T0223143

Lecture Notes in Mathematics

Edited by J.-M. Morel, F. Takens and B. Teissier

Editorial Policy for Multi-Author Publications: Summer Schools / Intensive Courses

1. Lecture Notes aim to report new developments in all areas of mathematics and their applications – quickly, informally and at a high level. Mathematical texts analysing new developments in modelling and numerical simulation are welcome. Manuscripts should be reasonably self-contained and rounded off. Thus they may, and often will, present not only results of the author but also related work by other people. They should provide sufficient motivation, examples and applications. There should also be an introduction making the text comprehensible to a wider audience. This clearly distinguishes Lecture Notes from journal articles or technical reports which normally are very concise. Articles intended for a journal but too long to be accepted by most journals, usually do not have this „lecture notes" character.

2. In general SUMMER SCHOOLS and other similar INTENSIVE COURSES are held to present mathematical topics that are close to the frontiers of recent research to an audience at the beginning or intermediate graduate level, who may want to continue with this area of work, for a thesis or later. This makes demands on the didactic aspects of the presentation. Because the subjects of such schools are advanced, there often exists no textbook, and so ideally, the publication resulting from such a school could be a first approximation to such a textbook.

   Usually several authors are involved in the writing, so it is not always simple to obtain a unified approach to the presentation.

   For prospective publication in LNM, the resulting manuscript should not be just a collection of course notes, each of which has been developed by an individual author with little or no co-ordination with the others, and with little or no common concept. The subject matter should dictate the structure of the book, and the authorship of each part or chapter should take secondary importance. Of course the choice of authors is crucial to the quality of the material at the school and in the book, and the intention here is not to belittle their impact, but simply to say that the book should be planned to be written by these authors jointly, and not just assembled as a result of what these authors happen to submit.

   This represents considerable preparatory work (as it is imperative to ensure that the authors know these criteria before they invest work on a manuscript), and also considerable editing work afterwards, to get the book into final shape. Still it is the form that holds the most promise of a successful book that will be used by its intended audience, rather than yet another volume of proceedings for the library shelf.

3. Manuscripts should be submitted (preferably in duplicate) either to Springer's mathematics editorial in Heidelberg, or to one of the series editors (with a copy to Springer). Volume editors are expected to arrange for the refereeing, to the usual scientific standards, of the individual contributions. If the resulting reports can be forwarded to us (series editors or Springer) this is very helpful. If no reports are forwarded or if other questions remain unclear in respect of homogeneity etc, the series editors may wish to consult external referees for an overall evaluation of the volume. A final decision to publish can be made only on the basis of the complete manuscript; however a preliminary decision can be based on a pre-final or incomplete manuscript. The strict minimum amount of material that will be considered should include a detailed outline describing the planned contents of each chapter.

   Volume editors and authors should be aware that incomplete or insufficiently close to final manuscripts almost always result in longer evaluation times. They should also be aware that parallel submission of their manuscript to another publisher while under consideration for LNM will in general lead to immediate rejection.

Continued on inside back-cover

Lecture Notes in Mathematics 1911

Editors:
J.-M. Morel, Cachan
F. Takens, Groningen
B. Teissier, Paris

FONDAZIONE CIME

ROBERTO CONTI

CENTRO INTERNAZIONALE MATEMATICO ESTIVO
INTERNATIONAL MATHEMATICAL SUMMER CENTER

C.I.M.E. means Centro Internazionale Matematico Estivo, that is, International Mathematical Summer Center. Conceived in the early fifties, it was born in 1954 and made welcome by the world mathematical community where it remains in good health and spirit. Many mathematicians from all over the world have been involved in a way or another in C.I.M.E.'s activities during the past years.

So they already know what the C.I.M.E. is all about. For the benefit of future potential users and co-operators the main purposes and the functioning of the Centre may be summarized as follows: every year, during the summer, Sessions (three or four as a rule) on different themes from pure and applied mathematics are offered by application to mathematicians from all countries. Each session is generally based on three or four main courses $(24-30$ hours over a period of $6\text{-}8$ working days) held from specialists of international renown, plus a certain number of seminars.

A C.I.M.E. Session, therefore, is neither a Symposium, nor just a School, but maybe a blend of both. The aim is that of bringing to the attention of younger researchers the origins, later developments, and perspectives of some branch of live mathematics.

The topics of the courses are generally of international resonance and the participation of the courses cover the expertise of different countries and continents. Such combination, gave an excellent opportunity to young participants to be acquainted with the most advance research in the topics of the courses and the possibility of an interchange with the world famous specialists. The full immersion atmosphere of the courses and the daily exchange among participants are a first building brick in the edifice of international collaboration in mathematical research.

C.I.M.E. Director
Pietro ZECCA
Dipartimento di Energetica "S. Stecco"
Università di Firenze
Via S. Marta, 3
50139 Florence
Italy
e-mail: zecca@unifi.it

C.I.M.E. Secretary
Elvira MASCOLO
Dipartimento di Matematica
Università di Firenze
viale G.B. Morgagni 67/A
50134 Florence
Italy
e-mail: mascolo@math.unifi.it

For more information see CIME's homepage: http://www.cime.unifi.it

CIME's activity is supported by:

– Istituto Nazionale di Alta Mathematica "F. Severi"
– Ministero dell'Istruzione, dell'Università e delle Ricerca
– Ministero degli Affari Esteri, Direzione Generale per la Promozione e la Cooperazione, Ufficio V

This CIME course was partially supported by: HyKE a Research Training Network (RTN) financed by the European Union in the 5th Framework Programme "Improving the Human Potential" (1HP). Project Reference: Contract Number: HPRN-CT-2002-00282

Alberto Bressan · Denis Serre
Mark Williams · Kevin Zumbrun

Hyperbolic Systems of Balance Laws

Lectures given at the
C.I.M.E. Summer School
held in Cetraro, Italy,
July 14–21, 2003

Editor: Pierangelo Marcati

 Springer

FONDAZIONE
CIME
ROBERTO CONTI

Editor and Authors

Alberto Bressan
Department of Mathematics
McAllister Building, Room 201
Penn State University
University Park
PA 16802, USA
e-mail: bressan@math.psu.edu

Kevin Zumbrun
Department of Mathematics
Rawles Hall 225
Indiana University
Bloomington
IN 47405, USA
e-mail: kzumbrun@indiana.edu

Denis Serre
Unité de mathématiques pures et
    appliquées
Ecole Normale Supérieure de Lyon
(UMR 5669 CNRS)
46, allée d'Italie
69364 Lyon Cedex 07, France
e-mail: serre@umpa.ens-lyon.fr

Pierangelo Marcati
Department of Pure
    and Applied Mathematics
University of L'Aquila
Via Vetoio – Loc. Coppito
67100 L'Aquila, Italy
e-mail: marcati@univaq.it
URL: *http://univaq.it/~marcati*

Mark Williams
Department of Mathematics
CB 3250
University of North Carolina
Chapel Hill
NC 27599, USA
e-mail: williams@email.unc.edu

Library of Congress Control Number: 2007925049

Mathematics Subject Classification (2000): 35L60, 35L65, 35L67, 35B35, 76L05, 35L50, 65M06

ISSN print edition: 0075-8434
ISSN electronic edition: 1617-9692

ISBN 978-3-540-72186-4 Springer Berlin Heidelberg New York

DOI 10.1007/978-3-540-72187-1

Springer is a part of Springer Science+Business Media
springer.com
© Springer-Verlag Berlin Heidelberg 2007

Typesetting by the authors and SPi using a Springer LATEX package
Cover design: WMXDesign GmbH, Heidelberg

Printed on acid-free paper   SPIN: 12055890   41/SPi   5 4 3 2 1 0

Preface

This volume includes the lecture notes delivered at the CIME Course "Hyperbolic Systems of Balance Laws" held July 14-21, 2003 in Cetraro (Cosenza, Italy). The present volume includes lectures notes by A. Bressan, D. Serre, K. Zumbrun and M. Williams and an appendix by A. Bressan on the center manifold theorem. These are among the "hot topics" in this field and can be of great interest, not only to professional mathematicians, but also for physicists and engineers.

The concept of hyperbolic systems of balance laws was introduced by the works of natural philosophers of the eighteenth century, predominantly L. Euler (1755), and has over the past one hundred and fifty years become the natural framework for the study of gas dynamics and, more broadly, of continuum physics. During this period of time great personalities like Stokes, Challis, Riemann, Rankine, Hugoniot, Lord Rayleigh and later Prandtl, Hadamard, H. Lewy, G.I. Taylor and many others wrote several fundamental papers, thus laying the groundwork for the further development of the mathematical theory. However the first part of the past century did not see much activity on the part of mathematicians in this field and it was only during the Second World War, in connection with the Manhattan Project, that associated research received a great impetus.

Many important scientists like J. Von Neumann, R. Courant, K.O. Friedrichs, H. Bethe and Ya. Zeldowich became interested in this field and proposed many new key concepts, the influence of which remains very great to the present day.

Immediately after the Second World War there was a considerable development in mathematical theory, with key results being obtained by a new generation of great mathematicians like S.K. Godunov, P. Lax, F. John, C. Morawetz and O. Oleinik, who led the field until the mid 1960s, when J. Glimm published an outstanding paper which marked the most important breakthrough in the history of this field. Glimm was able to prove the global existence of general systems in one space dimension, with small BV data. This result introduced a new approach to nonlinear wave interaction, but the proof was not fully deterministic. Tai-Ping Liu was later able to remove the probabilistic part of the proof, thus making it completely deterministic.

The relation between hyperbolic balance laws and continuum physics is not covered in any of the lectures in the present volume, but was the core topic of a series of lectures delivered in the Cetraro School by C. Dafermos entitled "Conservation Laws on Continuum Mechanics." In his wonderful monograph, published by Springer-Verlag in the Grundlehren der Mathematischen Wissenschaften, vol. 325, Dafermos provides an extremely thorough account of the most relevant aspects of the theory of hyperbolic conservation laws and systematically develops their ties to classical mechanics.

The notes by Alberto Bressan in this volume are intended to provide a self-contained presentation of recent results on hyperbolic conservation laws, based on the vanishing viscosity approach. Glimm's aforementioned theory was based on the construction of partially smooth approximating solutions with a locally self-similar structure. In order to get a uniform bound in BV norm, interaction potential was a crucial tool, an idea Glimm borrowed from physics. This potential, though a nonlinear functional, displays quadratic behaviour and decreases with time, provided the initial data have a small total variation.

In the 1990s Bressan and Tai-Ping Liu, together with various collaborators, completed this theory by proving the continuous dependence on initial data. The relations with the theory of compressible fluids have raised, since the very beginning of the theory, the question whether the inviscid solutions are in practice the same as the solutions with low viscosity. Although this fact had been established for various specific situations, it was only very recently that Bianchini and Bressan discovered a way to prove that, if the total variation of initial data remains sufficiently small, then the solutions of a viscous system of conservation laws converge to the solutions of the inviscid system, as long as the viscosity tends to zero. This approach allows the stability results obtained using the previous theories to be generalized.

The results are based on various technical steps, which in the present lecture notes Bressan describes in great detail, making a remarkable effort to make this difficult subject also accessible to non-specialists and young doctoral students. The notes of D. Serre cover the existence and stability of discrete shock profiles, another very exciting topic which, since the 1940s, has greatly interested applied mathematicians, including Von Neumann, Godounov and Lax, who were motivated by the need for efficient numerical codes to approximate the solutions of compressible fluid systems, including situations where shocks are present. It was immediately clear to them that a number of challenging and difficult mathematical problems needed to be solved. Partial differential equations are often approximated by finite difference schemes. The consistency and stability of a given scheme are usually studied through a linearization along elementary solutions such as constants, travelling waves, and shocks.

The study of the existence and stability of travelling waves faces significant difficulties; for example, existence may fail in rather natural situations because of small divisors problems. Ties to many branches of mathematics, ranging from dynamical systems to arithmetic number theory, prove to be relevant in this field. Serre's notes greatly emphasize this interdisciplinary aspect. His lecture notes not only provide a very useful and comprehensive introduction to this specific topic, but moreover propose a class of truly interesting and challenging problems in modern spectral theory. The analysis of the vanishing viscosity limit is far from being fully understood in the multidimensional setting. It is, in any case, important to understand the presence of stable and unstable modes along boundaries and shock profiles, where the most relevant linear and nonlinear phenomena take place. As such, the stability of viscous shock waves was the main focus of the lectures delivered by M. Williams and K. Zumbrun. This topic started, for the inviscid case, with the pioneering papers of Kreiss, Osher, Rauch and Majda. Later Metivier brought into the field a number of far-reaching ideas from microlocal analysis, in particular the paradifferential calculus introduced by Bony. The stability condition is expressed in terms of the so-called Kreiss-Lopatinskiĭ determinant. The viscous case can benefit from many of these ideas, but new tools are also needed.

Linearizing the system about a given profile (made stationary by Galilean invariance), and taking the Laplace transform in time and the Fourier transform in the hyperplane orthogonal to the direction of propagation, allows the formulation of an eigenvalue equation for a differential operator with variable coefficients. A necessary condition for the viscous profile to be stable is that these eigenvalue equations do not have (nontrivial) solutions. The Evans function technique provides a means to quantify this criterion.

But some rather subtle issues, in particular regarding regular dependence on parameters, call for a cautious approach. This was understood in a celebrated paper by J. Alexander, Gardner and C. Jones. Necessary stability conditions are expressed in terms of:

(1) the transversality of the connection in the travelling wave ODE, and

(2) the Kreiss-Lopatinskiĭ condition, which is known to ensure weak inviscid stability. The argument relies on the low frequency behaviour of the Evans function. Unlike the Kreiss-Lopatinskiĭ determinant Δ, encoding the linearized stability of the inviscid shock, the Evans function is not explicitly computable. But Zumbrun and Serre's result shows that the Evans function is tangent to Δ in the low frequency limit. Kevin Zumbrun's lectures focused on the planar stability for viscous shock waves in systems with real viscosity. His course provided an extensive overview of the technical tools and central concepts involved. He took great care to make such a difficult matter comparatively simple and approachable to an audience of young mathematicians.

M. Williams' course focused on the short time existence of curved multidimensional viscous shocks and the related small viscosity limit. It provided an accessible account of the main ideas and methods, trying to avoid most of the technical difficulties connected with the use of paradifferential calculus. His final lecture introduced the analysis of long time stability for planar viscous shocks. A fairly complete list of references to books and research articles is included at the bottom of all of the lecture notes.

The course was attended by several young mathematicians from various European countries, who worked very hard during the whole period of the Summer School. However they were also able to enjoy the beautiful hosting facility provided by CIME, in the paradise-like sea resort of Cetraro, under the Calabrian sun.

This course was organized with the collaboration and financial support of the European Network on Hyperbolic and Kinetic Equations (HyKE).

I would like to express my gratitude to the CIME Foundation, to the CIME Director Prof. Pietro Zecca and to the CIME Board Secretary Prof. Elvira Mascolo, for their invaluable help and support, and for the tremendous efforts they have invested to return the CIME Courses to their traditional greatness.

Pierangelo Marcati

Contents

BV Solutions to Hyperbolic Systems by Vanishing Viscosity

Alberto Bressan

Department of Mathematics, Penn State University
University Park, Pa. 16802 U.S.A.
bressan@math.psu.edu

1 Introduction

The aim of these notes is to provide a self-contained presentation of recent results on hyperbolic systems of conservation laws, based on the vanishing viscosity approach.

A system of conservation laws in one space dimension takes the form

$$u_t + f(u)_x = 0 \,. \tag{1.1}$$

Here $u = (u_1, \ldots, u_n)$ is the vector of **conserved quantities** while the components of $f = (f_1, \ldots, f_n)$ are called the **fluxes**. Integrating (1.1) over a fixed interval $[a, b]$ we find

$$\frac{d}{dt} \int_a^b u(t, x) \, dx \;=\; \int_a^b u_t(t, x) \, dx \;=\; -\int_a^b f(u(t, x))_x \, dx$$
$$= f(u(t, a)) - f(u(t, b)) \;=\; [\text{inflow at } a] - [\text{outflow at } b] \,.$$

Each component of the vector u thus represents a quantity which is neither created nor destroyed: its total amount inside any given interval $[a, b]$ can change only because of the flow across boundary points.

Systems of the form (1.1) are commonly used to express the fundamental balance laws of continuum physics, when small viscosity or dissipation effects are neglected. For a comprehensive discussion of conservation laws and their derivation from basic principles of physics we refer to the book of Dafermos [D1].

Smooth solutions of (1.1) satisfy the equivalent quasilinear system

$$u_t + A(u)u_x = 0 \,, \tag{1.2}$$

where $A(u) \doteq Df(u)$ is the Jacobian matrix of first order partial derivatives of f. We notice, however, that if u has a jump at some point x_0, then the left hand side of (1.2) contains a product of the discontinuous function $x \mapsto A(u(x))$

with the distributional derivative u_x, which in this case contains a Dirac mass at the point x_0. In general, such a product is not well defined. The quasilinear system (1.2) is thus meaningful only within a class of continuous functions. On the other hand, working with the equation in divergence form (1.1) allows us to consider discontinuous solutions as well, interpreted in distributional sense. We say that a locally integrable function $u = u(t, x)$ is a **weak solution** of (1.1) if $t \mapsto u(t, \cdot)$ is continuous as a map with values in $\mathbf{L}^1_{\text{loc}}$ and moreover

$$\iint \{u\phi_t + f(u)\phi_x\} \, dx \, dt \; = \; 0 \tag{1.3}$$

for every differentiable function with compact support $\phi \in \mathcal{C}^1_c$.

The above system is called **strictly hyperbolic** if each matrix $A(u) \doteq Df(u)$ has n real, distinct eigenvalues $\lambda_1(u) < \cdots < \lambda_n(u)$. One can then find dual bases of left and right eigenvectors of $A(u)$, denoted by $l_1(u), \ldots, l_n(u)$ and $r_1(u), \ldots, r_n(u)$, normalized according to

$$|r_i| = 1, \qquad l_i \cdot r_j = \begin{cases} 1 & if \quad i = j, \\ 0 & if \quad i \neq j. \end{cases} \tag{1.4}$$

To appreciate the effect of the non-linearity, consider first the case of a linear system with constant coefficients

$$u_t + Au_x = 0. \tag{1.5}$$

Call $\lambda_1 < \cdots < \lambda_n$ the eigenvalues of the matrix A, and let l_i, r_i be the corresponding left and right eigenvectors as in (1.4). The general solution of (1.5) can be written as a superposition of independent linear waves:

$$u(t, x) = \sum_i \phi_i(x - \lambda_i t)r_i, \qquad \phi_i(y) \doteq l_i \cdot u(0, y).$$

Notice that here the solution is completely decoupled along the eigenspaces of A, and each component travels with constant speed, given by the corresponding eigenvalue of A.

In the nonlinear case (1.2) where the matrix A depends on the state u, new features will appear in the solutions.

(i) Since the eigenvalues λ_i now depend on u, the shape of the various components in the solution will vary in time (fig. 1). Rarefaction waves will decay, and compression waves will become steeper, possibly leading to shock formation in finite time.

(ii) Since the eigenvectors r_i also depend on u, nontrivial interactions between different waves will occur (fig. 2). The strength of the interacting waves may change, and new waves of different families can be created, as a result of the interaction.

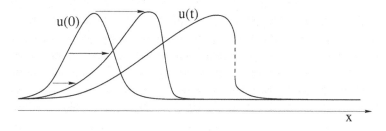

Fig. 1.

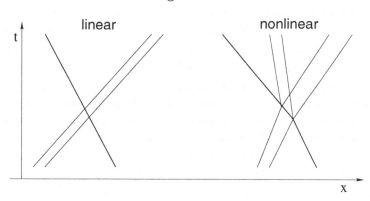

Fig. 2.

The strong nonlinearity of the equations and the lack of regularity of solutions, also due to the absence of second order terms that could provide a smoothing effect, account for most of the difficulties encountered in a rigorous mathematical analysis of the system (1.1). It is well known that the main techniques of abstract functional analysis do not apply in this context. Solutions cannot be represented as fixed points of continuous transformations, or in variational form, as critical points of suitable functionals. Dealing with vector valued functions, comparison principles based on upper or lower solutions cannot be used. Moreover, the theory of accretive operators and contractive nonlinear semigroups works well in the scalar case [C], but does not apply to systems. For the above reasons, the theory of hyperbolic conservation laws has largely developed by *ad hoc* methods, along two main lines.

1. The *BV* setting, considered by Glimm [G]. Solutions are here constructed within a space of functions with bounded variation, controlling the *BV* norm by a wave interaction functional.

2. The $\mathbf{L}^\infty$ setting, considered by Tartar and DiPerna [DP2], based on weak convergence and a compensated compactness argument.

Both approaches yield results on the global existence of weak solutions. However, the method of compensated compactness appears to be suitable only for 2×2 systems. Moreover, it is only in the BV setting that the well-posedness of the Cauchy problem could recently be proved, as well as the stability and convergence of vanishing viscosity approximations. Throughout the following we thus restrict ourselves to the study of BV solutions, referring to [DP2] or [Se] for the alternative approach based on compensated compactness.

Since the pioneering work of Glimm, the basic building block toward the construction and the analysis of more general solutions has been provided by the **Riemann problem**, i.e. the initial value problem with piecewise constant data

$$u(0, x) = \bar{u}(x) = \begin{cases} u^- & if \quad x < 0, \\ u^+ & if \quad x > 0. \end{cases} \tag{1.6}$$

This was first introduced by B.Riemann (1860) in the context of isentropic gas dynamics. A century later, P.Lax [Lx] and T.P.Liu [L1] solved the Riemann problem for more general $n \times n$ systems. The new approach of Bianchini [Bi] now applies to all strictly hyperbolic systems, not necessarily in conservation form. Solutions are always found in the self-similar form $u(t, x) = U(x/t)$. The central position taken by the Riemann problem is related to a symmetry of the equations (1.1). If $u = u(t, x)$ is a solution of (1.1), then for any $\theta > 0$ the function

$$u^\theta(t, x) \doteq u(\theta t, \theta x)$$

provides another solution. The solutions which are invariant under these rescalings of the independent variables are precisely those which correspond to some Riemann data (1.6).

For a general Cauchy problem, both the Glimm scheme [G] and the method of front tracking [D1], [DP1], [B1], [BaJ], [HR] yield approximate solutions of a general Cauchy problem by piecing together a large number of Riemann solutions. For initial data with small total variation, this approach is successful because one can provide a uniform a priori bound on the amount of new waves produced by nonlinear interactions, and hence on the total variation of the solution. It is safe to say that, in the context of weak solutions with small total variation, nearly all results on the existence, uniqueness, continuous dependence and qualitative behavior have relied on a careful analysis of the Riemann problem.

In [BiB] a substantially different perspective has emerged from the study of vanishing viscosity approximations. Solutions of (1.1) are here obtained as limits for $\varepsilon \to 0$ of solutions to the parabolic problems

$$u_t^\varepsilon + A(u^\varepsilon)u_x^\varepsilon = \varepsilon u_{xx}^\varepsilon \tag{1.7}$$

with $A(u) \doteq Df(u)$. This approach is very natural and has been considered since the 1950's. However, complete results had been obtained only in the scalar case [O], [K]. For general $n \times n$ systems, the main difficulty lies in establishing the compactness of the approximating sequence. We observe that $u^\varepsilon(t, x)$ solves (1.7) if and only if $u^\varepsilon(t, x) = u(t/\varepsilon, x/\varepsilon)$ for some function u which satisfies

$$u_t + A(u)u_x = u_{xx}.$$ (1.8)

In the analysis of vanishing viscosity approximations, the key step is to derive a priori estimates on the total variation and on the stability of solutions of (1.8). For this parabolic system, the rescaling $(t, x) \mapsto (\theta t, \theta x)$ no longer determines a symmetry. Hence the Riemann data no longer hold a privileged position. The role of basic building block is now taken by the **viscous traveling profiles**, i.e. solutions of the form

$$u(t, x) = U(x - \lambda t).$$

Of course, the function U must then satisfy the second order O.D.E.

$$U'' = (A(U) - \lambda)U'.$$

In this new approach, the profile $u(\cdot)$ of a viscous solution is viewed locally as a superposition of viscous traveling waves. More precisely, let a smooth function $u : \mathbb{R} \mapsto \mathbb{R}^n$ be given. At each point x, looking at the second order jet (u, u_x, u_{xx}) we seek traveling profiles $U_1, \ldots, U_n$ such that

$$U_i'' = (A(U_i) - \sigma_i)U_i'$$ (1.9)

for some speed σ_i close to the characteristic speed λ_i, and moreover

$$U_i(x) = u(x) \qquad i = 1, \ldots, n,$$ (1.10)

$$\sum_i U_i'(x) = u_x(x), \qquad \sum_i U_i''(x) = u_{xx}(x).$$ (1.11)

It turns out that this decomposition is unique provided that the traveling profiles are chosen within suitable center manifolds. We let $\tilde{r}_i$ be the unit vector parallel to U_i', so that $U_i' = v_i \tilde{r}_i$ for some scalar v_i. One can show that $\tilde{r}_i$ remains close to the eigenvector $r_i(u)$ of the Jacobian matrix $A(u) \doteq Df(u)$, but $\tilde{r}_i \neq r_i(u)$ in general. The first equation in (1.11) now yields the decomposition

$$u_x = \sum_i v_i \tilde{r}_i.$$ (1.12)

If $u = u(t, x)$ is a solution of (1.8), we can think of v_i as the density of i-waves in u. The remarkable fact is that these components satisfy a system of evolution equations

$$v_{i,t} + (\tilde{\lambda}_i v_i)_x - v_{i,xx} = \phi_i \qquad i = 1, \ldots, n,$$ (1.13)

where the source terms ϕ_i on the right hand side are INTEGRABLE over the whole domain $\{x \in \mathbb{R}, \ t > 0\}$. Indeed, we can think of the sources ϕ_i as new waves produced by interactions between viscous waves. Their total strength is controlled by means of viscous interaction functionals, somewhat similar to the one introduced by Glimm in [G] to study the hyperbolic case. Since the left hand side of (1.13) is in conservation form and the vectors $\tilde{r}_i$ have unit length, for an arbitrarily large time t we obtain the bound

$$\|u_x(t)\|_{\mathbf{L}^1} \leq \sum_i \|v_i(t)\|_{\mathbf{L}^1} \leq \sum_i \left(\|v_i(t_0)\| + \int_{t_0}^t \int |\phi_i(s,x)| \, dx ds \right). \quad (1.14)$$

This argument yields global BV bounds and stability estimates for viscous solutions. In turn, letting $\varepsilon \to 0$ in (1.7), a standard compactness argument yields the convergence of u^ε to a weak solution u of (1.1).

The plan of these notes is as follows. In Section 2 we briefly review the basic theory of hyperbolic systems of conservation laws: shock and rarefaction waves, entropies, the Liu admissibility conditions and the Riemann problem. For initial data with small total variation, we also recall the main results concerning the existence, uniqueness and stability of solutions to the general Cauchy problem. Section 3 contains the statement of the main new results on vanishing viscosity limits and an outline of the proof. In Section 4 we derive some preliminary estimates which can be obtained by standard parabolic techniques, representing a viscous solution in terms of convolutions with a heat kernel. Section 5 discusses in detail the local decomposition of a solution as superposition of viscous traveling waves. The evolution equations (1.13) for the components v_i and the strength of the source terms ϕ_i are then studied in Section 6. This will provide the crucial estimate on the total variation of viscous solutions, uniformly in time. In Section 7 we briefly examine the stability and some other properties of viscous solutions. The existence, uniqueness and stability of vanishing viscosity limits are then discussed in Section 8.

2 Review of Hyperbolic Conservation Laws

In most of this section, we shall consider a strictly hyperbolic system of conservation laws satisfying the additional hypothesis

(H) For each $i = 1, \ldots, n$, the i-th field is either **genuinely nonlinear**, so that $D\lambda_i(u) \cdot r_i(u) > 0$ for all u, or **linearly degenerate**, with $D\lambda_i(u) \cdot r_i(u) = 0$ for all u.

We observe that the i-th characteristic field is genuinely nonlinear iff the eigenvalue λ_i is strictly increasing along each integral curve of the corresponding field of eigenvectors r_i. It is linearly degenerate when λ_i is constant along each such curve.

2.1 Centered Rarefaction Waves.

We begin by studying a special type of solutions, in the form of centered rarefaction waves. Fix a state $u_0 \in \mathbb{R}^n$ and an index $i \in \{1, \ldots, n\}$. Let $r_i(u)$ be a field of i-eigenvectors of the Jacobian matrix $A(u) = Df(u)$. The integral curve of the vector field r_i through the point u_0 is called the i-**rarefaction curve** through u_0. It is obtained by solving the initial value problem for the O.D.E. in state space:

$$\frac{du}{ds} = r_i(u), \qquad u(0) = u_0. \tag{2.1}$$

We shall denote this curve as

$$s \mapsto R_i(s)(u_0).$$

Clearly, the parametrization depends on the choice of the eigenvectors r_i. In particular, if we impose the normalization $|r_i(u)| \equiv 1$, then the rarefaction curve R_i will be parametrized by arc-length.

Let the i-th field be genuinely nonlinear. Given two states u^-, u^+, assume that u^+ lies on the positive i-rarefaction curve through u^-, i.e. $u^+ = R_i(\sigma)(u^-)$ for some $\sigma > 0$. For each $s \in [0, \sigma]$, define

$$\lambda_i(s) = \lambda_i\big(R_i(s)(u^-)\big).$$

By genuine nonlinearity, the map $s \mapsto \lambda_i(s)$ is strictly increasing. Hence, for every $\lambda \in \big[\lambda_i(u^-), \lambda_i(u^+)\big]$, there is a unique value $s \in [0, \sigma]$ such that $\lambda = \lambda_i(s)$. We claim that, for $t > 0$, the function

$$u(t, x) = \begin{cases} u^- & if & x/t < \lambda_i(u^-), \\ R_i(s)(u^-) & if & x/t = \lambda_i(s) \in \big[\lambda_i(u^-), \lambda_i(u^+)\big], \\ u^+ & if & x/t > \lambda_i(u^+), \end{cases} \tag{2.2}$$

is a continuous solution of the Riemann problem (1.1), (1.6). Indeed, by construction it follows

$$\lim_{t \to 0+} \big\|u(t, \cdot) - \bar{u}\big\|_{\mathbf{L}^1} = 0.$$

Moreover, the equation (1.2) is trivially satisfied in the sectors where $x/t < \lambda_i(u^-)$ or $x/t > \lambda_i(u^+)$, since here $u_t = u_x = 0$. Next, consider what happens in the intermediate sector $\{\lambda_i(u^-) < x/t < \lambda_i(u^+)\}$. Since u is constant along each ray through the origin $\{x/t = c\}$, we have

$$u_t(t, x) + \frac{x}{t} u_x(t, x) = 0.$$

Our construction at (2.2) now yields $x/t = \lambda_i\big(u(t, x)\big)$. Moreover, u_x is parallel to $r_i(u)$, hence it is an eigenvector of the Jacobian matrix $A(u)$ with eigenvalue $\lambda_i(u)$. This implies $u_t + A(u)u_x = 0$, proving our claim. Notice that the assumption $\sigma > 0$ is essential for the validity of this construction. In the opposite case $\sigma < 0$, the definition (2.2) would yield a triple-valued function in the region where $x/t \in \,\,]\lambda_i(u^+), \lambda_i(u^-)[$.

2.2 Shocks and Admissibility Conditions.

The simplest type of discontinuous solution of (1.1) is given by a **shock**:

$$U(t, x) = \begin{cases} u^+ & if & x > \lambda t, \\ u^- & if & x < \lambda t, \end{cases} \tag{2.3}$$

for some left and right states $u^-, u^+ \in \mathbb{R}^n$ and a speed $\lambda \in \mathbb{R}$. Using the divergence theorem, one checks that the identity (1.2) is satisfied for all test functions ϕ if and only if the following Rankine-Hugoniot conditions hold:

$$\lambda (u^+ - u^-) = f(u^+) - f(u^-). \tag{2.4}$$

As usual, we denote by $A(u) = Df(u)$ the $n \times n$ Jacobian matrix of f at u. For any u, v, consider the averaged matrix

$$A(u, v) \doteq \int_0^1 A\big(\theta u + (1 - \theta)v\big) \, d\theta$$

and call $\lambda_i(u, v)$, $i = 1, \ldots, n$ its eigenvalues. We can then write (2.4) in the equivalent form

$$\lambda (u^+ - u^-) = \int_0^1 Df\big(\theta u^+ + (1 - \theta)u^-\big) \cdot (u^+ - u^-) \, d\theta = A(u^+, u^-) \cdot (u^+ - u^-). \tag{2.5}$$

In other words, the Rankine-Hugoniot conditions hold if and only if the jump $u^+ - u^-$ is an eigenvector of the averaged matrix $A(u^+, u^-)$ and the speed λ coincides with the corresponding eigenvalue.

The representation (2.5) is very useful for studying the set of pairs (u^-, u^+) that can be connected by a shock. Given a left state $u^- = u_0$, for each $i = 1, \ldots, n$ one can prove that there exists a one-parameter curve of right states $u^+ = S_i(s)(u_0)$ which satisfy the Rankine-Hugoniot conditions (2.4) for a suitable speed $\lambda = \lambda_i(s)$. This curve can also be parametrized by arc-length:

$$s \mapsto S_i(s)(u_0).$$

It has a second order tangency with the corresponding rarefaction curve R_i at the point u_0 (see fig. 3).

To study the behavior of a general weak solution at a point of discontinuity, we first recall some definitions.

A function $u = u(t, x)$ has an **approximate jump discontinuity** at the point (τ, ξ) if there exists vectors $u^+ \neq u^-$ and a speed λ such that

$$\lim_{r \to 0+} \frac{1}{r^2} \int_{-r}^r \int_{-r}^r \big|u(\tau + t, \, \xi + x) - U(t, x)\big| \, dx \, dt = 0, \tag{2.6}$$

with U as in (2.3). We say that u is **approximately continuous** at the point (τ, ξ) if the above limit hold with $u^+ = u^-$ (and λ arbitrary).

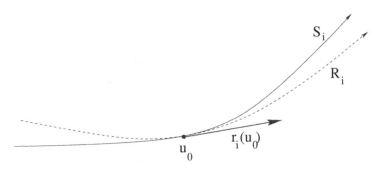

Fig. 3.

If u is now a solution of the system of conservation laws (1.1) having an approximate jump at a point (τ, ξ), one can prove that the states u^-, u^+ and the speed λ again satisfy the Rankine-Hugoniot conditions (see Theorem 4.1 in [B3]).

In order to achieve uniqueness of solutions to initial value problems, one needs to supplement the conservation equations (1.1) with additional admissibility conditions, to be satisfied at points of jump. We recall here two basic approaches.

A continuously differentiable function $\eta : \mathbb{R}^n \mapsto \mathbb{R}$ is called an **entropy** for the system of conservation laws (1.1), with **entropy flux** $q : \mathbb{R}^n \mapsto \mathbb{R}$, if it satisfies the identity

$$D\eta(u) \cdot Df(u) = Dq(u). \tag{2.7}$$

An immediate consequence of (2.7) is that, if $u = u(t, x)$ is a C^1 solution of (1.1), then

$$\eta(u)_t + q(u)_x = 0. \tag{2.8}$$

For a smooth solution u, not only the quantities $u_1, \ldots, u_n$ are conserved, but the additional conservation law (2.8) holds as well. However one should be aware that, for a discontinuous solution u, the quantity $\eta(u)$ may not be conserved. A standard admissibility condition for weak solutions can now be formulated as follows.

Let η be a convex entropy for the system (1.1), with entropy flux q. A weak solution u is **entropy-admissible** if

$$\eta(u)_t + q(u)_x \leq 0 \tag{2.9}$$

in distribution sense, i.e.

$$\iint \left\{ \eta(u)\varphi_t + q(u)\varphi_x \right\} \, dx \, dt \geq 0 \tag{2.10}$$

for every function $\varphi \geq 0$, continuously differentiable with compact support.

In analogy with (2.4), if u is an entropy admissible solution, at every point of approximate jump one can show that (2.9) implies

$$\lambda\big[\eta(u^+) - \eta(u^-)\big] \geq q(u^+) - q(u^-). \qquad (2.11)$$

The above admissibility condition can be useful only if some nontrivial convex entropy exists. This is not always the case. Indeed, according to (2.7), to construct an entropy we must solve a system of n equations for the two scalar functions η, q. This system is overdetermined whenever $n \geq 3$. In continuum physics, however, one finds various systems of conservation laws endowed with significant entropies.

An alternative admissibility condition, due to Liu [L2], is particularly useful because it can be applied to any system, even in the absence of nontrivial entropies. To understand its meaning, consider first a scalar conservation law. In this case, any two states u^-, u^+ satisfy the equations (2.4), provided that we choose the speed

$$\lambda = \frac{f(u^+) - f(u^-)}{u^+ - u^-}. \qquad (2.12)$$

To test the stability of the shock (2.3) w.r.t. a small perturbation, fix any intermediate state $u^* \in [u^-, u^+]$. We can slightly perturb the initial data by splitting the initial jump in two smaller jumps (fig. 4), say

$$u(0, x) = \begin{cases} u^- & if \quad x < 0, \\ u^* & if \quad 0 < x < \varepsilon, \\ u^+ & if \quad x > \varepsilon. \end{cases}$$

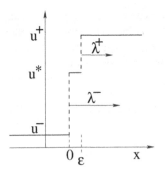

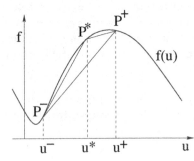

Fig. 4.   Fig. 5

Clearly, the original shock can be stable only if the speed λ^- of the jump behind is faster than the speed λ^+ of the jump ahead, i.e.

$$\lambda^- \doteq \frac{f(u^*) - f(u^-)}{u^* - u^-} \geq \frac{f(u^+) - f(u^*)}{u^+ - u^*} = \lambda^+ . \tag{2.13}$$

In fig. 5, the first speed corresponds to the slope of the segment $P^- P^*$ while the second speed is the slope of $P^* P^+$. On the other hand, by (2.12) the speed λ of the original shock is the slope of the segment $P^- P^+$. Therefore the inequality (2.13) holds if and only if

$$\frac{f(u^*) - f(u^-)}{u^* - u^-} \geq \frac{f(u^+) - f(u^-)}{u^+ - u^-} . \tag{2.14}$$

The condition (2.13) cannot be extended to the vector valued case, because in general one cannot find intermediate states u^* such that both couples (u^-, u^*) and (u^*, u^+) are connected by a shock. However, the condition (2.14) has a straightforward generalization to $n \times n$ systems. Denote by $s \mapsto S_i(s)(u^-)$ the i-shock curve through the point u^-, and let $\lambda_i(s)$ be the speed of the shock connecting u^- with $S_i(s)(u^-)$. Now consider any right state

$$u^+ = S_i(\sigma)(u^-)$$

on this shock curve. Following [L2], we say that the shock (2.3) satisfies the **Liu admissibility condition** if

$$\lambda_i(s) \geq \lambda_i(\sigma) \qquad \text{for all} \quad s \in [0, \sigma] \tag{2.15}$$

(or for all $s \in [\sigma, 0]$, in case where $\sigma < 0$). In other words, if $u^* = S_i(s)(u^-)$ is any intermediate state along the shock curve through u^-, the speed of the shock (u^-, u^*) should be larger or equal to the speed of the original shock (u^-, u^+).

2.3 Solution of the Riemann Problem.

Let the assumptions (H) hold, so that each characteristic field is either genuinely nonlinear or linearly degenerate. Relying on the previous analysis, the solution of the general Riemann problem

$$u_t + f(u)_x = 0, \qquad u(0, x) = \begin{cases} u^- & \text{if} \quad x < 0, \\ u^+ & \text{if} \quad x > 0. \end{cases} \tag{2.16}$$

can be obtained by finding intermediate states $w_0 = u^-, w_1, \ldots, w_n = u^+$ such that each pair of adjacent states w_{i-1}, w_i can be connected by an i-shock, or by a centered rarefaction i-wave. By the implicit function theorem, this can always be done provided that the two states u^-, u^+ are sufficiently close.

Following [Lx], the complete solution is now obtained by piecing together the solutions of the n Riemann problems

$$u_t + f(u)_x = 0, \qquad u(0, x) = \begin{cases} \omega_{i-1} & \text{if} \quad x < 0, \\ \omega_i & \text{if} \quad x > 0, \end{cases} \qquad (2.17)$$

on different sectors of the t-x plane. A typical situation is illustrated in fig. 6.

If the assumption (H) is removed, one can still solve the Riemann problem in terms of $n+1$ constant states $\omega_0, \ldots, \omega_n$. However, in this case each couple of states ω_{i-1}, ω_i can be connected by a large number of adjacent shocks and rarefaction waves (fig. 7). The analysis of the Riemann problem thus becomes far more complicated [L1].

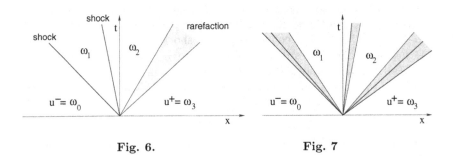

Fig. 6.   Fig. 7

2.4 Glimm and Front Tracking Approximations.

Approximate solutions for the general Cauchy problem can be constructed by patching together solutions of several Riemann problems. In the Glimm scheme (fig. 8), one works with a fixed grid in the t-x plane, with mesh sizes Δt, Δx. At time $t = 0$ the initial data is approximated by a piecewise constant function, with jumps at grid points. Solving the corresponding Riemann problems, a solution is constructed up to a time Δt sufficiently small so that waves emerging from different nodes do not interact. At time $t_1 \doteq \Delta t$, we replace the solution $u(\Delta t, \cdot)$ by a piecewise constant function having jumps exactly at grid points. Solving the new Riemann problems at every one of these points, one can prolong the solution to the next time interval $[\Delta t, 2\Delta t]$. At time $t_2 \doteq 2\Delta t$, the solution is again approximated by a piecewise constant functions with jumps exactly at grid points, etc... A key ingredient of the Glimm scheme is the restarting procedure. At each time $t_j \doteq j\,\Delta t$, a natural way to approximate a BV function with a piecewise constant one is by taking its average value on each subinterval $J_k \doteq [x_{k-1}, x_k]$. However, this procedure may generate an arbitrarily large amount of oscillations. Instead, the Glimm scheme is based on random sampling: a point y_k is selected at random inside each interval J_k and the old value $u(t_j-, y_k)$ at this particular point is taken as the new value of $u(t_j, x)$ for all $x \in J_k$. An excellent introduction to the Glimm scheme can be found in the book by J. Smoller [Sm].

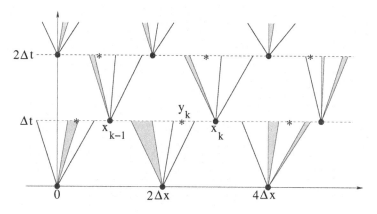

Fig. 8.

An alternative technique for constructing approximate solutions is by wave-front tracking (fig. 9). This method was introduced by Dafermos [D1] in the scalar case and later developed by various authors [DP1], [B1], [BaJ]. It now provides an efficient tool in the study of general $n \times n$ systems of conservation laws, both for theoretical and numerical purposes [B3], [HR].

The initial data is here approximated with a piecewise constant function, and each Riemann problem is solved approximately, within the class of piecewise constant functions. In particular, if the exact solution contains a centered rarefaction, this must be approximated by a *rarefaction fan*, containing several small jumps. At the first time t_1 where two fronts interact, the new Riemann problem is again approximately solved by a piecewise constant function. The solution is then prolonged up to the second interaction time t_2, where the new Riemann problem is solved, etc...

Comparing this method with the previous one of Glimm, we see that in the Glimm scheme one specifies a priori the nodal points where the the Riemann problems are to be solved. On the other hand, in a solution constructed by wave-front tracking the locations of the jumps and of the interaction points depend on the solution itself. No restarting procedure is needed and the map $t \mapsto u(t, \cdot)$ is thus continuous with values in $\mathbf{L}^1_{\mathrm{loc}}$.

In the end, both algorithms produce a sequence of approximate solutions, whose convergence relies on a compactness argument based on uniform bounds on the total variation. We sketch the main idea involved in these a priori BV bounds. Consider a piecewise constant function $u : \mathbb{R} \mapsto \mathbb{R}^n$, say with jumps at points $x_1 < x_2 < \cdots < x_N$. Call σ_α the amplitude of the jump at x_α. The *total strength of waves* is then defined as

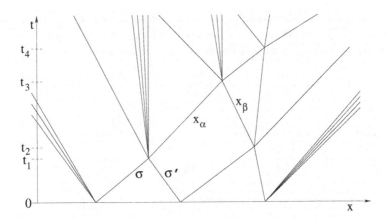

Fig. 9.

$$V(u) \doteq \sum_{\alpha} |\sigma_\alpha|. \tag{2.18}$$

Clearly, this is an equivalent way to measure the total variation. Along a solution $u = u(t, x)$ constructed by front tracking, the quantity $V(t) = V(u(t, \cdot))$ may well increase at interaction times. To provide global a priori bounds, following [G] one introduces a *wave interaction potential*, defined as

$$Q(u) = \sum_{(\alpha,\beta)\in\mathcal{A}} |\sigma_\alpha \sigma_\beta|, \tag{2.19}$$

where the summation runs over the set $\mathcal{A}$ of all couples of approaching waves. Roughly speaking, we say that two wave-fronts located at $x_\alpha < x_\beta$ are *approaching* if the one at x_α has a faster speed than the one at x_β (hence the two fronts are expected to collide at a future time). Now consider a time τ where two incoming wave-fronts interact, say with strengths σ, σ' (for example, take $\tau = t_1$ in fig. 9). The difference between the outgoing waves emerging from the interaction and the two incoming waves σ, σ' is of magnitude $\mathcal{O}(1) \cdot |\sigma\sigma'|$. On the other hand, after time τ the two incoming waves are no longer approaching. This accounts for the decrease of the functional Q in (2.19) by the amount $|\sigma\sigma'|$. Observing that the new waves generated by the interaction could approach all other fronts, the change in the functionals V, Q across the interaction time τ is estimated as

$$\Delta V(\tau) = \mathcal{O}(1) \cdot |\sigma\sigma'|, \qquad \Delta Q(\tau) = -|\sigma\sigma'| + \mathcal{O}(1) \cdot |\sigma\sigma'| V(\tau-).$$

If the initial data has small total variation, for a suitable constant C_0 the quantity

$$\Upsilon(t) \doteq V(u(t, \cdot)) + C_0 Q(u(t, \cdot))$$

is monotone decreasing in time. This argument provides a uniform BV bound on all approximate solutions. Using Helly's compactness theorem, one obtains the strong convergence in $\mathbf{L}^1_{\text{loc}}$ of a subsequence of approximate solutions, and hence the existence of a weak solution.

Theorem 1. *Consider the Cauchy problem*

$$u_t + f(u)_x = 0, \qquad u(0, x) = \bar{u}(x), \qquad (2.20)$$

for a system of conservation laws with a smooth flux f defined in a neighborhood of the origin. Assume that the system is strictly hyperbolic and satisfies the conditions (H). Then, for a sufficiently small $\delta > 0$ the following holds. For every initial condition $u(0, x) = \bar{u}(x)$ with

$$\|\bar{u}\|_{\mathbf{L}^\infty} < \delta, \qquad Tot.\,Var.\{\bar{u}\} < \delta, \qquad (2.21)$$

the Cauchy problem has a weak solution, defined for all times $t \geq 0$, obtained as limit of front tracking approximations.

2.5 A Semigroup of Solutions.

For solutions obtained as limits of Glimm or front tracking approximations, the uniqueness and the continuous dependence on the initial data were first established in [BC] for systems of two equations and then in [BCP] for $n \times n$ systems. Relying on some major new ideas introduced by Liu and Yang in [LY1], [LY2] a much simpler proof could then be worked out in [BLY]. Here the key step is the construction of a Lyapunov functional $\Phi(u, v)$, equivalent to the $\mathbf{L}^1$ distance

$$\frac{1}{C}\|u - v\|_{\mathbf{L}^1} \leq \Phi(u, v) \leq C\|u - v\|_{\mathbf{L}^1},$$

and almost decreasing in time along each pair of ε-approximate solutions constructed by front tracking

$$\frac{d}{dt}\Phi\big(u(t),\, v(t)\big) \leq \mathcal{O}(1) \cdot \varepsilon.$$

A comprehensive treatment of the above existence-uniqueness theory for BV solutions can be found in [B3], [HR]. The main result is as follows.

Theorem 2. *With the same assumptions as Theorem 1, the weak solution of the Cauchy problem (2.20) obtained as limit of Glimm or front tracking approximations is unique and satisfies the Liu entropy conditions. Adopting a semigroup notation, this solution can be written as*

$$u(t, \cdot) = S_t\bar{u}.$$

The semigroup $S : \mathcal{D} \times [0, \infty[\mapsto \mathcal{D}$ can be defined on a closed domain $\mathcal{D} \subset \mathbf{L}^1$ containing all functions with sufficiently small total variation, It is uniformly Lipschitz continuous w.r.t. both time and the initial data:

$$\left\|S_t\bar{u} - S_s\bar{v}\right\|_{\mathbf{L}^1} \le L\left\|\bar{u} - \bar{v}\right\|_{\mathbf{L}^1} + L'\left|t - s\right|. \tag{2.22}$$

We refer to the flow generated by a system of conservation laws as a *Riemann semigroup*, because it is entirely determined by specifying how Riemann problems are solved. As proved in [B2], if two semigroups S, S' yield the same solutions to all Riemann problems, then they coincide, up to the choice of their domains.

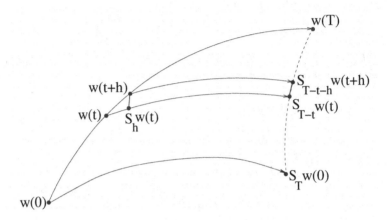

Fig. 10.

From (2.22) one can deduce the error bound (fig. 10)

$$\left\|w(T) - S_T w(0)\right\|_{\mathbf{L}^1} \le L \cdot \int_0^T \left\{ \liminf_{h\to 0+} \frac{\left\|w(t+h) - S_h w(t)\right\|_{\mathbf{L}^1}}{h} \right\} dt, \tag{2.23}$$

valid for every Lipschitz continuous map $w : [0, T] \mapsto \mathcal{D}$ taking values inside the domain of the semigroup. We can think of $t \mapsto w(t)$ as an approximate solution of (1.1), while $t \mapsto S_t w(0)$ is the exact solution having the same initial data. The distance $\left\|w(T) - S_T w(0)\right\|_{\mathbf{L}^1}$ is clearly bounded by the length of the curve $\gamma : [0, T] \mapsto \mathbf{L}^1$, with $\gamma(t) \doteq S_{T-t}w(t)$. According to (2.23), this length is estimated by the integral of an *instantaneous error rate*, amplified by the Lipschitz constant L of the semigroup.

Using (2.23), one can estimate the distance between a front tracking approximation and the corresponding exact solution. For approximate solutions constructed by the Glimm scheme, a direct application of this same formula is not possible because of the additional errors introduced by the restarting procedure at each time $t_k \doteq k\,\Delta t$. However, relying on a careful analysis of

Liu [L3], one can construct a front tracking approximate solution having the same initial and terminal values as the Glimm solution. By this technique, in [BM] the authors proved the estimate

$$\lim_{\Delta x \to 0} \frac{\left\| u^{\text{Glimm}}(\tau, \cdot) - u^{\text{exact}}(\tau, \cdot) \right\|_{\mathbf{L}^1}}{\sqrt{\Delta x} \cdot |\ln \Delta x|} = 0 \qquad (2.24)$$

for every $\tau > 0$. In other words, letting the mesh sizes $\Delta x, \Delta t \to 0$ while keeping their ratio $\Delta x / \Delta t$ constant, the $\mathbf{L}^1$ norm of the error in the Glimm approximate solution tends to zero at a rate slightly slower than $\sqrt{\Delta x}$.

2.6 Uniqueness and Characterization of Entropy Weak Solutions

The stability result stated in Theorem 1 refers to a special class of weak solutions: those obtained as limits of Glimm or front tracking approximations. It is desirable to have a uniqueness theorem valid for general weak solutions, without reference to any particular constructive procedure. Results in this direction were proved in, [BLF], [BG], [BLw]. They are all based on the error formula (2.23). In the proofs, one considers a weak solution $u = u(t, x)$ of the Cauchy problem

$$u_t + f(u)_x = 0, \qquad u(0, x) = \bar{u}(x). \qquad (2.25)$$

Assuming that u satisfies suitable entropy and regularity conditions, one shows that

$$\liminf_{h \to 0+} \frac{\left\| u(t + h) - S_h u(t) \right\|_{\mathbf{L}^1}}{h} = 0 \qquad (2.26)$$

at almost every time t. By (2.23), u thus coincides with the semigroup trajectory $t \mapsto S_t u(0) = S_t \bar{u}$ obtained as limit of front tracking approximations. Of course, this implies uniqueness. We state here the main result in [BG]. A function $u = u(t, x)$ is said to have **tame oscillation** if there exist constants $C', \beta > 0$ such that the following holds. For every $\tau > 0$ and $a < b$, the oscillation of u over the triangle

$$\Delta^{\tau}_{a,b} \doteq \left\{ (s, y) ; \ s \geq t, \ a + \beta(s - \tau) < y < b - \beta(s - \tau) \right\}$$

satisfies

$$\text{Osc.}\{u; \Delta^{\tau}_{a,b}\} \leq C' \cdot \text{Tot.Var.}\{u(\tau, \cdot); \ ]a, b[\}. \qquad (2.27)$$

We recall that the oscillation of a u on a set Δ is defined as

$$\text{Osc.}\{u; \Delta\} \doteq \sup_{(s,y),\, (s',y') \in \Delta} \left| u(s, y) - u(s', y') \right|.$$

Theorem 3. *In the same setting as Theorem 1, let $u = u(t, x)$ be a weak solution of the Cauchy problem (2.20) satisfying the tame oscillation condition*

(2.27) and the Liu admissibility condition (2.15) at every point of approximate jump. Then u coincides with the corresponding semigroup trajectory, i.e.

$$u(t, \cdot) = S_t \bar{u} \qquad (2.28)$$

for all $t \geq 0$. In particular, the solution that satisfies the above conditions is unique.

An additional characterization of these admissible weak solutions, based on local integral estimates, was given in [B2]. To state this result, we need to introduce some notations. Given a function $u = u(t, x)$ and a point (τ, ξ), we denote by $U^{\sharp}_{(u;\tau,\xi)}$ the solution of the Riemann problem with initial data

$$u^- = \lim_{x \to \xi-} u(\tau, x), \qquad u^+ = \lim_{x \to \xi+} u(\tau, x). \qquad (2.29)$$

In addition, we define $U^{\flat}_{(u;\tau,\xi)}$ as the solution of the linear hyperbolic Cauchy problem with constant coefficients

$$w_t + A^{\flat} w_x = 0, \qquad w(0, x) = u(\tau, x). \qquad (2.30)$$

Here $A^{\flat} \doteq A\big(u(\tau, \xi)\big)$. Observe that (2.30) is obtained from the quasilinear system (1.2) by "freezing" the coefficients of the matrix $A(u)$ at the point (τ, ξ) and choosing $u(\tau)$ as initial data. A notion of "good solution" can now be introduced, by locally comparing a function u with the self-similar solution of a Riemann problem and with the solution of a linear hyperbolic system with constant coefficients. More precisely, we say that a function $u = u(t, x)$ is a **viscosity solution** of the system (1.1) if $t \mapsto u(t, \cdot)$ is continuous as a map with values into $\mathbf{L}^1_{loc}$, and moreover the following integral estimates hold.

(i) At every point (τ, ξ), for every $\beta' > 0$ one has

$$\lim_{h \to 0+} \frac{1}{h} \int_{-\beta' h}^{\beta' h} \left| u(\tau + h, \, \xi + x) - U^{\sharp}_{(u;\tau,\xi)}(h, \, x) \right| dx = 0. \qquad (2.31)$$

(ii) There exist constants $C, \beta > 0$ such that, for every $\tau \geq 0$ and $a < \xi < b$, one has

$$\limsup_{h \to 0+} \frac{1}{h} \int_{a+\beta h}^{b-\beta h} \left| u(\tau + h, \, x) - U^{\flat}_{(u;\tau,\xi)}(h, x) \right| dx$$
$$\leq C \cdot \Big(\text{Tot.Var.}\{u(\tau); \,]a, \, b[\,\} \Big)^2. \qquad (2.32)$$

The above definition was introduced in [B2], with the expectation that the "viscosity solutions" in the above integral sense should precisely coincide with the limits of vanishing viscosity approximations (1.7). This conjecture was proved only many years later in [BiB]. The main result in [B2] shows that the integral estimates (2.31)–(2.32) completely characterize the semigroup trajectories.

Theorem 4. *Let $S : \mathcal{D} \times [0, \infty[\times \mathcal{D}$ be the semigroup generated by the system of conservation laws (1.1), whose trajectories are limits of front tracking approximations. A function $u : [0, T] \mapsto \mathcal{D}$ is a viscosity solution of (1.1) if and only if $u(t) = S_t u(0)$ for all $t \in [0, T]$.*

3 The Vanishing Viscosity Approach

In view of the previous uniqueness and stability results, it is natural to expect that the entropy weak solutions of the hyperbolic system (1.1) should coincide with the unique limits of solutions to the parabolic system

$$u_t^\varepsilon + f(u^\varepsilon)_x = \varepsilon \, u_{xx}^\varepsilon \qquad (3.1)$$

letting the viscosity coefficient $\varepsilon \to 0$. For smooth solutions, this convergence is easy to show. However, one should keep in mind that a weak solution of the hyperbolic system (1.1) in general is only a function with bounded variation, possibly with a countable number of discontinuities. In this case, as the smooth functions u^ε approach the discontinuous solution u, near points of jump their gradients u_x^ε must tend to infinity (fig. 11), while their second derivatives u_{xx}^ε become even more singular. Therefore, establishing the convergence $u^\varepsilon \to u$ is a highly nontrivial matter. In earlier literature, results in this direction relied on three different approaches:

1 - Comparison principles for parabolic equations. For a scalar conservation law, the existence, uniqueness and global stability of vanishing viscosity solutions was first established by Oleinik [O] in one space dimension. The famous paper by Kruzhkov [K] covers the more general class of $\mathbf{L}^\infty$ solutions and is also valid in several space dimensions.

2 - Singular perturbations. This technique was developed by Goodman and Xin [GX], and covers the case where the limit solution u is piecewise smooth, with a finite number of non-interacting, entropy admissible shocks. See also [Yu] and [Ro], for further results in this direction.

3 - Compensated compactness. With this approach, introduced by Tartar and DiPerna [DP2], one first considers a weakly convergent subsequence $u^\varepsilon \rightharpoonup u$. For a class of 2×2 systems, one can show that this weak limit u actually provides a distributional solution to the nonlinear system (1.1). The proof relies on a compensated compactness argument, based on the representation of the weak limit in terms of Young measures, which must reduce to a Dirac mass due to the presence of a large family of entropies.

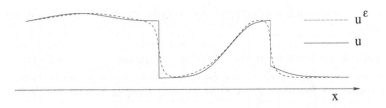

Fig. 11.

Since the hyperbolic Cauchy problem is known to be well posed within a space of functions with small total variation, it seems natural to develop a theory of vanishing viscosity approximations within the same space BV. This was indeed accomplished in [BiB], in the more general framework of nonlinear hyperbolic systems not necessarily in conservation form. The only assumptions needed here are the strict hyperbolicity of the system and the small total variation of the initial data.

Theorem 5. *Consider the Cauchy problem for the hyperbolic system with viscosity*

$$u_t^\varepsilon + A(u^\varepsilon)u_x^\varepsilon = \varepsilon\, u_{xx}^\varepsilon \qquad\qquad u^\varepsilon(0,x) = \bar u(x)\,. \qquad (3.2)$$

Assume that the matrices $A(u)$ are strictly hyperbolic, smoothly depending on u in a neighborhood of the origin. Then there exist constants C, L, L' and $\delta > 0$ such that the following holds. If

$$Tot.\,Var.\{\bar u\} < \delta\,, \qquad\qquad \|\bar u\|_{\mathbf{L}^\infty} < \delta, \qquad (3.3)$$

then for each $\varepsilon > 0$ the Cauchy problem $(3.2)_\varepsilon$ has a unique solution u^ε, defined for all $t \geq 0$. Adopting a semigroup notation, this will be written as $t \mapsto u^\varepsilon(t,\cdot) \doteq S_t^\varepsilon \bar u$. In addition, one has

BV bounds:
$$Tot.\,Var.\{S_t^\varepsilon \bar u\} \leq C\, Tot.\,Var.\{\bar u\}\,. \qquad (3.4)$$

L¹ stability:
$$\left\| S_t^\varepsilon \bar u - S_t^\varepsilon \bar v \right\|_{\mathbf{L}^1} \leq L \left\| \bar u - \bar v \right\|_{\mathbf{L}^1}, \qquad (3.5)$$

$$\left\| S_t^\varepsilon \bar u - S_s^\varepsilon \bar u \right\|_{\mathbf{L}^1} \leq L' \left(|t - s| + \left| \sqrt{\varepsilon t} - \sqrt{\varepsilon s} \right| \right)\,. \qquad (3.6)$$

Convergence: *As $\varepsilon \to 0+$, the solutions u^ε converge to the trajectories of a semigroup S such that*

$$\left\| S_t \bar u - S_s \bar v \right\|_{\mathbf{L}^1} \leq L \left\| \bar u - \bar v \right\|_{\mathbf{L}^1} + L' \, |t - s|\,. \qquad (3.7)$$

These vanishing viscosity limits can be regarded as the unique **vanishing viscosity solutions** *of the hyperbolic Cauchy problem*

$$u_t + A(u)u_x = 0, \qquad\qquad u(0,x) = \bar u(x)\,. \qquad (3.8)$$

In the conservative case $A(u) = Df(u)$, every vanishing viscosity solution is a weak solution of

$$u_t + f(u)_x = 0, \qquad\qquad u(0, x) = \bar{u}(x), \qquad\qquad (3.9)$$

satisfying the Liu admissibility conditions.

Assuming, in addition, that each field is genuinely nonlinear or linearly degenerate, the vanishing viscosity solutions coincide with the unique limits of Glimm and front tracking approximations.

In the genuinely nonlinear case, an estimate on the rate of convergence of these viscous approximations was provided in [BY]:

Theorem 6. *For the strictly hyperbolic system of conservation laws (3.9), assume that every characteristic field is genuinely nonlinear. At any time $t > 0$, the difference between the corresponding solutions of (3.2) and (3.9) can be estimated as*

$$\left\| u^\varepsilon(t, \cdot) - u(t, \cdot) \right\|_{\mathbf{L}^1} = \mathcal{O}(1) \cdot (1 + t)\sqrt{\varepsilon} |\ln \varepsilon| \ \text{Tot.Var.}\{\bar{u}\} .$$

The following remarks relate Theorem 5 with previous literature.

(i) Concerning the global existence of weak solutions to the system of conservation laws (3.9), Theorem 5 contains the famous result of Glimm [G]. It also slightly extends the existence theorem of Liu [L4], which does not require the assumption (H). Its main novelty lies in the case where the system (3.8) is non-conservative. However, one should be aware that for non-conservative systems the limit solution might change if we choose a viscosity matrix different from the identity, say

$$u_t + A(u)u_x = \varepsilon \left(B(u)u_x \right)_x .$$

The analysis of viscous solutions, with a more general viscosity matrix $B(u)$, is still a wide open problem.

(ii) Concerning the uniform stability of entropy weak solutions, the results previously available for $n \times n$ hyperbolic systems [BC1], [BCP], [BLY] always required the assumption (H). For 2×2 systems, this condition was somewhat relaxed in [AM]. On the other hand, Theorem 5 established this uniform stability for completely general $n \times n$ strictly hyperbolic systems, without any reference to the assumption (H).

(iii) For the viscous system (1.10), previous results in [L5], [SX], [SZ], [Yu] have established the stability of special types of solutions, such as traveling viscous shocks or viscous rarefactions, w.r.t. suitably small perturbations. Taking $\varepsilon = 1$, our present theorem yields at once the uniform Lipschitz stability of all viscous solutions with sufficiently small total variation, w.r.t. the

$\mathbf{L}^1$ distance. On the other hand, our approach has not yet been used to study the stability and convergence of viscous solutions with large total variation. For a result in this direction using singular perturbations, see [R].

We give below a general outline of the proof of Theorem 5. The main steps will then be discussed in more detail in the following sections.

1. As a preliminary, we observe that u^ε is a solution of (3.2) if and only if the rescaled function $u(t, x) \doteq u^\varepsilon(\varepsilon t, \varepsilon x)$ is a solution of the parabolic system with unit viscosity

$$u_t + A(u)u_x = u_{xx}, \tag{3.10}$$

with initial data $u(0, x) = \bar{u}(\varepsilon x)$. Clearly, the stretching of the space variable has no effect on the total variation. Notice however that the values of u^ε on a fixed time interval $[0, T]$ correspond to the values of u on the much longer time interval $[0, T/\varepsilon]$. To obtain the desired BV bounds for the viscous solutions u^ε, we can confine all our analysis to solutions of (3.10), but we need estimates uniformly valid for all times $t \geq 0$, depending only on the total variation of the initial data $\bar{u}$.

2. Consider a solution of (3.10), with initial data

$$u(0, x) = \bar{u}(x) \tag{3.11}$$

having small total variation, say

$$\text{Tot.Var.}\{\bar{u}\} \leq \delta_0.$$

Using standard parabolic estimates, one can then prove that the Cauchy problem (3.10)–(3.11) has a unique solution, defined on an initial time interval $[0, \hat{t}]$ with $\hat{t} \approx \delta_0^{-2}$. For $t \in [0, \hat{t}]$ the total variation remains small:

$$\text{Tot.Var.}\{u(t)\} = \|u_x(t)\|_{\mathbf{L}^1} = \mathcal{O}(1) \cdot \delta_0,$$

while all higher derivatives decay quickly. In particular

$$\|u_{xx}(t)\|_{\mathbf{L}^1} = \mathcal{O}(1) \cdot \frac{\delta_0}{\sqrt{t}}, \qquad \|u_{xxx}(t)\|_{\mathbf{L}^1} = \mathcal{O}(1) \cdot \frac{\delta_0}{t}.$$

As long as the total variation remains small, one can prolong the solution also for larger times $t > \hat{t}$. In this case, uniform bounds on higher derivatives remain valid.

3. To establish uniform bounds on the total variation, valid for arbitrarily large times, we decompose the gradient along a basis of unit vectors $\tilde{r}_i = \tilde{r}_i(u, u_x, u_{xx})$, say

$$u_x = \sum_i v_i \tilde{r}_i \tag{3.12}$$

We then derive an evolution equation for these gradient components, of the form

$$v_{i,t} + (\tilde{\lambda}_i v_i)_x - v_{i,xx} = \phi_i \qquad\qquad i = 1, \ldots, n, \qquad (3.13)$$

4. A careful examination of the source terms ϕ_i in (3.13) shows that they arise mainly because of interactions among different viscous waves. Their total strength can thus be estimated in terms of suitable interaction potentials. This will allow us to prove the implication

$$\|u_x(t)\|_{\mathbf{L}^1} \leq \delta_0 \quad \text{for all } t \in [\hat{t}, T], \qquad \sum_i \int_{\hat{t}}^T \int |\phi_i(t, x)| \, dx \, dt \leq \delta_0$$

$$\implies \quad \int_{\hat{t}}^T \int |\phi_i(t, x)| \, dx \, dt = \mathcal{O}(1) \cdot \delta_0^2 \qquad i = 1, \ldots, n.$$

$$(3.14)$$

Since the left hand side of (3.13) is in conservation form, one has

$$\|u_x(t)\|_{\mathbf{L}^1} \leq \sum_i \|v_i(t)\|_{\mathbf{L}^1} \leq \sum_i \left(\|v_i(\hat{t})\| + \int_{\hat{t}}^t \int |\phi_i(s, x)| \, dx \, ds \right). \quad (3.15)$$

By the implication (3.14), for δ_0 sufficiently small we obtain uniform bounds on the total strength of the source terms ϕ_i, and hence on the BV norm of $u(t, \cdot)$.

5. Similar techniques can also be applied to a solution $z = z(t, x)$ of the variational equation

$$z_t + \big[DA(u) \cdot z \big] u_x + A(u) z_x = z_{xx}, \qquad (3.16)$$

which describes the evolution of a first order perturbation to a solution u of (3.11). Assuming that the total variation of u remains small, one proves an estimate of the form

$$\|z(t, \cdot)\|_{\mathbf{L}^1} \leq L \|z(0, \cdot)\|_{\mathbf{L}^1} \qquad \text{for all } t \geq 0, \qquad (3.17)$$

valid for every solution u of (3.10) having small total variation and every $\mathbf{L}^1$ solution of the corresponding system (3.16).

6. Relying on (3.17), a standard homotopy argument yields the Lipschitz continuity of the flow of (3.10) w.r.t. the initial data, uniformly in time. Indeed, let any two solutions u, v of (3.10) be given (fig. 12). We can connect them by a smooth path of solutions u^θ, whose initial data satisfy

$$u^\theta(0, x) \doteq \theta u(0, x) + (1 - \theta) v(0, x), \qquad\qquad \theta \in [0, 1].$$

The distance $\|u(t, \cdot) - v(t, \cdot)\|_{\mathbf{L}^1}$ at any later time $t > 0$ is clearly bounded by the length of the path $\theta \mapsto u^\theta(t)$. In turn, this can be computed by integrating the norm of a tangent vector. Calling $z^\theta \doteq du^\theta/d\theta$, each vector z^θ is a solution of the corresponding equation (3.19), with u replaced by u^θ. Using (3.18) we thus obtain

$$\left\| u(t,\cdot) - v(t,\cdot) \right\|_{\mathbf{L}^1} \leq \int_0^1 \left\| \tfrac{d}{d\theta} u^\theta(t) \right\|_{\mathbf{L}^1} d\theta = \int_0^1 \left\| z^\theta(t) \right\|_{\mathbf{L}^1} d\theta$$

$$\leq L \int_0^1 \left\| z^\theta(0) \right\|_{\mathbf{L}^1} d\theta = L \left\| u(0,\cdot) - v(0,\cdot) \right\|_{\mathbf{L}^1}. \tag{3.18}$$

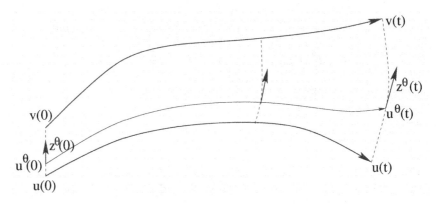

Fig. 12.

7. By the simple rescaling of coordinates $t \mapsto \varepsilon t$, $x \mapsto \varepsilon x$, all of the above estimates remain valid for solutions u^ε of the system $(3.2)_\varepsilon$. In particular this yields the bounds (3.4) and (3.5).

8. By a compactness argument, these BV bounds imply the existence of a strong limit $u^{\varepsilon_m} \to u$ in $\mathbf{L}^1_{\mathrm{loc}}$, at least for some subsequence $\varepsilon_m \to 0$. In the conservative case where $A = Df$, it is now easy to show that this limit u provides a weak solution to the Cauchy problem (3.9).

9. At this stage, it only remains to prove that the limit is unique, i.e. it does not depend on the choice of the sequence $\varepsilon_m \to 0$. For a system in conservative form, and with the standard assumption (H) that each field is either genuinely nonlinear or linearly degenerate, we can apply the uniqueness theorem in [BG] and conclude that the limit of vanishing viscosity approximations coincides with the limit of Glimm and of front tracking approximations.

10. To handle the general non-conservative case, some additional work is required. We first consider Riemann initial data and show that in this special case the vanishing viscosity solution is unique and can be accurately described. Then we prove that any weak solution obtained as limit vanishing viscosity approximations is also a "viscosity solution" in the sense that it satisfies the local integral estimates (2.31)–(2.32), where $U^\sharp$ is now the unique non-conservative solution of a Riemann problem. By an argument introduced in [B2], a Lipschitz semigroup is completely determined as soon as one specifies its local behavior for piecewise constant initial data. Characterizing its trajectories as "viscosity solutions" we thus obtain the uniqueness of the semigroup of vanishing viscosity limits.

4 Parabolic Estimates

Following a standard approach, one can write the parabolic system (3.11) in the form

$$u_t - u_{xx} = -A(u)u_x \qquad (4.1)$$

regarding the hyperbolic term $A(u)u_x$ as a first order perturbation of the heat equation. The general solution of (4.1) can then be represented as

$$u(t) = G(t) * u(0) - \int_0^t G(t-s) * \left[A(u(s))u_x(s) \right] ds$$

in terms of convolutions with the Gauss kernel

$$G(t,x) = \frac{1}{2\sqrt{\pi t}} e^{-x^2/4t}. \qquad (4.2)$$

From the above integral formula, one can derive local existence, uniqueness and regularity estimates for solutions of (4.1). Since we shall be dealing with a solution $u = u(t,x)$ having small total variation, a more effective representation is the following. Consider the state

$$u^* \doteq \lim_{x \to -\infty} u(t,x), \qquad (4.3)$$

which is independent of time. We then define the matrix $A^* \doteq A(u^*)$ and let λ_i^*, r_i^*, l_i^* be the corresponding eigenvalues and right and left eigenvectors, normalized as in (1.4). It will be convenient to use "$\bullet$" to denote a directional derivative, so that $z \bullet A(u) \doteq DA(u) \cdot z$ indicates the derivative of the matrix valued function $u \mapsto A(u)$ in the direction of the vector z. The systems (3.10) and (3.16) can now be written respectively as

$$u_t + A^* u_x - u_{xx} = \left(A^* - A(u) \right) u_x, \qquad (4.4)$$
$$z_t + A^* z_x - z_{xx} = \left(A^* - A(u) \right) z_x - \left(z \bullet A(u) \right) u_x. \qquad (4.5)$$

Observe that, if u is a solution of (4.4), then $z = u_x$ is a particular solution of the variational equation (4.5). Therefore, as soon as one proves an a priori bound on z_x or z_{xx}, the same estimate will be valid also for the corresponding derivatives u_{xx}, u_{xxx}.

In both of the equations (4.4), (4.5), we regard the right hand side as a perturbation of the linear parabolic system with constant coefficients

$$w_t + A^* w_x - w_{xx} = 0 \,. \tag{4.6}$$

We denote by G^* the Green kernel for (4.6), so that any solution admits the integral representation

$$w(t,x) = \int G^*(t,\, x - y)\, w(0,y)\, dy \,.$$

The matrix valued function G^* can be explicitly computed. If w solves (4.6), then the i-th component $w_i \doteq l_i^* \cdot w$ satisfies the scalar equation

$$w_{i,t} + \lambda_i^* w_{i,x} - w_{i,xx} = 0 \,.$$

Therefore $w_i(t) = G_i^*(t) * w_i(0)$, where

$$G_i^*(t,x) = \frac{1}{2\sqrt{\pi t}} \exp\left\{ -\frac{(x - \lambda_i^* t)^2}{4t} \right\} \,.$$

It is now clear that this Green kernel $G^* = G^*(t,x)$ satisfies the bounds

$$\left\| G^*(t) \right\|_{\mathbf{L}^1} \leq \kappa \,, \qquad \left\| G_x^*(t) \right\|_{\mathbf{L}^1} \leq \frac{\kappa}{\sqrt{t}} \,, \qquad \left\| G_{xx}^*(t) \right\|_{\mathbf{L}^1} \leq \frac{\kappa}{t} \,, \tag{4.7}$$

for some constant κ and all $t > 0$.

In the following, we consider an initial data $u(0,\cdot)$ having small total variation, but possibly discontinuous. We shall prove the local existence of solutions and some estimates on the decay of higher order derivatives. To get a feeling on this rate of decay, let us first take a look at the most elementary case.

Example 4.1. The solution to the Cauchy problem for the heat equation

$$u_t - u_{xx} = 0 \,, \qquad\qquad u(0,x) = \begin{cases} 0 & if \quad x < 0, \\ \delta_0 & if \quad x > 0, \end{cases}$$

is computed explicitly as

$$u(t,x) = \delta_0 \int_0^\infty G(t, x - y)\, dy \,.$$

Observing that the Gauss kernel (4.2) satisfies

$$G(t,x) = t^{-1/2}\, G(1,\, t^{-1/2} x),$$

for every $k \geq 1$ one obtains the estimates

$$\left\| \frac{\partial^{k-1}}{\partial x^{k-1}} G(t) \right\|_{\mathbf{L}^\infty} \leq \left\| \frac{\partial^k}{\partial x^k} G(t) \right\|_{\mathbf{L}^1} = \mathcal{O}(1) \cdot t^{-k/2} \,.$$

In the present example we have $u_x(t, x) = \delta_0 G(t, x)$. Therefore

$$\left\|u_x(t)\right\|_{\mathbf{L}^\infty} \leq \left\|u_{xx}(t)\right\|_{\mathbf{L}^1} = \mathcal{O}(1) \cdot \frac{\delta_0}{\sqrt{t}}, \tag{4.8}$$

$$\left\|u_{xx}(t)\right\|_{\mathbf{L}^\infty} \leq \left\|u_{xxx}(t)\right\|_{\mathbf{L}^1} = \mathcal{O}(1) \cdot \frac{\delta_0}{t}, \tag{4.9}$$

$$\left\|u_{xxx}(t)\right\|_{\mathbf{L}^\infty} = \mathcal{O}(1) \cdot \frac{\delta_0}{t\sqrt{t}}. \tag{4.10}$$

In the remainder of this chapter, our analysis will show that the same decay rates hold for solutions of the perturbed equation (4.4), restricted to some initial interval $[0, \hat{t}]$. More precisely, let δ_0 measure the order of magnitude of the total variation in the initial data, so that

$$\text{Tot.Var.}\{u_x(0, \cdot)\} = \mathcal{O}(1) \cdot \delta_0. \tag{4.11}$$

Then:

- There exists an initial interval $[0, \hat{t}]$, with $\hat{t} = \mathcal{O}(1) \cdot \delta_0^{-2}$ on which the solution of (4.4) is well defined. Its derivatives decay according to the estimates (4.8)–(4.10).
- As long as the total variation remains small, say

$$\left\|u_x(t)\right\|_{\mathbf{L}^1} \leq \delta_0, \tag{4.12}$$

the solution can be prolonged in time. In this case, for $t > \hat{t}$ the higher derivatives satisfy the bounds

$$\left\|u_x(t)\right\|_{\mathbf{L}^\infty}, \quad \left\|u_{xx}(t)\right\|_{\mathbf{L}^1} = \mathcal{O}(1) \cdot \delta_0^2, \tag{4.13}$$

$$\left\|u_{xx}(t)\right\|_{\mathbf{L}^\infty}, \quad \left\|u_{xxx}(t)\right\|_{\mathbf{L}^1} = \mathcal{O}(1) \cdot \delta_0^3, \tag{4.14}$$

$$\left\|u_{xxx}(t)\right\|_{\mathbf{L}^\infty} = \mathcal{O}(1) \cdot \delta_0^4. \tag{4.15}$$

The situation is summarized in fig. 13. All the estimates drawn here as solid lines will be obtained by standard parabolic-type arguments. If any breakdown occurs in the solution, the first estimate that fails must be the one in (4.12), concerning the total variation. Toward a proof of global existence of solutions, the most difficult step is to show that (4.12) remains valid for all times $t \in [\hat{t}, \infty[$ (the broken line in fig. 13). This will require hyperbolic-type estimates, based on the local decomposition of the gradient u_x as a sum of traveling waves, and on a careful analysis of all interaction terms. In the remainder of this section we work out a proof of local existence and a priori estimates for solutions of (4.4) and (4.5).

Proposition 4.2 (Local existence). *For $\delta_0 > 0$ suitably small, the following holds. Consider initial data*

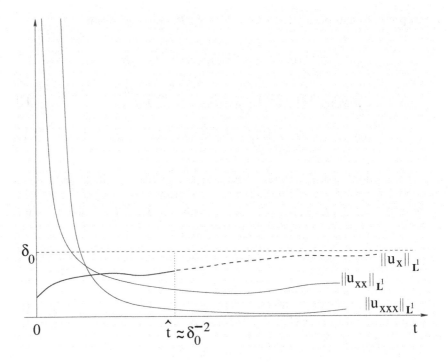

Fig. 13.

$$u(0, x) = \bar{u}(x), \qquad z(0, x) = \bar{z}(x). \qquad (4.16)$$

such that

$$Tot.\,Var.\{\bar{u}\} \le \frac{\delta_0}{4\kappa}, \qquad \bar{z} \in \mathbf{L}^1. \qquad (4.17)$$

Then the equations (4.4), (4.5) have unique solutions $u = u(t, x)$, $z = z(t, x)$
defined on the time interval $t \in [0, 1]$. *Moreover*

$$\left\| u_x(t) \right\|_{\mathbf{L}^1} \le \frac{\delta_0}{2}, \qquad \left\| z(t) \right\|_{\mathbf{L}^1} \le \frac{\delta_0}{2} \qquad \text{for all} \quad t \in [0, 1]. \qquad (4.18)$$

Proof. The couple (u, u_x), consisting of the solution of (4.2) together with
its derivative, will be obtained as the unique fixed point of a contractive
transformation. For notational convenience, we assume here

$$u^* \doteq \lim_{x \to -\infty} \bar{u}(x) = 0.$$

Of course this is not restrictive, since it can always be achieved by a translation
of coordinates. Consider the space E of all mappings

$$(u, z) : [0, 1] \mapsto \mathbf{L}^\infty \times \mathbf{L}^1,$$

with norm

$$\big\|(u,z)\big\|_E \doteq \sup_t \, \max\Big\{\big\|u(t)\big\|_{\mathbf{L}^\infty}, \ \big\|z(t)\big\|_{\mathbf{L}^1}\Big\}.$$

On E we define the transformation $T(u,z) = (\hat{u}, \hat{z})$ by setting

$$\hat{u}(t) \doteq G^*(t) * \bar{u} + \int_0^t G^*(t-s) * \big[A^* - A(u(s))\big] z(s) \, ds\,,$$

$$\hat{z}(t) \doteq G_x^*(t) * \bar{u} + \int_0^t G_x^*(t-s) * \big[A^* - A(u(s))\big] z(s) \, ds\,.$$

Of course, the above definition implies $\hat{z} = \hat{u}_x$. The assumption Tot.Var.$\{\bar{u}\} \le \delta_0/4\kappa$ together with the bounds (4.7) yields

$$\big\|G^*(t) * \bar{u}\big\|_{\mathbf{L}^\infty} \le \big\|G_x^*(t) * \bar{u}\big\|_{\mathbf{L}^1} \le \frac{\delta_0}{4}\,. \tag{4.19}$$

Moreover, if

$$\big\|u(s)\big\|_{\mathbf{L}^\infty} \le \frac{\delta_0}{2}\,, \qquad \big\|z(s)\big\|_{\mathbf{L}^1} \le \frac{\delta_0}{2} \qquad \text{for all} \quad s \in [0,1]\,,$$

then

$$\left\|\int_0^t G^*(t-s) * \big[A^* - A(u(s))\big] z(s) \, ds\right\|_{\mathbf{L}^\infty}$$
$$\le \int_0^t \kappa \, \|DA\| \, \big\|u(s)\big\|_{\mathbf{L}^\infty} \big\|z(s)\big\|_{\mathbf{L}^1} \, ds$$
$$\le t\kappa\kappa_A \frac{\delta_0^2}{4}$$

and similarly

$$\left\|\int_0^t G_x^*(t-s) * \big[A^* - A(u(s))\big] z(s) \, ds\right\|_{\mathbf{L}^1}$$
$$\le \int_0^t \frac{\kappa}{\sqrt{t-s}} \, \|DA\| \, \big\|u(s)\big\|_{\mathbf{L}^\infty} \big\|z(s)\big\|_{\mathbf{L}^1} \, ds$$
$$\le 2\sqrt{t}\,\kappa\kappa_A \frac{\delta_0^2}{4}\,.$$

Combining the above estimates with (4.19), we see that the transformation T maps the domain

$$\mathcal{D} \doteq \Big\{(u,z) : [0,1] \mapsto \mathbf{L}^\infty \times \mathbf{L}^1\,; \ \ \big\|u(t)\big\|_{\mathbf{L}^\infty}, \ \big\|z(t)\big\|_{\mathbf{L}^1} \le \delta_0/2 \ \text{ for all } \ t \in [0,1]\Big\}$$

into itself. To prove that T is a strict contraction, we compute the difference $T(u,z) - T(u',z')$. The norm of the first component is estimated as

$$\|\hat{u} - \hat{u}'\|_{\mathbf{L}^\infty} = \left\| \int_0^t G^*(t-s) * \Big\{ \big[A^* - A(u(s)) \big] \big(z(s) - z'(s) \big) \right.$$
$$\left. + \big[A(u'(s)) - A(u(s)) \big] z'(s) \Big\} ds \right\|_{\mathbf{L}^\infty}$$
$$\leq \int_0^t \|G^*(t-s)\|_{\mathbf{L}^\infty} \Big\{ \|DA\| \|u(s)\|_{\mathbf{L}^\infty} \|z(s) - z'(s)\|_{\mathbf{L}^1}$$
$$+ \|DA\| \|u(s) - u'(s)\|_{\mathbf{L}^\infty} \|z'(s)\|_{\mathbf{L}^1} \Big\} ds$$
$$\leq 2\sqrt{t} \kappa \kappa_A \delta_0 \|(u - u', \ z - z')\|_E .$$

An entirely similar computation yields

$$\|\hat{z}(t) - \hat{z}'(t)\|_{\mathbf{L}^1} \leq 2\sqrt{t} \kappa \kappa_A \delta_0 \|(u - u', \ z - z')\|_E .$$

Therefore, for δ_0 small enough, the map $\mathcal{T}$ is a strict contraction.

By the contraction mapping theorem, a unique fixed point exists, inside the domain $\mathcal{D}$. Clearly, this provides the solution of (4.4) with the prescribed initial data.

Having constructed a solution u of (4.4), we now prove the existence of a solution z of the linearized variational system (4.5), with initial data $\bar{z} \in \mathbf{L}^1$. Consider the space E' of all mappings $z : [0,1] \mapsto \mathcal{W}^{1,1}$ with norm

$$\|z\|_{E'} \doteq \sup_t \max \Big\{ \|z(t)\|_{\mathbf{L}^1} , \ \sqrt{t} \|z_x(t)\|_{\mathbf{L}^1} \Big\} .$$

On E' we define the transformation $\mathcal{T}(z) = \hat{z}$ by setting

$$\hat{z}(t) \doteq G^*(t) * \bar{z} + \int_0^t G^*(t-s) * \Big\{ \big[A^* - A(u(s)) \big] z_x(s) - \big[z(s) \bullet A(u(s)) \big] u_x(s) \Big\} ds .$$

The bounds (4.7) now yield

$$\|G^*(t) * \bar{z}\|_{\mathbf{L}^1} \leq \kappa \|\bar{z}\|_{\mathbf{L}^1} , \qquad \|G_x^*(t) * \bar{z}\|_{\mathbf{L}^1} \leq \frac{\kappa}{\sqrt{t}} \|\bar{z}\|_{\mathbf{L}^1} . \qquad (4.20)$$

Moreover, using the identity

$$\int_0^t \frac{1}{\sqrt{s(t-s)}} ds = \int_0^1 \frac{1}{\sqrt{\sigma(1-\sigma)}} d\sigma = \pi < 4 \qquad (4.21)$$

one obtains the bound

$$\left\| \int_0^t G_x^*(t-s) * \Big\{ \big[A^* - A(u(s)) \big] z_x(s) - \big[z(s) \bullet A(u(s)) \big] u_x(s) \Big\} ds \right\|_{\mathbf{L}^1}$$
$$\leq \int_0^t \frac{\kappa}{\sqrt{t-s}} \Big\{ \|DA\| \|u(s)\|_{\mathbf{L}^\infty} \|z_x(s)\|_{\mathbf{L}^1} + \|DA\| \|z_x(s)\|_{\mathbf{L}^1} \|u_x(s)\|_{\mathbf{L}^1} \Big\}$$
$$\leq 4\kappa \, \kappa_A \delta_0 \|z\|_{E'} .$$

Assuming that δ_0 is small enough, it is now clear that the linear transformation $\mathcal{T}'$ is a strict contraction in the space E'. Hence it has a unique fixed point, which provides the desired solution of (4.5). $\qquad \bullet$

Having established the local existence of a solution, we now study the decay of its higher order derivatives.

Proposition 4.3. *Let u, z be solutions of the systems (4.4) and (4.5), satisfying the bounds*

$$\left\|u_x(t)\right\|_{\mathbf{L}^1} \le \delta_0, \qquad\qquad \left\|z(t)\right\|_{\mathbf{L}^1} \le \delta_0, \qquad\qquad (4.22)$$

for some constant $\delta_0 < 1$ and all $t \in [0, \hat{t}]$, where

$$\hat{t} \doteq \left(\frac{1}{400\kappa\,\kappa_A\,\delta_0}\right)^2, \qquad \kappa_A \doteq \sup_u\left(\|DA\| + \|D^2 A\|\right) \qquad (4.23)$$

and κ is the constant in (4.7). Then for $t \in [0, \hat{t}]$ the following estimates hold.

$$\left\|u_{xx}(t)\right\|_{\mathbf{L}^1}, \ \left\|z_x(t)\right\|_{\mathbf{L}^1} \le \frac{2\kappa\delta_0}{\sqrt{t}}, \qquad\qquad (4.24)$$

$$\left\|u_{xxx}(t)\right\|_{\mathbf{L}^1}, \ \left\|z_{xx}(t)\right\|_{\mathbf{L}^1} \le \frac{5\kappa^2\delta_0}{t}, \qquad\qquad (4.25)$$

$$\left\|u_{xxx}(t)\right\|_{\mathbf{L}^\infty}, \ \left\|z_{xx}(t)\right\|_{\mathbf{L}^\infty} \le \frac{16\kappa^3\delta_0}{t\sqrt{t}}. \qquad\qquad (4.26)$$

Proof. We begin with (4.24). The function z_x can be represented in terms of convolutions with the Green kernel G^*, as

$$z_x(t) = G_x^*(t) * z(0) + \int_0^t G_x^*(t-s) * \left\{\left(A^* - A(u)\right)z_x(s) - \left(z \bullet A(u)\right)u_x(s)\right\} ds. \qquad (4.27)$$

Using (4.7) and (4.22) we obtain

$$\left\|\int_0^t G_x^*(t-s) * \left\{\left(A^* - A(u)\right)z_x(s) - \left(z \bullet A(u)\right)u_x(s)\right\} ds\right\|_{\mathbf{L}^1}$$
$$\le \int_0^t \left\|G_x^*(t-s)\right\|_{\mathbf{L}^1} \cdot \left\{\left\|u_x(s)\right\|_{\mathbf{L}^1}\left\|DA\right\|_{\mathbf{L}^\infty}\left\|z_x(s)\right\|_{\mathbf{L}^1}\right.$$
$$\left. + \left\|z(s)\right\|_{\mathbf{L}^\infty}\left\|DA\right\|_{\mathbf{L}^\infty}\left\|u_x(s)\right\|_{\mathbf{L}^1}\right\} ds$$
$$\le 2\delta_0\kappa\left\|DA\right\|_{\mathbf{L}^\infty} \cdot \int_0^t \frac{1}{\sqrt{t-s}}\left\|z_x(s)\right\|_{\mathbf{L}^1} ds.$$

Consider first the case of smooth initial data. We shall argue by contradiction. If (4.24) fails for some $t \le \hat{t}$, by continuity will be a first time $\tau < \hat{t}$ at which (4.24) is satisfied as an equality. In this case, recalling the identity (4.21) we compute

$$\left\|z_x(\tau)\right\|_{\mathbf{L}^1} \le \frac{\kappa}{\sqrt{\tau}}\delta_0 + 2\kappa\delta_0\left\|DA\right\|_{\mathbf{L}^\infty} \cdot \int_0^\tau \frac{1}{\sqrt{\tau-s}}\frac{2\delta_0\kappa}{\sqrt{s}} ds$$
$$< \frac{\kappa\delta_0}{\sqrt{\tau}} + 16\kappa^2\kappa_A\delta_0^2\left\|DA\right\|_{\mathbf{L}^\infty} \le \frac{2\kappa\delta_0}{\sqrt{\tau}},$$

reaching a contradiction. Hence, (4.24) must be satisfied as a strict inequality for all $t \in [0, \hat{t}]$.

A similar technique is used to establish (4.25). Indeed, we can write

$$z_{xx}(t) = G_x^*(t/2) * z_x(t/2)$$
$$- \int_{t/2}^{t} G_x^*(t - s) * \left\{ (z \bullet A(u)) u_x(s) + (A(u) - A^*) z_x(s) \right\}_x ds .$$

$$(4.28)$$

We prove (4.16) first in the case $z_{xx} = u_{xxx}$, then in the general case. If (4.25) is satisfied as an equality at a first time $\tau < \hat{t}$, using (4.7), (4.24) and recalling the definitions of the constants $\hat{t}, \kappa_A$ in (4.23), we compute

$$\left\| z_{xx}(\tau) \right\|_{\mathbf{L}^1} \leq \frac{\kappa}{\sqrt{\tau/2}} \cdot \frac{2\kappa\delta_0}{\sqrt{\tau/2}} + \int_{\tau/2}^{\tau} \frac{\kappa}{\sqrt{\tau - s}} \cdot \left\{ \left\| z_x \bullet A(u) u_x(s) \right\|_{\mathbf{L}^1} \right.$$
$$+ \left\| z \bullet (u_x \bullet A(u)) u_x(s) \right\|_{\mathbf{L}^1} + \left\| z \bullet A(u) u_{xx}(s) \right\|_{\mathbf{L}^1}$$
$$+ \left\| u_x \bullet A(u) z_x(s) \right\|_{\mathbf{L}^1} + \left. \left\| (A(u) - A^*) z_{xx}(s) \right\|_{\mathbf{L}^1} \right\} ds$$

$$\leq \frac{2\kappa^2\delta_0}{\tau/2} + \int_{\tau/2}^{\tau} \frac{\kappa}{\sqrt{\tau - s}} \cdot \left\{ \delta_0 \|DA\|_{\mathbf{L}^\infty} \left\| z_{xx}(s) \right\|_{\mathbf{L}^1} \right.$$
$$+ \delta_0 \|D^2 A\|_{\mathbf{L}^\infty} \left\| u_{xx}(s) \right\|_{\mathbf{L}^1}^2 + \delta_0 \|DA\|_{\mathbf{L}^\infty} \left\| u_{xxx}(s) \right\|_{\mathbf{L}^1}$$
$$+ \delta_0 \|DA\|_{\mathbf{L}^\infty} \left\| z_{xx}(s) \right\|_{\mathbf{L}^1} + \left. \delta_0 \|DA\|_{\mathbf{L}^\infty} \left\| z_{xx}(s) \right\|_{\mathbf{L}^1} \right\} ds$$

$$\leq \frac{4\kappa^2\delta_0}{\tau} + \kappa\delta_0 \left(4\kappa^2\delta_0^2 \|D^2 A\|_{\mathbf{L}^\infty} + 20\kappa^2\delta_0 \|DA\|_{\mathbf{L}^\infty} \right) \int_{\tau/2}^{\tau} \frac{1}{s\sqrt{\tau - s}} ds$$

$$< \frac{4\kappa^2\delta_0}{\tau} + 20\kappa^3\kappa_A\delta_0^2 \cdot \frac{4}{\sqrt{\tau/2}} < \frac{5\kappa^2\delta_0}{\tau} .$$

Again, we conclude that at time τ the bound (4.16) is still satisfied as a strict inequality, thus reaching a contradiction.

Finally, assume that $\tau < \hat{t}$ is the first time where the bound (4.26) is satisfied as an equality. Using (4.7) and the previous bounds (4.24)–(4.25), from the representation (4.28) we now obtain

$$\left\| z_{xx}(\tau) \right\|_{L^\infty} \leq \frac{\kappa}{\sqrt{\tau/2}} \cdot \frac{5\kappa^2\delta_0}{\tau/2} + \int_{\tau/2}^{\tau} \frac{\kappa}{\sqrt{\tau - s}} \cdot \left\{ \left\| z_x \bullet A(u) u_x(s) \right\|_{L^\infty} \right.$$
$$+ \left\| z \bullet (u_x \bullet A(u)) u_x(s) \right\|_{L^\infty} + \left\| z \bullet A(u) u_{xx}(s) \right\|_{L^\infty}$$
$$+ \left\| u_x \bullet A(u) z_x(s) \right\|_{L^\infty} + \left. \left\| (A(u) - A^*) z_{xx}(s) \right\|_{L^\infty} \right\} ds$$

$$leq \frac{10\sqrt{2}\kappa^3\delta_0}{\tau\sqrt{\tau}} + \left(8\kappa^4\delta_0^3 \|D^2 A\| + 46\kappa^4\delta_0^2 \|DA\| \right) \int_{\tau/2}^{\tau} \frac{1}{s^{3/2}\sqrt{\tau - s}} ds$$

$$\leq \frac{15\kappa^3\delta_0}{\tau\sqrt{\tau}} + 46\kappa^4\kappa_A\delta_0^2 \cdot \frac{4}{\tau/2} < \frac{16\kappa^3\delta_0}{\tau\sqrt{\tau}} .$$

Therefore, the bound (4.26) must be satisfied as a strict inequality for all $t \leq \hat{t}$.   •

Corollary 4.4. *In the same setting as Proposition 4.2, assume that the bounds (4.12) hold on a larger interval* $[0, T]$. *Then for all* $t \in [\hat{t}, T]$ *there holds*

$$\left\|u_{xx}(t)\right\|_{\mathbf{L}^1}, \ \left\|u_x(t)\right\|_{\mathbf{L}^\infty}, \ \left\|z_x(t)\right\|_{\mathbf{L}^1} = \mathcal{O}(1) \cdot \delta_0^2, \qquad (4.29)$$

$$\left\|u_{xxx}(t)\right\|_{\mathbf{L}^1}, \ \left\|u_{xx}(t)\right\|_{\mathbf{L}^\infty}, \ \left\|z_{xx}(t)\right\|_{\mathbf{L}^1} = \mathcal{O}(1) \cdot \delta_0^3, \qquad (4.30)$$

$$\left\|u_{xxx}(t)\right\|_{\mathbf{L}^\infty}, \ \left\|z_{xx}(t)\right\|_{\mathbf{L}^\infty} = \mathcal{O}(1) \cdot \delta_0^4. \qquad (4.31)$$

Proof. This follows by applying Proposition 4.3 on the interval $[t - \hat{t}, \ t]$.   ●

The next result provides a first estimate on the time interval where the solution u of (4.4) exists and remain small.

Proposition 4.5. *For* $\delta_0 > 0$ *sufficiently small, the following holds. Consider initial data* $\bar{u}, \bar{z}$ *as in (4.16), satisfying the bounds*

$$\mathrm{Tot.\,Var.}\{\bar{u}\} \le \frac{\delta_0}{4\kappa}, \qquad \qquad \|\bar{z}\|_{\mathbf{L}^1} \le \frac{\delta_0}{4\kappa}. \qquad (4.32)$$

Then the systems (4.4), (4.5) admit unique solutions $u = u(t, x)$, $z = z(t, x)$ *defined on the whole interval* $[0, \hat{t}]$, *with* $\hat{t}$ *defined at (4.23). Moreover, one has*

$$\left\|u_x(t)\right\|_{\mathbf{L}^1} \le \frac{\delta_0}{2}, \qquad \qquad \left\|z(t)\right\|_{\mathbf{L}^1} \le \frac{\delta_0}{2} \qquad \text{for all} \quad t \in \,]0, \hat{t}]. \quad (4.33)$$

Proof. Because of the local existence result proved in Proposition 4.2, it suffices to check that the bounds (4.32) remain valid for all $t \le \hat{t}$. The solution of (4.5) can be written in the form

$$z(t) = G^*(t) * \bar{z} + \int_0^t G^*(t - s) * \left\{ (A^* - A(u)) z_x(s) - (z \bullet A(u)) u_x(s) \right\} ds.$$

As before, we first establish the result for $z = u_x$, then for a general solution z of (4.5). Assume that there exists a first time $\tau < \hat{t}$ where the bound in (4.17) is satisfied as an equality. Using (4.7) and (4.24) and recalling the definition of $\hat{t}$ at (4.23), from the above identity we obtain

$$\left\|z(\tau)\right\|_{\mathbf{L}^1} \le \kappa \left\|\bar{z}\right\|_{\mathbf{L}^1} + \int_0^\tau \kappa \cdot \left\{ \|DA\| \left\|u_x(s)\right\|_{\mathbf{L}^1} \left\|z_x(s)\right\|_{\mathbf{L}^1} \right.$$
$$\left. + \|DA\| \left\|z_x(s)\right\|_{\mathbf{L}^1} \left\|u_x(s)\right\|_{\mathbf{L}^1} \right\} ds$$
$$\le \frac{\kappa\delta_0}{4\kappa} + \int_0^\tau \frac{2\kappa\delta_0^2}{\sqrt{s}} \|DA\| \, ds$$
$$\le \frac{\delta_0}{4} + 4\kappa \, \kappa_A \delta_0^2 \sqrt{\tau} < \frac{\delta_0}{2},$$

reaching a contradiction.   ●

To simplify the proof, we assumed here the same bounds on the functions u_x and z. However, observing that z solves a linear homogeneous equation, similar estimates can be immediately derived without any restriction on the initial size of $\bar{z} \in \mathbf{L}^1$. In particular, from Proposition 4.5 it follows

Corollary 4.6. *Let $u = u(t,x)$, $z = z(t,x)$ be solutions of (4.4), (4.5) respectively, with $\left\|u_x(0)\right\|_{\mathbf{L}^1} \leq \delta_0/4\kappa$. Then u, z are well defined on the whole interval $[0,\hat{t}]$ in (4.13), and satisfy*

$$\left\|u_x(t)\right\|_{\mathbf{L}^1} \leq 2\kappa\left\|u_x(0)\right\|_{\mathbf{L}^1}, \quad \left\|z(t)\right\|_{\mathbf{L}^1} \leq 2\kappa\left\|z(0)\right\|_{\mathbf{L}^1}, \quad \text{for all } t \leq \hat{t}. \tag{4.34}$$

5 Decomposition by Traveling Wave Profiles

Given a smooth function $u : \mathbb{R} \mapsto \mathbb{R}^n$, at each point x we want to decompose the gradient u_x as a sum of gradients of viscous traveling waves, say

$$u_x(x) = \sum_i v_i \tilde{r}_i = \sum_i U_i'(x). \tag{5.1}$$

We recall that, for the parabolic system

$$u_t + A(u)u_x = u_{xx}, \tag{5.2}$$

a traveling wave profile is a solution of the special form $u(t,x) = U(x - \sigma t)$. For the application that we have in mind, we can assume that the function u has small total variation, and that all of its derivatives are uniformly small. However, in order to come up with a unique decomposition of the form (1.10)–(1.11), we still face two major obstacles.

1. The system (1.9)–(1.11) is highly underdetermined. Indeed, by (1.9)–(1.10) each traveling profile U_i provides a solution to the Cauchy problem

$$U_i'' = \bigl(A(U_i) - \sigma_i\bigr)U_i', \qquad U_i(x) = u(x). \tag{5.3}$$

Since this is a second order equation, we can still choose the derivative $U'(x) \in \mathbb{R}^n$ arbitrarily, as well as the speed σ_i. The general solution of (5.3) thus depends on $n + 1$ scalar parameters. Summing over all n families of waves, we obtain $n(n + 1)$ free parameters. On the other hand, the system

$$\sum_i U_i'(x) = u_x(x), \qquad \sum_i U_i''(x) = u_{xx}(x) \tag{5.4}$$

consists of only $n + n$ scalar equations. When $n \geq 2$, these are not enough to uniquely determine the traveling profiles $U_1, \ldots, U_n$.

2. For particular values of u_x, u_{xx}, the equations (5.3)–(5.4) may not have any solution at all. This can already be seen for a scalar conservation law

$$u_t + \lambda(u)u_x = u_{xx}.$$

In this case, the above equations reduce to

$$U'' = \big(\lambda(U) - \sigma\big)U', \qquad\qquad U(x) = u(x),$$

$$U'(x) = u_x(x), \qquad\qquad U''(x) = u_{xx}(x).$$

The solution is now unique. The only choice for the speed is

$$\sigma = \lambda\big(u(x)\big) - \frac{u_{xx}(x)}{u_x(x)}.$$

The smallness assumption on u_x, u_{xx} is not of much use here. If $u_x = 0$ while $u_{xx} \neq 0$, no solution is found. This typically occurs at points where u has a local maximum or minimum.

To overcome the first obstacle, we shall consider not the family of all traveling profiles, but only certain subfamilies. More precisely, for each $i = 1, \ldots, n$, our decomposition will involve only those traveling profiles U_i which lie on a suitable center manifold.

The second difficulty will be removed by introducing a cutoff function. In a normal situation, our decomposition will be uniquely determined by (5.3)–(5.4). In the exceptional cases where the cutoff function is active, only the first equation in (5.4) will be satisfied, allowing for some error in the second equation.

5.1. Construction of a Center Manifold

To carry out our program, we begin by selecting certain families of traveling waves, depending on the correct number of parameters to fit the data. Observing that the system (5.4) consists of $n + n$ scalar equations, this number of parameters is easy to guess. To achieve a unique solution, through any given state u we need to construct n families of traveling wave profiles U_i, each depending on two scalar parameters. The construction will be achieved by an application of the center manifold theorem.

Traveling waves for the parabolic system (5.2) correspond to (possibly unbounded) solutions of

$$\big(A(U) - \sigma\big)U' = U''. \qquad\qquad (5.5)$$

We write (5.5) as a first order system on the space $\mathbb{R}^n \times \mathbb{R}^n \times \mathbb{R}$:

$$\begin{cases} \dot{u} = v, \\ \dot{v} = \big(A(u) - \sigma\big)v, \\ \dot{\sigma} = 0. \end{cases} \tag{5.6}$$

Let a state u^* be given and fix an index $i \in \{1, \ldots, n\}$. Linearizing (5.6) at the equilibrium point $P^* \doteq (u^*, 0, \lambda_i(u^*))$ we obtain the linear system

$$\begin{cases} \dot{u} = v, \\ \dot{v} = \big(A(u^*) - \lambda_i(u^*)\big)v, \\ \dot{\sigma} = 0. \end{cases} \tag{5.7}$$

Let $\{r_1^*, \ldots, r_n^*\}$ and $\{l_1^*, \ldots, l_n^*\}$ be dual bases of right and left eigenvectors of $A(u^*)$ normalized as in (1.4). We call $(V_1, \ldots, V_n)$ the coordinates of a vector $v \in \mathbb{R}^n$ w.r.t. this basis, so that

$$|r_i^*| = 1, \qquad v = \sum_j V_j r_j^*, \qquad V_j \doteq l_j^* \cdot v.$$

The center subspace $\mathcal{N}$ for (5.7) (see fig. 14) consists of all vectors $(u, v, \sigma) \in \mathbb{R}^n \times \mathbb{R}^n \times \mathbb{R}$ such that

$$V_j = 0 \qquad \text{for all } j \neq i, \tag{5.8}$$

and therefore has dimension $n + 2$. By the center manifold theorem (see the Appendix), there exists a smooth manifold $\mathcal{M} \subset \mathbb{R}^{n+n+1}$, tangent to $\mathcal{N}$ at the stationary point P^*, which is locally invariant under the flow of (5.6). This manifold has dimension $n + 2$ and can be locally defined by the $n - 1$ equations

$$V_j = \varphi_j(u, V_i, \sigma) \qquad j \neq i. \tag{5.9}$$

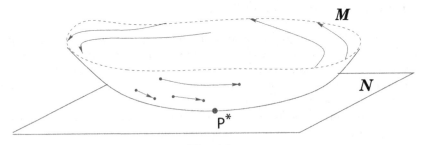

Fig. 14.

We seek a set of coordinates on this center manifold, involving $n + 2$ free parameters. As a preliminary, observe that a generic point on the center subspace $\mathcal{N}$ can be described as

$$P = (u, v, \sigma) = (u, v_i r_i^*, \sigma).$$

We can thus regard $(u, v_i, \sigma) \in \mathbb{R}^{n+2}$ as coordinates of the point P. Notice that $v_i = \pm |v|$ is the signed strength of the vector v, while σ is the wave speed. Similarly, we will show that a generic point on the center manifold $\mathcal{M}$ can be written as

$$P = (u, v, \sigma) = (u, v_i \tilde{r}_i, \sigma),$$

where again v_i is the signed strength of v and $\tilde{r}_i$ is a unit vector parallel to v. In general, this unit vector is not constant as in the linear case, but depends on u, v_i, σ. All the information about the center manifold is thus encoded in the map $(u, v_i, \sigma) \mapsto \tilde{r}_i(u, v_i, \sigma)$.

We can assume that the $n - 1$ smooth scalar functions φ_j in (5.9) are defined on the domain

$$\mathcal{D} \doteq \left\{ (u, V_i, \sigma_i) ; \ |u - u^*| < \epsilon, \ |V_i| < \epsilon, \ |\sigma_i - \lambda_i(u^*)| < \epsilon \right\} \subset \mathbb{R}^{n+2}, \quad (5.10)$$

for some small $\epsilon > 0$. The tangency condition implies

$$\varphi_j(u, V_i, \sigma) = \mathcal{O}(1) \cdot \left(|u - u^*|^2 + |V_i|^2 + |\sigma - \lambda_i(u^*)|^2 \right). \quad (5.11)$$

By construction, every trajectory

$$t \mapsto P(t) \doteq \big(u(t), v(t), \sigma(t) \big)$$

of (5.6), which remains within a small neighborhood of the point $P^* \doteq \big(u^*, 0, \lambda_i(u^*) \big)$ for all $t \in \mathbb{R}$, must lie entirely on the manifold $\mathcal{M}$. In particular, $\mathcal{M}$ contains all viscous i-shock profiles joining a pair of states u^-, u^+ sufficiently close to u^*. Moreover, all equilibrium points $(u, 0, \sigma)$ with $|u - u^*| < \epsilon$ and $|\sigma - \lambda_i(u^*)| < \epsilon$ must lie on $\mathcal{M}$. Hence

$$\varphi_j(u, 0, \sigma) = 0 \qquad \text{for all } j \neq i. \quad (5.12)$$

By (5.12) and the smoothness of the functions φ_j, we can "factor out" the component V_i and write

$$\varphi_j(u, V_i, \sigma) = \psi_j(u, V_i, \sigma) \cdot V_i,$$

for suitable smooth functions ψ_j. From (5.11) it follows

$$\psi_j \to 0 \qquad \text{as} \qquad (u, V_i, \sigma) \to \big(u^*, 0, \lambda_i(u^*) \big). \quad (5.13)$$

On the manifold $\mathcal{M}$ we thus have

$$v = \sum_k V_k r_k^* = V_i \cdot \left(r_i^* + \sum_{j \neq i} \psi_j(u, V_i, \sigma) r_j^* \right) \doteq V_i \, r_i^\sharp(u, V_i, \sigma). \quad (5.14)$$

By (5.13), the function $r^\sharp$ defined by the last equality in (5.14) satisfies

$$r_i^\sharp(u, V_i, \sigma) \to r_i^* \qquad \text{as} \qquad (u, V_i, \sigma) \to \big(u^*, 0, \lambda_i(u^*) \big). \quad (5.15)$$

Notice that, for a traveling profile U_i of the i-th family, the derivative $v = U_i'$ should be "almost parallel" to the eigenvector r_i^*. This is indeed confirmed by (5.15).

We can now define the new variable

$$v_i = v_i(u, V_i, \sigma) \doteq V_i \cdot \left| r_i^\sharp(u, V_i, \sigma) \right|. \tag{5.16}$$

As (u, V_i, σ) ranges in a small neighborhood of $(u^*, 0, \lambda_i(u^*))$, by (5.15) the vector $r_i^\sharp$ remains close to the unit vector $r_i^* \doteq r_i(u^*)$. In particular, its length $|r_i^\sharp|$ remains uniformly positive. Hence the transformation $V_i \longleftrightarrow v_i$ is invertible and smooth. We can thus reparametrize the center manifold $\mathcal{M}$ in terms of the variables $(u, v_i, \sigma) \in \mathbb{R}^n \times \mathbb{R} \times \mathbb{R}$. Moreover, we define the unit vector

$$\tilde{r}_i(u, v_i, \sigma) \doteq \frac{r_i^\sharp}{|r_i^\sharp|}. \tag{5.17}$$

Observe that $\tilde{r}_i$ is also a smooth function of its arguments. With the above definitions, in alternative to (5.9) we can express the manifold $\mathcal{M}$ by means of the equation

$$v = v_i \tilde{r}_i(u, v_i, \sigma). \tag{5.18}$$

The above construction of a center manifold can be repeated for every $i = 1, \ldots, n$. We thus obtain n center manifolds $\mathcal{M}_i \subset \mathbb{R}^{n+n+1}$ and vector functions $\tilde{r}_i = \tilde{r}_i(u, v_i, \sigma_i)$ such that

$$|\tilde{r}_i| \equiv 1, \tag{5.19}$$

$$\mathcal{M}_i = \left\{ (u, v, \sigma_i) ; \quad v = v_i \tilde{r}_i(u, v_i, \sigma_i) \right\}, \tag{5.20}$$

as $(u, v_i, \sigma_i) \in \mathbb{R}^n \times \mathbb{R} \times \mathbb{R}$ ranges in a neighborhood of $(u^*, 0, \lambda_i(u^*))$.

Next, we derive some identities for future use. The partial derivatives of $\tilde{r}_i = \tilde{r}_i(u, v_i, \sigma_i)$ w.r.t. its arguments will be written as

$$\tilde{r}_{i,u} \doteq \frac{\partial}{\partial u} \tilde{r}_i, \qquad \tilde{r}_{i,v} \doteq \frac{\partial}{\partial v_i} \tilde{r}_i, \qquad \tilde{r}_{i,\sigma} \doteq \frac{\partial}{\partial \sigma_i} \tilde{r}_i.$$

Of course $\tilde{r}_{i,u}$ is an $n \times n$ matrix, while $\tilde{r}_{i,v}$, $\tilde{r}_{i,\sigma}$ are n-vectors. Higher order derivatives are denoted as $\tilde{r}_{i,u\sigma}$, $\tilde{r}_{i,\sigma\sigma} \ldots$ We claim that

$$\tilde{r}_i(u, 0, \sigma_i) = r_i(u) \qquad \text{for all } u, \sigma_i. \tag{5.21}$$

Indeed, consider again the equation for a viscous traveling i-wave:

$$u_{xx} = \left(A(u) - \sigma_i \right) u_x. \tag{5.22}$$

For a solution contained in the center manifold, taking the derivative w.r.t. x of

$$u_x = v = v_i \tilde{r}_i(u, v_i, \sigma_i) \tag{5.23}$$

and using (5.22) one finds

$$v_{i,x} \tilde{r}_i + v_i \tilde{r}_{i,x} = \big(A(u) - \sigma_i\big) v_i \tilde{r}_i . \tag{5.24}$$

Since $|\tilde{r}_i| \equiv 1$, the vector $\tilde{r}_i$ is perpendicular to its derivative $\tilde{r}_{i,x}$. Taking the inner product of (5.24) with $\tilde{r}_i$ we thus obtain

$$v_{i,x} = (\tilde{\lambda}_i - \sigma_i) v_i , \tag{5.25}$$

where we defined the "generalized eigenvalue" $\tilde{\lambda}_i = \tilde{\lambda}_i(u, v_i, \sigma_i)$ as the inner product

$$\tilde{\lambda}_i \doteq \langle \tilde{r}_i , \ A(u)\tilde{r}_i \rangle . \tag{5.26}$$

Writing out the derivative

$$\tilde{r}_{i,x} = \tilde{r}_{i,u} u_x + \tilde{r}_{i,v} v_{i,x} + \tilde{r}_{i,\sigma} \sigma_x$$

and using (5.25), (5.23) and the identity $\sigma_x = 0$ clearly valid for a traveling wave, from (5.24) we finally obtain

$$(\tilde{\lambda}_i - \sigma_i) v_i \tilde{r}_i + v_i \big(\tilde{r}_{i,u} \tilde{r}_i v_i + \tilde{r}_{i,v}(\tilde{\lambda}_i - \sigma_i) v_i\big) = \big(A(u) - \sigma_i\big) v_i \tilde{r}_i . \tag{5.27}$$

This yields a fundamental identity satisfied by our "generalized eigenvectors" $\tilde{r}_i$, namely

$$\big(A(u) - \tilde{\lambda}_i\big) \tilde{r}_i = v_i \big(\tilde{r}_{i,u} \tilde{r}_i + \tilde{r}_{i,v}(\tilde{\lambda}_i - \sigma_i)\big) . \tag{5.28}$$

Notice how (5.28) replaces the familiar identity

$$\big(A(u) - \lambda_i\big) r_i = 0 \tag{5.29}$$

satisfied by the usual eigenvectors and eigenvalues of the matrix $A(u)$. The presence of some small terms on the right hand side of (5.28) is of great importance. Indeed, in the evolution equations (3.13), these terms achieve a crucial cancellation with other source terms that would otherwise not be integrable.

Comparing (5.28) with (5.29) we see that, as $v_i \to 0$, the unit vector $\tilde{r}_i(u, v_i, \sigma_i)$ approaches an eigenvector of the matrix $A(u)$, while $\tilde{\lambda}_i(u, v_i, \sigma_i)$ approaches the corresponding eigenvalue. By continuity, this establishes our claim (5.21).

By (5.21), when $v_i = 0$ the vector $\tilde{r}_i$ does not depend on the speed σ_i, hence $\tilde{r}_{i,\sigma}(u, 0, \sigma_i) = 0$. In turn, by the smoothness of the vector field $\tilde{r}_i$ we also have

$$\begin{aligned} \tilde{r}_i(u, v_i, \sigma_i) - r_i(u) &= \mathcal{O}(1) \cdot v_i, & \tilde{r}_{i,\sigma} &= \mathcal{O}(1) \cdot v_i, \\ \tilde{r}_{i,u\sigma} &= \mathcal{O}(1) \cdot v_i, & \tilde{r}_{i,\sigma\sigma} &= \mathcal{O}(1) \cdot v_i. \end{aligned} \tag{5.30}$$

Observing that the vectors $\tilde{r}_{i,v}$ and $\tilde{r}_{i,\sigma}$ are both perpendicular to $\tilde{r}_i$, from (5.28) and the definition (5.26) we deduce

$$\tilde{\lambda}_i(u, v_i, \sigma_i) - \lambda_i(u) = \mathcal{O}(1) \cdot v_i, \qquad \tilde{\lambda}_{i,\sigma} = \mathcal{O}(1) \cdot v_i. \qquad (5.31)$$

We conclude our analysis of trajectories on the center manifold by proving the identity

$$-\sigma_i v_{i,x} + (\tilde{\lambda}_i v_i)_x - v_{i,xx} = 0, \qquad (5.32)$$

valid for a traveling wave profile. Toward this goal, we first differentiate (5.24) and obtain

$$v_{i,xx} \tilde{r}_i + 2 v_{i,x} \tilde{r}_{i,x} + v_i \tilde{r}_{i,xx} = \big(A(u) v_i \tilde{r}_i\big)_x - \sigma_i v_{i,x} \tilde{r}_i - \sigma_i v_i \tilde{r}_{i,x}. \qquad (5.33)$$

From the identities

$$\langle \tilde{r}_i, \tilde{r}_{i,x} \rangle = 0, \qquad \langle \tilde{r}_i, \tilde{r}_{i,xx} \rangle = -\langle \tilde{r}_{i,x}, \tilde{r}_{i,x} \rangle,$$

taking the inner product of (5.24) with $\tilde{r}_{i,x}$ we find

$$\langle \tilde{r}_i, \tilde{r}_{i,xx} \rangle v_i = -\langle \tilde{r}_{i,x}, A(u) \tilde{r}_i \rangle v_i. \qquad (5.34)$$

On the other hand, taking the inner product of (5.33) with $\tilde{r}_i$ we find

$$v_{i,xx} + \langle \tilde{r}_i, \tilde{r}_{i,xx} \rangle v_i = \langle \tilde{r}_i, (A(u) \tilde{r}_i v_i)_x \rangle - \sigma_i v_{i,x}.$$

By (5.34), this yields (5.32).

The significance of the identity (5.32) can be explained as follows. Consider a very special solution of the parabolic system (5.2), say

$$u = u(t, x) = U_i(x - \sigma_i t)$$

consisting of a traveling i-wave on the center manifold. The corresponding decomposition (5.1) will then contain one single term:

$$u_x(t, x) = v_i(t, x) \tilde{r}_i(t, x) = U_i'(x - \sigma_i t),$$

while $v_j \equiv 0$ for all other components $j \neq i$. We seek the evolution equation satisfied by the scalar component v_i. Since we are dealing with a traveling wave, we have

$$v_{i,t} + \sigma_i v_{i,x} = 0.$$

Inserting this in (5.32) we thus conclude

$$v_{i,t} + (\tilde{\lambda}_i v_i)_x - v_{i,xx} = 0. \qquad (5.35)$$

In other words, in connection with a traveling wave profile, the source terms ϕ_i in (3.13) vanish identically.

5.2. Wave Decomposition

Let $u : \mathbb{R} \mapsto \mathbb{R}^n$ be a smooth function with small total variation. At each point x, we seek a decomposition of the gradient u_x in the form (5.1), where $\tilde{r}_i = \tilde{r}_i(u, v_i, \sigma_i)$ are the vectors defining the center manifold at (5.20). To uniquely determine the $\tilde{r}_i$, we must first define the wave strengths v_i and speeds σ_i in terms of u, u_x, u_{xx}.

Consider first the special case where u is precisely the profile of a viscous traveling wave of the j-th family (contained in the center manifold $\mathcal{M}_j$). In this case, our decomposition should clearly contain one single component:

$$u_x = v_j \tilde{r}_j(u, v_j, \sigma_j). \tag{5.36}$$

It is easy to guess what v_j, σ_j should be. Indeed, since by construction $|\tilde{r}_j| = 1$, the quantity

$$v_j = \pm |u_x|$$

is the signed strength of the wave. Notice also that for a traveling wave the vectors u_x and u_t are always parallel, because $u_t = -\sigma_j u_x$ where σ_j is the speed of the wave. We can thus write

$$u_t = u_{xx} - A(u)u_x = \omega_j \tilde{r}_j(u, v_j, \sigma_j) \tag{5.37}$$

for some scalar ω_j. The speed of the wave is now obtained as $\sigma_j = -\omega_j/v_j$.

Motivated by the previous analysis, as a first attempt we define

$$u_t \doteq u_{xx} - A(u)u_x \tag{5.38}$$

and try to find scalar quantities v_i, ω_i such that

$$\begin{cases} u_x = \sum_i v_i \, \tilde{r}_i(u, v_i, \sigma_i), \\ u_t = \sum_i \omega_i \, \tilde{r}_i(u, v_i, \sigma_i), \end{cases} \qquad \sigma_i = -\frac{\omega_i}{v_i}. \tag{5.39}$$

The trouble with (5.39) is that the vectors $\tilde{r}_i$ are defined only for speeds σ_i close to the i-th characteristic speed $\lambda_i^* \doteq \lambda_i(u^*)$. However, when $u_x \approx 0$ one has $v_i \approx 0$ and the ratio ω_i/v_i may become arbitrarily large.

To overcome this problem, we introduce a cutoff function (fig. 15). Fix $\delta_1 > 0$ sufficiently small. Define a smooth odd function $\theta : \mathbb{R} \mapsto [-2\delta_1, 2\delta_1]$ such that

$$\theta(s) = \begin{cases} s & if \ |s| \leq \delta_1 \\ 0 & if \ |s| \geq 3\delta_1 \end{cases} \qquad |\theta'| \leq 1, \quad |\theta''| \leq 4/\delta_1. \tag{5.40}$$

We now rewrite (5.39) in terms of the new variable w_i, related to ω_i by $w_i \doteq \omega_i - \lambda_i^* v_i$. We require that σ_i coincides with $-\omega_i/v_i$ only when this ratio is sufficiently close to $\lambda_i^* \doteq \lambda_i(u^*)$. Our basic equations thus take the form

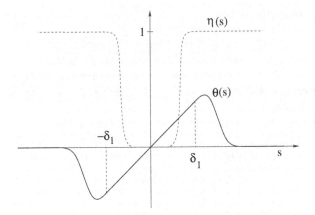

Fig. 15.

$$\begin{cases} u_x = \sum_i v_i \, \tilde{r}_i(u, v_i, \sigma_i), \\ u_t = \sum_i (w_i - \lambda_i^* v_i) \, \tilde{r}_i(u, v_i, \sigma_i), \end{cases} \tag{5.41}$$

where

$$u_t = u_{xx} - A(u)u_x, \qquad \sigma_i = \lambda_i^* - \theta\left(\frac{w_i}{v_i}\right). \tag{5.42}$$

Notice that σ_i is not well defined when $v_i = w_i = 0$. However, recalling (5.21), in this case we have $\tilde{r}_i = r_i(u)$ regardless of σ_i. Hence the two equations in (5.41) are still meaningful.

Remark 5.1. By construction, the vectors $v_i \tilde{r}_i$ are the gradients of viscous traveling waves U_i such that

$$U_i(x) = u(x), \qquad U_i'(x) = v_i \tilde{r}_i, \qquad U_i'' = \big(A(u) - \sigma_i\big)U_i'.$$

From the first equation in (5.41) it follows

$$u_x(x) = \sum_i U_i'(x).$$

If $\sigma_i = \lambda_i^* - w_i/v_i$ for all $i = 1, \dots, n$, i.e. if none of the cutoff functions is active, then

$$u_{xx}(x) = u_t + A(u)u_x = \sum_i (w_i - \lambda_i^* v_i)\tilde{r}_i + A(u)\sum_i v_i \tilde{r}_i$$

$$= \sum_i \big(A(u) - \sigma_i\big)v_i \tilde{r}_i = \sum_i U_i''(x).$$

In this case, both of the equalities in (5.4) hold. Notice however that the second equality in (5.4) may fail if $|w_i/v_i| > \delta_1$ for some i.

We now show that the system of equations (5.41)–(5.42) uniquely defines the components v_i, w_i. Moreover, we study the smoothness of these components, as functions of u, u_x, u_{xx}.

Lemma 5.2. *For $|u - u^*|$, $|u_x|$ and $|u_{xx}|$ sufficiently small, the system of $2n$ equations (5.41) has a unique solution $(v, w) = (v_1, \ldots, v_n, w_1, \ldots, w_n)$. The map $(u, u_x, u_{xx}) \mapsto (v, w)$ is smooth outside the n manifolds $\mathcal{Z}_i \doteq \{(v, w); \ v_i = w_i = 0\}$; moreover it is $C^{1,1}$, i.e. continuously differentiable with Lipschitz continuous derivatives on a whole neighborhood of the point $(u^*, 0, 0)$.*

Proof. Consider the mapping $\Lambda : \mathbb{R}^n \times \mathbb{R}^n \times \mathbb{R}^n \mapsto \mathbb{R}^{2n}$ defined by

$$\Lambda(u, v, w) \doteq \sum_{i=1}^n \Lambda_i(u, v_i, w_i), \tag{5.43}$$

$$\Lambda_i(u, v_i, w_i) \doteq \begin{pmatrix} v_i \, \tilde{r}_i\big(u, \ v_i, \ \lambda_i^* - \theta(w_i/v_i)\big) \\ (w_i - \lambda_i^* v_i) \, \tilde{r}_i\big(u, \ v_i, \ \lambda_i^* - \theta(w_i/v_i)\big) \end{pmatrix}. \tag{5.44}$$

Given u close to u^* and (u_x, u_{xx}) in a neighborhood of $(0,0) \in \mathbb{R}^{n+n}$, we need to find vectors v, w such that

$$\Lambda(u, v, w) = \big(u_x, \ u_{xx} - A(u)u_x\big). \tag{5.45}$$

This will be achieved by applying the implicit function theorem.

Observe that each map Λ_i is well defined and continuous also when $v_i = 0$, because in this case (5.21) implies $\tilde{r}_i = r_i(u)$. Computing the Jacobian matrix of partial derivatives w.r.t. (v_i, w_i) we find

$$\frac{\partial \Lambda_i}{\partial(v_i, w_i)} = \begin{pmatrix} \tilde{r}_i, & 0 \\ -\lambda_i^* \tilde{r}_i, & \tilde{r}_i \end{pmatrix} +$$

$$\begin{pmatrix} v_i \tilde{r}_{i,v} + (w_i/v_i)\theta_i' \tilde{r}_{i,\sigma}, & -\theta_i' \tilde{r}_{i,\sigma} \\ w_i \tilde{r}_{i,v} - \lambda_i^* v_i \tilde{r}_{i,v} - \lambda_i^* (w_i/v_i)\theta_i' \tilde{r}_{i,\sigma} + (w_i/v_i)^2 \theta_i' \tilde{r}_{i,\sigma}, & \lambda_i^* \theta_i' \tilde{r}_{i,\sigma} - (w_i/v_i)\theta_i' \tilde{r}_{i,\sigma} \end{pmatrix}$$
$$\doteq B_i + \tilde{B}_i .$$
$$\tag{5.46}$$

Here and throughout the following, by θ_i, θ_i' we denote the function θ and its derivative, evaluated at the point $s = w_i/v_i$. Notice that the matrix valued function $\tilde{B}_i$ is well defined and continuous also when $v_i = 0$. Indeed, either $|w_i/v_i| > 2\delta_1$, in which case $\theta_i' = 0$, or else $|w_i| \leq 2\delta_1|v_i|$. In this second case, by (5.30) we have $\tilde{r}_{i,\sigma} = \mathcal{O}(1) \cdot v_i$. In all cases we have the estimate

$$\tilde{B}_i(u, v_i, w_i) = \mathcal{O}(1) \cdot v_i . \tag{5.47}$$

According to (5.46), we can split the Jacobian matrix of partial derivatives of Λ as the sum of two matrices:

$$\frac{\partial \Lambda}{\partial (v, w)} = B(u, v, w) + \widetilde{B}(u, v, w). \tag{5.48}$$

For (v, w) small, B has a uniformly bounded inverse, while $\widetilde{B} \to 0$ as $(v, w) \to 0$. Since $\Lambda(u, 0, 0) = (0, 0) \in \mathbb{R}^{2n}$, by the implicit function theorem we conclude that the map $(v, w) \mapsto \Lambda(u, v, w)$ is $\mathcal{C}^1$ and invertible in a neighborhood of the origin, for $|u - u^*|$ suitably small. We shall write Λ^{-1} for this inverse mapping. In other words,

$$\Lambda^{-1}(u, u_x, u_t) = (v, w) \quad \text{if and only if} \quad \Lambda(u, v, w) = (u_x, u_t).$$

Having proved the existence and uniqueness of the decomposition, it now remains to study its regularity. Since the cutoff function θ vanishes for $|s| \geq 3\delta_1$, it is clear that each Λ_i is smooth outside the manifold $\mathcal{Z}_i \doteq \{(v, w) ; v_i = w_i = 0\}$ having codimension 2. We shall not write out the second derivatives of the maps Λ_i explicitly. However, it is clear that

$$\frac{\partial^2 \Lambda}{\partial v_i \partial v_j} = \frac{\partial^2 \Lambda}{\partial v_i \partial w_j} = \frac{\partial^2 \Lambda}{\partial w_i \partial w_j} = 0 \quad \text{if} \quad i \neq j. \tag{5.49}$$

Moreover, recalling that $\theta_i \equiv 0$ for $|w_i/v_i| > 2\delta_i$ and using the bounds (5.30), we have the estimates

$$\frac{\partial^2 \Lambda}{\partial v_i^2}, \ \frac{\partial^2 \Lambda}{\partial v_i \partial w_i}, \ \frac{\partial^2 \Lambda}{\partial w_i^2} = \mathcal{O}(1). \tag{5.50}$$

In other words, all second derivatives exist and are uniformly bounded outside the n manifolds $\mathcal{Z}_i$. Therefore, Λ is continuously differentiable with Lipschitz continuous first derivatives on a whole neighborhood of the point $(u^*, 0, 0)$. Clearly the same holds for the inverse mapping Λ^{-1}.   ●

Remark 5.3. By possibly performing a linear transformation of variables, we can assume that the matrix $A(u^*)$ is diagonal, hence its eigenvectors $r_1^*, \ldots r_n^*$ form an orthonormal basis:

$$\langle r_i^*, r_j^* \rangle = \delta_{ij}. \tag{5.51}$$

Observing that

$$\left| \tilde{r}_i(u, v_i, \sigma_i) - r_i^* \right| = \mathcal{O}(1) \cdot \left(|u - u^*| + |v_i| \right), \tag{5.52}$$

from (5.21) and the above assumption we deduce

$$\langle \tilde{r}_i(u, v_i, \sigma_i), \tilde{r}_j(u, v_j, \sigma_j) \rangle = \delta_{ij} + \mathcal{O}(1) \cdot \left(|u - u^*| + |v_i| + |v_j| \right)$$
$$= \delta_{ij} + \mathcal{O}(1) \cdot \delta_0, \tag{5.53}$$

$$\langle \tilde{r}_i, A(u)\tilde{r}_j \rangle = \mathcal{O}(1) \cdot \delta_0 \qquad j \neq i. \tag{5.54}$$

Another useful consequence of (5.51)–(5.52) is the following. Choosing the bound on the total variation $\delta_0 > 0$ small enough, the vectors $\tilde{r}_i$ will remain close the an orthonormal basis. Hence the decomposition (5.1) will satisfy

$$|u_x| \leq \sum_i |v_i| \leq 2\sqrt{n}|u_x|. \tag{5.55}$$

5.3. Bounds on Derivative Components

Relying on the bounds on the derivatives of u stated in Corollary 4.4, in this last section we derive some analogous estimates, valid for the components v_i, w_i and their derivatives.

Lemma 5.4. *In the same setting as Proposition 4.2, assume that the bound*

$$\left\|u_x(t)\right\|_{\mathbf{L}^1} \leq \delta_0$$

holds on a larger time interval $[0, T]$. Then for all $t \in [\hat{t}, T]$, the decomposition (5.41) is well defined. The components v_i, w_i satisfy the estimates

$$\left\|v_i(t)\right\|_{\mathbf{L}^1}, \ \left\|w_i(t)\right\|_{\mathbf{L}^1} = \mathcal{O}(1) \cdot \delta_0, \tag{5.56}$$

$$\left\|v_i(t)\right\|_{\mathbf{L}^\infty}, \ \left\|w_i(t)\right\|_{\mathbf{L}^\infty}, \left\|v_{i,x}(t)\right\|_{\mathbf{L}^1}, \ \left\|w_{i,x}(t)\right\|_{\mathbf{L}^1} = \mathcal{O}(1) \cdot \delta_0^2, \tag{5.57}$$

$$\left\|v_{i,x}(t)\right\|_{\mathbf{L}^\infty}, \ \left\|w_{i,x}(t)\right\|_{\mathbf{L}^\infty} = \mathcal{O}(1) \cdot \delta_0^3. \tag{5.58}$$

Proof. By Lemma 5.2, in a neighborhood of the origin the map $(v, w) \mapsto \Lambda(u, v, w)$ in (5.43) is well defined, locally invertible, and continuously differentiable with Lipschitz continuous derivatives. Hence, for $\delta_0 > 0$ suitably small, the $\mathbf{L}^\infty$ bounds in (4.29) and (4.30) guarantee that the decomposition (5.41) is well defined. From the identity (5.45) we deduce

$$v_i, w_i = \mathcal{O}(1) \cdot \left(|u_x| + |u_{xx}|\right).$$

By (4.12) and (4.29)–(4.30) this yields the $\mathbf{L}^1$ bounds in (5.56) and the $\mathbf{L}^\infty$ bounds in (5.57). Differentiating (5.45) w.r.t. x we obtain

$$\frac{\partial \Lambda}{\partial u} u_x + \frac{\partial \Lambda}{\partial(v, w)}(v_x, w_x) = \left(u_{xx}, \ u_{xxx} - A(u)u_{xx} - \left(u_x \bullet A(u)\right)u_x\right). \tag{5.59}$$

Using the estimate

$$\frac{\partial \Lambda}{\partial u} = \mathcal{O}(1) \cdot \left(|v| + |w|\right),$$

since the derivative $\partial \Lambda / \partial(v, w)$ has bounded inverse, from (5.59) we deduce

$$(v_x, w_x) = \mathcal{O}(1) \cdot \left(|u_{xx}| + |u_{xxx}| + |u_x|^2 + |u_x|(|v| + |w|)\right).$$

Together with the bounds (4.29)–(4.31), this yields the remaining $\mathbf{L}^1$ estimates in (5.57) and the $\mathbf{L}^\infty$ estimates in (5.58). $\qquad \bullet$

Corollary 5.5. *With the same assumptions as Lemma 5.2, one has the estimates*

$$\left\|\tilde{r}_{i,x}(t)\right\|_{\mathbf{L}^1}, \quad \left\|\tilde{\lambda}_{i,x}(t)\right\|_{\mathbf{L}^1} = \mathcal{O}(1) \cdot \delta_0, \tag{5.60}$$

$$\left\|\tilde{r}_{i,x}(t)\right\|_{\mathbf{L}^\infty}, \quad \left\|\tilde{\lambda}_{i,x}(t)\right\|_{\mathbf{L}^\infty} = \mathcal{O}(1) \cdot \delta_0^2. \tag{5.61}$$

Proof. Recalling (5.30) we have

$$
\begin{aligned}
\tilde{r}_{i,x} &= \sum_j v_j \tilde{r}_{i,u} \tilde{r}_j + v_{i,x} \tilde{r}_{i,v} + \sigma_{i,x} \tilde{r}_{i,\sigma} \\
&= \mathcal{O}(1) \cdot \sum_j |v_j| + \mathcal{O}(1) \cdot |v_{i,x}| + \mathcal{O}(1) \cdot |\sigma_{i,x}| \, |v_i| \\
&= \mathcal{O}(1) \cdot \sum_j |v_j| + \mathcal{O}(1) \cdot \left(|v_{i,x}| + |w_{i,x}| \right).
\end{aligned}
$$

The estimate for $\tilde{r}_{i,x}$ thus follows from the corresponding bounds in Lemma 5.2. Clearly $\tilde{\lambda}_{i,x} = \mathcal{O}(1) \cdot |\tilde{r}_{i,x}|$, hence it satisfies the same bounds. ●

In the remaining part of this section we shall examine the relations between w_i and $v_{i,x}$. Due to the presence of a cutoff function, the results will depend on the ratio $|w_i/v_i|$. It is convenient to recall here the bounds

$$|\tilde{\lambda}_i - \lambda_i^*| = \mathcal{O}(1) \cdot |\tilde{r}_i - r_i^*| = \mathcal{O}(1) \cdot \delta_0, \qquad |v_i|, \; |w_i| = \mathcal{O}(1) \cdot \delta_0^2, \tag{5.62}$$

which follow from (5.52) and (5.57). Moreover, one should keep in mind our choice of the constants

$$0 < \delta_0 \ll \delta_1 \ll 1. \tag{5.63}$$

From the basic relations (5.41) and the identity $u_t + A(u)u_x = u_{xx}$ it follows

$$
\begin{aligned}
\sum_i w_i \tilde{r}_i &+ \sum_i \left(A(u) - \lambda_i^* \right) v_i \tilde{r}_i \\
&= \sum_i v_{i,x} \tilde{r}_i + \sum_{ij} v_i \tilde{r}_{i,u} \, v_j \tilde{r}_j + \sum_i v_i \tilde{r}_{i,v} v_{i,x} + \sum_i v_i \tilde{r}_{i,\sigma} \sigma_{i,x}.
\end{aligned}
$$

Taking the inner product with $\tilde{r}_i$ and recalling that $\tilde{r}_i$ has unit norm and is thus orthogonal to its derivatives $\tilde{r}_{i,v}$, $\tilde{r}_{i,u}\tilde{r}_j$, we obtain

$$w_i + (\tilde{\lambda}_i - \lambda_i^*)v_i = v_{i,x} + \Theta_i, \tag{5.64}$$

where

$$
\begin{aligned}
\Theta_i &\doteq \sum_{j \neq i} \left\langle \tilde{r}_i, \; (\lambda_j^* - A(u))\tilde{r}_j \right\rangle v_j + \sum_{j \neq i} \sum_k \left\langle \tilde{r}_i, \; \tilde{r}_{j,u}\tilde{r}_k \right\rangle v_j v_k \\
&\quad + \sum_{j \neq i} \left\langle \tilde{r}_i, \; \tilde{r}_{j,v} \right\rangle v_j v_{j,x} + \sum_{j \neq i} \left\langle \tilde{r}_i, \; \tilde{r}_{j,\sigma} \right\rangle v_j \sigma_{j,x} + \sum_{j \neq i} \left\langle \tilde{r}_i, \; \tilde{r}_j \right\rangle (v_{j,x} - w_j) \\
&= \mathcal{O}(1) \cdot \delta_0 \sum_{j \neq i} \left(|v_j| + |w_j - v_{j,x}| \right).
\end{aligned}
\tag{5.65}
$$

The above estimate on Θ_i is obtained using (5.54) together with the $\mathbf{L}^\infty$ bounds in (5.62) and the bound $\tilde{r}_{j,\sigma} = \mathcal{O}(1) \cdot v_i$ in (5.30). Summing (5.64) over $i = 1, \ldots, n$ and using (5.61)–(5.62), from (5.64) we obtain

$$\sum_i |w_i - v_{i,x}| = \mathcal{O}(1) \cdot \delta_0 \sum_j |v_j|. \tag{5.66}$$

This yields the implications

$$|w_i| < 3\delta_1|v_i| \qquad \Longrightarrow \qquad v_{i,x} = \mathcal{O}(1) \cdot v_i + \mathcal{O}(1) \cdot \delta_0 \sum_{j \neq i} |v_j|, \tag{5.67}$$

$$|w_i| > \delta_1|v_i| \qquad \Longrightarrow \qquad v_i = \mathcal{O}(1) \cdot v_{i,x} + \mathcal{O}(1) \cdot \delta_0 \sum_{j \neq i} |v_j|. \tag{5.68}$$

The following technical lemma will be used in the estimate of the sources due to cut-off terms.

Lemma 5.6. *If* $|w_i/v_i| \geq 3\delta_1/5$, *then*

$$|w_i| \leq 2|v_{i,x}| + \mathcal{O}(1) \cdot \delta_0 \sum_{j \neq i} |v_j|, \quad |v_i| \leq \frac{5}{2\delta_1}|v_{i,x}| + \mathcal{O}(1) \cdot \delta_0 \sum_{j \neq i} |v_j|. \tag{5.69}$$

On the other hand, if $|w_i/v_i| \leq 4\delta_1/5$, *then*

$$|v_{i,x}| \leq \delta_1|v_i| + \mathcal{O}(1) \cdot \delta_0 \sum_{j \neq i} |v_j|. \tag{5.70}$$

Proof. The analysis is based on the formula (5.64). By (5.62)–(5.63), from the condition $|w_i/v_i| \geq 3\delta_1/5$ two cases can arise. On one hand, if

$$|\Theta_i| \leq \frac{\delta_1}{10}|v_i|, \tag{5.71}$$

then

$$|v_{i,x}| \geq \frac{3\delta_1}{5}|v_i| - \mathcal{O}(1) \cdot \delta_0|v_i| - \frac{\delta_1}{10}|v_i| \geq \frac{2\delta_1}{5}|v_i|,$$

and hence

$$|v_i| \leq \frac{5}{2\delta_1}|v_{i,x}|, \qquad |w_i| \leq |v_{i,x}| + \mathcal{O}(1) \cdot \delta_0|v_i| + \frac{\delta_1}{10}|v_i| \leq 2|v_{i,x}|. \tag{5.72}$$

On the other hand, if (5.71) fails, using (5.65)–(5.66) and then (5.62) we find

$$|v_i| = \mathcal{O}(1) \cdot \delta_0 \sum_{j \neq i} |v_j|, \qquad |\tilde{\lambda}_i - \lambda_i^*| \, |v_i| = \mathcal{O}(1) \cdot \delta_0^2 \sum_{j \neq i} |v_j|. \tag{5.73}$$

In both cases, the estimates in (5.69) hold.

Next, if $|w_i/v_i| \leq 4\delta_1/5$, from (5.64)–(5.66) we deduce

$$|v_{i,x}| \leq \frac{4\delta_1}{5}|v_i| + \mathcal{O}(1) \cdot \delta_0 |v_i| + \mathcal{O}(1) \cdot \delta_0 \sum_{j \neq i} |v_j|\,.$$

By (5.63) we can assume that $\mathcal{O}(1) \cdot \delta_0 \leq \delta_1/10$. If (5.71) holds, we now conclude $|v_{i,x}| \leq \delta_1|v_i|$. If (5.71) fails, then (5.73) is valid. In both cases we have (5.70). ●

6 Interaction of Viscous Waves

Let $u = u(t, x)$ be a solution of the system

$$u_t + A(u)u_x = u_{xx}\,. \tag{6.1}$$

According to the analysis of the previous chapter, at each point we can decompose the gradient u_x in terms of viscous traveling waves. Assuming that u_x, u_{xx} remain sufficiently small, by Lemma 5.2 the corresponding components v_i, w_i in (5.41) are then well defined. The equations governing the evolution of these $2n$ components can be written in the form

$$\begin{cases} v_{i,t} + (\tilde{\lambda}_i v_i)_x - v_{i,xx} = \phi_i\,, \\ w_{i,t} + (\tilde{\lambda}_i w_i)_x - w_{i,xx} = \psi_i\,, \end{cases} \tag{6.2}$$

with $\tilde{\lambda}_i \doteq \langle \tilde{r}_i, A(u)\tilde{r}_i \rangle$, as in (5.26). Here ϕ_i, ψ_i are appropriate functions involving u, v, w and their first first order derivatives. Writing out their explicit form requires lengthy calculations. We shall only outline the basic procedure.

Differentiating (6.1) one finds that the vector $(u_x, u_t) = \Lambda(u, v, w)$ satisfies the evolution equation

$$\begin{pmatrix} u_x \\ u_t \end{pmatrix}_t + \left(\begin{bmatrix} A(u) & 0 \\ 0 & A(u) \end{bmatrix} \begin{pmatrix} u_x \\ u_t \end{pmatrix} \right)_x - \begin{pmatrix} u_x \\ u_t \end{pmatrix}_{xx}$$

$$= \begin{pmatrix} 0 \\ (u_x \bullet A(u))u_t - (u_t \bullet A(u))u_x \end{pmatrix}\,. \tag{6.3}$$

Observe that, in the conservative case $A(u) = Df(u)$, the right hand side vanishes because

$$(u_x \bullet A(u))u_t = (u_t \bullet A(u))u_x = D^2 f(u)(u_x \otimes u_t)\,.$$

For notational convenience, we introduce the variable $z \doteq (v, w)$ and write $\tilde{\lambda}$ for the $2n \times 2n$ diagonal matrix with entries $\tilde{\lambda}_i$ defined at (5.26):

$$\tilde{\lambda} \doteq \begin{pmatrix} \mathrm{diag}(\tilde{\lambda}_i) & 0 \\ 0 & \mathrm{diag}(\tilde{\lambda}_i) \end{pmatrix}\,.$$

From (6.3) it now follows

$$\frac{\partial \Lambda}{\partial u} u_t + \frac{\partial \Lambda}{\partial z} \begin{pmatrix} v \\ w \end{pmatrix}_t + \left(\begin{bmatrix} A(u) & 0 \\ 0 & A(u) \end{bmatrix} \Lambda \right)_x - \frac{\partial \Lambda}{\partial z} \begin{pmatrix} v \\ w \end{pmatrix}_{xx} - \frac{\partial \Lambda}{\partial u} u_{xx}$$

$$- \frac{\partial^2 \Lambda}{\partial u^{[2]}} (u_x \otimes u_x) - \frac{\partial^2 \Lambda}{\partial z^{[2]}} \cdot \begin{pmatrix} v_x \\ w_x \end{pmatrix} \otimes \begin{pmatrix} v_x \\ w_x \end{pmatrix} - 2 \frac{\partial^2 \Lambda}{\partial u \, \partial z} u_x \otimes \begin{pmatrix} v_x \\ w_x \end{pmatrix}$$

$$= \begin{pmatrix} 0 \\ (u_x \bullet A(u)) u_t - (u_t \bullet A(u)) u_x \end{pmatrix} .$$

Therefore,

$$\frac{\partial \Lambda}{\partial z} \left[\begin{pmatrix} v \\ w \end{pmatrix}_t + \left(\tilde{\lambda} \begin{pmatrix} v \\ w \end{pmatrix} \right)_x - \begin{pmatrix} v \\ w \end{pmatrix}_{xx} \right]$$

$$= \frac{\partial \Lambda}{\partial z} \left(\tilde{\lambda} \begin{pmatrix} v \\ w \end{pmatrix} \right)_x - \left(\begin{bmatrix} A(u) & 0 \\ 0 & A(u) \end{bmatrix} \Lambda \right)_x$$

$$+ \frac{\partial \Lambda}{\partial u} A(u) u_x + \begin{pmatrix} 0 \\ (u_x \bullet A(u)) u_t - (u_t \bullet A(u)) u_x \end{pmatrix}$$

$$+ \frac{\partial^2 \Lambda}{\partial u^{[2]}} (u_x \otimes u_x) + \frac{\partial^2 \Lambda}{\partial z^{[2]}} \cdot \begin{pmatrix} v_x \\ w_x \end{pmatrix} \otimes \begin{pmatrix} v_x \\ w_x \end{pmatrix} + 2 \frac{\partial^2 \Lambda}{\partial u \, \partial z} u_x \otimes \begin{pmatrix} v_x \\ w_x \end{pmatrix}$$

$$\doteq E$$

(6.4)

Multiplying both sides of (6.4) by the inverse of the matrix differential $\partial \Lambda / \partial z$, one obtains an expression for the source terms ϕ_i, ψ_i in (6.2). For our purposes, however, we only need an upper bound for the norms $\|\phi_i\|_{\mathbf{L}^1}$ and $\|\psi_i\|_{\mathbf{L}^1}$. In this direction we observe that, since $\partial \Lambda / \partial z$ has uniformly bounded inverse, one has the bounds

$$\phi_i = \mathcal{O}(1) \cdot |E|, \qquad \psi_i = \mathcal{O}(1) \cdot |E|, \qquad i = 1, \dots, n. \qquad (6.5)$$

The right hand side of (6.4) can be computed by differentiating the map Λ in (5.43) and using the identities (5.39) and (5.28).

Before giving a precise estimate on the size of the source terms, we provide an intuitive explanation of how they arise. Consider first the special case where u is precisely one of the traveling wave profiles on the center manifold (fig. 16a), say $u(t, x) = U_j(x - \sigma_j t)$. We then have

$$u_x = v_j \tilde{r}_j, \qquad u_t = (w_j - \lambda_j^* v_j) \tilde{r}_j, \qquad v_i = w_i = 0 \quad \text{for } i \neq j,$$

and therefore

$$\begin{cases} v_{i,t} + (\tilde{\lambda}_i v_i)_x - v_{i,xx} = 0, \\ w_{i,t} + (\tilde{\lambda}_i w_i)_x - w_{i,xx} = 0. \end{cases}$$

Indeed, this is obvious when $i \neq j$. The identity $\phi_j = 0$ was proved in (5.35), while the relation $w_j = (\lambda_j^* - \sigma_j) v_j$ implies $\psi_j = 0$.

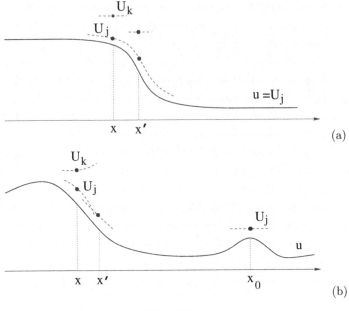

Fig. 16.

Next, consider the case of a general solution $u = u(t, x)$. The sources on the right hand sides of (6.2) arise for three different reasons (fig. 16b).

1. The ratio $|w_j/v_j|$ is large and hence the cutoff function θ in (5.40) is active. Typically, this will happen near a point x_0 where $u_x = 0$ but $u_t = u_{xx} \neq 0$. In this case the identity (5.35) fails because of a "wrong" choice of the speed: $\sigma_j \neq \lambda_j^* - (w_j/v_j)$. The difference in the speed is $|\theta_j - (w_j/v_j)|$, where θ_j denotes the value of θ at $s = w_j/v_j$. In the corresponding source terms, a detailed analysis will show that this cutoff error always enters multiplied by a factor $v_j v_{j,x}$ or by $v_j w_{j,x}$. Its strength is thus of order $\mathcal{O}(1) \cdot \left(|v_{j,x}| + |w_{j,x}|\right) \left|w_j - \theta_j v_j\right|$.

2. Waves of two different families $j \neq k$ are present at a given point x. These will produce quadratic source terms, due to transversal interactions. The strength of these terms is estimated by the product of different components of v, w and of their first order derivatives. For example: $|v_j v_k|$, $|v_j w_k|$, $|v_{j,x} w_k| \cdots$

3. Since the decomposition (5.39) is defined pointwise, it may well happen that the traveling j-wave profile U_j at a point x is not the same as the profile U_j at a nearby point x'. This is the case whenever these two traveling waves have different speeds. It is the rate of change in this speed, i.e. $\sigma_{j,x}$, that determines the infinitesimal interaction between nearby waves of the same family. A detailed analysis will show that the corresponding source terms can only be linear or quadratic w.r.t. $\sigma_{j,x}$, with the square of the strength of the wave always appearing as a factor. These terms can thus be estimated as $\mathcal{O}(1) \cdot v_j^2 \sigma_{j,x} + \mathcal{O}(1) \cdot v_j^2 \sigma_{j,x}^2$. Observing that $\sigma_{j,x} = (w_j/v_j)_x \theta_j'$ and $\theta_j' = 0$ when $|w_j/v_j| \geq 3$, this motivates the presence of the remaining terms in the following lemma.

Lemma 6.1. *The source terms in (6.2) satisfy the bounds*

$$\phi_i, \ \psi_i = \mathcal{O}(1) \cdot \sum_j \left(|v_{j,x}| + |w_{j,x}| \right) \cdot |w_j - \theta_j v_j| \qquad \text{(cutoff error)}$$

$$+ \mathcal{O}(1) \cdot \sum_j |v_{j,x} w_j - v_j w_{j,x}| \qquad \text{(change in speed, linear)}$$

$$+ \mathcal{O}(1) \cdot \sum_j \left| v_j \left(\frac{w_j}{v_j} \right)_x \right|^2 \cdot \chi_{\{|w_j/v_j| < 3\delta_1\}}$$

$$\text{(change in speed, quadratic)}$$

$$+ \mathcal{O}(1) \cdot \sum_{j \neq k} \left(|v_j v_k| + |v_{j,x} v_k| + |v_j w_k| + |v_{j,x} w_k| + |v_j w_{k,x}| + |w_j w_k| \right)$$

$$\text{(interaction of waves of different families)}$$

A proof of this Lemma, involving lengthy computations, is given in [BiB].

6.1. Transversal Wave Interactions

In this subsection we establish an a priori bound on the total amount of interactions between waves of different families. More precisely, let $u = u(t, x)$ be a solution of the parabolic system (6.1) and assume that

$$\|u_x(t)\|_{\mathbf{L}^1} \leq \delta_0 \qquad t \in [0, T]. \tag{6.6}$$

In this case, for $t \geq \hat{t}$, by Corollary 4.3 all higher derivatives will be suitably small and we can thus define the components v_i, w_i according to (5.39). These will satisfy the linear evolution equation (6.2), with source terms ϕ_i, ψ_i estimated in Lemma 6.1. Assuming that

$$\int_{\hat{t}}^T \int \left\{ |\phi_i(t, x)| + |\psi_i(t, x)| \right\} dx \, dt \leq \delta_0, \qquad i = 1, \ldots, n, \tag{6.7}$$

and relying on the bounds (5.56)–(5.58) on v_i, w_i and their derivatives, we shall prove the sharper estimate

$$\int_{\hat{t}}^T \int \sum_{j \neq k} \left(|v_j v_k| + |v_{j,x} v_k| + |v_j w_k| + |v_{j,x} w_k| + |w_{j} w_{k,x}| + |w_j w_k| \right) dx \, dt = \mathcal{O}(1) \cdot \delta_0^2 . \tag{6.8}$$

As a preliminary, we establish a more general estimate on solutions of two independent linear parabolic equations, with strictly different drifts.

Lemma 6.2. *Let* $z, z^\sharp$ *be solutions of the two independent scalar equations*

$$\begin{cases} z_t + (\lambda(t,x)z)_x - z_{xx} = \varphi(t,x), \\ z^\sharp_t + (\lambda^\sharp(t,x)z^\sharp)_x - z^\sharp_{xx} = \varphi^\sharp(t,x), \end{cases} \tag{6.9}$$

defined for $t \in [0,T]$. *Assume that*

$$\inf_{t,x} \lambda^\sharp(t,x) - \sup_{t,x} \lambda(t,x) \geq c > 0. \tag{6.10}$$

Then

$$\int_0^T \int |z(t,x)|\,|z^\sharp(t,x)|\,dx\,dt$$
$$\leq \frac{1}{c}\left(\int |z(0,x)|\,dx + \int_0^T \int |\varphi(t,x)|\,dx\,dt\right) \tag{6.11}$$
$$\left(\int |z^\sharp(0,x)|\,dx + \int_0^T \int |\varphi^\sharp(t,x)|\,dx\,dt\right).$$

Proof. We consider first the homogeneous case, where $\varphi = \varphi^\sharp = 0$. Define the interaction potential

$$Q(z, z^\sharp) \doteq \iint K(x-y)\,|z(x)|\,|z^\sharp(y)|\,dx\,dy, \tag{6.12}$$

with

$$K(s) \doteq \begin{cases} 1/c & \text{if} \quad s \geq 0, \\ 1/c \cdot e^{cs/2} & \text{if} \quad s < 0. \end{cases} \tag{6.13}$$

Computing the distributional derivatives of the kernel K we find that $cK' - 2K''$ is precisely the Dirac distribution, i.e. a unit mass at the origin. A direct computations now yields

$$\frac{d}{dt}Q(z(t), z^\sharp(t))$$
$$= \frac{d}{dt}\iint K(x-y)|z(x)|\,|z^\sharp(y)|\,dx\,dy$$
$$= \iint K(x-y)\Big\{(z_{xx} - (\lambda z)_x)\,\mathrm{sgn}z(x)|z^\sharp(y)|$$
$$\qquad + |z(x)|(z^\sharp_{yy} - (\lambda^\sharp z^\sharp)_y)\,\mathrm{sgn}z^\sharp(y)\Big\}\,dx\,dy$$
$$= \iint K'(x-y)\Big\{\lambda|z(x)|\,|z^\sharp(y)| - \lambda^\sharp|z(x)|\,|z^\sharp(y)|\Big\}\,dx\,dy$$
$$\qquad + \iint K''(x-y)\Big\{|z(x)|\,|z^\sharp(y)| + |z(x)|\,|z^\sharp(y)|\Big\}\,dx\,dy$$
$$\leq -\iint (cK' - 2K'')|z(x)|\,|z^\sharp(y)|\,dx\,dy$$
$$= -\int |z(x)|\,|z^\sharp(x)|\,dx$$

Therefore

$$\int_0^T \int |z(t,x)| \, |z^\sharp(t,x)| \, dx \, dt \leq Q\big(z(0),\, z^\sharp(0)\big) \leq \frac{1}{c} \|z(0)\|_{\mathbf{L^1}} \|z^\sharp(0)\|_{\mathbf{L^1}},$$

(6.14)

proving the lemma in the homogeneous case.

To handle the general case, call $\Gamma, \Gamma^\sharp$ the Green functions for the corresponding linear homogeneous systems. The general solution of (6.9) can thus be written in the form

$$\begin{cases} z(t,x) = \displaystyle\int \Gamma(t,x,0,y)z(0,y)dy + \int_0^t \int \Gamma(t,x,s,y)\varphi(s,y)dy \, ds, \\[2mm] z^\sharp(t,x) = \displaystyle\int \Gamma^\sharp(t,x,0,y)z^\sharp(0,y)dy + \int_0^t \int \Gamma^\sharp(t,x,s,y)\varphi^\sharp(s,y)dy \, ds. \end{cases}$$

(6.15)

From (6.14) it follows

$$\int_{\max\{s,s'\}}^T \int \Gamma(t,x,s,y) \cdot \Gamma^\sharp(t,x,s',y') \, dx \, dt \leq \frac{1}{c} \qquad (6.16)$$

for every couple of initial points (s,y) and (s',y'). The estimate (6.11) now follows from (6.16) and the representation formula (6.15).   ●

Remark 6.3. Exactly the same estimate (6.11) would be true also for a system without viscosity. In particular, if

$$z_t + \big(\lambda(t,x)z\big)_x = 0, \qquad z_t^\sharp + \big(\lambda^\sharp(t,x)z^\sharp\big)_x = 0,$$

and if the speeds satisfy the gap condition (6.10), then

$$\frac{d}{dt}\left[\frac{1}{c}\iint_{x<y} \big|z^\sharp(t,x)z(t,y)\big| \, dx \, dy\right] \leq -\int \big|z(t,x)\big| \, \big|z^\sharp(t,x)\big| \, dx.$$

Notice that the left hand side can be written as $(d/dt)Q(z, z^\sharp)$, with Q as in (6.12) but now

$$K(s) \doteq \begin{cases} 1/c & if \quad s < 0, \\ 0 & if \quad s > 0. \end{cases}$$

In the case where viscosity is present, our definition (6.12)–(6.13) thus provides a natural counterpart to the Glimm interaction potential between waves of different families, introduced in [G] for strictly hyperbolic systems.

Lemma 6.2 allows us to estimate the integral of the terms $|v_i v_k|$, $|v_j w_k|$ and $|w_j w_k|$ in (6.8). We now work toward an estimate of the remaining terms $|v_{j,x}v_k|$, $|v_{j,x}w_k|$ and $|v_j w_{k,x}|$, containing one derivative w.r.t. x.

Lemma 6.4. Let $z, z^\sharp$ be solutions of (6.9) and assume that (6.10) holds, together with the estimates

$$\int_0^T \int |\varphi(t,x)| \, dx \, dt \le \delta_0, \qquad \int_0^T \int |\varphi^\sharp(t,x)| \, dx \, dt \le \delta_0, \qquad (6.17)$$

$$\left\| z(t) \right\|_{\mathbf{L}^1}, \; \left\| z^\sharp(t) \right\|_{\mathbf{L}^1} \le \delta_0, \qquad \left\| z_x(t) \right\|_{\mathbf{L}^1}, \; \left\| z^\sharp(t) \right\|_{\mathbf{L}^\infty} \le C^* \delta_0^2, \qquad (6.18)$$

$$\left\| \lambda_x(t) \right\|_{\mathbf{L}^\infty}, \; \left\| \lambda_x(t) \right\|_{\mathbf{L}^1} \le C^* \delta_0, \qquad \lim_{x \to -\infty} \lambda(t,x) = 0 \qquad (6.19)$$

for all $t \in [0,T]$. Then one has the bound

$$\int_0^T \int |z_x(t,x)| \, |z^\sharp(t,x)| \, dx \, dt = \mathcal{O}(1) \cdot \delta_0^2. \qquad (6.20)$$

Proof. The left hand side of (6.20) is clearly bounded by the quantity

$$\mathcal{I}(T) \doteq \sup_{(\tau,\xi) \in [0,T] \times \mathbb{R}} \int_0^{T-\tau} \int |z_x(t,x) z^\sharp(t+\tau, x+\xi)| \, dx \, dt \le (C^* \delta_0^2)^2 \cdot T,$$

the last inequality being a consequence of (6.18). For $t > 1$ we can write z_x in the form

$$z_x(t,x) = \int G_x(1,y) z(t-1,\, x-y) \, dy + \int_0^1 \!\! \int G_x(s,y) \big[\varphi - (\lambda z)_x \big] (t-s,\, x-y) \, dy \, ds,$$

where $G(t,x) \doteq \exp\{-x^2/4t\}/2\sqrt{\pi t}$ is the standard heat kernel. Using (6.11) we obtain

$$\int_1^{T-\tau} \int \left| z_x(t,x)\, z^\sharp(t+\tau, x+\xi) \right| dx\, dt$$

$$\leq \int_1^{T-\tau} \iint \left| G_x(1,y)\, z(t-1,\, x-y)\, z^\sharp(t+\tau, x+\xi) \right| dy\, dx\, dt$$

$$+ \int_1^{T-\tau} \iint_0^1 \int \|\lambda_x\|_{\mathbf{L}^\infty} \left| G_x(s,y)\, z(t-s,\, x-y)\, z^\sharp(t+\tau, x+\xi) \right| dy\, ds\, dx\, dt$$

$$+ \int_0^{T-\tau} \iint_0^1 \int \|\lambda\|_{\mathbf{L}^\infty} \left| G_x(s,y)\, z_x(t-s,\, x-y)\, z^\sharp(t+\tau, x+\xi) \right| dy\, ds\, dx\, dt$$

$$+ \int_1^{T-\tau} \iint_{t-1}^t \int \left| G_x(t-s,\, x-y)\, \varphi(s,y)\, z^\sharp(t+\tau, x+\xi) \right| dy\, ds\, dx\, dt$$

$$\leq \left(\int \left| G_x(1,y) \right| dy + \|\lambda_x\|_{\mathbf{L}^\infty} \int_0^1 \int \left| G_x(s,y) \right| dy\, ds \right)$$

$$\cdot \sup_{s,y,\tau,\xi} \left(\int_1^{T-\tau} \int \left| z(t-s,\, x-y) \right| \left| z^\sharp(t+\tau, x+\xi) \right| dx\, dt \right)$$

$$+ \left(\|\lambda\|_{\mathbf{L}^\infty} \cdot \int_0^1 \int \left| G_x(s,y) \right| dy\, ds \right)$$

$$\cdot \left(\sup_{s,y,\tau,\xi} \int_1^{T-\tau} \int \left| z_x(t-s,\, x-y) \right| \left| z^\sharp(t+\tau, x+\xi) \right| dx\, dt \right)$$

$$+ \|z^\sharp\|_{\mathbf{L}^\infty} \cdot \int_0^1 \int \left| G_x(s,y) \right| ds\, dy \cdot \int_0^T \int \left| \varphi(t,x) \right| dx\, dt$$

$$\leq \left(\frac{1}{\sqrt{\pi}} + \|\lambda_x\|_{\mathbf{L}^\infty} \frac{2}{\sqrt{\pi}} \right) \frac{4\delta_0^2}{c} + \|\lambda\|_{\mathbf{L}^\infty} \frac{2}{\sqrt{\pi}} \mathcal{I}(T) + C^* \delta_0^2 \frac{2}{\sqrt{\pi}} \delta_0 \,.$$

$$(6.21)$$

On the initial time interval $[0,1]$, by (6.18) one has

$$\int_0^1 \int \left| z_x(t,x)\, z^\sharp(t+\tau,\, x+\xi) \right| dx\, dt \leq \int_0^1 \|z_x(t)\|_{\mathbf{L}^1} \|z^\sharp(t+\tau)\|_{\mathbf{L}^\infty} dt \leq (C^* \delta_0^2)^2 \,.$$

$$(6.22)$$

Moreover, (6.19) implies

$$\|\lambda\|_{\mathbf{L}^\infty} \leq \|\lambda_x\|_{\mathbf{L}^1} \leq C^* \delta_0 \ll 1 \,.$$

From (6.21) and (6.22) it thus follows

$$\mathcal{I}(T) \leq (C^* \delta_0^2)^2 + \frac{4\delta_0^2}{c} + \frac{1}{2}\mathcal{I}(T) + C^* \delta_0^3 \,.$$

For δ_0 sufficiently small, this implies $\mathcal{I}(T) \leq 9\delta_0^2/c$, proving the lemma.   •

Using the two previous lemmas we now prove the estimate (6.8). Setting $z \doteq v_j$, $z^\sharp \doteq v_k$, $\lambda \doteq \tilde{\lambda}_j$, $\lambda^\sharp \doteq \tilde{\lambda}_k$, an application of Lemma 6.2 yields the desired bound on the integral of $|v_j v_k|$. Moreover, Lemma 6.4 allows us to estimate the integral of $|v_{j,x} v_k|$. Notice that the assumptions in (6.18) are a consequence of (5.56)–(5.57), while (6.19) follows from (5.60)–(5.61). The simplifying condition $\lambda(t, -\infty) = 0$ in (6.19) can be easily achieved, using a new space coordinate $x' \doteq x - \lambda_j^* t$.

The other terms $|v_j w_k|$, $|w_j w_k|$, $|v_{j,x} w_k|$ and $|v_j w_{k,x}|$ are handled similarly.

6.3. Interaction of Waves of the Same Family

We now study the interaction of viscous waves of the same family. As in the previous section, let $u = u(t, x)$ be a solution of the parabolic system (6.1) whose total variation remains bounded according to (6.6). Assume that the components v_i, w_i satisfy the evolution equation (6.2), with source terms ϕ_i, ψ_i bounded as in (6.7). Using the bounds (5.56)–(5.58), for each $i = 1, \ldots, n$ we shall prove the estimates

$$\int_{\hat{t}}^{T} \int |w_{i,x} v_i - w_i v_{i,x}| \, dx \, dt = \mathcal{O}(1) \cdot \delta_0^2, \tag{6.23}$$

$$\int_{\hat{t}}^{T} \int_{|w_i/v_i| < 3\delta_1} |v_i|^2 \left| \left(\frac{w_i}{v_i} \right)_x \right|^2 dx \, dt = \mathcal{O}(1) \cdot \delta_0^3. \tag{6.24}$$

The above integrals will be controlled in terms of two functionals, related to shortening curves. Consider a parametrized curve in the plane $\gamma : \mathbb{R} \mapsto \mathbb{R}^2$. Assuming that γ is sufficiently smooth, its **length** is computed by

$$\mathcal{L}(\gamma) \doteq \int |\gamma_x(x)| \, dx. \tag{6.25}$$

We can also define the **area** functional as the integral of a wedge product:

$$\mathcal{A}(\gamma) \doteq \frac{1}{2} \iint_{x < y} |\gamma_x(x) \wedge \gamma_x(y)| \, dx \, dy. \tag{6.26}$$

To understand its geometrical meaning, observe that if γ is a closed curve, the integral

$$\frac{1}{2} \int \gamma(y) \wedge \gamma_x(y) \, dy = \frac{1}{2} \iint_{x < y} \gamma_x(x) \wedge \gamma_x(y) \, dx \, dy$$

yields the sum of the areas of the regions enclosed by the curve γ, multiplied by the corresponding winding number. In general, the quantity $\mathcal{A}(\gamma)$ provides an upper bound for the area of the convex hull of γ.

Let now $\gamma = \gamma(t, x)$ be a planar curve which evolves in time, according to the vector equation

$$\gamma_t + \lambda \gamma_x = \gamma_{xx}. \tag{6.27}$$

Here $\lambda = \lambda(t, x)$ is a sufficiently smooth scalar function. It is then clear that the length $\mathcal{L}(\gamma(t))$ of the curve is a decreasing function of time. One can show that also the area functional $\mathcal{A}(\gamma(t))$ is monotonically decreasing. Moreover, the amount of decrease dominates the area swept by the curve during its motion. An intuitive way to see this is the following. In the special case where γ is a polygonal line, with vertices at the points $P_0, \ldots, P_m$, the integral in (6.26) reduces to a sum:

$$\mathcal{A}(\gamma) = \frac{1}{2} \sum_{i<j} |\mathbf{v}_i \wedge \mathbf{v}_j|, \qquad \mathbf{v}_i \doteq P_i - P_{i-1}.$$

If we now replace γ by a new curve γ' obtained by replacing two consecutive edges $\mathbf{v}_h, \mathbf{v}_k$ by one single edge (fig. 17b), the area between γ and γ' is precisely $|\mathbf{v}_h \wedge \mathbf{v}_k|/2$, while an easy computation yields

$$\mathcal{A}(\gamma') \leq \mathcal{A}(\gamma) - \frac{1}{2} |\mathbf{v}_h \wedge \mathbf{v}_k|.$$

The estimate on the area swept by a smooth curve (fig. 17a) is now obtained by approximating a shortening curve γ by a sequence of polygons, each obtained from the previous one by replacing two consecutive edges by a single segment.

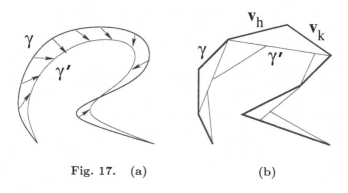

Fig. 17.   (a)   (b)

We shall apply the previous geometric considerations toward a proof of the estimates of (6.23)–(6.24). Let v, w be two scalar functions, satisfying

$$\begin{aligned} v_t + (\lambda v)_x - v_{xx} &= \phi, \\ w_t + (\lambda w)_x - w_{xx} &= \psi. \end{aligned} \qquad (6.28)$$

Define the planar curve γ by setting

$$\gamma(t, x) = \left(\int_{-\infty}^x v(t, y) dy, \ \int_{-\infty}^x w(t, y) dy \right). \qquad (6.29)$$

Integrating (6.27) w.r.t. x, one finds the corresponding evolution equation for γ:

$$\gamma_t + \lambda \gamma_x - \gamma_{xx} = \Phi(t,x) \doteq \left(\int_{-\infty}^x \phi(t,y)dy, \quad \int_{-\infty}^x \psi(t,y)dy \right). \qquad (6.30)$$

In particular, if no sources were present, the motion of the curve would reduce to (6.27). At each fixed time t, we now define the **Length Functional** as

$$\mathcal{L}(t) = \mathcal{L}(\gamma(t)) = \int \sqrt{v^2(t,x) + w^2(t,x)} \, dx \qquad (6.31)$$

and the **Area Functional** as

$$\mathcal{A}(t) = \mathcal{A}(\gamma(t)) = \frac{1}{2} \iint_{x<y} \left| v(t,x)w(t,y) - v(t,y)w(t,x) \right| dx \, dy. \qquad (6.32)$$

We now estimate the time derivative of the above functionals, in the general case when sources are present.

Lemma 6.5 *Let v, w be solutions of (6.2), defined for $t \in [0,T]$. For each t, assume that the maps $x \mapsto v(t,x)$ $x \mapsto w(t,x)$ and $x \mapsto \lambda(t,x)$ are $C^{1,1}$, i.e. continuously differentiable with Lipschitz derivative. Then the corresponding area functional (6.32) satisfies*

$$\frac{d}{dt} \mathcal{A}(t) \leq - \int \left| v_x(t,x)w(t,x) - v(t,x)w_x(t,x) \right| dx$$
$$+ \|v(t)\|_{L^1} \|\psi(t)\|_{L^1} + \|w(t)\|_{L^1} \|\phi(t)\|_{L^1}. \qquad (6.33)$$

Proof. In the following, given a curve γ, at each point x where $\gamma_x \neq 0$ we define the unit normal $\mathbf{n} = \mathbf{n}(x)$ oriented so that $\gamma_x(x) \wedge \mathbf{n} = |\gamma_x(x)| > 0$. For every vector $\mathbf{v} \in \mathbb{R}^2$ this implies

$$\gamma_x(x) \wedge \mathbf{v} = |\gamma_x(x)| \langle \mathbf{n}, \mathbf{v} \rangle.$$

Given a unit vector $\mathbf{n}$, we shall also consider the projection of γ along $\mathbf{n}$, namely

$$y \mapsto \chi^{\mathbf{n}}(y) \doteq \langle \mathbf{n}, \gamma(y) \rangle.$$

If $\gamma = \gamma(t,x)$ is any smooth curve evolving in time, the time derivative of the area functional in (6.26) can be computed as

$$\begin{aligned} \frac{d\mathcal{A}}{dt} &= \frac{1}{2} \iint_{x<y} \mathrm{sgn}(\gamma_x(x) \wedge \gamma_x(y)) \left\{ \gamma_{xt}(x) \wedge \gamma_x(y) + \gamma_x(x) \wedge \gamma_{xt}(y) \right\} dx \, dy \\ &= \frac{1}{2} \iint \mathrm{sgn}(\gamma_x(x) \wedge \gamma_x(y)) \left\{ \gamma_x(x) \wedge \gamma_{xt}(y) \right\} dy \, dx \\ &= \frac{1}{2} \int |\gamma_x(x)| \left(\frac{d}{dt} \int |\langle \mathbf{n}, \gamma_x(y) \rangle| dy \right) dx \\ &= \frac{1}{2} \int |\gamma_x(x)| \left(\int \mathrm{sgn} \langle \mathbf{n}, \gamma_x(y) \rangle \cdot \langle \mathbf{n}, \gamma_{xt}(y) \rangle dy \right) dx \\ &= \frac{1}{2} \int |\gamma_x(x)| \frac{d}{dt} \left(\mathrm{Tot.Var.} \{\chi^{\mathbf{n}}\} \right) dx \end{aligned}$$

For each x, we are here choosing the unit normal $\mathbf{n} = \mathbf{n}(x)$ to the curve γ at the point x. To compute the derivative of the total variation, assume that the function $y \mapsto \chi^{\mathbf{n}(x)}(y)$ has a finite number of local maxima and minima, say attained at the points (fig. 18)

$$y_{-p} < \cdots < y_{-1} < y_0 = x < y_1 < \cdots < y_q .$$

Assume, in addition, that its derivative $d\chi^{\mathbf{n}}/dy$ changes sign across every such point. Then

$$\frac{d}{dt}\left(\text{Tot.Var.}\{\chi^{\mathbf{n}}\}\right) = -\text{sgn}\left\langle \mathbf{n},\, \gamma_{xx}(x)\right\rangle \cdot 2 \sum_{-p\leq\alpha\leq q} (-1)^\alpha\left\langle \mathbf{n},\, \gamma_t(y_\alpha)\right\rangle. \quad (6.34)$$

Notice the sign factor in front of (6.34). If the inner product $\left\langle \mathbf{n},\, \gamma_{xx}(x)\right\rangle$ is positive, the even indices α correspond to local minima and the odd indices to local maxima. The opposite is true if the inner product is negative. Using (6.34) and observing that

$$\left\langle \mathbf{n},\, \gamma_x(x)\right\rangle = 0, \qquad \text{sgn}\left\langle \mathbf{n},\, \gamma_{xx}(y_\alpha)\right\rangle = (-1)^\alpha \cdot \text{sgn}\left\langle \mathbf{n},\, \gamma_{xx}(x)\right\rangle,$$

we obtain

$$\begin{aligned}
\frac{d\mathcal{A}}{dt} &= -\int \left|\gamma_x(x)\right|\text{sgn}\left\langle \mathbf{n},\, \gamma_{xx}(x)\right\rangle \cdot \left(\sum_{-p\leq\alpha\leq q}(-1)^\alpha\left\langle \mathbf{n},\, \gamma_t(y_\alpha)\right\rangle\right) dx \\
&\leq -\int \sum_\alpha \left|\gamma_x(x)\wedge\gamma_{xx}(y_\alpha)\right| dx + \int \left|\gamma_x(x)\right| \cdot \left|\sum_\alpha(-1)^\alpha\left\langle \mathbf{n},\, \varPhi(y_\alpha)\right\rangle\right| dx \\
&\leq -\int \left|\gamma_x(x)\wedge\gamma_{xx}(x)\right| dx + \iint \left|\gamma_x(x)\wedge\varPhi_x(y)\right| dy\,dx .
\end{aligned}$$
$$(6.35)$$

If we specialize this formula to the case where γ is the curve in (6.29) and $\varPhi$ the function in (6.30), we find

$$\begin{aligned}
\frac{d\mathcal{A}}{dt} &\leq -\int \left|\binom{v_x}{w_x}\wedge\binom{v_x(x)}{w_x(x)}\right| dx \\
&\quad + \iint \left|\binom{v_x}{w_x}\wedge\binom{\phi(y)}{\psi(y)}\right| dy\,dx ,
\end{aligned}$$

from which (6.33) clearly follows. Notice that, by an approximation argument, we can assume that the functions $\chi^{\mathbf{n}(x)}$ have the required regularity for almost every $(t, x) \in \mathbb{R}^2$.   ●

Lemma 6.6. *Together with the hypotheses of Lemma 6.5, at a fixed time t assume that $\gamma_x(t, x) \neq 0$ for every x. Then*

$$\frac{d}{dt}\mathcal{L}(t) \leq -\frac{1}{(1+9\delta_1^2)^{3/2}}\int_{|w/v|\leq 3\delta_1} |v|\left|\left(\frac{w}{v}\right)_x\right|^2 dx + \|\phi(t)\|_{\mathbf{L}^1} + \|\psi(t)\|_{\mathbf{L}^1} .$$
$$(6.36)$$

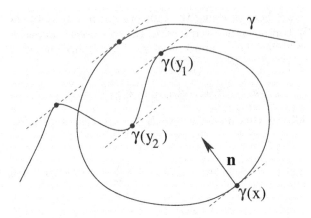

Fig. 18.

Proof. As a preliminary, recalling that

$$\gamma_x = (v, w), \qquad \gamma_{xt} + (\lambda \gamma_x)_x - \gamma_{xxx} = (\phi, \psi),$$

we derive the identities

$$|\gamma_{xx}|^2 |\gamma_x|^2 - \langle \gamma_x, \gamma_{xx} \rangle^2 = (v_x^2 + w_x^2)(v^2 + w^2) - (vv_x + ww_x)^2$$
$$= (vw_x - v_x w)^2 = v^4 |(w/v)_x|^2,$$

$$\frac{|v|^3}{|\gamma_x|^3} = \frac{1}{\left(1 + (w/v)^2\right)^{3/2}} \cdot$$

Thanks to the assumption that γ_x never vanishes, we can now integrate by parts and obtain

$$\frac{d}{dt} \mathcal{L}(t)$$

$$= \int \frac{\langle \gamma_x, \gamma_{xt} \rangle}{\sqrt{\langle \gamma_x, \gamma_x \rangle}} \, dx = \int \left\{ \frac{\langle \gamma_x, \gamma_{xxx} \rangle}{|\gamma_x|} - \frac{\langle \gamma_x, (\lambda \gamma_x)_x \rangle}{|\gamma_x|} + \frac{\langle \gamma_x, (\phi, \psi) \rangle}{|\gamma_x|} \right\} dx$$

$$= \int \left\{ |\gamma_x|_{xx} - (\lambda |\gamma_x|)_x - \frac{|\gamma_{xx}|^2 - \langle \gamma_x/|\gamma_x|, \gamma_{xx} \rangle^2}{|\gamma_x|} \right\} dx + \int \frac{\langle \gamma_x, (\phi, \psi) \rangle}{|\gamma_x|} \, dx$$

$$\leq - \int \frac{|v| |(w/v)_x|^2}{\left(1 + (w/v)^2\right)^{3/2}} \, dx + \|\phi(t)\|_{\mathbf{L}^1} + \|\psi(t)\|_{\mathbf{L}^1}.$$

Since the integrand is non-negative, the last inequality clearly implies (6.36).

•

Remark 6.7. Let $u = u(t, x)$ be a solution to a scalar, viscous conservation law

$$u_t + f(u)_x - u_{xx} = 0\,,$$

and consider the planar curve $\gamma \doteq (u,\ f(u) - u_x)$ whose components are respectively the conserved quantity and the flux (fig. 19). If $\lambda \doteq f'$, the components $v \doteq u_x$ and $w \doteq -u_t$ evolve according to (6.27). Defining the speed $s(x) \doteq -u_t(x)/u_x(x)$, the area functional $\mathcal{A}(\gamma)$ in (6.36) can now be written as

$$
\begin{aligned}
\mathcal{A}(\gamma) &= \frac{1}{2} \iint_{x<y} \left| u_x(x) u_t(y) - u_t(x) u_x(y) \right| dx\, dy \\
&= \frac{1}{2} \iint_{x<y} \left| u_x(x)\, dx \right| \cdot \left| u_x(y)\, dy \right| \cdot \left| s(x) - s(y) \right| \\
&= \frac{1}{2} \iint_{x<y} [\text{wave at } x] \times [\text{wave at } y] \times [\text{difference in speeds}]\,.
\end{aligned}
$$

It now becomes clear that the area functional can be regarded as an interaction potential between waves of the same family. In the case where viscosity is present, this provides a counterpart to the interaction functional introduced by Liu in [L4] for general hyperbolic systems.

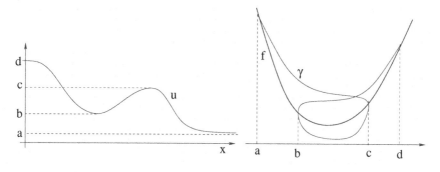

Fig. 19.

Remark 6.8. In the case where u is precisely a viscous traveling wave, the curve $\gamma = (u,\ f(u) - u_x)$ reduces to a segment. Assume now that the flux f is genuinely nonlinear, say with $f'' \geq c > 0$. Consider a solution u which initially consists of two viscous traveling waves, far apart from each other (fig. 20). To fix the ideas, let the first wave join a left state a with a middle state b, and the second wave join the middle state b with a right state c, with $a > b > c$. The strength of the two waves can be measured as $s = a - b$, $s' = b - c$. The corresponding curve γ is approximately given by two segments, joining the points $P \doteq (a, f(a))$, $Q \doteq (b, f(b))$, $P' \doteq (c, f(c))$. After a long time $\tau \gg 0$, the two shocks will interact, merging into one single viscous shock.

The curve $\gamma(\tau)$ will then reduce to one single segment, joining P with P'. The area swept by the curve is approximately the area of the triangle PQP'. The assumption of genuine nonlinearity implies

$$\text{area swept} = \mathcal{O}(1) \cdot |ss'| \left(|s| + |s'| \right).$$

In this case, the decrease in the area functional is of cubic order w.r.t. the strengths s, s' of the interacting waves. This is indeed the correct order of magnitude needed to control the strength of new waves generated by the interaction in the genuinely nonlinear case. It is remarkable that this area functional gives the correct order of magnitude of waves generated by an interaction also for a general flux function f, not necessarily convex.

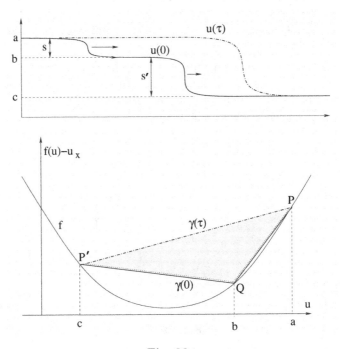

Fig. 20.

We now return to the proof of the crucial estimates (6.23)–(6.24). Recalling that the components v_i, w_i satisfy the equations (6.2), we can apply the previous lemmas with $v \doteq v_i$, $w \doteq w_i$, $\lambda \doteq \tilde{\lambda}_i$, $\phi \doteq \phi_i$, $\psi \doteq \psi_i$, calling $\mathcal{L}_i$ and $\mathcal{A}_i$ the corresponding length and area functionals. For $t \in [\hat{t}, T]$, the bounds (5.56)–(5.57) yield

$$\mathcal{A}_i(t) \leq \left\| v_i(t) \right\|_{\mathbf{L}^\infty} \cdot \left\| w_i(t) \right\|_{\mathbf{L}^1} = \mathcal{O}(1) \cdot \delta_0^3, \tag{6.37}$$

$$\mathcal{L}_i(t) \leq \left\|v_i(t)\right\|_{\mathbf{L}^1} + \left\|w_i(t)\right\|_{\mathbf{L}^1} = \mathcal{O}(1) \cdot \delta_0. \tag{6.38}$$

Using (6.33) we now obtain

$$\int_{\hat{t}}^T \int \left| w_{i,x} v_i - w_i v_{i,x} \right| dx \, dt$$

$$\leq \int_{\hat{t}}^T \left| \frac{d}{dt} \mathcal{A}_i(t) \right| dt + \int_{\hat{t}}^T \left(\left\|v_i(t)\right\|_{L^1} \left\|\psi_i(t)\right\|_{L^1} + \left\|w_i(t)\right\|_{L^1} \left\|\phi_i(t)\right\|_{L^1} \right) dt$$

$$\leq \mathcal{A}_i(\hat{t}) + \sup_{t \in [\hat{t},T]} \left(\left\|v_i(t)\right\|_{\mathbf{L}^1} + \left\|w_i(t)\right\|_{\mathbf{L}^1} \right) \cdot \int_{\hat{t}}^T \int \left(\left|\phi_i(t,x)\right| + \left|\psi_i(t,x)\right| \right) dx \, dt$$

$$= \mathcal{O}(1) \cdot \delta_0^2, \tag{6.39}$$

proving (6.23). To establish (6.24), we first observe that, by an approximation argument, it is not restrictive to assume that the set of points in the t-x plane where $v_{i,x}(t,x) = w_{i,x}(t,x) = 0$ is at most countable. In this case, for almost every $t \in [\hat{t}, T]$ the inequality (6.36) is valid, and hence

$$\int_{\hat{t}}^T \int_{|w_i/v_i|<3\delta_1} |v_i| \left| \left(\frac{w_i}{v_i}\right)_x \right|^2 dx \, dt$$

$$\leq \int_{\hat{t}}^T \left| \frac{d}{dt} \mathcal{L}_i(t) \right| dt + \int_{\hat{t}}^T \left(\left\|\phi_i(t)\right\|_{L^1} + \left\|\psi_i(t)\right\|_{L^1} \right) dt \tag{6.40}$$

$$\leq \mathcal{L}_i(\hat{t}) + \int_{\hat{t}}^T \int \left(\left|\phi_i(t,x)\right| + \left|\psi_i(t,x)\right| \right) dx \, dt$$

$$= \mathcal{O}(1) \cdot \delta_0.$$

The estimate (6.24) is now a consequence of (6.40) and of the bound $\left\|v_i(t)\right\|_{\mathbf{L}^\infty} = \mathcal{O}(1) \cdot \delta_0^3$, given at (5.57). $\bullet$

6.4. Bounds on Cutoff Terms

In this section we provide an estimate on the total strength of source terms due to the cutoff function. In the same setting as the two previous sections, in particular assuming that (6.7) holds, we claim that

$$\int_{\hat{t}}^T \int \left(\left|v_{i,x}\right| + \left|w_{i,x}\right| \right) \left|w_i - \theta_i v_i\right| dx \, dt = \mathcal{O}(1) \cdot \delta_0^2. \tag{6.41}$$

We recall that $\theta_i \doteq \theta(w_i/v_i)$, where θ is the cut-off function introduced at (5.40). Notice that the integrand can be $\neq 0$ only when $|w_i/v_i| > \delta_1$.

Consider another cut-off function $\eta : \mathbb{R} \mapsto [0,1]$ such that (fig. 15)

$$\eta(s) = \begin{cases} 0 & if & |s| \leq 3\delta_1/5, \\ 1 & if & |s| \geq 4\delta_1/5. \end{cases} \tag{6.42}$$

We can assume that η is a smooth even function, such that

$$|\eta'| \le 21/\delta_1, \qquad\qquad |\eta''| \le 101/\delta_1^2.$$

For convenience, we shall write $\eta_i \doteq \eta(w_i/v_i)$.

Toward a proof of the estimate (6.41), we first reduce the integrand to a more tractable expression. Since the term $|w_i - \theta_i v_i|$ vanishes when $|w_i/v_i| \le \delta_1$, and is $\le |w_i|$ otherwise, by (5.69) we always have the bound

$$|w_i - \theta_i v_i| \le |\eta_i w_i| \le \eta_i \left(2|v_{i,x}| + \mathcal{O}(1) \cdot \delta_0 \sum_{j \ne i} |v_j| \right).$$

Therefore

$$
\begin{aligned}
\big(|v_{i,x}| + |w_{i,x}|\big) &\cdot |w_i - \theta_i v_i| \\
&\le \big(|v_{i,x}| + |w_{i,x}|\big)\eta_i \left(2|v_{i,x}| + \mathcal{O}(1) \cdot \delta_0 \sum_{j \ne i} |v_j| \right) \\
&\le 2\eta_i v_{i,x}^2 + 2\eta_i |v_{i,x} w_{i,x}| + \sum_{j \ne i} \big(|v_j v_{i,x}| + |v_j w_{i,x}| \big) \\
&\le 3\eta_i v_{i,x}^2 + \eta_i w_{i,x}^2 + \sum_{j \ne i} \big(|v_j v_{i,x}| + |v_j w_{i,x}| \big).
\end{aligned}
\tag{6.43}
$$

Since we already proved the bounds (6.8) on the integrals of transversal terms, to prove (6.41) we only need to consider the integrals of $v_{i,x}^2$ and $w_{i,x}^2$, in the region where $\eta_i \ne 0$. In both cases, energy type estimates can be used.

We work out in detail the bound for $v_{i,x}^2$. Multiplying the first equation in (6.1) by $\eta_i v_i$ and integrating by parts, we obtain

$$
\begin{aligned}
\int \eta_i v_i \phi_i \, dx &= \int \Big\{ \eta_i v_i v_{i,t} + \eta_i v_i (\tilde{\lambda}_i v_i)_x - \eta_i v_i v_{i,xx} \Big\} \, dx \\
&= \int \Big\{ \eta_i (v_i^2/2)_t - \eta_i \tilde{\lambda}_i v_i v_{i,x} - \eta_{i,x} \tilde{\lambda}_i v_i^2 + \eta_i v_{i,x}^2 + \eta_{i,x} v_{i,x} v_i \Big\} \, dx \\
&= \int \Big\{ (\eta_i v_i^2/2)_t + (\tilde{\lambda}_i \eta_i)_x (v_i^2/2) - (\eta_{i,t} + 2\tilde{\lambda}_i \eta_{i,x} - \eta_{i,xx})(v_i^2/2) \\
&\qquad\quad + \eta_i v_{i,x}^2 + 2\eta_{i,x} v_i v_{i,x} \Big\} \, dx.
\end{aligned}
$$

Therefore

$$
\begin{aligned}
\int \eta_i v_{i,x}^2 \, dx = {}& -\frac{d}{dt}\left[\int \eta_i v_i^2 / 2 \, dx \right] + \int (\eta_{i,t} + \tilde{\lambda}_i \eta_{i,x} - \eta_{i,xx})(v_i^2/2) \, dx \\
& - \int \tilde{\lambda}_{i,x} \eta_i (v_i^2/2) \, dx - 2 \int \eta_{i,x} v_i v_{i,x} \, dx + \int \eta_i v_i \phi_i \, dx.
\end{aligned}
\tag{6.44}
$$

A direct computation yields

$$\eta_{i,t} + \tilde{\lambda}_i \eta_{i,x} - \eta_{i,xx}$$

$$= \eta_i'\left(\frac{w_{i,t}}{v_i} - \frac{v_{i,t}w_i}{v_i^2}\right) + \tilde{\lambda}_i \eta_i'\left(\frac{w_{i,x}}{v_i} - \frac{v_{i,x}w_i}{v_i^2}\right)$$

$$-\eta_i''\left(\frac{w_i}{v_i}\right)_x^2 - \eta_i'\left(\frac{w_{i,xx}}{v_i} - \frac{v_{i,xx}w_i}{v_i^2} - 2\frac{v_{i,x}w_{i,x}}{v_i^2} + 2\frac{v_{i,x}^2 w_i}{v_i^3}\right) \qquad (6.45)$$

$$= \left[\eta_i'\left(w_{i,t} + (\tilde{\lambda}_i w_i)_x - w_{i,xx}\right)/v_i - \eta_i' w_i\left(v_{i,t} + (\tilde{\lambda}_i v_i)_x - v_{i,xx}\right)/v_i^2\right]$$

$$+2v_{i,x}\eta_i'/v_i \cdot (w_i/v_i)_x - \eta_i''(w_i/v_i)_x^2$$

$$= \eta_i'\left(\frac{\psi_i}{v_i} - \frac{w_i}{v_i}\frac{\phi_i}{v_i}\right) + 2\eta_i'\frac{v_{i,x}}{v_i}\left(\frac{w_i}{v_i}\right)_x - \eta_i''\left(\frac{w_i}{v_i}\right)_x^2.$$

Since $\tilde{\lambda}_{i,x} = (\tilde{\lambda}_i - \lambda_i^*)_x$, integrating by parts and using the second estimate in (5.69) one obtains

$$\left|\int \tilde{\lambda}_{i,x}\eta_i(v_i^2/2)\,dx\right|$$

$$= \left|\int (\tilde{\lambda}_i - \lambda_i^*)\left(\eta_{i,x}v_i^2/2 + \eta_i v_i v_{i,x}\right)dx\right|$$

$$\leq \|\tilde{\lambda}_i - \lambda_i^*\|_{\mathbf{L}^\infty} \cdot \left\{\frac{1}{2}\int |\eta_i'|\,|w_{i,x}v_i - v_{i,x}w_i|\,dx + \frac{5}{2\delta_1}\int \eta_i v_{i,x}^2\,dx\right.$$

$$\left. +\mathcal{O}(1)\cdot\delta_0 \int \sum_{j\neq i}|v_{i,x}v_j|\,dx\right\} \qquad (6.46)$$

$$\leq \int |w_{i,x}v_i - v_{i,x}w_i|\,dx + \frac{1}{2}\int \eta_i v_{i,x}^2\,dx + \delta_0 \int \sum_{j\neq i}|v_{i,x}v_j|\,dx.$$

Indeed, $|\tilde{\lambda}_i - \lambda_i^*| = \mathcal{O}(1)\cdot\delta_0 \ll \delta_1$. Using (6.45)–(6.46) in (6.44) we now obtain

$$\frac{1}{2}\int \eta_i v_{i,x}^2\,dx \leq -\frac{d}{dt}\left[\int \frac{\eta_i v_i^2}{2}\,dx\right]$$

$$+\frac{1}{2}\int |\eta_i'|\left(|v_i\psi_i| + |w_i\phi_i|\right)dx + \int \left|\eta_i'v_iv_{i,x}\left(\frac{w_i}{v_i}\right)_x\right|dx$$

$$+\frac{1}{2}\int \left|\eta_i''v_i^2\left(\frac{w_i}{v_i}\right)_x^2\right|dx + \int |w_{i,x}v_i - w_iv_{i,x}|\,dx \qquad (6.47)$$

$$+\delta_0 \int \sum_{j\neq i}|v_{i,x}v_j|\,dx + 2\int |\eta_{i,x}v_iv_{i,x}|\,dx + \int |v_i\phi_i|\,dx.$$

Recalling the definition of η_i, on regions where $\eta_i' \neq 0$ one has $|w_i/v_i| < \delta_1$, hence the bound (5.70) holds. In turn, this implies

$$\left|\eta_{i,x}v_iv_{i,x}\right| = \left|\eta_i'v_iv_{i,x}\left(\frac{w_i}{v_i}\right)_x\right|$$

$$\leq 2\left|\delta_1\eta_i'v_i^2\left(\frac{w_i}{v_i}\right)_x\right| + \mathcal{O}(1)\cdot\delta_0\sum_{j\neq i}\left|\eta_i'v_iv_j\left(\frac{w_i}{v_i}\right)_x\right|$$

$$\leq 2\left|\delta_1\eta_i'\right|\left|w_{i,x}v_i - w_iv_{i,x}\right| + \mathcal{O}(1)\cdot\delta_0|\eta_i'|\sum_{j\neq i}\left(\left|v_jw_{i,x}\right| + \left|v_jv_{i,x}\right|\left|\frac{w_i}{v_i}\right|\right).$$

(6.48)

Using the bounds (5.56)–(5.57), (6.7)–(6.8) and (6.23)–(6.24), from (6.47) we conclude

$$\int_{\hat t}^T\int \eta_i\,v_{i,x}^2\,dx\,dt \leq \int \eta_i v_i^2(\hat t, x)\,dx + \mathcal{O}(1)\cdot\int_{\hat t}^T\int\left(|v_i\psi_i| + |w_i\phi_i|\right)dx\,dt$$

$$+ \mathcal{O}(1)\cdot\int_{\hat t}^T\int\left|w_{i,x}v_i - w_iv_{i,x}\right|dx\,dt$$

$$+ \mathcal{O}(1)\cdot\delta_0\int_{\hat t}^T\int\sum_{j\neq i}\left(|v_jw_{i,x}| + |v_jv_{i,x}|\right)dx\,dt$$

$$+ \mathcal{O}(1)\cdot\int_{\hat t}^T\int_{|w_i/v_i|<\delta_1}\left|v_i(w_i/v_i)_x\right|^2dx\,dt$$

$$+ 2\delta_0\int_{\hat t}^T\int\sum_{j\neq i}|v_{i,x}v_j|\,dx\,dt + 2\int_{\hat t}^T\int|v_i\phi_i|\,dx\,dt$$

$$= \mathcal{O}(1)\cdot\delta_0^2\,.$$

(6.49)

In a similar way (see [BiB]) one establishes the bound

$$\int_{\hat t}^T\int \eta_i\,w_{i,x}^2\,dx\,dt = \mathcal{O}(1)\cdot\delta_0^2\,.$$

(6.50)

6.5. Proof of the BV Estimates

We now conclude the proof of the uniform BV bounds, for solutions of the Cauchy problem

$$u_t + A(u)u_x = u_{xx}\,, \qquad u(0, x) = \bar u(x)\,. \tag{6.51}$$

Assume that the initial data $\bar u : \mathbb{R} \mapsto \mathbb{R}^n$ has suitably small total variation, so that

$$\text{Tot.Var.}\{\bar u\} \leq \frac{\delta_0}{8\sqrt{n}\,\kappa}\,, \qquad \lim_{x\to-\infty}\bar u(x) = u^*\,. \tag{6.52}$$

We recall that κ is the constant introduced at (4.7), related to the Green kernel G^* of the linearized equation (4.6). This constant actually depends on the matrix $A(u^*)$, but it is clear that it remains uniformly bounded when u^* varies in a compact subset of $\mathbb{R}^n$.

An application of Proposition 4.5 yields the existence of the solution to the Cauchy problem (6.51) on an initial interval $[0, \hat{t}]$, satisfying the bound

$$\left\| u_x(\hat{t}) \right\|_{\mathbf{L}^1} \leq \frac{\delta_0}{4\sqrt{n}} \,. \tag{6.53}$$

This solution can be prolonged in time as long as its total variation remains small. By the previous analysis, for $t \geq \hat{t}$ we can decompose the vectors u_x, u_x along a basis of unit vectors $\tilde{r}_i$, as in (5.41). The corresponding components v_i, w_i then satisfy the system of evolution equations (6.2). Define the time

$$T \doteq \sup \left\{ \tau \, ; \quad \sum_i \int_{\hat{t}}^{\tau} \int |\phi_i(t, x)| + |\psi_i(t, x)| \, dx \, dt \leq \frac{\delta_0}{2} \right\}. \tag{6.54}$$

If $T < \infty$, a contradiction is obtained as follows. By (5.55) and (6.53), for all $t \in [\hat{t}, T]$ one has

$$\left\| u_x(t) \right\|_{\mathbf{L}^1} \leq \sum_i \left\| v_i(t) \right\|_{\mathbf{L}^1}$$

$$\leq \sum_i \left(\left\| v_i(\hat{t}) \right\|_{\mathbf{L}^1} + \int_{\hat{t}}^{T} \int |\phi_i(t, x)| \, dx \, dt \right)$$

$$\leq 2\sqrt{n} \left\| u_x(\hat{t}) \right\|_{\mathbf{L}^1} + \frac{\delta_0}{2} \leq \delta_0 \,.$$

Using Lemma 6.1 and the bounds (6.7), (6.23), (6.24) and (6.41) we now obtain

$$\sum_i \int_{\hat{t}}^{T} \int |\phi_i(t, x)| + |\psi_i(t, x)| \, dx \, dt = \mathcal{O}(1) \cdot \delta_0^2 \, < \, \frac{\delta_0}{2}, \tag{6.55}$$

provided that δ_0 was chosen suitably small. Therefore T cannot be a supremum. This contradiction shows that the total variation remains $< \delta_0$ for all $t \in [\hat{t}, \infty[$. In particular, the solution u is globally defined.

7 Stability of Viscous Solutions

By the homotopy argument (3.18), to show the global stability of every small BV solution of the viscous system (3.10) it suffices to prove the bound

$$\left\| z(t, \cdot) \right\|_{\mathbf{L}^1} \leq L \left\| z(0, \cdot) \right\|_{\mathbf{L}^1} \tag{7.1}$$

on first order perturbations. Toward this goal, consider a solution $u = u(t, x)$ of (3.10). The linearized evolution equation (3.16) for a first order perturbation can be written as

$$z_t + \big(A(u)z\big)_x - z_{xx} = \big(u_x \bullet A(u)\big)z_x - \big(z \bullet A(u)\big)u_x. \qquad (7.2)$$

In connection with with (7.2), we introduce the quantity $\Upsilon \doteq z_x - A(u)z$, related to the flux of z. Differentiating (3.10) we derive an evolution equation for Υ:

$$
\begin{aligned}
\Upsilon_t + \big(A(u)\Upsilon\big)_x - \Upsilon_{xx} = & \Big[\big(u_x \bullet A(u)\big)z - \big(z \bullet A(u)\big)u_x\Big]_x \\
& -A(u)\Big[\big(u_x \bullet A(u)\big)z - \big(z \bullet A(u)\big)u_x\Big] \qquad (7.3) \\
& +\big(u_x \bullet A(u)\big)\Upsilon - \big(u_t \bullet A(u)\big)z.
\end{aligned}
$$

Notice that $z = u_x$ and $\Upsilon = u_t$ yield a particular solution to (7.2)–(7.3). In this special case, (7.1) reduces to the bound on the total variation,

In general, the estimate (7.1) will still be obtained following a similar strategy as for the bounds on the total variation. By (4.34) we already know that the desired estimate holds on the initial time interval $[0, \hat{t}]$. To obtain a uniform bound valid for all $t > 0$, we decompose the vector z along a basis of unit vectors $\tilde{r}_i$ and derive an evolution equation for these scalar components. At first sight, it looks promising to write

$$z = \sum_i z_i \tilde{r}_i(u, v_i, \sigma_i),$$

where $\tilde{r}_1, \ldots, \tilde{r}_n$ are the same vectors used in the decomposition of u_x at (5.41)–(5.42). Unfortunately, this choice leads to an evolution equation for the components z_i with non-integrable source terms. Instead, we shall use a different basis of unit vectors, depending not only on the reference solution u but also on the perturbation z. As shown in [BiB], a decomposition leading to successful estimates is

$$
\begin{cases}
z = \sum_i h_i \tilde{r}_i\big(u, v_i, \lambda_i^* - \theta(g_i/h_i)\big), \\
\Upsilon = \sum_i (g_i - \lambda_i^* h_i)\tilde{r}_i\big(u, v_i, \lambda_i^* - \theta(g_i/h_i)\big),
\end{cases} \qquad (7.4)
$$

where θ is the cutoff function introduced at (5.40). Here $v_1, \ldots, v_n$ are the wave strengths in the decomposition (5.41) of the gradient u_x. The next result, analogous to Lemma 5.2, provides the existence and regularity of the decomposition (7.4). For a proof, see [BiB].

Lemma 7.1 Let $|u - u^*|$ and $|v|$ be sufficiently small. Then for all $z, \Upsilon \in \mathbb{R}^n$ the system of $2n$ equations (7.4) has a unique solution $(h_1, \ldots, h_n, g_1, \ldots, g_n)$. The map $(z, \Upsilon) \mapsto (h, g)$ is Lipschitz continuous. Moreover, it is smooth outside the n manifolds $\widehat{\mathcal{N}}_i \doteq \{h_i = g_i = 0\}$.

Remark 7.2. If we had chosen a basis of unit vectors $\{\tilde{r}_1, \ldots, \tilde{r}_n\}$ depending only on the reference solution u, the map $(z, \Upsilon) \mapsto (h, g)$ would be linear.

This approach seems very natural, because z, Υ both satisfy linear evolution equations. However, it would not achieve the desired estimates (7.1). On the other hand, our decomposition (7.4) yields a mapping $(z, \Upsilon) \mapsto (h, g)$ which is homogeneous of degree one, but not linear. Indeed, the unit vectors $\tilde{r}_i$ are here related to viscous traveling waves U_i whose signed strength v_i depends on the reference solution u, but whose speed $\sigma_i = \theta(h_i/g_i)$ depends on the perturbation z.

Differentiating (7.2)–(7.3) and using (7.4) we obtain a system of evolution equations for the scalar components h_i, g_i :

$$\begin{cases} h_{i,t} + (\tilde{\lambda}_i h_i)_x - h_{i,xx} = \hat{\phi}_i, \\ g_{i,t} + (\tilde{\lambda}_i g_i)_x - g_{i,xx} = \hat{\psi}_i. \end{cases} \qquad (7.5)$$

Since the left hand sides are in conservation form, we have

$$\left\| z(t,\cdot) \right\|_{\mathbf{L}^1} \leq \sum_i \left\| h_i(t) \right\|_{\mathbf{L}^1} \leq \sum_i \left(\left\| h_i(\hat{t}) \right\| + \int_{\hat{t}}^t \int \left| \hat{\phi}_i(s,x) \right| dx ds \right). \quad (7.6)$$

Relying on the same techniques described in Sections 6–7, a careful analysis of the source terms $\hat{\phi}_i, \hat{\psi}_i$ eventually leads to the bounds (7.1). For all details we refer to [BiB].

7.2. Time Dependence.

The continuity w.r.t. time of a solution of the viscous system (3.10) easily follows from the bounds (4.24) and (4.29). Indeed

$$\left\| u_t(t) \right\|_{\mathbf{L}^1} \leq \mathcal{O}(1) \cdot \left\| u_x(t) \right\|_{\mathbf{L}^1} + \left\| u_{xx}(t) \right\|_{\mathbf{L}^1}$$
$$= \mathcal{O}(1) \cdot \delta_0 + \mathcal{O}(1) \cdot \max\left\{ \frac{\delta_0}{\sqrt{t}}, \, \delta_0^2 \right\}.$$

For a suitable constant $L' = \mathcal{O}(1) \cdot \text{Tot.Var.}\{u(0,\cdot)\}$, the above implies

$$\left\| u(t) - u(s) \right\|_{\mathbf{L}^1} \leq v(t,x) = \int_s^t \left\| u_t(t) \right\|_{\mathbf{L}^1} dt \leq L' \left(|t-s| + \left| \sqrt{t} - \sqrt{s} \right| \right). \quad (7.7)$$

7.3. Propagation Speed.

It is well known that, for a parabolic equation, a localized perturbation can be instantly propagated over the whole domain. However, one can show that the amount of perturbation which is transported with large speed is actually very small.

Example 7.3. Fix a constant $\lambda > 0$ and consider the scalar equation

$$u_t + \lambda u_x = u_{xx} . \tag{7.8}$$

Consider an initial data such that $u(0,x) = \bar{u}(x) = 0$ for all $x > 0$. In general, for $t > 0$ one may well find a solution with $u(t,x) \neq 0$ for every $x \in \mathbb{R}$. However, we claim that the values of $u(t,x)$ are exponentially small for large x. Indeed, our assumption implies

$$\left|u(x,0)\right| \leq \|\bar{u}\|_{\mathbf{L}^\infty} \cdot e^{-x} \doteq \bar{v}(x) .$$

One easily checks that the solution of (7.8) with the above initial data $\bar{v}$ is

$$v(t,x) = \|\bar{u}\|_{\mathbf{L}^\infty} \cdot e^{(\lambda+1)t-x} .$$

Applying a standard comparison argument to both u and $-u$, we thus obtain

$$\left|u(t,x)\right| \leq \|\bar{u}\|_{\mathbf{L}^\infty} \cdot e^{(\lambda+1)t-x} . \tag{7.9}$$

The estimate (7.9) is rather crude. However, it already shows that for $x > (\lambda+1)t$ the size of the solution u is exponentially small.

Estimates of the same type can be proved, more generally, on the difference between any two solutions of the viscous system (3.10). The following result will be useful in order to prove the finite propagation speed of solutions obtained as vanishing viscosity limits.

Lemma 7.4. *For some constants $\alpha, \beta > 0$ the following holds. Let u, v be solutions of (3.10) with small total variation, whose initial data satisfy*

$$u(0,x) = v(0,x) \qquad x \in [a,b] . \tag{7.10}$$

Then one has

$$\left|u(t,x) - v(t,x)\right| \leq \|u(0) - v(0)\|_{\mathbf{L}^\infty} \cdot \left(\alpha e^{\beta t - (x-a)} + \alpha e^{\beta t + (x-b)}\right). \tag{7.11}$$

A proof can be found in [BiB].

8 The Vanishing Viscosity Limit

All the previous analysis has been concerned with solutions of the parabolic system (3.10) with unit viscosity. These results can be easily restated in connection with the Cauchy problem

$$u_t^\varepsilon + A(u^\varepsilon)u_x^\varepsilon = \varepsilon\, u_{xx}^\varepsilon , \qquad u^\varepsilon(0,x) = \bar{u}(x) \tag{8.1}$$

for any $\varepsilon > 0$. Indeed, a function u^ε is a solution of (8.1) if and only if

$$u^\varepsilon(t, x) = u(t/\varepsilon, \, x/\varepsilon), \tag{8.2}$$

where u is the solution of

$$u_t + A(u)u_x = u_{xx}, \qquad u(0, x) = \bar{u}(\varepsilon x). \tag{8.3}$$

Since the rescaling (8.2) does not change the total variation, our earlier analysis yields the first part of Theorem 5. Namely, for every initial data $\bar{u}$ with sufficiently small total variation, the corresponding solution $u^\varepsilon(t) \doteq S_t^\varepsilon \bar{u}$ is well defined for all times $t \geq 0$. Recalling (3.18) and (7.7), the bounds (3.4)–(3.6) are derived as follows.

$$\text{Tot.Var.}\{u^\varepsilon(t)\} = \text{Tot.Var.}\{u(t/\varepsilon)\} \leq C \,\text{Tot.Var.}\{\bar{u}\}, \tag{8.4}$$

$$\left\|u^\varepsilon(t) - v^\varepsilon(t)\right\|_{\mathbf{L}^1} = \varepsilon \left\|u(t) - v(t)\right\|_{\mathbf{L}^1} \leq \varepsilon L \left\|u(0) - v(0)\right\|_{\mathbf{L}^1} = \varepsilon L \frac{1}{\varepsilon}\|\bar{u} - \bar{v}\|_{\mathbf{L}^1}, \tag{8.5}$$

$$\left\|u^\varepsilon(t) - u^\varepsilon(s)\right\|_{\mathbf{L}^1} \leq \varepsilon \left\|u(t/\varepsilon) - u(s/\varepsilon)\right\|_{\mathbf{L}^1} \leq \varepsilon L' \left(\left|\frac{t}{\varepsilon} - \frac{s}{\varepsilon}\right| + \left|\sqrt{\frac{t}{\varepsilon}} - \sqrt{\frac{s}{\varepsilon}}\right| \right). \tag{8.6}$$

Moreover, if $\bar{u}(x) = \bar{v}(x)$ for $x \in [a, b]$, then by (7.11) the corresponding solutions $u^\varepsilon, v^\varepsilon$ of (8.1) satisfy

$$\begin{aligned}
&\left|u^\varepsilon(t, x) - v^\varepsilon(t, x)\right| \\
&\leq \|\bar{u} - \bar{v}\|_{\mathbf{L}^\infty} \cdot \left\{ \alpha \exp\left(\frac{\beta t - (x - a)}{\varepsilon}\right) + \alpha \exp\left(\frac{\beta t + (x - b)}{\varepsilon}\right) \right\}.
\end{aligned} \tag{8.7}$$

We now consider the vanishing viscosity limit. Call $\mathcal{U} \subset \mathbf{L}^1_{\text{loc}}$ the set of all functions $\bar{u} : \mathbb{R} \mapsto \mathbb{R}^n$ with small total variation, satisfying (3.3). For each $t \geq 0$ and every initial condition $\bar{u} \in \mathcal{U}$, call $S_t^\varepsilon \bar{u} \doteq u^\varepsilon(t, \cdot)$ the corresponding solution of (8.1). Thanks to the uniform BV bounds (8.4), we can apply Helly's compactness theorem and obtain a sequence $\varepsilon_\nu \to 0$ such that

$$\lim_{\nu \to \infty} u^{\varepsilon_\nu}(t, \cdot) = u(t, \cdot) \qquad \text{in} \quad \mathbf{L}^1_{\text{loc}}. \tag{8.8}$$

holds for some BV function $u(t, \cdot)$. By extracting further subsequences and then using a standard diagonalization procedure, we can assume that the limit in (8.8) exists for all rational times t and all solutions u^ε with initial data in a countable dense set $\mathcal{U}^* \subset \mathcal{U}$. Adopting a semigroup notation, we thus define

$$S_t \bar{u} \doteq \lim_{m \to \infty} S_t^{\varepsilon_m} \bar{u} \qquad \text{in} \quad \mathbf{L}^1_{\text{loc}}, \tag{8.9}$$

for some particular subsequence $\varepsilon_m \to 0$. By the uniform continuity of the maps $(t, \bar{u}) \mapsto u^\varepsilon(t, \cdot) \doteq S_t^\varepsilon \bar{u}$, stated in (8.5)–(8.6), the set of couples $(t, \bar{u})$ for which the limit (8.9) exists must be closed in $\mathbb{R}_+ \times \mathcal{U}$. Therefore, this limit is well defined for all $\bar{u} \in \mathcal{U}$ and $t \geq 0$. Notice that (8.4) and (8.9) imply a uniform BV bound on the vanishing viscosity limits:

$$\text{Tot.Var.}\{S_t \bar{u}\} \leq \limsup_{m \to \infty} \text{Tot.Var.}\{u^{\varepsilon m}(t)\} \leq C \,\text{Tot.Var.}\{\bar{u}\}\,. \qquad (8.10)$$

To complete the proof of Theorem 5, we need to show that the map S defined at (8.9) is a semigroup, satisfies the continuity properties (3.7) and does not depend on the choice of the subsequence $\{\varepsilon_m\}$. We give here a proof in the special case of a conservative system satisfying the additional hypotheses (H).

1. **(Continuous dependence)** Let S be the map defined by (8.9). Then

$$\left\| S_t \bar{u} - S_t \bar{v} \right\|_{\mathbf{L}^1} = \sup_{r>0} \int_{-r}^{r} \left| (S_t \bar{u})(x) - (S_t \bar{v})(x) \right| dx\,.$$

For every $r > 0$, the convergence in $\mathbf{L}^1_{\text{loc}}$ implies

$$\int_{-r}^{r} \left| (S_t \bar{u})(x) - (S_t \bar{v})(x) \right| dx$$
$$= \lim_{m \to \infty} \int_{-r}^{r} \left| (S_t^{\varepsilon m} \bar{u})(x) - (S_t^{\varepsilon m} \bar{v})(x) \right| dx \leq L \left\| \bar{u} - \bar{v} \right\|_{\mathbf{L}^1}\,.$$

because of (8.5). This yields the Lipschitz continuous dependence w.r.t. the initial data:

$$\left\| S_t \bar{u} - S_t \bar{v} \right\|_{\mathbf{L}^1} \leq L \left\| \bar{u} - \bar{v} \right\|_{\mathbf{L}^1}\,. \qquad (8.11)$$

The continuous dependence w.r.t. time is proved in a similar way. By (8.6), for every $r > 0$ we have

$$\int_{-r}^{r} \left| (S_t \bar{v})(x) - (S_s \bar{v})(x) \right| dx$$
$$= \lim_{m \to \infty} \int_{-r}^{r} \left| (S_t^{\varepsilon m} \bar{v})(x) - (S_s^{\varepsilon m} \bar{v})(x) \right| dx$$
$$= \lim_{m \to \infty} \varepsilon_m L' \left(\left| \frac{t}{\varepsilon_m} - \frac{s}{\varepsilon_m} \right| + \left| \sqrt{\frac{t}{\varepsilon_m}} - \sqrt{\frac{s}{\varepsilon_m}} \right| \right)$$
$$= L' |t - s|\,.$$

Hence

$$\left\| S_t \bar{v} - S_s \bar{v} \right\|_{\mathbf{L}^1} \leq L' |t - s|\,. \qquad (8.12)$$

From (8.11) and (8.12) it now follows (3.7).

2. **(Finite propagation speed)** Consider any interval $[a, b]$ and two initial data $\bar{u}, \bar{v}$, with $\bar{u}(x) = \bar{v}(x)$ for $x \in [a, b]$. By (8.7), for every $t \geq 0$ and $x \in]a + \beta t, b - \beta t[$ one has

$$\left| \left(S_t \bar{u} \right)(x) - \left(S_t \bar{v} \right)(x) \right|$$

$$\leq \limsup_{m \to \infty} \left| \left(S_t^{\varepsilon_m} \bar{u} \right)(x) - \left(S_t^{\varepsilon_m} \bar{v} \right)(x) \right|$$

$$\leq \lim_{m \to \infty} \| \bar{u} - \bar{v} \|_{\mathbf{L}^\infty} \cdot \left\{ \alpha \exp \left(\frac{\beta t - (x - a)}{\varepsilon_m} \right) + \alpha \exp \left(\frac{\beta t + (x - b)}{\varepsilon_m} \right) \right\}$$

$$= 0 \, .$$

$$(8.13)$$

In other words, the restriction of the function $S_t \bar{u} \in \mathbf{L}^1_{\mathrm{loc}}$ to a given interval $[a', b']$ depends only on the values of the initial data $\bar{u}$ on the interval $[a' - \beta t, \, b' + \beta t]$. Using (8.13), we now prove a sharper version of the continuous dependence estimate (8.11), namely

$$\int_a^b \left| \left(S_t \bar{u} \right)(x) - \left(S_t \bar{v} \right)(x) \right| dx \leq L \cdot \int_{a - \beta t}^{b + \beta t} \left| \bar{u}(x) - \bar{v}(x) \right| dx \, , \qquad (8.14)$$

valid for every $\bar{u}, \bar{v}$ and $t \geq 0$. For this purpose, define the auxiliary function

$$\bar{w}(x) = \begin{cases} \bar{u}(x) & \text{if} \quad x \in [a - \beta t, \, b + \beta t] \, , \\ \bar{v}(x) & \text{if} \quad x \notin [a - \beta t, \, b + \beta t] \, . \end{cases}$$

Using the finite propagation speed, we now have

$$\int_a^b \left| \left(S_t \bar{u} \right)(x) - \left(S_t \bar{v} \right)(x) \right| dx = \int_a^b \left| \left(S_t \bar{w} \right)(x) - \left(S_t \bar{v} \right)(x) \right| dx$$

$$\leq L \, \| \bar{w} - \bar{v} \|_{\mathbf{L}^1} = L \cdot \int_{a - \beta t}^{b + \beta t} \left| \bar{u}(x) - \bar{v}(x) \right| dx \, .$$

3. (Semigroup property) We claim that the map $(t, \bar{u}) \mapsto S_t \bar{u}$ is a semigroup, i.e.

$$S_0 \bar{u} = \bar{u} \, , \qquad S_s S_t \bar{u} = S_{s+t} \bar{u} \, . \qquad (8.15)$$

Since every S^ε is a semigroup, the first equality in (8.15) is a trivial consequence of the definition (8.9). To prove the second equality, we observe that

$$S_{s+t} \bar{u} = \lim_{m \to \infty} S_s^{\varepsilon_m} S_t^{\varepsilon_m} \bar{u} \, , \qquad S_s S_t \bar{u} = \lim_{m \to \infty} S_s^{\varepsilon_m} S_t \bar{u} \, . \qquad (8.16)$$

We can assume $s > 0$. Fix any $r > 0$ and consider the function

$$u_m^\sharp(x) \doteq \begin{cases} \left(S_t \bar{u} \right)(x) & \text{if} \quad |x| > r + 2\beta s \, , \\ \left(S_t^{\varepsilon_m} \bar{u} \right)(x) & \text{if} \quad |x| < r + 2\beta s \, . \end{cases}$$

Observing that $S_t^{\varepsilon_m} \bar{u} \to S_t \bar{u}$ in $\mathbf{L}^1_{\mathrm{loc}}$ and hence $u_m^\sharp \to S_t \bar{u}$ in $\mathbf{L}^1$, we can use (8.7) and (8.5) and obtain

$$\limsup_{m\to\infty} \int_{-r}^{r} \left| \left(S_s^{\varepsilon_m} S_t^{\varepsilon_m} \bar{u}\right)(x) - \left(S_s^{\varepsilon_m} S_t \bar{u}\right)(x) \right| dx$$

$$\leq \lim_{m\to\infty} 2r \cdot \sup_{|x|<r} \left| \left(S_s^{\varepsilon_m} S_t^{\varepsilon_m} \bar{u}\right)(x) - \left(S_s^{\varepsilon_m} u_m^\sharp\right)(x) \right| + \lim_{m\to\infty} \left\| S_s^{\varepsilon_m} u_m^\sharp - S_s^{\varepsilon_m} S_t \bar{u} \right\|_{\mathbf{L}^1}$$

$$\leq \lim_{m\to\infty} 2r \left\| S_t^{\varepsilon_m} \bar{u} - u_m^\sharp \right\|_{\mathbf{L}^\infty} \cdot 2\alpha e^{-\beta s/\varepsilon_m} + \lim_{m\to\infty} L \cdot \left\| u_m^\sharp - S_t \bar{u} \right\|_{\mathbf{L}^1}$$

$$= 0 \,.$$

By (8.16), this proves the second identity in (8.15).

4. (Tame Oscillation) Next, we prove that every function $u(t,x) = \left(S_t \bar{u}\right)(x)$ satisfies the tame oscillation property introduced at (2.27). Indeed, let a, b, τ be given, together with an initial data $\bar{u}$. By the semigroup property, it is not restrictive to assume $\tau = 0$. Consider the auxiliary initial condition

$$\bar{v}(x) \doteq \begin{cases} \bar{u}(x) & \text{if} \quad a < x < b, \\ \bar{u}(a+) & \text{if} \quad x \leq a, \\ \bar{u}(b-) & \text{if} \quad x \geq b, \end{cases}$$

and call $v(t,x) \doteq \left(S_t \bar{v}\right)(x)$ the corresponding trajectory.

Observe that

$$\lim_{x\to-\infty} v(t,x) = \bar{u}(a+)$$

for every $t \geq 0$. Using (8.4) and the finite propagation speed, we can thus write

$$\text{Osc.}\{u \,;\, \Delta_{a,b}^\tau\} = \text{Osc.}\{v \,;\, \Delta_{a,b}^\tau\} \leq 2\sup_t \left(\text{Tot.Var.}\{S_t \bar{v}\}\right)$$

$$\leq 2C \cdot \text{Tot.Var.}\{\bar{v}\} = 2C \cdot \text{Tot.Var.}\{u(\tau) \,;\,]a, b[\} \,,$$

proving (2.27) with $C' = 2C$.

5. (Conservation equations) Assume that the system (8.1) is in conservation form, i.e. $A(u) = Df(u)$ for some flux function f. In this special case, we claim that every vanishing viscosity limit is a weak solution of the system of conservation laws (1.1). Indeed, if ϕ is a $\mathcal{C}^2$ function with compact support contained in the half plane $\{x \in \mathbb{R}, \, t > 0\}$, integrating by parts we find

$$\int\!\!\int \left[u \phi_t + f(u)\phi_x\right] dx \, dt = \lim_{m\to\infty} \int\!\!\int \left[u^{\varepsilon_m} \phi_t + f(u^{\varepsilon_m})\phi_x\right] dx \, dt$$

$$= -\lim_{m\to\infty} \int\!\!\int \left[u_t^{\varepsilon_m} \phi + f(u^{\varepsilon_m})_x \phi\right] dx \, dt = -\lim_{m\to\infty} \int\!\!\int \varepsilon_m u_{xx}^{\varepsilon_m} \phi \, dx \, dt$$

$$= -\lim_{m\to\infty} \int\!\!\int \varepsilon_m u^{\varepsilon_m} \phi_{xx} \, dx \, dt = 0 \,.$$

An approximation argument shows that the identity (1.3) still holds if we only assume $\phi \in \mathcal{C}_c^1$.

6. (Approximate jumps) From the uniform bound on the total variation and the Lipschitz continuity w.r.t. time, it follows that each function $u(t,x) =$

$(S_t \bar{u})(x)$ is a BV function, jointly w.r.t. the two variables t, x. In particular, an application of Theorem 2.6 in [B3] yields the existence of a set of times $\mathcal{N} \subset \mathbb{R}_+$ of measure zero such that, for every $(\tau, \xi) \in \mathbb{R}_+ \times \mathbb{R}$ with $\tau \notin \mathcal{N}$, the following holds. Calling

$$u^- \doteq \lim_{x \to \xi-} u(\tau, x), \qquad u^+ \doteq \lim_{x \to \xi+} u(\tau, x),$$

there exists a finite speed λ such that the function

$$U(t, x) \doteq \begin{cases} u^- & if \quad x < \lambda t, \\ u^+ & if \quad x > \lambda t, \end{cases}$$

for every constant $\kappa > 0$ satisfies

$$\lim_{r \to 0+} \frac{1}{r^2} \int_{-r}^{r} \int_{-\kappa r}^{\kappa r} \left| u(\tau + t,\ \xi + x) - U(t, x) \right| dx\, dt = 0,$$

$$\lim_{r \to 0+} \frac{1}{r} \int_{-\kappa r}^{\kappa r} \left| u(\tau + r,\ \xi + x) - U(r, x) \right| dx = 0.$$

In the case where $u^- \neq u^+$, we say that (τ, ξ) is a point of *approximate jump* for the function u. On the other hand, if $u^- = u^+$ (and hence λ can be chosen arbitrarily), we say that u is *approximately continuous* at (τ, ξ). The above result can thus be restated as follows: with the exception of a null set $\mathcal{N}$ of "interaction times", the solution u is either approximately continuous or has an approximate jump discontinuity at each point (τ, ξ).

7. (Shock conditions) Assume again that the system is in conservative form. Consider a semigroup trajectory $u(t, \cdot) = S_t \bar{u}$ and a point (τ, ξ) where u has an approximate jump. Since u is a weak solution, the states u^-, u^+ and the speed λ in (8.19) must satisfy the Rankine-Hugoniot equations (2.4). If u is a limit of vanishing viscosity approximations, the same is true of the solution U in (8.19). In particular (see [D2]), the *Liu admissibility condition* (2.15) must hold.

8. (Uniqueness) Assume that the system is in conservation form and that each characteristic field is either linearly degenerate or genuinely nonlinear. By the previous steps, the semigroup trajectory $u(t, \cdot) = S_t \bar{u}$ provides a weak solution to the Cauchy problem (1.1)–(1.2) which satisfies the Tame Oscillation and the Lax shock conditions. By a theorem in [BG], [B3], such a weak solution is unique and coincides with the limit of front tracking approximations. In particular, it does not depend on the choice of the subsequence $\{\varepsilon_m\}$:

$$S_t \bar{u} = \lim_{\varepsilon \to 0+} S_t^\varepsilon \bar{u},$$

i.e. the same limit actually holds over all real values of ε.

The above discussion has established Theorem 5 in the special case where the system is in conservation form and each characteristic field is either linearly degenerate or genuinely nonlinear. For a proof in the general case of a non-conservative hyperbolic system we again refer to [BiB].

References

[AM]   F. Ancona and A. Marson, Well posedness for general 2×2 systems of conservation laws, *Memoir Amer. Math. Soc.*, **801**, 2004.

[BaJ]   P. Baiti and H. K. Jenssen, On the front tracking algorithm, *J. Math. Anal. Appl.* **217** (1998), 395–404.

[Bi]   S. Bianchini, On the Riemann problem for non-conservative hyperbolic systems, *Arch. Rat. Mach. Anal.* **166** (2003), 1–26.

[BiB]   S. Bianchini and A. Bressan, Vanishing viscosity solutions to nonlinear hyperbolic systems, *Annals of Mathematics*, **161** (2005), 223–342.

[B1]   A. Bressan, Global solutions to systems of conservation laws by wave-front tracking, *J. Math. Anal. Appl.* **170** (1992), 414–432.

[B2]   A. Bressan, The unique limit of the Glimm scheme, *Arch. Rational Mech. Anal.* **130** (1995), 205–230.

[B3]   A. Bressan, *Hyperbolic Systems of Conservation Laws. The One Dimensional Cauchy Problem*. Oxford University Press, 2000.

[BC1]   A. Bressan and R. M. Colombo, The semigroup generated by 2×2 conservation laws, *Arch. Rational Mech. Anal.* **133** (1995), 1–75.

[BCP]   A. Bressan, G. Crasta and B. Piccoli, Well posedness of the Cauchy problem for $n \times n$ conservation laws, *Amer. Math. Soc. Memoir* **694** (2000).

[BG]   A. Bressan and P. Goatin, Oleinik type estimates and uniqueness for $n \times n$ conservation laws, *J. Diff. Equat.* **156** (1999), 26–49.

[BLF]   A. Bressan and P. LeFloch, Uniqueness of weak solutions to systems of conservation laws, *Arch. Rat. Mech. Anal.* **140** (1997), 301–317.

[BLw]   A. Bressan and M. Lewicka, A uniqueness condition for hyperbolic systems of conservation laws, *Discr. Cont. Dynam. Syst.* **6** (2000), 673–682.

[BLY]   A. Bressan, T. P. Liu and T. Yang, L^1 stability estimates for $n \times n$ conservation laws, *Arch. Rational Mech. Anal.* **149** (1999), 1–22.

[BM]   A. Bressan and A. Marson, Error bounds for a deterministic version of the Glimm scheme, *Arch. Rational Mech. Anal.* **142** (1998), 155–176.

[BY]   A. Bressan and T. Yang, On the rate of convergence of vanishing viscosity approximations, *Comm. Pure Appl. Math* **57** (2004), 1075–1109.

[C]   M. Crandall, The semigroup approach to first-order quasilinear equations in several space variables, *Israel J. Math.* **12** (1972), 108–132.

[D1]   C. Dafermos, Polygonal approximations of solutions of the initial value problem for a conservation law, *J. Math. Anal. Appl.* **38** (1972), 33–41.

[D2]   C. Dafermos, *Hyperbolic Conservation Laws in Continuum Physics*, Springer-Verlag, Berlin 1999.

[DP1]   R. DiPerna, Global existence of solutions to nonlinear hyperbolic systems of conservation laws, *J. Diff. Equat.* **20** (1976), 187–212.

[DP2]   R. DiPerna, Convergence of approximate solutions to conservation laws, *Arch. Rational Mech. Anal.* **82** (1983), 27–70.

[G]   J. Glimm, Solutions in the large for nonlinear hyperbolic systems of equations, *Comm. Pure Appl. Math.* **18** (1965), 697–715.

[GX]   J. Goodman and Z. Xin, Viscous limits for piecewise smooth solutions to systems of conservation laws, *Arch. Rational Mech. Anal.* **121** (1992), 235–265.

[HR]   H. Holden and N. H. Risebro, *Front Tracking for Hyperbolic Conservation Laws*, Springer Verlag, New York 2002.

[K]   S. Kruzhkov, First order quasilinear equations with several space variables, *Math. USSR Sbornik* **10** (1970), 217–243.

[J]   H. K. Jenssen, Blowup for systems of conservation laws, *SIAM J. Math. Anal.* **31** (2000), 894–908.

[Lx]   P. Lax, Hyperbolic systems of conservation laws II, *Comm. Pure Appl. Math.* **10** (1957), 537–566.

[L1]   T. P. Liu, The Riemann problem for general systems of conservation laws, *J. Diff. Equat.* **18** (1975), 218–234.

[L2]   T. P. Liu, The entropy condition and the admissibility of shocks, *J. Math. Anal. Appl.* **53** (1976), 78–88.

[L3]   T. P. Liu, The deterministic version of the Glimm scheme, *Comm. Math. Phys.* **57** (1977), 135–148.

[L4]   T. P. Liu, Admissible solutions of hyperbolic conservation laws, *Amer. Math. Soc. Memoir* **240** (1981).

[L5]   T. P. Liu, Nonlinear stability of shock waves for viscous conservation laws, *Amer. Math. Soc. Memoir* **328** (1986).

[LY1]   T. P. Liu and T. Yang, A new entropy functional for scalar conservation laws, *Comm. Pure Appl. Math.* **52** (1999), 1427–1442.

[LY2]   T. P. Liu and T. Yang, L^1 stability for 2×2 systems of hyperbolic conservation laws, *J. Amer. Math. Soc.* **12** (1999), 729–774.

[O]   O. Oleinik, Discontinuous solutions of nonlinear differential equations (1957), *Amer. Math. Soc. Translations* **26**, 95–172.

[R]   F. Rousset, Viscous approximation of strong shocks of systems of conservation laws, *SIAM J. Math. Anal.* 35 (2003), 492–519.

[Sc]   S. Schochet, Sufficient conditions for local existence via Glimm's scheme for large BV data, *J. Differential Equations* **89** (1991), 317–354.

[Se]   D. Serre, *Systems of Conservation Laws I, II*, Cambridge University Press, 2000.

[Sm]   J. Smoller, *Shock Waves and Reaction-Diffusion Equations*, Springer-Verlag, New York, 1983.

[SX]   A. Szepessy and Z. Xin, Nonlinear stability of viscous shocks, *Arch. Rational Mech. Anal.* **122** (1993), 53–103.

[SZ]   A. Szepessy and K. Zumbrun, Stability of rarefaction waves in viscous media, *Arch. Rational Mech. Anal.* **133** (1996), 249–298.

[V]   A. Vanderbauwhede, Centre manifolds, normal forms and elementary bifurcations, *Dynamics Reported, Vol. 2* (1989), 89–169.

[Yu]   S. H. Yu, Zero-dissipation limit of solutions with shocks for systems of hyperbolic conservation laws, *Arch. Rational Mech. Anal.* **146** (1999), 275–370.

Discrete Shock Profiles: Existence and Stability

Denis Serre *

ENS Lyon, UMPA (UMR 5669 CNRS),
46, allée d'Italie, F-69364 Lyon Cedex 07 FRANCE
serre@umpa.ens-lyon.fr

Summary. Partial differential equations are often approximated by finite difference schemes. The consistency and stability of a given scheme are usually studied through a linearization along elementary solutions, for instance constants. So long as time-dependent problems are concerned, one may also ask for the behaviour of schemes about traveling waves. A rather complete study was made by Chow & al. [12] in the context of fronts in reaction-diffusion equations, for instance KPP equation; see also the monograph by Fiedler & Scheurle [17] for different aspects of the same problem. We address here similar questions in the context of hyperbolic systems of conservation laws. Besides constants, traveling waves may be either linear waves, corresponding to a linear characteristic field, or simple discontinuities such as shock waves of various kinds: Lax shocks, under-compressive shocks, over-compressive ones, anti-Lax ones.

Up to some extent, the approximation of hyperbolic systems of conservation laws by conservative finite difference schemes displays features that are similar to those encountered in other kinds of approximations; like viscosity and relaxation, schemes often display some kind of dissipation and/or dispersion. However, as far as the existence and stability of traveling waves are concerned, one faces significant differences. For instance, the existence may fail in rather natural situations because of small divisors problems. That such problems are present or not depend on arithmetical properties of the dimensionless parameter $\eta := s\Delta t/\Delta x$ where s is the shock velocity. The simplest case is the "rational one", but even then, difference schemes and viscosity yield completely different patterns, for instance in the existence of under-compressive shock profiles.

We present here an overview of the existence and stability theory. We wish to clarify observations that were made after numerical experiments. For instance, it is a part of the folklore that large steady shocks may be unstable under the Godunov scheme. This turns out to be true, as proved in [11]. Also, it is well-known that the Lax–Wendroff scheme fixes every steady discontinuities, even those which are irrelevant, for instance anti-Lax shocks. In both cases, the Evans function is a powerful tool in the study of stability. Besides stability questions, the existence

---

* This research was done from one part in accomplishment of the European IHP project "HYKE", contract # HPRN-CT-2002-00282.

theory is extremely complicated, as mentionned above, and is by far incomplete. Since difficulties arise form the fact that essential spectrum cannot be pushed away from the imaginary axis in general, we also present some ideas in the easier context of reaction-diffusion equations, and discuss whether they might be applied, or not, to systems of conservation laws.

New Results:

Although this text represents primarily lecture notes, it contains a small amount of new results. As a matter of fact, it is impossible to think seriously to an active research topic without asking some new questions or filling some gaps. The original contributions of this paper are the following:

- A counter-example to (2.9) is build in the case of the Lax–Wendroff scheme,
- An analysis of the generic dynamics that holds in the reaction-diffusion case shows that the situation for conservation laws should be completely different,
- We give a general construction of the Evans function when the scheme displays some numerical viscosity, following the careful analysis by P. Godillon [21],
- In particular, many calculations on examples are carried out in full details,
- We generalize the results of the seminal paper [11] for steady shock profiles of the Godunov scheme,
- We interpret the function Y in the light of the expected asymptotic behaviour.

Acknowledgements:

The author is happy to thank S. Benzoni–Gavage, A. Bressan, H.–K. Jenssen and K. Zumbrun for their valuable comments and questions. He is also indebted to P. Marcati, who organized a high level summer school in the lovely place of Grand Hotel San Michele at Cetraro, for his suggestion to lecture about discrete shock profiles.

Introduction

Let
$$\partial_t u + \partial_x f(u) = 0, \quad x \in \mathbb{R},\, t > 0 \tag{0.1}$$

be a system of conservation laws, where $f : \mathcal{U} \to \mathbb{R}^n$ is a given smooth flux, and $\mathcal{U}$, the phase space, is a convex open subset of $\mathbb{R}^n$. We shall assume a hyperbolicity property in the vicinity of two distinguished points u^l and u^r that we call the *left* and *right* states. Hyperbolicity means that the differential (or Jacobian matrix) $\mathcal{D}f$ is diagonalizable with real eigenvalues. We shall specialize later on this assumption. At least, we wish to allow some complicated pattern as under-compressive shock waves, and this is why we do not require that hyperbolicity holds everywhere in $\mathcal{U}$, but only in neighbourhoods of $u^{l,r}$.

We study in this text the behaviour of numerical approximations of (0.1). The approximations that we consider are conservative finite difference schemes. The system is supplemented with an initial datum
$$u(x,0) = a(x), \quad x \in \mathbb{R}. \tag{0.2}$$

The likely solution is approximated along the staggered grid
$$\Delta x \mathbb{Z} \times \Delta t \mathbb{N}.$$

The grid points are labelled $(x_j, t_m) := (j\Delta x, m\Delta t)$ and the corresponding discrete unknowns are u_j^m. A conservative one-step difference scheme is defined by an iteration of the form
$$u_j^{m+1} = u_j^m + \sigma(F_{j-1/2}^m - F_{j+1/2}^m), \quad \sigma := \Delta t / \Delta x. \tag{0.3}$$

The numerical flux $F_{j-1/2}^m$ is a $\mathcal{C}^1$ function of the following $p+q+1$ quantities:
$$\sigma,\, u_{j-p}^m, \ldots, u_{j+q-1}^m,$$

where p, q are non-negative integers. One speaks of a $(p+q+1)$-points scheme, because u_j^{m+1} is determined by the $p+q+1$ previous states $(u_{j-p}, \ldots, u_{j+q})$. For instance, the simple case $p = q = 1$ corresponds to three-point schemes:
$$u_j^{m+1} = u_j^m + \sigma(F(\sigma; u_{j-1}^m, u_j^m) - F(\sigma; u_j^m, u_{j+1}^m)).$$

The schemes known as Godunov's, Lax-Friedrichs' and Lax-Wendroff's are three-point conservative difference schemes. We refer to the books by Leveque [36] or Godlewski and Raviart [GR] for the basic notions about conservative finite difference schemes.

Consistency

The consistency is the property that whenever a sequence of approximate solutions converges to some function $u(x,t)$ in a strong enough topology as the mesh size tends to zero, then u must be a solution of (0.1). In some sense, it tells that a scheme is faithful to the system.

Of course, we must be cautious when considering such sequences, since there are two mesh sizes, one for the space variable and another one for the time variable. For well-known stability reasons that rely upon the fact that the system is first-order in both space and time variables, we shall always keep fixed the *mesh ratio* $\sigma = \Delta t / \Delta x$. Then, letting $\Delta x \to 0^+$ will make Δt converging too.

A standard consistency condition that deals with data close to constants is the identity

$$F(\sigma; v, \ldots, v) = f(v), \quad v \in \mathcal{U}. \tag{0.4}$$

However, since the most interesting solutions of (0.1) have jumps (say shock waves) of significant strength, it is also desirable to study traveling waves of (0.3) that correspond to such simple patterns. The fact that a given scheme admits a "stable" traveling wave for every "admissible" shock wave, and only for them, may be regarded as an enhanced consistency property of the scheme. Whence the topic of this paper.

A traveling wave from u^l to u^r will be called a *discrete shock profile* (short-coming: DSP), although we shall see in a moment that this terminology might often be confusing. A DSP is a special solution of (0.3), of the form

$$u_j^m = U(j - ms\sigma), \quad j \in \mathbb{Z}, m \in \mathbb{N}, \tag{0.5}$$

with the boundary conditions

$$U(-\infty) = u^l, \quad U(+\infty) = u^r. \tag{0.6}$$

The parameter $s \in \mathbb{R}$ is called the wave velocity, since the above formula really means that the value of u^{app} at the grid point (x_j, t_m) equals

$$U\left(\frac{x_j - st_m}{\Delta x}\right).$$

The function U is called the "shock profile". We see on the formula above that the thickness of the numerical shock layer is of order Δx. Hence u^{app} approaches a simple wave of the form

$$u^{exact}(x, t) = \begin{cases} u^l, x < st, \\ u^r, x > st. \end{cases}$$

The fact that u^{app}, defined in (0.5), be a traveling wave of the scheme, translates into the following functional-difference equation:

$$\begin{aligned} U(y - \eta) &= U(y) + \sigma\{F(\sigma; U(y - p), \ldots, U(y + q - 1)) \\ &\quad - F(\sigma; U(y - p + 1), \ldots, U(y + q))\} \\ &=: G(\sigma; U(y - p), \ldots, U(y + q)), \end{aligned} \tag{0.7}$$

where

$$\eta := s\sigma = s\frac{\Delta t}{\Delta x}. \tag{0.8}$$

We emphasize that, since s is a velocity, η is a dimensionless parameter. It is sometimes called the "grid velocity", because it represents the average number of meshes that the shock advances during one time step. Because of obvious stability reasons (the information in the numerical scheme needs to travel at a velocity higher than that of the wave), one always need CFL-like inequalities:

$$-q < \eta < p. \tag{0.9}$$

For instance, $|\eta| < 1$ in three-point schemes. In upwind schemes with $p = 1$ and $q = 0$, one must have $0 < \eta < 1$.

Equation (0.7) shows that the domain of the profile U must be invariant under the translations by units and multiples of η. Hence a minimal domain of U is the additive group $\mathbb{Z} + \eta\mathbb{Z}$. When $\eta = r/\ell$ is rational, we may content ourselves with the discrete domain $\ell^{-1}\mathbb{Z}$; we call this the *rational case*. However, when η is irrational (the *irrational case*), $\mathbb{Z} + \eta\mathbb{Z}$ becomes dense in $\mathbb{R}$. In particular, it is not discrete at all, despite the terminology "discrete shock profiles". In this latter case, we impose that U be defined in the whole real line $\mathbb{R}$. It is thus a "continuous discrete" profile!

We remark that the profile equation (0.7) can be "integrated" once, as noticed in the rational case by S. Benzoni [4]. To explain what is going on, let us separate the rational case from the irrational one. If $\eta = r/\ell \in \mathbb{Q}$, with $r \geq 0$ for instance, we rewrite

$$U(y) - U(y-\eta) = U(y) + \cdots + U\left(y - \frac{r-1}{\ell}\right) - U\left(y - \frac{1}{\ell}\right) - \cdots - U\left(y - \frac{r}{\ell}\right).$$

We decompose in a similar way the F-difference, and deduce that the quantity

$$\sum_{j=0}^{r-1} U\left(y - \frac{j}{\ell}\right) - \sigma \sum_{k=1}^{\ell} F\left(\sigma; U\left(y - p + \frac{k}{\ell}\right), \ldots, U\left(y + q - 1 + \frac{k}{\ell}\right)\right)$$

is constant upon a shift by $1/\ell$. Hence it remains constant over the domain $\ell^{-1}\mathbb{Z}$. We identify its value by letting $y \to \pm\infty$:

$$\sum_{j=0}^{r-1} U\left(y - \frac{j}{\ell}\right) - \sigma \sum_{k=1}^{\ell} F\left(\sigma; U\left(y - p + \frac{k}{\ell}\right), \ldots\right) = ru^{l,r} - \sigma\ell f(u^{l,r}). \tag{0.10}$$

Hereabove, we made use of (0.4). As a by-product, we obtain the necessary condition

$$f(u^r) - f(u^l) = \frac{\eta}{\sigma}(u^r - u^l).$$

This is nothing but the well-known *Rankine–Hugoniot condition*, which tells that u^l and u^r are connected by a discontinuity of velocity s that is a distributional solution of (0.1):

$$[f(u)] = s[u]. \tag{0.11}$$

In the irrational case, we make the natural assumption that the profile $U :$ $\mathbb{R} \to \mathcal{U}$ is continuous. Then the quantity

$$x \mapsto \sigma \int_x^{x+1} F(\sigma; U(y-p), \dots, U(y+q-1)) \, dy - \int_{x-\eta}^x U(y) \, dy$$

is constant, according to (0.7). Computing its value at both ends $\pm\infty$, we obtain

$$\sigma \int_x^{x+1} F(\sigma; U(y-p), \dots, U(y+q-1)) \, dy$$
$$- \int_{x-\eta}^x U(y) \, dy \equiv \sigma f(u^{l,r}) - \eta u^{l,r}. \tag{0.12}$$

Whence, once again, the Rankine–Hugoniot condition. Notice that identity (0.12) holds true for every parameter η so long as the profile is defined on the whole line $\mathbb{R}$, since we did not use irrationality.

Discrete *vs* Viscous Profiles

A viscous approximation of (0.1) has the form

$$\partial_t u + \partial_x f(u) = \epsilon \partial_x (B(u) \partial_x u), \quad x \in \mathbb{R}, \, t > 0, \tag{0.13}$$

where B is a smooth tensor field that makes the system parabolic[2]. Given the left and right states, a viscous profile is a traveling wave

$$u(x,t) = U_v \left(\frac{x - st}{\epsilon} \right)$$

that satisfies the boundary conditions

$$U_v(-\infty) = u^l, \quad U_v(+\infty) = u^r.$$

The profile equation is a differential equation of second order:

$$(B(U_v)U_v')' = f(U_v)' - sU_v'.$$

As in the discrete case, it can be integrated once as

$$B(U_v)U_v' = F(U_v) - f(u^l) - s(U_v - u^l). \tag{0.14}$$

The constant of integration was determined by using the value of U at $-\infty$. Using its value at $+\infty$, we would have ended with

---

[2] Incompletely parabolic system may be considered and are even welcome because of their practical relevance.

$$B(U_v)U_v' = f(U_v) - f(u^r) - s(U_v - u^r).$$

Equating both values of U_v', we obtain the Rankine–Hugoniot condition (0.11) as a necessary condition, as in the discrete case.

Notice that if a given shock wave admits a viscous and a discrete shock profiles, both have the same velocity, which is that of the shock itself, according to the Rankine–Hugoniot condition. Although the profiles differ from each other in general[3], they have a similar structure, each one depending on $x - st$, scaled by the small parameter ϵ or Δx. The fact that first order schemes display viscous features indicates that the same tools will enter into play. We shall see however that the situation is by far more complicated in the discrete case.

Distinct Profiles

A first significant difference from the viscous case is the parametrization of the set of profiles for a given shock wave.

Let us begin with the most common case of a Lax p-shock, say of small amplitude. Assuming that the p-th characteristic field of (0.1) is simple and genuinely nonlinear, it has been well-known since M. Foy [18] that there exists a unique small viscous shock profile U_v (given the viscosity tensor B), up to a translation of the "space" variable y. In other words, the set of viscous profiles is homeomorphic, and even isomorphic, to the real line. This result remains true when one drops the nonlinearity assumption, provided the shock satisfies Liu's E-condition [39].

As far as discrete shock profiles are concerned, the situation is pretty much different, and even depends on whether the parameter η is rational or not, as well as on the nature of the shock. The simplest pattern, which is also the closest to the viscous case, is when η is rational and the shock is a Lax shock. Under the same assumptions as above[4], one proves (see below) that the set of small profiles with domain $\ell^{-1}\mathbb{Z}$ is homeomorphic to the real line. A translation of the parameter by a multiple of $1/\ell$ corresponds to a space-time translation of the approximate solution. However, a translation by a shift less than $1/\ell$ changes significantly the profile. When η is irrational, the situation is even worse. It may happen that the set of discrete profiles is void, or that it is rough, in a sense that has not been understood in full details by now.

By contrast, the case of undercompressive shock waves is really amazing. It is well-known that, since viscous profiles correspond to heteroclinic orbits, they are not structurally stable in this case. Hence the set of undercompressive shocks that admit a viscous profile must be of codimension one at least; on the other hand, its elements still admit a one-parameter family of profiles,

---

[3] Similarly, viscous profiles associated to distinct viscosity tensors differ from each other, and discrete profiles associated to distinct schemes also do.

[4] But smallness is now that of $\ell[u]$ where ℓ is the denominator of η.

because of the translational invariance. The discrete case is precisely the opposite. A discrete profile is, generically, structurally stable, although isolated (see Sect. 1.2.) Hence, the set of undercompressive shocks that admit a discrete profile usually has a non-void interior, while the set of profiles of such a shock is usually discrete! A questionable conclusion is that, whenever some under-compressive shock is needed for the well-posedness of some system of conservation laws, every reasonable conservative difference scheme is inconsistent ! As a matter of fact, either this shock does not admit a DSP for some rational choice of η, and the scheme fails to approach this admissible wave, or it does, but then there exist a DSP for every nearby under-compressive shock, although we know that their admissibility is a codimension-one condition. We point out that asking for *stable* DSPs instead of any kind of DSP does not remove the contradiction.

We summarize the picture in Table 1.

Table 1. Existence of shock profiles for hyperbolic systems of conservation laws; G: *generic*, NG: *non-generic*, FD: *finite difference scheme*.

$\sharp$ incoming char.	viscosity	FD, $\eta \in \mathbb{Q}$	FD, $\eta \notin \mathbb{Q}$
n+2 (overcompressive)	2-parameter family G.	2-parameter family G.	?
n+1 (Lax shock)	1-parameter family G.	1-parameter family G.	Diophantine case OK
n (undercompressive)	1-parameter family NG.	discrete set G.	?
less than n ($n-1$: anti-Lax)	1-parameter family highly NG.	discrete set NG.	?

1 Existence Theory Rational Case

The rational case is the easiest one. To fix the notations, we set $\eta = r/\ell$ with $\ell \in \mathbb{N}^*$, $r \in \mathbb{Z}$ and $gcd(r, \ell) = 1$. Up to a symmetry in the space variable, we may suppose that r is non-negative.

The existence theory is, to a large extent, an application of the theory of discrete dynamical systems. For instance, it uses such objects as stable and unstable manifolds, center manifolds, transversality and structural stability. For most of these concepts, we refer to the excellent book by Palis and de Melo [47]. Concerning the center manifold Theorem, the reader may consult the book by Hirsch & al. [29] or the Appendix by A. Bressan in the present volume.

1.1 Steady Lax Shocks

For Lax shocks, the main result is the existence of a one-parameter family of DSP's whenever the strength of the shock is small enough. It is due to Majda and Ralston [43]. Unfortunately, the smallness is that of $\ell[u^l - u^r]$, so that the result cannot be extended to the irrational case. We shall see later on that this weakness is inherent to the existence problem, thus is not due to the method employed.

The method uses the center manifold theory for dynamical systems. To present it, we begin with the simplest situation, that of steady shock waves $(s = 0)$. Then the Rankine–Hugoniot condition reads

$$f(u^r) = f(u^l). \tag{1.1}$$

A DSP for a steady shock is itself steady: Since $\eta = 0$, the "integration" of (0.7) gives the implicit induction[5] in $u_j := U(j)$

$$F(u_{j-p+1}, \dots, u_{j+q}) = f(u^{r,l}) =: \bar{f}. \tag{1.2}$$

In this section, we specialize to the schemes whose numerical fluxes $F(u_{-p+1}, \dots, u_q)$ are invertible with respect to their extreme arguments u_{-p+1} and u_q, at least within relevant states. This is the case for the Lax–Friedrichs and Lax–Wendroff schemes:

- $F_{LF}(a, b) = \frac{1}{2}(f(a) + f(b)) + \frac{a-b}{2\sigma}$. The CFL (Courant–Friedrichs–Levy) condition[6]

$$\sigma \, \rho(Df(u)) < 1 \tag{1.3}$$

  ensures that $\mathcal{D}_a F_{LF}$ and $\mathcal{D}_b F_{LF}$ are non-singular for relevant values of state. From Implicit Function Theorem, the flux is invertible with respect to both a and b.
- $F_{LW}(a, b) = \frac{1}{2}(f(a) + f(b)) + \frac{\sigma}{2}Df_m(f(a) - f(b))$. The subscript m, for "middle", means that we use an interpolated value of Df. In the genuine Lax–Wendroff scheme, one puts $Df_m = Df\left(\frac{a+b}{2}\right)$. The choice $Df_m = \frac{1}{2}(Df(a) + Df(b))$ is reasonable too. One also may take

$$\mathcal{D}f_m = \int_0^1 Df(\tau a + (1 - \tau)b) \, d\tau,$$

  a matrix that plays an important role in the Osher scheme, due to the following equivalent form of the Rankine–Hugoniot condition:

$$\left(\int_0^1 Df(\tau a + (1 - \tau)b) \, d\tau - s \right)(b - a) = 0.$$

---

[5] Since the mesh ratio is kept fixed, we drop the argument σ in the numerical flux.
[6] ρ denotes the spectral radius of an endomorphism.

For most of the present analysis, the choice of $\mathcal{D}f_m$ is not important. Here, the CFL condition (same as above) ensures the invertibility, but only when b is close to a. As a matter of fact,

$$\mathcal{D}_a F_{LW}(u,u) = \frac{1}{2}\mathcal{D}f(u)(I_n + \sigma\mathcal{D}f(u)),$$

$$\mathcal{D}_b F_{LW}(u,u) = \frac{1}{2}\mathcal{D}f(u)(I_n - \sigma\mathcal{D}f(u)).$$

When the numerical flux $F(u_{-p+1}, \ldots, u_q)$ is invertible with respect to its extreme arguments u_{-p+1} and u_q, say in some convex open set that contains u^r and u^l, we may rewrite (1.2) as a discrete dynamical system of the form

$$u_{j+q} = G(u_{j-p+1}, \ldots, u_{j+q-1}), \quad j \in \mathbb{Z}, \tag{1.4}$$

where G is a diffeomorphism, defined implicitly by

$$F(u_{1-p}, \ldots, u_{q-1}, G(u_{1-p}, \ldots, u_{q-1})) = \bar{f}. \tag{1.5}$$

Remark that

$$G(u^r, \ldots, u^r) = u^r, \quad G(u^l, \ldots, u^l) = u^l. \tag{1.6}$$

In order to apply the Center Manifold Theorem (denoted by CMT below), we fix a state u^* that is sonic: the k-th eigenvalue λ_k vanishes at u^*. We assume however[7] that λ_k is a simple eigenvalue that satisfies

$$\mathcal{D}\lambda_k(u^*)r_k(u^*) \neq 0, \tag{1.7}$$

a property that has been called *genuine nonlinearity* since Lax [35].

Under Assumption (1.7), the equation $\lambda_k(u) = 0$ defines a smooth hypersurface Σ that is transversal to $r_k(u^*)$ at u^*. When $u \in \Sigma$, the tangent space to the graph of f contains a horizontal vector. Hence the graph folds above Σ: $f(\Sigma)$ is a smooth hypersurface that splits a neighbourhood O^* of $f^* := f(u^*)$ in $\mathbb{R}^n$ into two components O_0 and O_2. When $z \in O_0$, the equation $f(u) = z$ has no solution, while if $z \in O_2$, it has two solutions, say $u(z)$ and $u'(z)$. When $u \in \Sigma$, we define $u' = u$. The map $u \mapsto u'$ is well-defined and smooth in some neighbourhood $U(u^*)$ of u^*. We leave the reader checking that the product $\lambda_k(u)\lambda_k(u')$ is strictly negative away from Σ.

We now let the point u^l vary in $U(u^*)$, and consider the dynamical system (1.2) with $\bar{f} := f(u^l)$. This means that G also depends on u^l in (1.6). The fact that u^l becomes a variable rather than a constant means that we replace this u^l in G by a v_j, and we add the trivial iteration $v_{j+1} = v_j$! Hence we have to deal with a one-step dynamical system in an open set of $\mathbb{R}^{(p+q)n}$:

$$(u_{j-p+2}, \ldots, u_{j+q}, v_{j+1})^T = H\left((u_{j-p+1}, \ldots, u_{j+q-1}, v_j)^T\right).$$

---

[7] We use standard notations: The eigenvalues are labelled in increasing order. The right and left eigenvectors of $\mathcal{D}f(u)$ are denoted $r_k(u)$ and $l_k(u)$ respectively. The latter and $\mathcal{D}\lambda_k(u)$ are to be viewed as linear forms.

The differential of H at $(u^*, \ldots, u^*)^T$ is the matrix

$$
\begin{pmatrix}
0_n & I_n & 0_n & \cdots & & 0_n \\
\vdots & \ddots & \ddots & & \ddots & \vdots \\
0_n & \cdots & 0_n & I_n & & 0_n \\
\mathcal{D}_{1-p}G^* & \cdots & \cdots & \mathcal{D}_{q-1}G^* & \mathcal{D}_v G^* \\
0_n & \cdots & \cdots & & 0_n & I_n
\end{pmatrix},
$$

where the star stands for the argument $(u^*, \ldots, u^*)^T$. This matrix, being blockwise triangular, admits the unity as an eigenvalue of order n at least. The rest of the spectrum is given by the equation

$$
\det \left(X^{p+q-1} I_n - X^{p+q-2} \mathcal{D}_{q-1} G^* - \cdots - \mathcal{D}_{1-p} G^* \right) = 0. \tag{1.8}
$$

Since we assume that $\mathcal{D}G^*$ is non-singular, the matrix is invertible and H is locally a diffeomorphism around $(u^*, \ldots, u^*)^T$. It turns out that $X = 1$ is again a root of (1.8). As a matter of fact, a differentiation of (1.5) gives

$$
\mathcal{D}_{1-p}F^* + \cdots + \mathcal{D}_{q-1}F^* + \mathcal{D}_q F^* (\mathcal{D}_{q-1}G^* + \cdots + \mathcal{D}_{1-p}G^*) = 0_n.
$$

On the other hand, the consistency implies

$$
\mathcal{D}_{1-p}F^* + \cdots + \mathcal{D}_q F^* = Df(u^*).
$$

An elimination gives the formula

$$
\mathcal{D}_q F^* (I_n - \mathcal{D}_{q-1}G^* - \cdots - \mathcal{D}_{1-p}G^*) = Df^*. \tag{1.9}
$$

Since $Df(u^*)$ is singular and $\mathcal{D}_q F^*$ where supposed invertible, we deduce that the parenthesis is singular too.

We now assume that (1.8) does not have any other root on the unit circle. In particular, the unity is a simple root. This is called in the litterature a *non-resonance condition*. It means that the unity is an eigenvalue of $\mathcal{D}H^*$, of multiplicity $n+1$, and that no other eigenvalue belongs to the unit circle. In this situation, it is well-known that there exists a smooth manifold Γ_c (the *center manifold*) in $\mathbb{R}^{(p+q)n}$, with the following properties:

1. Γ_c is locally invariant forward and backward under the dynamical system. This means that there exists a neighbourhood V of $(u^*, \ldots, u^*)^T$, such that whenever $w \in V \cap \Gamma_c$, then $H(w)$ and $H^{-1}(w)$ belong to Γ_c,
2. If a global orbit $(w_j)_{j \in \mathbb{Z}}$ is included in V, then it is included in Γ_c,
3. $\dim \Gamma_c = n+1$.

Notice than in general, Γ_c is not unique. However, the second point above is a partial uniqueness property. Also, two center manifolds are tangent at every order. In other words, in the description of Γ_c as a graph of some function, the jet of this function a $(u^*, \ldots, u^*)^T$ is unique.

Since a DSP can be viewed as a heteroclinic orbit of H, linking a point $(u^l, \ldots, u^l)$ to a point $(u^r, \ldots, u^r)$, the second point above is fundamental. It tells that such profiles, provided they keep a small amplitude, must run over Γ_c. This property also tells that Γ_c contains all the fixed points of H that are close to $(u^*, \ldots, u^*)$. Using (1.5), we see that these fixed points split into two families, namely

$$(u, \ldots, u, u), \quad u \in U(u^*)$$

and

$$(u, \ldots, u, u'), \quad u \in U(u^*),$$

with the notation used above. Each family is an n-dimensional submanifold of Γ_c, parametrized by its first component, as well as by the last one. These submanifolds intersect at points $(u, \ldots, u, u)$ where u runs over Σ. It turns out that this intersection is transverse because $du'r_k = -r_k \neq r_k$ at u^*. Next, the map H, restricted to Γ_c, is a local diffeomorphism that preserves the orientation, and the "vertical lines" $\{v = \text{cst}\}$ are invariant under H, by definition. On the other hand, the genuine nonlinearity implies that the manifolds of fixed points are transversal to the vertical lines (their tangent spaces at $(u^*, \ldots, u^*)$ are not vertical.) Hence every vertical line δ, close enough to $(u^*, \ldots, u^*)$, contains exactly one fixed point of each family. Since they must have the same last entry, they have the form

$$w := (u, \ldots, u, u), \quad w' := (u', \ldots, u', u).$$

Since H preserves δ and the orientation, the interval between w and w' along δ is invariant. At last, since there is no fixed point on it, this interval is a *continuous* heteroclinic orbit: For every point $w_0 \in (w, w')$, the whole orbit $(w_j := H^j(w_0))_{j \in \mathbb{Z}}$ stays in (w, w') and tends to the end points w and w' as j goes to $\pm\infty$. This exactly means that given $u \in U(u^*)$, there exists a one-parameter family of DSPs connecting either u to u', or the converse. An obvious stability analysis shows that the point $(u, \ldots, u, u)$ is unstable (in Γ_c) if and only if $\lambda_k(u) > 0$. Hence, given two points $u^{l,r}$ in $U(u^*)$, there exists a DSP from u^l to u^r if and only if $(u^l, u^r; \sigma = 0)$ is a steady shock satisfying the Lax shock condition

$$\lambda_k(u^r) \leq 0 \leq \lambda_k(u^l). \tag{1.10}$$

In case of equalities in (1.10), one has $u^r = u^l$, and the profile is constant.

We summarize this fundamental result:

Theorem 1.1 *Let u^* be given, as well as an eigenvalue λ_k that is simple and genuinely nonlinear at u^*. Assume that the numerical flux is locally invertible with respect to both its first and last arguments. Assume at last the non-resonnance condition.*

Then, given a steady k-discontinuity (u^l, u^r) in a neighbourhood of (u^*, u^*), their exist small amplitude DSPs if and only if this discontinuity is a k-shock (namely, it satisfies the Lax shock condition.)

Additionally, the set of small DSPs of the shock (u^l, u^r) is a connected curve in the following sense. Parametrizing a DSP by the $(p + q)$-uplet $(u(-p), \ldots, u(q - 1))$, the parameter runs over a curve connecting the points $(u^l, \ldots, u^l)$ and $(u^r, \ldots, u^r)$.

Remark.

The Lax–Friedrichs is resonant because the even grid, defined by $j + m \in 2\mathbb{Z}$ and the odd one (its complement) ignore each other. Hence each one is responsible for an eigenvalue $\lambda = 1$, thus the total multiplicity is 2 at least. Non-resonance can be recovered by restriction to the even grid for instance. However, this subgrid has not the same tensorial structure as the usual one. It is therefore convenient to iterate once the scheme, thus working on the subgrid $(j, m) \in 2\mathbb{Z} \times 2\mathbb{Z}$. The iterated scheme is still a conservative 3-point scheme, which reads

$$u_{2j}^{2m+2} = u_{2j}^{2m} + \sigma(F_{2j-1}^{2m} - F_{2j+1}^{2m}), \quad F_{2j-1}^{2m} = F_{LF2}(u_{2j-2}^{2m}, u_{2j}^{2m}),$$

with

$$F_{LF2}(a, b) := \frac{1}{4\sigma}(a - b) + \frac{1}{4}(f(a) + f(b)) + \frac{1}{2}f\left(\frac{a+b}{2} + \frac{\sigma}{2}(f(a) - f(b))\right).$$

This form of the Lax–Friedrichs is non-resonant under the CFL condition. This provides a context where Theorem 1.1 applies.

More Complex Situations

The genuine nonlinearity assumption served in the description of the number of solutions of the equation $f(u) = \bar{f}$, when $\bar{f}$ is given. When dropping it, we encounter more complicated patterns. The structurally stable ones are described by Thom's "theory of catastrophies". The length of the list of possible patterns increases with n. The simplest one is when $Df(u^*)$ is non-singular: Nothing happens, the equation has exactly one solution, the steady discontinuities are trivial ($u^r = u^l$). Next, the case analysed above: Except for points on Σ, each u close to u^* is either a left state (if $\lambda_k(u) > 0$) or a right state (if $\lambda_k(u) < 0$) of a Lax shock wave, and this shock admits a one-parameter family of DSPs.

When $D\lambda_k(u^*)r_k(u^*)$ vanishes (*linear degeneracy* at point u^*), let us assume that $\mathcal{D}(D\lambda_k r_k)$ does not vanish at u^*, and that it is not parallel to $\mathcal{D}\lambda_k$. Then the zero set of $D\lambda_k r_k$ is a hypersurface that is transversal to Σ. The number of solutions of $f(u) = \bar{f}$ is either 1 or 3, except for non-generic values[8] $\bar{f} \in f(\Sigma)$. Using the same procedure as above, we obtain the following result. On a typical vertical line δ, H has either 1 or 3 fixed points. In the former case, there cannot be non-trivial steady discontinuities. In the latter, let w, w', w''

---

[8] $f(\Sigma)$ is a smooth hypersurface.

be the fixed points, with w' in the middle; we denote by u, u', u'' their respective first components. Then either w' is an attractor or it is a repellor. In the attracting case, both shocks (u, u') and (u'', u') admit a one-parameter family of DSPs. In the repulsing case, both shocks (u', u) and (u', u'') admit a one-parameter family of DSPs.

More generally, the same kind of bifurcation analysis can be performed without any nonlinearity assumption. The general result is analogous to Theorem 1.1, except that the necessary and sufficient condition for the existence of DSPs is Liu's E-condition instead of Lax's. We recall Liu's E-condition: given a small shock[9] $(u^l, u^r; \sigma(u^l, u^r))$, we know that u^r belongs to a branch $H_k(u^l)$ of the Hugoniot curve based on u^l. Then the shock is admissible if there holds

$$\sigma(u^l, u^r) < \sigma(u^l, v), \tag{1.11}$$

for every $v \in H_k(u^l)$ that lies between u^l and u^r. Notice that one usually admits also the shocks that satisfy large inequalities in (1.11), but these do not admit a DSP. This borderline case is reminiscent to the notion of *contact discontinuities*.

Other Rational Values of η

Assume that $\eta = r/\ell$ is a non-zero rational number. The stategy follows closely that used in the steady case. The main difference is that the group $\mathbb{Z}/\ell\mathbb{Z}$ replaces $\mathbb{Z}$. Hence, we search for heteroclinic orbits of a dynamical system whose increment is $1/\ell$. This needs to multiply by ℓ the number of variables u. Hence the induction takes place in $\mathbb{R}^{(p+q-1)\ell n+n}$. For instance, in the not-too-much-complex case $\eta = 1/2$, (0.7) gives implicitly u_{j+q} in terms of $u_{j-p}, \ldots, u_{j+q-1}$ *and* $u_{j-1/2}$. Hence the one-step induction concerns the set of variables

$$u_{j-p}, u_{j-p+1/2}, u_{j-p+1}, \ldots, u_{j+q-1}, u_{j+q-1/2}, v_j.$$

Notice that the "trivial" induction is now $v_{j+1/2} = v_j$, since our increment has become $1/2$.

The rest of the analysis is really similar to that of the steady case. Of course, the non-resonnance condition is more involved, since we deal with a polynomial of higher degree $(p + q - 1)\ell n$.

Explicit Profiles for the Godunov Scheme

For some schemes of practical interest, the numerical flux might not be invertible with respect to its first and last arguments. This could be caused by the lack of numerical viscosity and/or the sonicity of some points. Typical examples are the Godunov schemes, whatever the states in consideration, or the Lax–Wendroff scheme near sonic shocks. It is remarquable that something

---

[9] $\sigma(a, b)$ denotes the velocity of a discontinuity with end states a and b.

can be said in both these cases. Since these are three-points schemes, we have $p = q = 1$ and the numerical flux involves only two points.

Let us begin with the Godunov scheme. The numerical flux is

$$F_{God}(a, b) = f(R(a, b)),$$

where $R(a, b)$ denotes the middle point (on the line $x/t = 0$) in the Riemann problem between the left state a and the right state b. When this Riemann problem involves a steady shock (and we are precisely instered in steady shocks), $R(a, b)$ is not well-defined but $f(R(a, b))$ is, thanks to the Rankine–Hugoniot condition. Hence the scheme is well-defined, although its numerical flux is not C^1 in general.

One immediately sees that the dependency of F_{God} on its first argument concerns only the waves of negative velocities. If one replaces a by another state $\hat{a}$ still belonging to the set of states that can be linked to $R(a, b)$ by backward waves, we do not modify $R(a, b)$, hence $F_{God}(a, b)$. This shows that $\mathcal{D}_a F_{God}$ is not one-to-one, except if $\mathcal{D}f(R(a, b))$ has a positive spectrum. Even in this latter case, which corresponds to $R(a, b) = a$, that is $F_{God}(a, b) = f(a)$, the flux is not invertible with respect to b.

Let us examine the DSPs of a steady Lax shock (u^l, u^r). The profile equation is

$$f(u_{j+1/2}) = \bar{f} := f(u^{l,r}), \quad u_{j+1/2} := R(u_j, u_{j+1}).$$

For the sake of simplicity, we make the realistic assumption[10] that the equation $f(u) = \bar{f}$ has only two solutions, namely u^l and u^r. Then the profile equation reduces to

$$u_{j+1/2} \in \{u^l, u^r\}.$$

Since u_j tends to u^l as $j \to -\infty$, and $u_{j+1/2}$ does too because of the continuity of the Riemann solver, we have $u_{j+1/2} = u^l$ for j less than some number A, and similarly $u_{j+1/2} = u^r$ for j larger than some B. In particular, there must be an index i such that $u_{i-1/2} = u^l$ and $u_{i+1/2} = u^r$. If there is more than one such index, then there must also exists an index m such that $u_{m-1/2} = u^r$ and $u_{m+1/2} = u^l$. However this is impossible in realistic problems, since it would allow a non-constant solution of the Riemann problem between u_m and itself[11]: Go from u_m to u^l by backward waves (this is possible since $R(u_m, u_{m+1}) = u^l$), then use the steady shock (u^l, u^r), then go from u^r to u_m by forward waves (this is possible since $R(u_{m-1}, u_m) = u^r$.) We deduce that the index i above is unique. With the same uniqueness argument as above, we find that $u_j = u^l$ for $j < i$ and $u_j = u^r$ for $j > i$. At last, the state u_i belongs

---

[10] This assumption corresponds more or less to genuine nonlinearity of the corresponding characteristic field.

[11] Such a non-constant solution violates the uniqueness of the Cauchy problem. Since a constant data is smooth, the uniqueness of a weak entropy solution is ensured when the system is endowed with a strictly convex entropy. See Dafermos [15], Chapter 5.

to the intersection $\mathcal{W}^-(u^r) \cap \mathcal{W}^+(u^l)$ of the following sets: $\mathcal{W}^-(u)$ is the set of states that can be linked to u by backward waves, and $\mathcal{W}^+(u)$ is the set of states to which u can be linked by forward waves. Assuming that (u^l, u^r) is a steady k-th shock, $\mathcal{W}^+(u^l)$ is a manifold of dimension $n - k + 1$ with a boundary that passes through u^r, and $\mathcal{W}^-(u^r)$ is a manifold of dimension k with a boundary that passes through u^l. These manifolds usually intersect transversally along a curve $\gamma(u^l, u^r)$ that links u^l to u^r. Each choice of a point $a \in \gamma(u^l, u^r)$ and of an index i such that $u_i = a$ yields a DSP for the steady shock (u^l, u^r), and a DSP determines uniquely the pair $(a, i) \in \gamma(u^l, u^r) \times \mathbb{Z}$, with the exception that (u^l, i) and $(u^r, i + 1)$ correspond to the same DSP. Therefore the set of DSPs is parametrized by $\gamma(u^l, u^r) \times \mathbb{Z}$, quotiented or glued by $(u^l, i) \sim (u^r, i + 1)$; see Figure 1. This quotient is homeomorphic to $\mathbb{R}$. Thus there is a one-parameter set of DSPs for a given steady Lax shock. In other words, there exists a *continuous* DSP. Notice however that it is not of class $\mathcal{C}^1$, since it displays a periodic array of kinks (see Figure 1.)

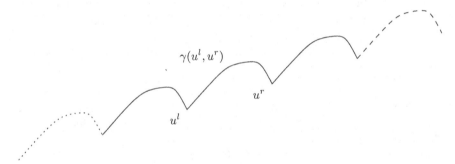

$\gamma(u^l, u^r)$

u^r

u^l

Fig. 1. The curve of Godunov DSPs (the "continuous" DSP) for a Lax shock.

For an extreme shock, that is $k = 1$ or $k = n$, the set $\gamma(u^l, u^r)$ has a pretty simple description. For instance, if $k = 1$, then $\mathcal{W}^+(u^r)$ is an open set whose boundary passes through u^l, while $\mathcal{W}^-(u^l)$ is the part of the 1-forward[12] wave wave curve $\mathcal{O}_f^1(u^l)$ originated from u^l, which stops at u^r because the shocks $(u^l, u; \sigma)$ with u beyond u^r have typically a negative velocity[13]. Thus $\gamma(u^l, u^r)$ is the part of $\mathcal{O}_f^1(u^l)$ bounded by u^l and u^r. Similarly, when $k = n$, $\gamma(u^l, u^r)$ is the part of the n-backward wave curve $\mathcal{O}_b^n(u^r)$ between u^l and u^r. When the first characteristic field is genuinely non-linear, this branch of $\mathcal{O}_f^1(u^l)$ is a part of the shock curve $\mathcal{S}_f^1(u^l)$. By symmetry, an analogous characterization

---

[12] When a simple or composite wave of the k-th family passes from a to b as $\xi = x/t$ increases, we say that b belongs to the k-forward wave curve of a, and that a belongs to the k-backward wave curve of b.

[13] Here, we assume for the sake of simplicity that the wave velocity is monotonous along the wave curve.

holds for $k = n$, with $\mathcal{S}_b^n(u^l)$ instead. Again, the subscripts f, b stand for *forward/backward*.

So to speak, the Godunov DSPs of steady shocks have a "compact support", because u_j is constant on each sides of a single index, though with two distinct constants. We remark that the existence of Godunov DSPs when η is a non-zero rational number is an open problem. Since F_{God} is not invertible with respect to its arguments, it does not follow from the previous section. But it cannot be solved in an explicit way as the case $\eta = 0$ could be. In particular, we do not expect that such a profile would be of compact support.

Open Problem 1.1 *Prove the existence of Godunov DSPs for Lax shocks with non-zero rational parameters.*

DSPs for Strong Steady Shocks Under the Lax–Wendroff Scheme

Given a pair (u^l, u^r) that satisfies the Rankine–Hugoniot condition at zero velocity, that is $f(u^l) = f(u^r)$, the Lax–Wendroff admits the trivial DSPs defined by $u_j \in \{u^l, u^r\}$ and $u_j = u^l$ for j less than some A, $u_j = u^r$ for j larger than some B. That is a very bad feature of this scheme, since there are much too many profiles, and that even non-admissible discontinuities admit DSPs. These are reasons why the Lax–Wendroff was not in favour. However, since these profiles might not be dynamically stable, and therefore not observable, the flaw might be ignored. On the other hand, the scheme has the advantage of being second-order.

We now restrict to a steady k-Lax shock and consider the "monotonous" profile $\bar{u}$ given by $u_j = u^l$ for $j \le 0$ and $u_j = u^r$ for $j \ge 1$. It is compactly supported. We shall show that there exists a one-parameter family of profiles close to this one, but which are not compactly supported. Of course, we assume the CFL condition, so that F_{LW} is invertible with respect to each of its arguments near (u^l, u^l) and (u^r, u^r). Therefore the profile equation (1.2) may be recast as a discrete dynamical system of the form (1.4), say

$$u_{j+1} = G(u_j), \tag{1.12}$$

as long as u_j and u_{j+1} stay in a neighbourhood of u^l. The matrix $\mathcal{D}G(u^l)$ can be computed by linearizing (1.2):

$$\mathcal{D}G(u^l) = (A(u^l) - I_n)^{-1}(A(u^l) + I_n),$$

where $A(u) := \sigma \mathcal{D}f(u)$ has real eigenvalues, of modulus less than one. Since $A(u^l)$ is non-singular, $\mathcal{D}G(u^l)$ is *hyperbolic* in the terminology of dynamical systems: It has no eigenvalue on the unit circle. Hence the trajectories of (1.12) that tend to u^l as $j \to -\infty$ describe locally the *unstable manifold* $W^u(u^l)$. This manifold is tangent at u^l to the vector space $E^u(u^l)$ spanned by the eigenvectors[14] of $\mathcal{D}G(u^l)$ corresponding to the eigenvalues outside the

---

[14] One should take in account the generalized eigenvectors in general, but here our $\mathcal{D}G(u^l)$ is diagonalisable.

unit circle. These are precisely the eigenvectors of $A(u^l)$ corresponding to its negative eigenvalues.

Similarly, the trajectories that tend to u^r as $j \to +\infty$ cover the stable manifold $W^s(u^r)$. The latter is tangent at u^r to the invariant subspace $E^s(u^r)$ of $A(u^r)$ corresponding to the positive eigenvalues[15].

We now remark that $F_{LW}(u^l, u^r) = \bar{f}$ and therefore there is a possibility to define a diffeomorphism from a neighbourhood $\mathcal{V}(u^l)$ to a neighbourhood $\mathcal{V}(u^r)$, provided that $\mathcal{D}_a F_{LW}(u^l, u^r)$ and $\mathcal{D}_b F_{LW}(u^l, u^r)$ be invertible. Since

$$\mathcal{D}_a F_{LW}(u^l, u^r) = \frac{1}{2\sigma}(I_n + A_m)A(u^l), \quad \mathcal{D}_b F_{LW}(u^l, u^r) = \frac{1}{2\sigma}(I_n - A_m)A(u^r),$$

it will be enough that $\rho(A_m) < 1$, a property that is usually implied by a suitable form of the CFL condition. Let us thus denote H this diffeomorphism. We have

$$H(u^l) = u^r, \quad \mathcal{D}H(u^l) = A(u^r)^{-1}(A_m - I_n)^{-1}(A_m + I_n)A(u^l). \qquad (1.13)$$

Since $(u^l, u^r; s = 0)$ is a k-Lax shock, the dimension of $W^u(u^l)$ is $n - k + 1$ and that of $W^s(u^r)$ is k. To build shock profiles in an as simple as possible way, we glue a trajectory on $W^s(u^l)$ to another one on $W^s(u^r)$, through a large jump that uses H. This means that we choose a u_1 in the intersection

$$W^s(u^r) \cap H\left(W^u(u^l)\right) \qquad (1.14)$$

and define $u_0 := H^{-1}(u_1)$. By construction, the backward trajectory of (1.12), originated from u_0, tends to u^l at $-\infty$, while the forward trajectory originated from u_1 tends to u^r.

There remains to study the intersection in (1.14). This is the point where we use the assumption that the shock is a Lax shock. The dimension of $W^u(u^l)$ is $n - k + 1$, while that of $H(W^s(u^r))$ equals that of $W^s(u^r)$, say k. If we assume that their tangent spaces span the ambient space $\mathbb{R}^n$ (a transversality assumption), this intersection is locally a smooth curve. Hence, generically, the trivial profile is embedded in a one-parameter smooth family of DSPs. Notice that, due to the special form of $\mathcal{D}H(u^l)$, the transversality amounts to saying that

$$(A_m + I_n)E^u(u^l) + (A_m - I_n)E^s(u^r) = \mathbb{R}^n, \qquad (1.15)$$

where the sum of dimensions is $n + 1$.

Remarks.

1. The eigenvalues of $\mathcal{D}G(u^{l,r})$ are negative. Therefore every DSP has an oscillating behaviour at infinity.

---

[15] The notation is a bit confusing, since the words *stable* and *unstable* refer to the dynamical system (1.12), but not to the differential equation $\dot{y} = A(u^{l,r})y$.

2. In Sect. 1.2, we study the case of undercompressive shocks that satisfy[16]

$$\lambda_k(u^{l,r}) < 0 < \lambda_{k+1}(u^{l,r}). \tag{1.16}$$

The sum of dimension equals n and a transversality property holds true at each point of the intersection, generically. In such a case, the intersection is discrete, and it is *structurally stable* in the sense that it persists under a small change of the steady shock. Hence there may be isolated DSPs that are perturbed smoothly when the shock is modified. This feature is in contrast with the situation for viscous shock profiles[17]. It will have an important consequence in the study of its stability.

Scalar Shocks Under Monotone Schemes

A scalar conservation law ($n = 1$) has the property of monotony, according to Kruzkhov's Theorem. For two initial data a and b that satisfy $a \leq b$ almost everywhere, the entropy solutions satisfy $S_t a \leq S_t b$. A difference scheme that is consistent with this property is said *monotone*. Monotone schemes are characterized by the implication

$$(u_{-p} \leq v_{-p}, \ldots, u_q \leq v_q) \Longrightarrow G(\sigma; u_{-p}, \ldots, u_q) \leq G(\sigma; v_{-p}, \ldots, v_q), \tag{1.17}$$

where G is the function defined in (0.7). When $p, q \geq 1$, monotonicity implies

$$\frac{\partial F}{\partial u_{-p}} \geq 0 \geq \frac{\partial F}{\partial u_{q-1}}. \tag{1.18}$$

When the monotonicity in (1.17) is strict and the inequalities in (1.18) are strict too, one says that the scheme is *strictly monotone*.

Remarks.

- In many cases, monotony is equivalent to the CFL condition.
- In particular, monotony might hold on a bounded interval only, for instance if f is not globally Lipschitz. Then monotony will make sense for every data with values in this interval.
- A monotone scheme is consistent with the L^1-contraction property of the Kruzkhov semi-group S_t. In particular, it is TVD (*total variation diminishing*).

When dealing with a strictly monotone scheme in an interval that contains $u^{l,r}$, ordering arguments were employed by G. Jennings [31] to prove that every Lax shock with a rational η admits a one-parameter family of DSPs, whatever the strength $|u^r - u^l|$ of the shock. For every u^* taken in (u^l, u^r) (or (u^r, u^l)),

---

[16] For instance, a phase transition in a Van der Waals gas.
[17] See the end of the introduction.

there exists a unique DSP with $u_0 = u^*$. This DSP is itself strictly monotone. The monotony and the uniqueness are consequences of the TVD property of scheme.

Jennings claimed that this result could be extended to irrational values of η. However, his density argument did not contain any detail. The question has been therefore considered as open for a long time. The gap was filled recently in [55], using the function Y described in Sect. 2.1. This method allows to relax also the strict monotony in (1.17) and the strict inequalities in (1.18). In particular, it handles the case of the Godunov scheme, for which Jennings' proof was powerless. We now have an as general as possible existence and uniqueness theorem, since only the monotony of the scheme over (u^l, u^r) is required. See Paragraph 2.1 for a rigorous statement.

1.2 Under-Compressive Shocks

We focus in this paragraph on undercompressive steady shock waves. By this, we mean that there are exactly n incoming characteristics and as many outgoing ones. Thus there exists an index k such that (1.16) holds true. For the sake of simplicity, we limit ourselves to 3-point schemes. In other words, the numerical flux involves only two points. We assume also that the profile equation (1.2) can be rewritten as a dynamical system (1.12) for an orientation preserving diffeomorphism G. We know a priori that u^l and u^r are fixed points of G. Let assume finally that these are hyperbolic points, meaning that $DG(u^{l,r})$ do not have eigenvalues on the unit circle. This usually follows from the CFL condition. Then the DSPs of the shock (u^l, u^r) are in one-to-one correspondance with the intersection $W^u(u^l) \cap W(u^r)$: Given a point u_0 in the intersection, it defines a unique trajectory, and this trajectory tends to u^l and u^r at infinity. In realistic cases, as the Lax–Friedrichs or the Lax–Wendroff schemes, the dimension of $W^u(u^l)$ is $n - k$ and that of $W^s(u^r)$ is k; we notice that the sum of the dimensions equals that of the ambient space.

Since there is no particular constraint on the respective positions of their tangent spaces along their intersection, they generically intersect along a discrete subset, transversally. It is a well-known fact that transversality ensures the persistence of an intersection under small perturbations of the manifolds, of class C^1. Hence a DSP depends continuously on the underlying undercompressive steady shock, so long as it corresponds to a transverse intersection.

When counting the number of DSPs of a given undercompressive shock, it is easier not to distiguish the profiles that differ only by a shift of the index. Two such profiles correspond to possibly distinct points u_0 but which belong to the same orbit.

The orbits may actually be classified according to a natural orientation. To simplify the exposition, let us consider the case $n = 2$ with $k = 1$. Then both $W^u(u^l)$ and $W^s(u^r)$ are planar curves, that are oriented according to the dynamics (1.12). This means that their tangent bundles are naturally oriented. Assume that these curves intersect transversally at some point u_0. Let X_0^u be a positively oriented tangent vector to $W^u(u^l)$ at u_0; we denote similarly X_0^s a positively oriented tangent vector to $W^s(u^r)$ at u_0. By assumption, the pair (X_0^u, X_0^s) is a basis. The images $u_j := G^{(j)}(u_0)$ are other intersection points, where the positively oriented tangent vectors may be taken as $X_j^{u,s} = \mathcal{D}G^{(j)}X_0^{u,s}$. There holds

$$\det(X_j^u, X_j^s) = \left(\det \mathcal{D}G^{(j)}\right) \det(X_0^u, X_0^s).$$

In particular, the intersection is transverse at every point of the orbit. Since G is orientation preserving, the sign of $\det(X_j^u, X_j^s)$, that is the orientation of the basis (X_j^u, X_j^s), does not depend on j. This analysis shows that, on the simplest possible figure 2, two consecutive intersection points correspond to distinct DSPs, because the orientation of the basis (X_j^u, X_j^s) changes.

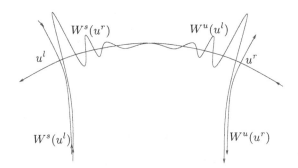

Fig. 2. The simplest transversally heteroclinic connection.

An Example from Reaction-Diffusion

The observation in the previous section is not specific to DSPs within conservation laws. It concerns more generally DSPs associated to standing waves of evolutionary PDEs, that is heteroclinic solutions of difference equations in $\mathbb{R}^n$, provided

$$\dim W^u(u^l) + \dim W^s(u^r) = n, \tag{1.19}$$

where $u^{l,r}$ are the extreme values of the wave.

Because it is pretty easy to make explicit calculations, we illustrate this principle with an example taken from reaction-diffusion modelling. The PDE is

$$\partial_t u = \Delta u - \phi'(u), \tag{1.20}$$

where $\phi : \mathbb{R} \to \mathbb{R}$ is an even double-wells potential that is super-linear at infinity. For instance, $\phi(u) = (u^2-1)^2$ works. The standing waves are solutions of the ODE

$$\frac{d^2u}{dx^2} = \phi'(u).$$ (1.21)

The wells $u^{l,r}$ (the zeroes of ϕ') are fixed points of the ODE, but there is also an heteroclinic solution from u^l to u^r, as well as a converse, using the symmetry $x \mapsto -x$ (we do not use the parity of ϕ here.)

Let us discretize the PDE in the simplest form:

$$\frac{u_j^{n+1} - u_j^n}{\Delta t} = \frac{u_{j+1}^n - 2u_j^n + u_{j-1}^n}{(\Delta x)^2} - \phi'(u_j^n).$$ (1.22)

One immediately sees that the equation for DSPs, namely

$$\frac{u_{j+1} - 2u_j + u_{j-1}}{(\Delta x)^2} = \phi'(u_j),$$ (1.23)

is the Euler–Lagrange equation of the functional

$$J[u] := \sum_{j \in \mathbb{Z}} \frac{1}{2\Delta x}(u_j - u_{j-1})^2 + \Delta x \sum_{j \in \mathbb{Z}} \phi(u_j).$$

Notice that J is the discrete version of the energy of the system,

$$J_{cont}[u] = \int_{\mathbb{R}} \left(\frac{1}{2}|\dot{u}|^2 + \phi(u)\right) dx.$$

An elementary calculations shows that

$$\dim W^u(u^l) = \dim W^s(u^r) = 1.$$

Since (1.23) is a two-step[18] induction, the ambient space is $\mathbb{R}^2$, hence (1.19) holds true.

Because ϕ hence J are even, the minimisation of J, subjected to the constraint $u(-\infty) = u^l$, $u(+\infty) = u^r$, together with one of the following parities

$$u_j = -u_{-j},$$ (1.24)

or

$$u_j = -u_{1-j},$$ (1.25)

yields to the same Euler-Lagrange equation (1.23), hence to DSPs of the standing wave[19]. Since the parities (1.24) and (1.25) are not compatible, we obtain two distinct DSPs for the same wave. One may checks that, according to the former paragraph, they have opposite "orientations".

---

[18] For a reaction-diffusion equation, there is no way to integrate once the profile equation for a standing wave.

[19] The fact that both minimization problems admit a solution is not too difficult, because each parity constraint breaks the translational invariance.

Homoclinic and Chaotic Orbits

An especially rich dynamics arises in the following situation:

- The dynamical system (1.12) has two hyperbolic fixed points $u^{l,r}$,
- The equality (1.19) holds true, whence the symmetric equality

$$\dim W^s(u^l) + \dim W^u(u^r) = n, \tag{1.26}$$

- There exist simultaneously a DSP from u^l to u^r, and another one from u^r to u^l. In other words, each pair $(W^u(u^l), W^s(u^r))$ and $(W^s(u^l), W^u(u^r))$ has a non-void intersection,
- These intersections are transverse.

Under these assumptions, one proves that there exists a homoclinic orbit to u^l (and one to u^r too) that is transverse. Actually, as an invariant set, $W^u(u^r)$ is attracting in the sense that all forward orbit starting near u^l, which does not belong to $W^s(u^r)$, is asymptotic to $W^u(u^r)$. Therefore, apart from its intersection points[20] with $W^s(u^r)$, $W^u(u^l)$ tends to cover a larger and larger portion of $W^u(u^r)$. Since the latter intersect transversally $W^s(u^l)$, it happens that $W^u(u^l)$ eventually intersects $W^s(u^l)$ transversally too (see Figure 3.)

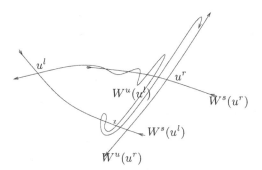

Fig. 3. The onset of Smale's "horse-shoe". The simplest of the infinitely many homoclinic orbits to u^l is that passing through z.

This situation is known to imply a Smale's *horse-shoe dynamics*: In some small neighbourhood of u^l, there exists a Cantor set K, invariant under some power G^N, on which the dynamics defined by G^N is conjugated to the *shift* on $\{0,1\}^{\mathbb{Z}}$. In K, the set of periodic points is dense (Birkhoff), and the periods may be arbitrarily large. The set K contains also an infinity of points homoclinic to u^l, whose orbits may be arbitrarily close to periodic ones, on intervals

---

[20] Because of these countable intersection points, $W^u(u^l)$ must fold countably many times. The asymptotic picture is extremely complicated and drawing it threatened Poincaré himself.

of arbitrarily large lengths. There is also an infinity of heteroclinic orbits[21] from u^l to u^r, and again they may be arbitrarily close to periodic orbits, on intervals of arbitrarily large lengths. Hence there are a lot of DSPs, both from u^l to u^r and in the reverse sense. Even stranger behaviours do arise. For instance, there exist orbits (under G) that oscillate in a choatic way between u^l and u^r; they stay in neighbourhoods of u^l or u^r during large intervals, and the dynamics of the interval lengths looks inpredictible. Let us mention at last that this situation is *structurally stable*, meaning that it persists under a small change of G, for instance under a small change of the steady shock (u^l, u^r).

An illustration of this pattern is the standing DSP in the reaction-diffusion equation (1.20). We have shown the existence of a DSP from u^l to u^r. The reflexion $j \mapsto -j$ gives the DSP from u^r to u^l. These intersections turn out to be transverse. The global pattern in this example is even more complicated than described above, because of the presence of an *elliptic* fixed point of G, namely the origin. The stability of the latter must be studied through the Kolmogorov–Arnold–Moser Theorem.

It is extremely exciting to compare the existence and the multiplicity of profiles, between the viscous approximation and a discretization of the same equation. Typically, they both yield dynamical systems, either discrete or continuous, for which the end points $u^{l,r}$ are hyperbolic points. The dimensions of the stable/unstable manifolds are usually the same in both approaches, because they depend only on properties of the underlying systems around $u^{l,r}$. The general picture is governed by the intersection $W^u(u^l) \cap W^s(u^r)$. When (1.19) holds true, this intersection is usually void in the continuous case (because the intersection may not be of dimension zero) and there does not exist a viscous profile. Even if there exists one, it is not structurally stable and it disappears once we perturb the data $u^{l,r}$ a little bit. On the other hand, in the discrete context, the intersection is not necessarily void, it is generically transverse, hence persistent under small perturbations. There even may happen that both pairs $(W^u(u^l), W^s(u^r))$ and $(W^s(u^l), W^u(u^r))$ intersect transversally. Again, this fact persists after perturbation, and it exhibits an extremely rich chaotic dynamics.

Exponentially Small Splitting

We finish this chapter by mentioning another amazing fact about transverse intersections under condition (1.19). In some cases, G depends on a small parameter (the mesh size Δx in a discretization). Though it does not happen for systems of conservation laws, because of the scale invariance $(x, t) \mapsto (\mu x, \mu t)$, this case is interesting enough to motivate a paragraph to this phenomenon.

Let us consider the reaction-diffusion equation (1.20) with the same assumptions as above. Because of the parity of ϕ, there is a standing wave

---

[21] The heteroclinic orbits do not intersect K.

between u^l and u^r. It is a heteroclinic orbit of (1.21) that is *not* structurally stable. On the other hand, we have seen that the discretization yields structurally stable orbits of (1.23). What is the mechanism under which this structural stability is lost as $\Delta x \to 0^+$? What is the allowable size of disturbances that do not alter the intersection?

We represent the dynamics in terms of u_j and $(u_j - u_{j-1})/\Delta x$. It is thus described by

$$G_{\Delta x}(u, v) = (u + \Delta x(v + \Delta x\, \phi'(u)), v + \Delta x\, \phi'(u)).$$

Let us consider the manifolds $W^u_{\Delta x}(u^l)$ and $W^s_{\Delta x}(u^r)$, which are exchanged under the symmetry $(u, v) \mapsto (-u, v)$. We limit our study to a finite portion of each manifold, between its extremity and the first vertical point $P = (0, v_0)$. This point is common to both manifolds. Its existence is ensured by that of a heteroclinic trajectory with the symmetry $u_{-j} = -u_j$. The intersection being transversal, the tangent spaces at P form a non-zero angle $\theta_{\Delta x}$. As Δx tends to zero, this bounded part of $W^u_{\Delta x}(u^l)$ (respectively $W^s_{\Delta x}(u^r)$) tends to the unstable manifold of u^l (respectively the stable manifold of u^r), associated to the differential equation (1.21). The convergence, which holds at least in the C^1-topology, is due to the stability and the consistency of the forward Euler scheme for ODEs. Since both manifolds coincide at the continuous level, we obtain that

$$\lim_{\Delta x \to 0^+} \theta_{\Delta x} = 0. \tag{1.27}$$

When trying to evaluate the smallness through an asymptotic expansion in powers of Δx, one faces the difficulty that all terms vanish, although $\theta_{\Delta x}$ is non-zero when $\Delta x > 0$. Actually, $\theta_{\Delta x}$ is *exponentially* small in terms of Δx. The leading exponential term may be computed explicitly, provided ϕ is analytic. See the memoir by Fiedler and Scheurle [17].

A drawback of such a smallness is that the structural stability of the heteroclinic connections is extremely weak. The size of disturbances that do not affect their persistence is exponentially small as well. In practice, the monotonous connection from u^l to u^r cannot be observed for small values of Δx, except if ϕ must a priori be even for some natural reason.

1.3 Conclusions

When the parameter $\eta := s\Delta t/\Delta x = r/\ell$ is rational, a DSP for a shock wave $(u^l, u^r; s)$ is a heteroclinic orbit of a dynamical system in $\mathbb{R}^{\ell n}$. For some difference schemes, for instance that of Godunov, this dynamical system is *implicit*, and DSPs must be studied on a case-to-case basis.

Those situations where the dynamics is explicit, of the form

$$(u_{j+1/\ell}, u_{j+2/\ell}, \ldots, u_{j+1}) = G(u_j, u_{j+1/\ell}, \ldots, u_{j+1-1/\ell}),$$

must be treated through the modern dynamical system theory. DSPs are in one-to-one correspondance with the intersection of the stable manifold $W^s(u^r)$

and the unstable manifold $W^u(u^l)$. The dimensions of these manifolds depend both on ℓ and on the number of characteristics that are incoming/outgoing with respect to the shock. Typical cases are the following:

- **Lax shock**. Usually, there holds

$$\dim W^u(u^l) + \dim W^s(u^r) = \ell n + 1.$$

The manifolds are likely to intersect transversally along a curve, in which case the set of DSPs is homeomorphic to the line. This kind of intersection can be proved rigorously for shocks of small amplitude through a bifurcation analysis that involves the Center Manifold Theorem. The existence of a small DSP is equivalent to Liu's E-condition.

- **Undercompressive shock**, with n incoming and as many ougoing characteristics. Usually, there holds

$$\dim W^u(u^l) + \dim W^s(u^r) = \ell n.$$

The manifolds are likely to intersect transversally along a discrete set. The set of DSPs is thus *discrete*. This behaviour contrasts with the viscous approximation, as does the fact that such DSPs persist under small disturbances of the shock data. Considerations about an orientation may be used to prove that some pairs of DSPs are not congruent.

- **Anti-Lax shock**. Usually, there holds

$$\dim W^u(u^l) + \dim W^s(u^r) = \ell n - 1.$$

It is unlikely that the manifolds intersect. That they do for special schemes (Lax–Wendroff) does not mean that the scheme is inconsistent; such a phenommenon should be regarded as similar to the fact that the shock wave solves the conservation laws in the distributional sense. As the entropy condition encodes a kind of instability of the latter, a dynamical instability might prevent observing an anti-Lax DSP.

2 Existence Theory the Irrational Case

The knowledge in the irrational case is by far lighter than in the rational case. This is mainly due to the fact the profile equation (0.7) is not any more a dynamical system in a finite dimensional space[22]. When dealing with a reaction-diffusion equation, the infinite-dimensional context is merely a technical difficulty that may be overcome with the help of functional analysis. We shall develop this aspect in Sect. 3.2. Unfortunately, as far as conservation laws are concerned, the infinite dimension is coupled with a small divisors

---

[22] Remark that the dimension ℓn of the ambiant space in the rational case increases unboundedly as the denominator ℓ tends to infinity.

problem that causes a lack of invertibility of the linearized operator. Again, this uncomfortable situation could be just a technical difficulty that has to be solved through new mathematical arguments. However, this expectation is hopeless, for there are some obstructions[23] to a smooth existence theory of DSPs. Therefore, the only existence result, due to T.-P. Liu and H.-S. Yu [41], is etablished under rather restrictive assumptions.

2.1 Obstructions

The Small Divisors Problem

A general strategy for the existence theory, which is expected to work for shocks of small amplitude, consists in bifurcating from the trivial profile $U = \text{cst} = u^*$. An important step in the procedure is to solve the linearized profile equation

$$Lv = h, \tag{2.1}$$

where

$$\begin{aligned}(Lv)(y) := {} & v(y) - v(y - \eta) + \sigma D_{-p}F^*(v(y - p) - v(y - p + 1)) \\ & + \cdots + \sigma D_{q-1}F^*(v(y + q - 1) - v(y + q)).\end{aligned}$$

Hereabove, D_jF^* denotes the derivative with respect to the j-th argument (with $-p \leq j \leq q - 1$), computed at the diagonal point $(u^*, \ldots, u^*)$.

Of course, we should be happy that L be a Fredholm operator in some Banach space, L^2 or L^∞ for instance. The L^2 case can be studied through a Fourier analysis since L has constant coefficients. One computes easily that

$$L[e^{i\xi y}w] = e^{i\xi y}M(\xi)w,$$

where the matrix M is given by

$$M(\xi) = (1 - e^{-i\eta\xi})I_n + \sigma(1 - e^{i\xi})\sum_{j=-p}^{q-1} e^{ij\xi}D_jF^*.$$

In the rational case $\eta = r/\ell$, $M(\xi)$ runs over a closed compact curve as $\xi \in \mathbb{R}$, which contains the null matrix, since $M(0) = 0_n$. Hence the Fredholm property does not hold (there are matrices of arbitrary small determinant.) However, parameters ξ that do not belong to $\ell\mathbb{Z}$ yield to invertible matrices, under a CFL property and a non-resonnance condition.

The situation is much worse when η is irrational. Since the set of pairs $(e^{-i\eta\xi}, e^{i\xi})$ is dense in $\mathcal{S}^1 \times \mathcal{S}^1$, the set of matrices $M(\xi)$ is dense in the set of matrices

$$M(\alpha, \omega) = (1 - \alpha)I_n + \sigma(1 - \omega)\sum_{j=-p}^{q-1} \omega^j D_jF^*,$$

---

[23] A counter example has been build by Baiti, Bressan and Jenssen for both the Godunov and the Lax-Friedrichs scheme. See [2] and Comm. Pure Appl. Math. 59 (2006), pp 1604–1638.

as α and ω run over $\mathcal{S}^1$. Since $M(1,1) = 0_n$, $M(\xi)^{-1}$ is not uniformly bounded for $|\xi| > \epsilon$ ($\epsilon > 0$), even though each inverse $M(\xi)^{-1}$ other than that of $M(0)$ exists. Therefore, when solving (2.1), we do not expect that the solution v, when it exists, be of the same regularity as the data h. There must hold a *loss of derivatives* in estimates. This loss will be *finite* if the growth of $M(\xi)^{-1}$ is at most algebraic, a fact that corresponds to "Diophantine" values of η (see Sect. 2.2 for this notion.) Roughly speaking, these numbers cannot be approximated too fast by rationals.

The Function Y

Given a DSP, namely a solution of (0.7) with $u(\pm\infty) = u^{l,r}$, we define a function $Y(x; h)$, where x is a point and h a shift:

$$Y(x;h) := \sum_{y \in x+\mathbb{Z}} (u(y+h) - u(y)).$$

The fact that the series converges is not systematic. It certainly does if u is of bounded total variation (we write $u \in BV(\mathbb{R})$). This regularity seems reasonable as it is implied by monotonicity properties. For instance, the monotone schemes for scalar equations produce monotone, hence BV, DSPs according to Jennings and [55]. Also, the viscous shock profiles, which are expected to mimic DSPs since first-order difference schemes have a lot in common with viscous perturbations, have a finite total variation too.

Recall that the domain $\mathcal{D}$ of the DSP is either $\ell^{-1}\mathbb{Z}$ if $\eta = r/\ell$ is rational, or $\mathbb{R}$ if it is irrational. The function Y is therefore defined whenever $x, h \in \mathcal{D}$. The most elementary property of Y is

$$Y(x; h+k) = Y(x+h; k) + Y(x; h). \tag{2.2}$$

The next one comes from ressumation. We have $Y(x; 1) = u^r - u^l =: [u]$ and more generally

$$Y(x; h+1) - Y(x; h) = [u]. \tag{2.3}$$

The last property, which ressembles very much the former, has a different origin since it involves the profile equation itself. Let us compute:

$$Y(x; -\eta) = \sigma \sum_{x+\mathbb{Z}} (F(y-1/2) - F(y+1/2)),$$

with $F(y+1/2) := F(u(y-p+1), \ldots, u(y+q))$. Resummation and consistency give

$$Y(x; -\eta) = \sigma(F(-\infty) - F(+\infty)) = \sigma(f(u^l) - f(u^r)).$$

Using the Rankine–Hugoniot condition, we end with

$$Y(x; -\eta) = -\eta[u]. \tag{2.4}$$

At last, combining (2.2), (2.3) and (2.4, we obtain the following statement.

Proposition 2.1 *Given a DSP of BV class, there holds*

$$Y(x; h) = h[u], \quad \text{for all } h \in \mathbb{Z} + \eta\mathbb{Z}. \tag{2.5}$$

Irrational Case:

When η is irrational, Y coincides on a dense subset with a continuous function. The coincidence extends to the whole line whenever the DSP is regular enough to imply the continuity of Y. This happens whenever the profile is absolutely continuous, since

$$|Y(x; k) - Y(x; h)| = \left| \sum_{x+\mathbb{Z}} (u(y + k) - u(y + h)) \right| \leq \sum_{x+\mathbb{Z}} |u(y + k) - u(y + h)|.$$

Corollary 2.1 *Assume that $\eta \notin \mathbb{Q}$. Given a DSP that is absolutely continuous, there holds*

$$Y(x; h) = h[u], \quad \text{for all } h \in \mathbb{R}. \tag{2.6}$$

Since the right-hand side in (2.5) does not depend on x, we feel free to drop this argument and write simply $Y(h)$. The corollary applies for instance to the scalar DSP associated to monotone schemes, since they are monotone and since one can prove their continuity (see [55]).

Rational Case:

If the shock is small enough and satisfies Liu's E-condition, we expect that the set of DSPs be a connected curve, even for $\eta \in \mathbb{Q}$. When the profile equation is explicit, we proved this fact for small shocks under rather general assumptions ; see for instance Theorem 1.1. A direct analysis yielded an analogous result for the Godunov scheme.

We parametrize the set of DSPs by the map

$$u \mapsto (u(-p), u(-p + 1/\ell), \ldots, u(q - 1/\ell)).$$

In the light of the case $p = q = 1$, $\eta = 0$, one may expect that

$$(u(-p), u(-p + 1/\ell), \ldots, u(q - 1/\ell)) \mapsto u_0$$

is one-to-one. Therefore, we identify this set to a curve in $\mathbb{R}^n$, whose ends are $u^{l,r}$. We then construct a "continuous DSP", by choosing a parametrization of this curve (*shift-invariant* parametrization), such that each section

$$u(x + \cdot) : \ell^{-1}\mathbb{Z} \to \mathbb{R}^n$$

is a DSP in the usual sense. We warn the reader that a continuous DSP is far from being unique, even up to shifts along $\mathbb{R}$. Given any continuous increasing function[24] ϕ that satisfies

---

[24] Such functions are commonly called circle diffeomorphisms.

$$\phi(x + 1/\ell) = \phi(x) + 1/\ell,$$

another continuous DSP is $v := u \circ \phi$.

Given a continuous DSP, the function Y makes sense for every real values of x and h. It satisfies (2.5); but since $\ell^{-1}\mathbb{Z}$ is not dense, that does not imply (2.6). As a matter of fact, since the DSPs corresponding to base points $u(x)$ and $u(y)$ are distinct when $x - y \notin \ell^{-1}\mathbb{Z}$, there is not special reason why (2.6) should hold. For instance, the composition by a circle diffeomorphism transforms Y in the following way:

$$Y_{u \circ \phi}(x; h) = Y_u(\phi(x); \phi(x + h) - \phi(x)). \tag{2.7}$$

Formula (2.7) shows that one can fit (2.6) if, and only if there holds

$$Y(x; h) \parallel [u], \tag{2.8}$$

together with a kind of monotonicity. As a matter of fact, defining $\alpha(x; h)$ by $Y(x; h) = \alpha(x; h)[u]$, it satisfies the same identity (2.2) as Y. That allows to choose a function ϕ in such a way that

$$\alpha(\phi(x); \phi(x + h) - \phi(x)) = h, \quad x, h \in \mathbb{R}.$$

Since two sections of a continuous DSP on sets of the form $x + \ell^{-1}\mathbb{Z}$ are nothing but two basic (genuinely discrete) profiles, property (2.8) amounts to saying that given any two discrete shock profiles v and w of the shock (u^l, u^r), there holds

$$\left(\sum_{\ell^{-1}\mathbb{Z}} (v_r - w_r) \right) \parallel [u] \tag{2.9}$$

The Limit Towards Rational Parameters.

Up to now, our analysis of DSPs concerned a single shock and a fixed mesh ratio σ. At a higher level, we shall consider the dependency of Y upon other data. For instance, the shock being fixed, one might let σ varying continuously so that the parameter η passes from irrational values to rational ones.

If, as an optimistic mathematician should expect, Y varies continuously with respect to σ, property (2.6) extends to rational values of η. According to the previous paragraph, this means that, when $\eta \in \mathbb{Q}$, any two DSPs satisfy (2.9). As mentionned above, there is no special reason why it should be true, except in two special situations described below. A counter-example is provided in the next paragraph. Hence we see a limit of the numerical approximation by conservative finite difference schemes: *In general, the DSPs do not depend in a smooth enough way on either the space variable y or other data like σ.*

Counter-Examples to (2.9)

We display here two counter-examples to (2.9), build from our calculations on the Godunov and the Lax–Wendroff schemes with $\eta = 0$. Hence, in both cases, we deal with standing DSPs.

Godunov Scheme.

The DSPs were constructed in full details in Sect. 1.1. For a k-Lax shock, a steady DSP between u^l and u^r differs from its end states only on one mesh. Up to a translation, one may restrict to DSPs that equal u^l for $j < 0$ and equal u^r for $j > 0$. The intermediate state is any point on the curve $\gamma(u^l, u^r)$. Given to such DSPs v and w, the left-hand side of (2.9) is $v_0 - w_0$. Hence property (2.9) exactly means that $\gamma(u^l, u^r)$ is the straight segment $[u^l, u^r]$.

Of course, there is no special reason why it should be. For instance, for an extreme shock ($k = 1$ or n) of a genuinely non-linear field, $\gamma(u^l, u^r)$ is a piece of a Hugoniot curve, and Hugoniot curves are usually not straight lines. For a 2×2 systems, where every shock is extreme, that $\gamma(u^l, u^r)$ is straight would mean that the system belongs to the rather narrow Temple class. We leave the reader checking that in full gas dynamics ($n = 3$), in which all shocks are extreme since the intermediate field is linearly degenerate, $\gamma(u^l, u^r)$ is never straight, hence (2.9) does not hold. This fact reveals a lack of existence or regularity of Godunov's DSPs as η tends to zero.

Lax–Wendroff Scheme.

We use here the calculations of Sect. 1.1. We consider a Lax shock that satisfies the transversality condition (1.15). Hence, passing through the "compactly supported" DSP (denoted by w, with $w_0 = u^l$ and $w_1 = u^r$), there exists a continuum of DSP that is parametrized by the points a in the intersection

$$W := W^u(u^l) \cap H^{-1}(W^s(u^r)).$$

Let $A_{l,r}$ denote the matrices $\sigma \mathcal{D}f(u^{l,r})$. Since $W^u(u^l)$ is tangent at u^l to $E^u(A_l)$, a point $a \in W$ decomposes in the following way:

$$a - u^l \sim \sum_{i \geq k} \mu_i^l R_i^l, \tag{2.10}$$

where the R_i^l are small (possibly vanishing) eigenvectors of A_l associated to its i-th eigenvalue λ_i^l (hence $0 < \lambda_i^l < 1$), and

$$\mu_i^l := \frac{\lambda_i^l + 1}{\lambda_i^l - 1} < -1.$$

Similarly, its image $b := H(a)$ splits into

$$b - u^r \sim \sum_{i \leq k} \mu_i^r R_i^r, \tag{2.11}$$

where the R_i^r are small eigenvectors of A_r associated to its i-th eigenvalue λ_i^r (hence $-1 < \lambda_i^r < 0$), and

$$\mu_i^r = \frac{\lambda_i^r + 1}{\lambda_i^r - 1} \in (-1, 0).$$

Let us denote by v the profile such that $v_0 = a$ and $v_1 = b$. Then the following formulæ hold:

$$v_j - u^l \sim \sum_{i \geq k} (\mu_i^l)^{j+1} R_i^l, \quad j \leq 0,$$

and

$$v_j - u^r \sim \sum_{i \leq k} (\mu_i^r)^j R_i^r, \quad j \geq 1.$$

Summing up, we obtain (we leave the reader checking that the sum of remainders is negligible as a tends to u^l)

$$\sum_{j \in \mathbb{Z}} (v_j - w_j) \sim \sum_{i \geq k} \frac{\mu_i^l}{\mu_i^l - 1} R_i^l + \sum_{i \leq k} \frac{\mu_i^r}{1 - \mu_i^r} R_i^r.$$

In other words,

$$\sum_{j \in \mathbb{Z}} (v_j - w_j) \sim (A_l - I_n)^{-1}(a - u^l) + (I_n - A_r)^{-1}(b - u^r).$$

On the other hand, b is linked to a by $b = H(a)$, which implies

$$b - u^r \sim \mathcal{D}H(u^l)(a - u^l) = (A_r)^{-1}(A_m - I_n)^{-1}(A_m + I_n)A_l(a - u^l),$$

where we have used (1.13). Therefore we conclude that

$$\sum_{j \in \mathbb{Z}} (v_j - w_j) \sim M_{LW}(a - u^l),$$

where M_{LW} stands for the matrix

$$(A_l - I_n)^{-1} + (I_n - A_r)^{-1}(A_r)^{-1}(A_m - I_n)^{-1}(A_m + I_n)A_l.$$

Remark that M_{LW} is non-singular in general.

When our choice a tends to u^l, the vector $a - u^l$ tends to be parallel to τ, the tangent vector at u^l to $\mathcal{W}$. Hence the direction of $\sum_{j \in \mathbb{Z}} (v_j - w_j)$ tends to that of $M_{LW}\tau$. Remark that τ is explicitly known: It is a generator of the intersection of $E^u(u^l)$, the tangent space to $W^u(u^l)$ at u^l, together with

$$(A_l)^{-1}(A_m + I_n)^{-1}(A_m - I_n)A_r E^s(u^r),$$

the tangent space to $H^{-1}(W^s(u^r))$ at u^r.

Let us explain the form of this counter-example. If $\sum_{j \in \mathbb{Z}} (v_j - w_j)$ was parallel to $[u]$ for every choice of $a \in \mathcal{W}$ close to u^l, then $M_{LW}\tau$ would be parallel to $[u]$ too. However, this introduces a rigid dependence between A_l, A_r and A_m on the one hand, since M_{LW} and τ depend only on these matrices, and the jump $[u]$ on the other hand. It is clear that we may adjust arbitrarily the matrices without changing neither $u^{l,r}$, nor $f(u^{l,r})$, hence keeping the same Lax shock.

Decoupled Systems.

In a recent work [2], Baiti, Bressan and Jenssen construct continuous DSPs for 2×2 systems ($n = 2$) for which the first equation is scalar and the flux in the second one involves both conserved quantities:

$$\partial_t u + \partial_x f(u) = 0, \quad \partial_t v + \partial_x g(u, v) = 0.$$

Considering a Lax shock associated to the velocity $f'(u)$, one knows that the first equation admits a continuous DSP for every value of η (see Section 1.1.) These authors then build a profile for the second equation, where the main issue is the source induced by the profile in the component u. This requires an accurate knowledge of a Green's function, which turns out to be very sensitive to the arithmetical properties of η. From this construction, they deduce that a non-trivial amount of total variation may escape at infinity when η tends to a rational value, thus preventing the function $\eta \mapsto Y(h; \eta)$ from being continuous.

The Lax–Friedrichs Scheme with an Almost Linear Flux

There exist however two situations where (2.9) holds true systematically. The first one is provided by the Lax–Friedrichs scheme when $n - 1$ components of the flux f are affine forms. We may assume that $f_1, \ldots, f_{n-1}$ are these forms, and that they are actually linear. We shall need at last the generic assumption that the forms $f_k - se_k^*$ are linearly independent, s being the shock speed and $e_k^* U$ the k-th component of the vector U. This is a very natural requirement, since the failure of independency would imply that the shock is characteristic on both sides!

From Rankine-Hugoniot condition, we learn that

$$f_k([u]) = [f_k(u)] = [se_k^* u] = se_k^*[u].$$

Hence the jump $[u] = u^r - u^l$ belongs to the intersection of the kernels of the forms $f_k - e_k^*$. Since these are independent, this intersection is a line and $[u]$ turns out to be a generator of this line.

Let us use the integrated form (0.10) of the profile equation. Because of the form of the Lax–Friedrichs scheme, and because of our assumption, the k-th component of F_{LF} is linear. Hence, given two DSPs v and w for the same shock with $\eta \in \mathbb{Q}$, and defining $z := v - w$, and making the difference, we obtain

$$\sum_{j=0}^{r-1} e_k^* z(y - j/\ell) - \sigma \sum_{m=1}^{\ell} F_{LF,k}(\sigma; z(y - p + m/\ell), \ldots) = 0,$$

for every $k = 1, \ldots, n - 1$ and every $y \in \ell^{-1}\mathbb{Z}$. Recall that

$$F_{LF}(a,b) = \frac{1}{2}(f(a) + f(b)) + \frac{a-b}{2\sigma}.$$

Using again the linearity of f_k and summing over $y \in \ell^{-1}\mathbb{Z}$, we derive

$$\ell\sigma f_k\left(\sum_{\ell^{-1}\mathbb{Z}} z(y)\right) = \frac{\sigma}{2}\sum_{\ell^{-1}\mathbb{Z}}\sum_{m=1}^{\ell}(f_k(z(y-1+m/\ell)) + f_k(z(y+m/\ell)))$$

$$= e_k^*\sum_{\ell^{-1}\mathbb{Z}}\left(\sum_{j=0}^{r-1}z(y-j/\ell)\right.$$

$$\left. + \frac{1}{2}\sum_{m=1}^{\ell}(z(y+m/\ell) - z(y-1+m/\ell))\right)$$

$$= re_k^*\sum_{\ell^{-1}\mathbb{Z}} z(y).$$

Since $r/(\ell\sigma) = s$, this amounts to saying that

$$\sum_{\ell^{-1}\mathbb{Z}} z(y) \in \bigcap_{k=1,\ldots,n-1} \ker(f_k - se_k^*).$$

Since the right-hand side is nothing but the line spanned by $[u]$, we deduce that

$$\sum_{\ell^{-1}\mathbb{Z}} z(y) \parallel [u],$$

which exactly means (2.9).

Examples:

There are at least two physically interesting systems with almost linear flux.

- The so-called p-system, where $n = 2$. It reads

$$\partial_t u_1 + \partial_x u_2 = 0, \quad \partial_t u_2 + \partial_x p(u_2) = 0. \tag{2.12}$$

  It is hyperbolic whenever $p' > 0$.
- The Euler equations for full gas dynamics, when the molecules consist of single atoms. Denoting the mass density, the velocity and the internal energy per unit mass by ϱ, z, e, the pressure of a mono-atomic gas is given by

$$p = \frac{2}{d}\varrho e, \tag{2.13}$$

  where d is the space dimension. In the one-dimensional setting, we have[25] $p = 2\varrho e$. The system reads

---

[25] Notice that the one-dimensional monoatomic gas is not the restriction of a three-dimensional one, for which there holds $p = 2\varrho e/3$.

$$\partial_t \varrho + \partial_x(\varrho z) = 0,$$
$$\partial_t(\varrho z) + \partial_x(\varrho z^2 + 2\varrho e) = 0,$$
$$\partial_t\left(\frac{1}{2}\varrho z^2 + \varrho e\right) + \partial_x\left((\frac{1}{2}\varrho z^2 + 3\varrho e)z\right) = 0.$$

Notice that the flux of momentum equals twice the specific total energy.
- The Saint-Venant system for shallow water, which is equivalent to the Euler equations with the law of state (2.13), when identifying the height of water with a mass density.

The Scalar Case

The other favourable case is obviously the scalar one, since two vectors in a one-dimensional space must be parallel. Therefore, the computation that involves the function Y does not suggest any kind of bad behaviour when η varies. Amazingly, the opposite happens, as we shall be able to use Y in order to pass from the rational case to the irrational one. We reproduce here the strategy employed in [55]. This analysis extends that of H. Fan [16], which only concerns the Godunov scheme when the flux f is convex.

Recall that when the scheme is monotone and the flux F is invertible with respect to its extreme arguments, Jennings [31] was able to prove the existence of a monotone "continuous" DSP in the rational case, for any given shock that satisfies the Oleinik admissibility condition[26]. The function Y is monotone in h, and we may normalize the DSP u in such a way that

$$Y(x; h) = h[u].$$

Since u is monotone, the sum that defines Y consists in terms of equal signs. Thus $|Y|$ is an upper bound of the modulus of each term. We deduce that

$$|u(x + h) - u(x)| \le |h[u]|, \quad x, h \in \mathbb{R}. \tag{2.14}$$

In other words, the normalized DSP is Lipschitz, uniformly in the space variable, and locally uniformly in the shock strength. Higher regularity is still an open problem in general, although it follows from dynamical systems theory in the rational case. Notice that in [55], it was also shown that, provided the shock is non-characteristic, its DSP has exponentially converging tails at infinity.

Of course, we may complete the normalization by assuming

$$u(0) = \frac{u^l + u^r}{2}.$$

Given a sequence of rational numbers η_m that tends to an irrational η, and a converging sequence of shocks $(u_m^l, u_m^r; s_m) \to (u^l, u^r; s)$ with $\eta_m = s_m\sigma$, the

---

[26] The Oleinik condition is nothing but Liu's E-condition, written for scalar shocks. However, we keep this terminology because Oleinik's pre-existed to Liu's.

Ascoli–Arzela's Theorem tells us that we may extract from the sequence of DSPs u_m a sub-sequence (still labelled by m, say) that converges uniformly on every compact interval. Let u be the limit, which is monotone. Passing to the limit in (0.7), we see that u satisfies the profile equation with parameter η. Since $\min(u^l, u^r) \leq u(y) \leq \max(u^l, u^r)$, u admits finite limits $u(\pm\infty)$. Thus it is a DSP for the shock $(u(-\infty), u(+\infty); s)$. There remains to check that $u(\pm\infty) = u^{l,r}$. To do this, we apply the integrated form (0.12) of the profile equation to the sequence $(U = u_m, u_m^l, u_m^r; \eta_m)$. Because of uniform convergence on compact intervals, we may pass to the limit and obtain that $(U = u, u^l, u^r; \eta)$ solves (0.12) too. Letting x going to $\pm\infty$ in (0.12), we conclude that $u(\pm\infty) = u^{l,r}$.

The same strategy can be employed to treat a non-strictly monotone scheme, by adding a small amount of strict monotonicity. For instance, we can prove the existence of continuous DSPs under the Godunov's scheme. To summarize this analysis, we have:

Theorem 2.1 *Let n equal one (scalar case) and assume that the difference scheme is monotone (not necessarily strictly) in some interval I. Then every shock $(u^l, u^r; s)$ satisfying the Oleinik condition admits a continuous DSP.*

Explicit DSPs for a Special Flux.

The calculations of this paragraph use a discrete version of the famous Hopf–Cole transformation. This idea is mainly due to P. Lax.

Let us fix the ratio $\sigma = \Delta t / \Delta x$ and the flux

$$f(u) := -\frac{2}{\sigma} \log \cosh \frac{u}{2}. \tag{2.15}$$

We consider the Lax–Friedrichs scheme. The discrete Hopf–Cole transformation is

$$u_j^m = \log \frac{z_{j+1}^m}{z_{j-1}^m}.$$

Under the special flux above, the scheme is tranformed into the linear induction

$$z_j^{m+1} = \frac{z_{j+1}^m + z_{j-1}^m}{2}. \tag{2.16}$$

Given a grid velocity $\eta \neq 0$, there exists a unique positive solution of

$$2a^{-\eta} = a^{-1} + a, \quad a \neq 1. \tag{2.17}$$

With this number, we may construct the traveling wave

$$z_j^m = 1 + a^{j - \eta m}$$

of (2.16). The corresponding traveling wave

$$U(y) = \log \frac{a^{y+1} + 1}{a^{y-1} + 1}$$

is a continuous discrete profile between the states $u = 0$ and $u = 2 \log a$ with grid velocity η. Conversely, given $\bar{u} \neq 0$, and s being given by the Rankine–Hugoniot equation

$$s = \frac{f(\bar{u}) - f(0)}{\bar{u}}, \qquad \eta := s\sigma,$$

then $a := \exp(\bar{u}/2)$ is a solution of (2.17). Whence a continuous DSP for the shock between 0 and $\bar{u}$. Notice that f is strictly concave, so that every discontinuity between a u^l and a u^r satisfies the Oleinik condition if and only if $u^l < u^r$.

Of course, there is no reason to specialize the value $u = 0$ as above. To find an explicit form of every DSPs associated to f, we just make a translation. Given an end point u^*, we use the variable $v := u - u^*$. We remark that $f(u)$ rewrites as

$$f^*(v) = -\frac{2}{\sigma} \log(\alpha e^{-v/2} + \gamma e^{v/2}), \qquad \alpha := \frac{1}{2} e^{-u^*/2}, \quad \gamma := \frac{1}{2} e^{u^*/2}.$$

Then the discrete Hopf–Cole transformation

$$v_j^m = \log \frac{z_{j+1}^m}{z_{j-1}^m}$$

yields the induction

$$(\alpha + \gamma) z_j^{m+1} = \alpha z_{j-1}^m + \gamma z_{j+1}^m. \tag{2.18}$$

Given an other end point $\bar{u} \neq u^*$, and s being given by the Rankine–Hugoniot relation

$$s = \frac{f(\bar{u}) - f(u^*)}{\bar{u} - u^*}, \qquad \eta := s\sigma,$$

there exists a unique positive solution of

$$(\alpha + \gamma) a^{-\eta} = \alpha a^{-1} + \gamma a, \qquad a \neq 1.$$

Then $Z(y) := 1 + a^y$ is a traveling wave solution of (2.18) that yields the continous DSP between u^* and $\bar{u}$

$$U(y) = u^* + \log \frac{a^{y+1} + 1}{a^{y-1} + 1}.$$

Remark that u^* can be either the left or the right state of the shock, so that the Oleinik condition be satisfied. In the former case, one has $a > 1$ while in the latter case $a < 1$.

We end this paragraph with the remark that the DSPs for the special flux (2.15) are C^∞ and even analytic.

2.2 The Approach by Liu and Yu

Since Liu & Yu's work [41] is extremely technical, it is not possible to give here more than a flavour of it. Roughly speaking, Liu and Yu attempted to overcome the small divisor problem by establishing accurate estimates for the linear problem associated to the operator L defined in Section 2.1. As mentionned in Section 2.1, this could become impossible when η is closer and closer to rational values. Hence there is a need of an assumption saying that η is not too well approximable by rational numbers.

Let us say that a real number x is *Diophantine* if there exists a positive constants $C = C(x)$ and $\nu = \nu(x) > 0$ such that

$$\left| x - \frac{r}{\ell} \right| > C \frac{1}{\ell^\nu}, \quad \text{for all } r \in \mathbb{Z}, \text{ for all } \ell \in \mathbb{N}^*, \quad (r, \ell) = 1. \qquad (2.19)$$

The infimum of exponents ν in (2.19) is the *measure of irrationality* of x. For irrational numbers, it cannot be less than two, since every number may be approximated at order two by the convergents of its continued fraction. The irrationality measure of an algebraic number is its degree (Liouville's Theorem.) Therefore algebraic numbers of degree two are the least approximable numbers. Actually, the irrationality measure of Lebesgue-almost every number equals two, and therefore we may suppose that η is such a number. In this respect, quadratic numbers like $\sqrt{2}/2$ or $(\sqrt{5}-1)/2$ are typical. This property also tells that almost every number is Diophantine and therefore it makes sense to prove a theorem under this assumption. The Diophantine property might be difficult to prove for some universal numbers; for instance, the irrationality measure of π is not larger than 8.0161 . . . (see M. Hata [28]). Its exact value is unknown and many upper bounds had been found since K. Mahler's 42 (in [42].) The knowledge is slightly better for $\zeta(3)$, with a bound 5.513891 by Rhin & Viola [48]. On the other hand, numbers that are not Diophantine have been known for a long time. A classical example is due to Liouville:

$$x = \sum_{s=1}^{\infty} 10^{-s!}.$$

The strategy followed by Liu and Yu is of PDE style. What they actually look for are *stable* DSPs. Thus one lets the difference scheme evolve from an initial data that is a presumably good approximation of the DSP. Then one proves that the solution admits a limit as the discrete time goes to infinity. The initial data is provided by a viscous shock profile, since the schemes that are relevant in the theorem are first-order and somehow dissipative. The estimates are first established for the Green's function of the linear operator L through Fourier analysis. A non-resonnance assumption ensures that $M(\xi)$ is non-singular for $\xi \neq 0$, while the Diophantine assumption tells that its inverse cannot grow too fast, hence is a multiplier between some Hilbert spaces. Estimating the Green's function needs a deep understanding of the wave propagation and the wave interaction. The most difficult part is that associated to the *crossing* waves, that is j-waves with $j \neq k$ if the shock is a k-Lax shock. This is the point that requires the Diophantine assumption. The fact that there are no crossing waves when $n = 1$ explains why no restriction is needed in the scalar case. One could wander whether that remains true for system with almost linear fluxes.

To summarize, the assumptions made by Liu & Yu are:

- The system is strictly hyperbolic, each characteristic field being either genuinely nonlinear or linearly degenerate.
- The scheme is dissipative, non-resonant and satisfies the CFL stability condition. Actually, the proof was carried out in the case of the Lax–Friedrichs scheme with a CFL number less than one fourth.
- The parameter η is Diophantine. Actually, the proof was carried out for $\eta = 1/\sqrt{8}$. As mentionned above, this value is typical.

Its conclusion is that, whenever a Lax shock $(u^l, u^r; s = \eta \Delta x/\Delta t)$ is weak enough, there exists a unique continuous DSP for that shock. Of course, the allowable shock strength depends on the constants $C(\eta)$ and $\nu(\eta)$. It deteriorates as $C(\eta)$ tends to zero or $\nu(\eta)$ tends to infinity, since the result cannot hold uniformly near rational parameters.

3 Semi-Discrete *vs* Discrete Traveling Waves

When discretizing an evolutionary PDE, the time unit Δt is constrained by the mesh length Δx. In first-order systems of conservation laws, the CFL condition tells that $\Delta t = \mathcal{O}(\Delta x)$, while in parabolic PDEs, we need $\Delta t = \mathcal{O}(\Delta x^2)$. In both cases, we may let Δt tend to zero while keeping Δx fixed. In the limit, we recover a *semi-discretization* of the PDE. For instance,

- For the reaction-diffusion equation (1.20), we consider

$$\frac{du_j}{dt} = \frac{u_{j+1} - 2u_j + u_{j+1}}{\Delta x^2} - \phi'(u_j), \quad j \in \mathbb{Z}, t > 0. \tag{3.1}$$

- For the system of conservation laws (0.1), we may have

$$\frac{du_j}{dt} = \frac{F_0(u_{j-1}, u_j) - F_0(u_j, u_{j+1})}{\Delta x}, \quad j \in \mathbb{Z}, t > 0, \tag{3.2}$$

where F_0 is the limit as $\sigma \to 0$ of a consistent flux. In particular, it satisfies $F(a, a) \equiv f(a)$. Interesting example are that of Godunov and Lax–Wendroff, but we point out that the Lax–Friedrichs scheme is not relevant since its flux is singular at $\sigma = 0$.

More generally, we write a semi-discretization in the abstract form

$$\frac{du}{dt} = G(u), \tag{3.3}$$

where G commutes with the shift operator S,

$$(Su)_j = u_{j-1}.$$

A system like (3.3) is called in the litterature a *Lattice Dynamical System*. It is an ODE in some Banach space, for instance $\mathcal{X} := \ell^\infty(\mathbb{Z}; \mathbb{R}^n)$. Its discrete counterpart

$$u^{m+1} = u^m + hG(u^m), \quad m \in \mathbb{N} \tag{3.4}$$

is called a *Coupled Map Lattice*. It is a full discretization of the underlying PDE.

3.1 Semi-Discrete Profiles

The notion of semi-discrete traveling waves[27] mimics that in the fully discrete case. It is a solution of (3.3) of the form

$$u_j(t) = \phi(j - ct).$$

The function ϕ is the profile. It satisfies the functional difference equation

$$c\frac{d\phi}{dy} + G(\phi)(y) = 0. \tag{3.5}$$

It usually has prescribed values at $\pm\infty$. We warn the reader that the left-hand side of (3.5) is not a function of $\phi(y)$ alone, but involves other values like $\phi(y \pm 1)$. Hence (3.5) is certainly not a differential equation in $\mathcal{X}$. The number c is the numerical velocity. It may be explicit, thus independent of the semi-discretization, in the case of shocks, because of the conservativity (we obtain $c = s/\Delta x$), but it is an unknown in the case of reaction-diffusion equations.

We shall not treat the difficult question of the existence and stability of semi-discrete traveling waves, since this course is devoted to the fully discrete case. As far as shock waves are concerned, we refer to the series of works by S. Benzoni and co-authors [4, 5, 6, 7, 8, 9]. The reaction-diffusion case has been considered first by B. Zinner [59, 26], and also by J. Mallet-Paret [44, 45]. Let us only say that the profile equation is transformed into a PDE in a space of the form $\mathcal{X}_1 = \mathcal{C}([0, 1]; \mathbb{R}^n)$. For weak waves, the existence may be attacked through a bifurcation analysis. This requires a kind of Center Manifold Theorem, but for a dynamical system that is not an ODE.

3.2 A Strategy Towards Fully Discrete Traveling Waves

Acknowledgement. Most of this section is taken from an elegant and concise presentation by S. Benzoni of the work of Chow & al. [12]. I thank her warmly for having given to me her careful notes.

---

[27] Semi-discrete profiles (SDSP) in case of shocks.

As seen above, there is a fundamental difficulty in the search of DSPs when η is close to rationals. On the other hand, there is no dimensionless parameter in the semi-discrete case, and therefore there is no restriction involving arithmetical properties. Whence the following idea. Given an unsteady SDSP, with suitable stability properties, prove that the traveling wave persists after full discretization, provided that the time mesh h is small enough. If so, there would exist DSPs for values of the parameter η in some interval. In particular, such a result would cover values that have bad arithmetical properties, contrary to Liu & Yu's Theorem.

More generally, we consider an unsteady traveling front p of a LDS, since p would immediately be a traveling wave of (3.4) if c vanished. Without loss of generality, we assume that $c > 0$. Introducing the "return time" $T = 1/c$, the property that p is a traveling wave is characterized by

$$p(t + T) = Sp(t), \tag{3.6}$$

and the profile is given by

$$\phi(y) := p_0(-yT).$$

Assume that G is differentiable enough and consider the linearized system around p:

$$\frac{du}{dt} = DG(p(t))u(t). \tag{3.7}$$

Denoting by $A(t; t_0)$ the solution operator of (3.7), we immediately have

$$A(T; 0)\dot{\phi} = S\dot{\phi}.$$

in other words, $\dot{\phi}$ is an eigenvector and $\zeta = 1$ is an eigenvalue of $R :=$ $S^{-1}A(T; 0)$. From $A(t + T; t)S = SA(t; t - T)$, we find

$$R^m = S^{-m}A(mT; 0).$$

Since S is an isometry in $\mathcal{X}$, this shows that the spectral properties of R encode the spectral stability of p. Whence the

Definition 3.1 *The traveling wave p is said* uniformly spectrally stable *if*

- *the spectrum of R lies in $\{\zeta \in \mathbb{C}; |\zeta| < 1\} \cup \{1\}$,*
- *the eigenvalue $\zeta = 1$ is simple and isolated in the spectrum of R.*

We warn the reader that there is no hope that a SDSP be uniformly spectrally stable. As a matter of fact, given the end values $u^{l,r} = \phi(\pm\infty)$, the essential spectrum of R is the union of the operators $R_{l,r}$ that correspond to the constant waves $p^{l,r} \equiv u^{l,r}$. Let us compute the latter for a LDS like (3.2). Rescaling the time variable, we may assume that Δx is the unit. The linearization around the constant state u^l yields the system

$$\frac{du_j}{dt} = \mathcal{D}_a F^l(u_{j-1} - u_j) + \mathcal{D}_b F^l(u_j - u_{j+1}).$$

This can be solved explicitly with the help of the discrete Fourier transform:

$$\frac{dv}{dt} = N(\xi)v, \quad \xi \in \mathbb{R}/2\pi\mathbb{Z},$$

with

$$N(\xi) := (e^{-i\xi} - 1)\mathcal{D}_a F^l + (1 - e^{i\xi})\mathcal{D}_b F^l.$$

In Fourier variable, the operator R_l is the multiplier

$$\hat{R}_l(\xi) = e^{-i\xi} e^{TN(\xi)}.$$

Since $R(\xi) = I_n$ and R is a continous function of ξ, we see that the essential spectrum reaches $\zeta = 1$.

The situation is much different in reaction-diffusion. The end states $u^{l,r}$ are stable equilibria, and the linearized system about $u \equiv u^l$, for instance, is strongly dissipative. We leave the reader to compute the essential spectrum of $R_{l,r}$ by the same method as above and prove that it is a compact subset of $\{\zeta \in \mathbb{C}; |\zeta| < 1\}$. This explains why the method developped by Chow & al. has been successful for reaction-diffusion equations but not for systems of conservation laws, so far. Their result writes as follows.

Theorem 3.1 (Chow, Mallet–Paret, Shen) *Assume that $p(t)$ is a uniformly spectrally stable traveling wave solution of the LDS (3.3) with speed $c \neq 0$, such that*

$$\lim_{t \to \pm\infty} \inf \|p(t) - p(0)\| > 0. \tag{3.8}$$

Then there exists a positive h_0 so that, for every $h \in (0, h_0)$, there exists a smooth one-dimensional manifold M_h, close to $M := \{p(\theta); \theta \in \mathbb{R}\}$ in $\mathcal{X}$, which is invariant under the CML (3.4). Moreover, M_h contains traveling wave solutions of (3.4), of speed ϱ_h close to ch.

Comment:

The condition (3.8) tells that the profile ϕ is not quasi-periodic or periodic. In particular, heteroclinic waves do satisfy it.

3.3 Sketch of Proof of Theorem 3.1

Let us take the quotient of $\mathcal{X}$ by the shift group

$$\mathcal{G} := \{S^j \, ; \, j \in \mathbb{Z}\}.$$

We warn the reader that the action of $\mathcal{G}$ is not faithful on $\mathcal{X}$. For instance, $\mathcal{G}$ has torsion on every periodic sequence. Therefore the resulting space $\mathcal{V}$ is not globally a Banach manifold. However, it is locally a Banach manifold near each point that is not periodic. When $q \in \mathcal{X}$ we shall write $\dot{q} \in \mathcal{V}$ for the class of q.

Since both the CML and the LDS commute with the shift, they define a discrete and a continuous dynamical systems on $\mathcal{V}$. Because of (3.6), $\dot{p}$ is a periodic solution of the latter. Denote by V_0 the curve described by $\dot{p}$. It is a closed compact curve along which $\mathcal{V}$ is a Banach manifold, thanks to Assumption (3.8).

Let us denote by $(E(t))_{t \geq 0}$ the solution operator of the LDS in $\mathcal{V}$. The manifold V_0 is invariant under E, and $E(T)$ acts trivially on V_0. The stability assumption tells actually that V_0 is a uniformly attracting cycle. In particular, it is *transversally hyperbolic*.

Similarly, let us denote by $(\Gamma_h^m)_{m \in \mathbb{N}}$ the solution operator of the CML on $\mathcal{V}$. Since the CML is a backward Euler discretization of the LDS, standard estimates in Numerical Analysis give

$$\Gamma_h^N - E(T) = \mathcal{O}(h),$$

where N is the integral part of T/h. Using now the persistence of normally hyperbolic invariant manifolds (see Bates & al. [3] for an infinite dimensional setting), we obtain that Γ_h^N possesses a unique attracting cycle V_h, close to V_0, provided that $h > 0$ is small enough. From uniqueness, and the fact that $\Gamma_h V_h$ is invariant under Γ_h^N and still close to V_0, we see that V_h is actually invariant under Γ_h.

Being close to V_0, the closed curve V_h can be parametrized by some parameter $\theta \in \mathbb{R}$, in a T-periodic way. The dynamics induced by Γ_h on V_h can be coded by a one-to-one map

$$\theta \mapsto \beta_h(\theta)$$

that satisfies $\beta_h(\theta + T) = \beta_h(\theta) + T$. Therefore, β_h is a circle diffeomorphism:

$$\beta_h : \mathbb{R}/T\mathbb{Z} \to \mathbb{R}/T\mathbb{Z}.$$

The map β_h is of class $\mathcal{C}^{r-1}$ if G is of class $\mathcal{C}^r$, and is close to the identity. Actually, a first order Taylor expansion gives

$$\beta_h(\theta) = \theta + h + \mathcal{O}(h^2). \tag{3.9}$$

The parametrization of V_h is a bit arbitrary. In practice, θ is the privileged coordinate of the transformation mentionned above. However, it has the bad feature that β_h is not a "rotation" in general. Actually, if β_h was a rotation $\theta \mapsto \theta + \varrho$, one would have immediately our discrete traveling wave, with a velocity $\varrho T/h$, and θ would stand for the time variable. Hence the last step consists in finding an alternate parametrization. To this end, we recall the notion, due to Poincaré, of the "rotation number" of a circle diffeomorphism β on $\mathbb{R}/T\mathbb{Z}$:

$$\varrho(\beta) := \frac{1}{T} \lim_{m \to +\infty} \frac{\beta^m(\theta)}{m}, \tag{3.10}$$

where the limit is uniform in θ. The rotation number is invariant under conjugacy. We simply denote in the sequel $\varrho_h := \varrho(\beta_h)$. Because of (3.9), there holds

$$\varrho_h \sim ch. \tag{3.11}$$

In other words, $h \mapsto \varrho_h$ is differentiable at the origin, with

$$\frac{d\varrho_h}{dh}(0) = c.$$

The Rational Case.

We recall that $\varrho(\beta)$ is rational if, and only if, β admits a finite orbit. In such a case, the length ℓ of the finite orbits of β is the denominator of $\varrho(\beta)$, written in reduced form r/ℓ. Assume now that $\varrho_h = r/\ell$ is rational. Let $\theta_0, \beta_h(\theta_0), \ldots$ be such an orbit. Then the points of V_h, parametrized by $\theta_0, \beta_h(\theta_0), \ldots$ form a periodic orbit of length l of the CML on $\mathcal{V}$. When lifted to $\mathcal{X}$, this gives rise to a traveling wave that is shifted by r meshes in ℓ times steps. Its grid velocity is therefore ϱ_h, while its space-time velocity is $s_h := \varrho_h/\sigma = \varrho_h(\Delta x)/h$. From (3.11), one has $s_h \sim c\Delta x$, the latter quantity being the space-time velocity of the semi-discrete front. We warn the reader that this periodic orbit defines a traveling front that is genuinely discrete. We also point out that in general β_h is not conjugated to the rotation $\theta \mapsto \theta + \varrho_h T$ since this would mean that *every* orbit on V_h be periodic. Hence there does not exist in general a "continuous" traveling front, since it would have to cover the whole loop V_h. See below for a more complete description.

The Irrational Case.

When $\varrho_h \notin \mathbb{Q}$, the situation changes completely since then β_h must be conjugated to the rotation $\tau \mapsto \tau + \varrho_h T$. In other words, there exists a homeomorphism Λ_h of $\mathbb{R}/T\mathbb{Z}$ such that

$$\beta_h(\Lambda_h(\tau)) = \Lambda_h(\tau + \varrho_h T),$$

provided that β_h is of class $\mathcal{C}^2$. Hence G of class $\mathcal{C}^3$ is enough. This is Denjoy's Theorem, for which we refer to Arnold's book [1]. This means that there is a parametrization of V_h for which the dynamics is simply the rotation $\tau \mapsto \tau + \varrho_h T$. Lifting back to the original variables, we obtain the manifold M_h that is invariant under the CML (3.4). The above analysis provides a parametrization of M_h for which one time step corresponds to a grid shift by ϱ_h. Whence a traveling wave of space-time velocity $s_h = \varrho_h/\sigma$. We warn the reader that Λ_h is not necessarily a diffeomorphism and hence the wave profile may not be of class $\mathcal{C}^1$.

The Richness of Discrete Dynamics

What we described above is the emerging part of the iceberg. One may ask a lot of questions about the dynamics. What is the most likely situation? What happens to other trajectories on V_h when ϱ_h is rational? When is the wave profile differentiable?...

First of all, let us kill a naive belief. Although the rationals are rare in the reals because of countability, it is likely that the rotation number be rational! As a matter of fact, if ϱ_h is rational, then there is a fixed point θ_0 for some power β_h^ℓ. Generically (in the C^1-topology), this fixed point is hyperbolic (one says that the corresponding periodic orbit is hyperbolic), meaning that

$$\frac{d\beta_h^\ell}{d\theta}(\theta_0) \neq 1.$$

Such a fixed point is either an attractor or a repellor. It persists under a small C^0-perturbation of β_h, in particular under a small change of h. Since the rationals are countable, one obtains that for generic one-parameter families of circle diffeomorphisms, for instance $h \mapsto \beta_h$, the preimage of each rational number under ϱ_h is a non-void interval. On the other hand, the preimage of an irrational number is usually a single point. Hence the graph of the continuous map $h \mapsto \varrho_h$ is a *devil stair-case*: The set of numbers h such that ϱ_h is irrational is a Cantor set, whose complement is dense.

This could make us thinking that the irrational situation is rare, and especially that the Lebesgue measure has no role in the theory. However, M. Herman proved that in such generic one-parameter families, the Cantor set

$$\{h \in (0, h_0); \, \varrho_h \in \mathbb{R} \setminus \mathbb{Q}\}$$

is of *positive* Lebesgue measure. This measure can even be pretty large, since it approaches the total length (here h_0) as the family of circle diffeomorphisms approaches that of rotations.

These observations provide a two-fold picture. On the one hand, ϱ_h is likely to be rational, since this is the case for time meshes h running over a dense subset. More precisely, the diffeomorphism admits hyperbolic periodic orbits for every h in a dense open subset whose complement is a Cantor set. Then there are only finitely many discrete traveling waves and no continuous traveling wave. On the other hand, ϱ_h is likely to be irrational, since this happens for time meshes h running over a set of positive Lebesgue measure. In the latter situation, there is a unique traveling wave, up to a shift, and it is a continuous one, instead of a discrete one.

Remarks:

- In the picture above, the pinning phenomenon happens on a set of rather small Lebesgue measure, because large powers of β_h are close to rotations (remember that the measure of the rationality set tends to zero as a family of circle diffeomorphisms approaches that of rotations.) As a matter of fact, given a large integer $L \gg 1$ and $h \in (1/L, 2/L)$, then $\|\Gamma_h^L - E(hL)\|_\infty = \mathcal{O}(h)$, showing that $\|\beta_h^L - R_{chL}\|_\infty = \mathcal{O}(h)$, where R_θ denotes the rotation of angle θ. This implies

$$|\varrho(\beta_h^L) - chL| = \mathcal{O}(h), \tag{3.12}$$

and therefore

$$|\varrho(\beta_h) - ch| = \mathcal{O}(h^2),$$

confirming (3.11). On another hand, (3.12) tells that the family $h \mapsto \beta_h^L$ (where $h \in (1/L, 2/L)$) is close to the rotations, and therefore its rationality set is likely to be Lebesgue-small. This is exactly saying that the rationality set of $h \mapsto \beta_h$ be Lebesgue-small.

- Since β_h is smooth, the order of regularity of the profile is that of the conjugation Λ_h. However, when ϱ_h is irrational, it is not always true that Λ_h be differentiable. General results in this direction depend on arithmetical properties of ϱ_h. On the one hand, in a class like $(\beta_h)_{h \in (0, h_0)}$, there must exist some parameters h for which ϱ_h is irrational but Λ_h is not differentiable, thanks to an argument of Herman. On the other hand, every $\mathcal{C}^\infty$-circle diffeomorphism with a *Diophantine* rotation number (see (2.19)) is conjugated to a rotation through a $\mathcal{C}^\infty$-diffeomorphism. This provides a sufficient condition for the wave profile to be $\mathcal{C}^\infty$. This result is sharp in the sense that, for non-Diophantine rotation numbers, the conjugation is generically less regular.

- We finish with a remark in the rational case. Generically, the periodic orbits of β_h are hyperbolic; half of them are attracting and the other ones are repelling. Both types alternate. Every other trajectory on V_h is asymptotic to an unstable periodic orbit in the past, and to a stable one in the future. At the level of the CML, this means that besides the unstable and stable discrete traveling waves, there also exist solutions of the CML, defined for all time steps $m \in \mathbb{Z}$, which tend to an unstable traveling wave as $m \to -\infty$, and to a stable one as $m \to +\infty$.

- There does not seem that Chow & al. [12] analyse the likeliness of rationality or irrationality of ϱ_h. Of course, all the comments made above need that some genericity be checked, a task that is always difficult.

What for DSPs now?

Because of the lack of stability in the sense of Definition 3.1, one does not know how to prove the existence of an invariant curve V_h. S. Benzoni, P. Huot & F. Rousset [9] derived an sharp result of linear stability, which is much weaker than uniform stability. One has not been able to use it yet in the present problem. If it existed, the pair (V_h, β_h) would have rather strange properties, hence would be highly non-generic. First of all, recall that when η ($\eta := sh/\Delta x$, s the shock speed) is rational and a Lax shock is small enough, then under reasonable assumptions, there exists a one-parameter family of discrete DSPs (Majda–Ralston Theorem.) Hence for rational rotation numbers, β_h would be conjugated to a rotation! This fact allows that ϱ_h be non-constant, even at rational points. Hence the devil stair-case picture does not hold any more. This is compatible with another observation: Because of the Rankine–Hugoniot and calculations of Section 2.1, the rotation number of β_h must be linear with respect to h.

The conclusion is that, despite an apparent similarity between the discretization of both types of PDEs, namely reaction-diffusion equations and systems of conservation laws, the expected behaviour of the discrete fronts are so much different that it is hopeless to mimic the analysis of the former to solve the latter. Remark however that not all difference schemes admit a limit as Δt tends to zero (for instance, the Lax–Friedrichs scheme does not), and therefore they cannot even be compared with a semi-discretization.

4 Stability Analysis: The Evans Function

The stability analysis of shock profiles is a fundamental issue. As a matter of fact, we shall see in a moment that an instability manifests itself on a very short time interval, of the order of Δt (discrete shock profiles) or of the viscosity parameter (viscous shock profiles.) Hence infinitesimal disturbances destroy the structure of the profile before it could have been observed. Therefore the only meaningful shock profiles are the stable ones.

Since the situation in the irrational case is covered by Liu–Yu's Theorem, we shall focus on DSPs with a rational parameter from now on. Say that $\eta = r/\ell$. Before defining our stability notion, we make the following observation. Iterating ℓ times the scheme (0.3), we obtain a formula of the same kind:

$$u_j^{m+\ell} = u_j^m + \sigma(\hat{F}_{j-1/2}^m - \hat{F}_{j+1/2}^m), \tag{4.1}$$

which is still conservative and involves $\ell(p+q+1)$ points in general. We rewrite (4.1) as a first order scheme of the form (0.3) by selecting $v^m := u^{\ell m}$. The new scheme is still consistent with (0.1). It has the advantage that it admits a DSP $\bar{v}$ such that $\bar{v}^m = S^{rm}\bar{v}^0$. Last, we consider the following modification:

$$w_j^{m+1} = w_{j-r}^m + \sigma(\hat{F}_{j-r-1/2}^m - \hat{F}_{j-r+1/2}^m). \tag{4.2}$$

The scheme (4.2) is consistent with the system

$$\partial_t u + \partial_x(f(u) - su) = 0.$$

It admits the standing DSP $\bar{w} := S^{-rm}\bar{v}^m$. Therefore we always may restrict to the case of a standing shock associated to a standing DSP. However, we have to pay a price by an increase of the number of variables in the numerical flux.

4.1 Spectral Stability

From now on, assume that (u^l, u^r) is a standing shock of (0.1), and that $\bar{u}$ is a standing DSP, that is a steady solution of (0.3) with $u(\pm\infty) = u^{l,r}$. The scheme is a $(p + q + 1)$- points one. Let

$$v^{m+1} = Lv^m \qquad (4.3)$$

be the linearization of (0.3) about $\bar{u}$. This is a linear system in $\mathcal{X} := \ell^2(\mathbb{Z})$. The linear operator L is bounded. The value $(Lv)_j$ involves the variables $v_{j-p}, \ldots, v_{j+q}$. The coefficients of L are not constant since they involve the values $\bar{u}_j$ of the profile. However, since $\bar{u}_j$ tends to constant values at infinities, we see that the coefficients of L are asymptotically constant. Hence L behaves in the far field as the constant coefficients operators $L_{l,r}$ obtained by linearizing (0.3) about u^l and u^r. The latter can be analyzed through a discrete Fourier transform.

Definition 4.1 *We say that $\bar{u}$ is a stable DSP if the spectrum of the operator L satisfies the following properties:*

- *It is contained in the set*

$$\{\zeta \in \mathbb{C}; |\zeta| < 1\} \cup \{1\},$$

- *If (u^l, u^r) is a Lax shock, $\zeta = 1$ is a simple eigenvalue, meaning that it is a simple root of the Evans function (see Sect. 4.3 for its construction),*
- *If (u^l, u^r) is an undercompressive shock, $\zeta = 1$ is not an eigenvalue, meaning that it is not a zero of the Evans function (see Sect. 4.3 for its construction).*

For an overcompressive shock wave (the number N of incoming characteristics is larger than $n + 1$), one should ask that $\zeta = 1$ is a zero of order $N - n$.

If the space $\mathcal{X}$ was finite-dimensional, a convenient tool for the study of the spectrum of the endomorphism L would be its characteristic polynomial. The *Evans function* plays more or less (and more less than more) the role of the characteristic polynomial in infinite dimension. As the latter, it is a holomorphic function. Of course, it cannot be a polynomial since we expect an infinite spectrum. It has three flaws:

- The Evans function is defined only on an open subset of the complex plane. As a matter of fact, one has not been able to define it within the essential spectrum[28],
- There is a lot of arbitrariness in the construction of the Evans function. Hence it is defined only up to the multiplication by a non-vanishing holomorphic function. Hence, the only intrinsic contents are its zeroes and their multiplicities,

---

[28] However, scattering techniques may handle some features of the essential spectrum. But both techniques remain separated so far.

- It is far from being explicit.

Consequently, the Evans function has been exploited so far through very robust analytic tools: Intermediate Value Theorem for continuous real functions (see [GZ, 11] and below), Rouché's Theorem of persistence of zeroes of holomorphic functions (see for instance [ZS]).

4.2 The Essential Spectrum of L

Recall that a complex number ζ lies in the *resolvant set* $\rho(L)$ if $L - \zeta$ is an isomorphism from $\mathcal{X}$ into itself. It is a spectral value otherwise. The set of spectral values is the *spectrum* of L and is denoted by $\sigma(L)$. When the kernel of $L - \zeta$ is non-trivial, ζ is an *eigenvalue* ; in particular, it is a spectral value. The essential spectrum $\sigma_{ess}(L)$ consists in those ζ such that $L - \zeta$ is not a Fredholm operator of index zero. By Fredholm theory, $\sigma_{ess}(L)$ is closed and $\rho(L)$ is open. The object of this paragraph is to show that, under natural assumptions about the numerical scheme, $\sigma_{ess}(L)$ is contained in the subset

$$\{\zeta \in \mathbb{C};\ |\zeta| < 1\} \cup \{1\},$$

and therefore, the unstable spectral values are either $\zeta = 1$ or isolated eigenvalues of finite multiplicities.

We thus consider the linear system

$$Lv = \zeta v + h, \tag{4.4}$$

where $h \in \mathcal{X}$ is given and v is searched in $\mathcal{X}$. For the sake of simplicity, we assume a three-points scheme. We have

$$(Lv)_j = (1 - \sigma \mathcal{D}_a F_{j+1/2} + \sigma \mathcal{D}_b F_{j-1/2})v_j + \sigma \mathcal{D}_a F_{j-1/2}v_{j-1} - \sigma \mathcal{D}_b F_{j+1/2}v_{j+1}, \tag{4.5}$$

where the subscript $j + 1/2$ means that the argument is the pair $(\bar{u}_j, \bar{u}_{j+1})$. Then (4.4) reads

$$(1 - \zeta - \sigma \mathcal{D}_a F_{j+1/2} + \sigma \mathcal{D}_b F_{j-1/2})v_j + \sigma \mathcal{D}_a F_{j-1/2}v_{j-1} - \sigma \mathcal{D}_b F_{j+1/2}v_{j+1} = h_j.$$

For the sake of simplicity, we assume that the matrices $\mathcal{D}_a F$ and $\mathcal{D}_b F$ are non-singular, otherwise the computations are a bit different, although they follow similar guidelines. This rules out the case of the Godunov's scheme, for which we refer to Bultelle & coll. [11]. Thanks to our assumption, we may rewrite this equation in matrix form

$$\begin{pmatrix} v_j \\ v_{j+1} \end{pmatrix} = A_j(\zeta) \begin{pmatrix} v_{j-1} \\ v_j \end{pmatrix} + B_j(\zeta)h_j. \tag{4.6}$$

The matrix $A_j(\zeta)$ is $2n \times 2n$, while $B_j(\zeta)$ is $2n \times n$. They depend holomorphically on ζ and have real entries when ζ is real. As j tends to $\pm\infty$, they admit limits $A^{\pm}(\zeta)$. The latter are associated to the linearized operators $L^{l,r}$ around the constant states $u^{l,r}$.

A basic assumption is that both u^l and u^r are stable under the difference scheme. In mathematical words, this means the following. Since $L^{l,r}$ have constant coefficients, they may be analyzed through the Fourier transform

$$\hat{v}(\xi) := \sum_{j \in \mathbb{Z}} e^{-ij\xi} v_j, \quad \xi \in \mathbb{R}/2\pi\mathbb{Z}.$$

There holds

$$\widehat{L^r v}(\xi) = K^r(\xi)\hat{v}(\xi),$$

where $K^r(\xi)$ is a matrix whose entries are trigonometric polynomials. The spectrum of L^r is purely essential and is the union of the spectra of the matrices $K^r(\xi)$ as ξ runs over the real line. Because of the consistency, there holds $K^r(0) = I_n$. The L^2-stability amounts to saying that there exists a positive[29] number θ such that, for every eigenvalue $\lambda(\xi)$ of $K^{l,r}(\xi)$, we have

$$|\lambda(\xi)| \leq 1 - \theta \sin^2 \frac{\xi}{2}. \tag{4.7}$$

In particular, the spectra of L^l and L^r are contained in some compact disk D, contained in

$$\{\zeta \in \mathbb{C}; |\zeta| < 1\} \cup \{1\}$$

and tangent to the latter at $\zeta = 1$.

Remark.

For the Lax–Friedrichs scheme, the spectra of $L^{l,r}$ also contain the point $\zeta = -1$ on the unit circle. This is due to the fact that, since u_j^{m+1} depends only on u_{j-1}^m and u_{j+1}^m, the grid $\mathbb{N} \times \mathbb{Z}$ is really the union of two independent grids, each one defined by the parity of $m+j$. We may avoid this spurious spectrum by iterating once more the scheme, thus by working on the rectangular subgrid $2\mathbb{N} \times 2\mathbb{Z}$. In many papers, this phenomenon is considered as a resonnance and authors prefer to eliminate it by replacing the average

$$\frac{1}{2}(u_{j-1}^m + u_{j+1}^m)$$

by

$$\frac{1-\nu}{2}(u_{j-1}^m + u_{j+1}^m) + \nu u_j^m, \quad \nu \in (0,1).$$

This choice yields the so-called "modified" Lax–Friedrichs scheme.

Let us denote by Ω the outer connected component of $\mathbb{C} \setminus D$. In particular every unstable ζ, that is $|\zeta| > 1$, belong to Ω. The stability property has

---

[29] The positivity of θ tells that the scheme displays some viscosity.

the effect that for every $\zeta \in \Omega$, the spectra of the matrices $A^{\pm}(\zeta)$ do not intersect the unit circle ; in other words, $A^{\pm}(\zeta)$ are hyperbolic matrices. With a continuity argument, we derive that the number of unstable[30] (respectively stable) eigenvalues of $A^{\pm}(\zeta)$ are constant as ζ runs over Ω. These numbers determine the "type" of $A^{\pm}(\zeta)$. The type is usually computed by letting ζ tend to infinity. There are several natural arguments in favour of the fact that the types of $A^{-}(\zeta)$ and of $A^{+}(\zeta)$ are equal. For instance, a continuum of states, including $u^{l,r}$, being stable. Or the fact that, along the diagonal $a = b$, the matrix $\sigma \mathcal{D}_a F$ (respectively $\sigma \mathcal{D}_b F$) is a polynomial in $\sigma \mathcal{D} f$, this polynomial mapping $(-1, 1)$ into $(0, +\infty)$ (resp. $(-\infty, 0)$.) The latter situation, which yields the type (n, n), occurs for instance in Lax–Friedrichs and Lax–Wendroff schemes.

For the sake of simplicity, we adopt the latter situation. Then the set of solutions of the homogeneous iteration

$$\begin{pmatrix} v_j \\ v_{j+1} \end{pmatrix} = A^{-}(\zeta) \begin{pmatrix} v_{j-1} \\ v_j \end{pmatrix}, \quad j \in \mathbb{Z} \tag{4.8}$$

that tend to zero as $j \to -\infty$ is an n-dimensional vector space, denoted by $\mathcal{E}^{-}(\zeta)$ and called the *unstable subspace* of $A^{-}(\zeta)$. It corresponds to the unstable eigenvalues of $A^{-}(\zeta)$. Similarly, there is the *stable subspace* of $A^{+}(\zeta)$, denoted by $\mathcal{E}^{+}(\zeta)$, with the property that v_j tends to zero as $j \to +\infty$. This constant coefficients situation is very simple, but we really are interested in the case where the matrices $A_j(\zeta)$ do depend on j. Since $A_j(\zeta)$ is asymptotic to $A^{\pm}(\zeta)$, the description above persists: The subspaces of solutions of the homogeneous induction

$$\begin{pmatrix} v_j \\ v_{j+1} \end{pmatrix} = A_j(\zeta) \begin{pmatrix} v_{j-1} \\ v_j \end{pmatrix} \tag{4.9}$$

that tend to zero either as $j \to -\infty$, or as $j \to +\infty$, are n-dimensional. We denote their traces at $j = 0$ by $E^{-}(\zeta)$ and $E^{+}(\zeta)$ respectively. We point out that the solutions that decay to zero at some infinity actually decay exponentially fast and therefore are square-summable.

It is clear that when ζ lies in Ω, it is an eigenvalue if and only if

$$E^{-}(\zeta) \cap E^{+}(\zeta) \neq \{0\}, \tag{4.10}$$

or equivalently

$$E^{-}(\zeta) + E^{+}(\zeta) \neq \mathbb{C}^{2n}. \tag{4.11}$$

---

[30] In this fully discrete context, an unstable (respectively stable) eigenvalue is always a complex number of modulus larger (resp. smaller) than one. However, the stability may refer to either the time evolution or to the space evolution. For instance, eigenvalues of L or $K(\xi)$ are associated to time evolution, while eigenvalues of $A(\zeta)$ are associated to spatial "evolution".

As a matter of fact, every solution of (4.9) is uniquely defined by its value at $j = 0$, a vector $V \in \mathbb{C}^n$. It is square-summable on $\mathbb{N}$, respectively $-\mathbb{N}$ if and only if V belongs to $E^+(\zeta)$, respectively $E^-(\zeta)$.

One actually shows that $L - \zeta$ is a Fredholm operator of index zero, as soon as $\zeta \in \Omega$. To prove that, we use the *Geometric Dichotomy*. This technique was developped by P. Godillon in her thesis [21], after the *Exponential Dichotomy*, coined by Coppel [14] in the context of differential equations. Define the fundamental matrix $(X_j(\zeta))_{j \in \mathbb{Z}}$ by

$$X_{j+1} = A_j X_j, \quad X_0 = I_{2n},$$

and choose projections $P^-(\zeta)$ on $E^-(\zeta)$ and $P^+(\zeta)$ on $E^+(\zeta)$. There exists a positive number $\alpha(\zeta)$, such that

$$\|X_j P^- X_k^{-1}\| \le c e^{\alpha(k-j)}, \, k \le j \le 0, \tag{4.12}$$

$$\|X_j (I_{2n} - P^-) X_k^{-1}\| \le c e^{\alpha(j-k)}, \, j \le k \le 0, \tag{4.13}$$

and

$$\|X_j P^+ X_k^{-1}\| \le c e^{\alpha(k-j)}, \, 0 \le k \le j, \tag{4.14}$$

$$\|X_j (I_{2n} - P^+) X_k^{-1}\| \le c e^{\alpha(j-k)}, \, 0 \le j \le k. \tag{4.15}$$

These projectors can be used to build solutions of the non-homogeneous problem (4.6). Writing $H \in \mathcal{X}$ for the source term, the expression

$$W_j := \sum_{k=-\infty}^{j} X_j P^- X_k^{-1} H_{k-1} - \sum_{k=j+1}^{1} X_j (I_{2n} - P^-) X_k^{-1} H_{k-1}, \quad j \le 0$$

is square-summable because of (4.12,4.13). As a matter of fact, the Young inequality for the convolution $\ell^1 * \ell^2$ tells that

$$\|W\|_{\mathcal{X}} \le \frac{c}{\sqrt{\alpha(\zeta)}} \|H\|_{\mathcal{X}}.$$

Similarly, we define

$$Z_j := \sum_{k=0}^{j} X_j P^+ X_k^{-1} H_{k-1} - \sum_{k=j+1}^{+\infty} X_j (I_{2n} - P^+) X_k^{-1} H_{k-1}, \quad j \ge 0,$$

which is square-summable with again

$$\|Z\|_{\mathcal{X}} \le \frac{c}{\sqrt{\alpha(\zeta)}} \|H\|_{\mathcal{X}}.$$

One checks immediately that

$$W_{j+1} - A_j W_j = H_j, \quad j \le 0 \tag{4.16}$$

and
$$Z_{j+1} - A_j Z_j = H_j, \quad j \geq 0. \tag{4.17}$$

The other solutions of (4.16) that are square-summable have the form

$$W_j + X_j V^-, \quad V^- \in E^-(\zeta).$$

Similarly, the solutions of (4.17) that are square-summable have the form

$$Z_j + X_j V^+, \quad V^+ \in E^+(\zeta).$$

Therefore, a solution of (4.6) exists if and only if there is a pair (V^-, V^+) in $E^-(\zeta) \times E^+(\zeta)$ such that

$$V^+ - V^- = W_0 - Z_0. \tag{4.18}$$

The right-hand side of (4.18) is a bounded linear function $\nu(H)$. Hence the range $R(L - \zeta)$, defined by the relation

$$\nu(H) \in E^-(\zeta) \times E^+(\zeta),$$

is closed, of finite codimension. Thus $L - \zeta$ is a Fredholm operator.

By continuity, the index of $L - \zeta$ is constant over Ω. For $|\zeta| > \|L\|$, it vanishes since $L - \zeta$ is invertible. Therefore we have proved:

Theorem 4.1 *For every ζ in Ω, $L - \zeta$ is a Fredholm operator of index zero. The eigenvalues of L in Ω are characterized by Property (4.10), or equivalently (4.11).*

When ζ is not an eigenvalue, we have

$$R(P^-) \oplus \ker P^+ = \mathbb{C}^{2n}.$$

Thus there exists a unique projection $P(\zeta)$ with $R(P) = E^-(\zeta)$ and $\ker P = E^+(\zeta)$. We may choose $P^- = I_{2n} - P^+ = P$. Then the formula

$$V_j = \sum_{k=-\infty}^{j} X_j P X_k^{-1} H_{k-1} - \sum_{k=j+1}^{+\infty} X_j (I_{2n} - P) X_k^{-1} H_{k-1}$$

provides the unique solution in $\mathcal{X}$ of the non-homogeneous problem (4.6).

4.3 Construction of the Evans Function

Recall that the set of subspaces of given dimension in $\mathbb{C}^{2n}$, called a *Grassmannian* manifold, is endowed with a complex structure. Since the operator $L - \zeta$ is (in the most obvious way) holomorphic with respect to ζ, the functions $\zeta \mapsto E^\pm(\zeta)$ are holomorphic on Ω. Following Kato ([Kat], Sect. 4.2), we may construct respective bases

$$\mathcal{B}^-(\zeta) = \{V_1^-(\zeta), \dots, V_n^-(\zeta)\}, \quad \mathcal{B}^+(\zeta) = \{V_1^+(\zeta), \dots, V_n^+(\zeta)\},$$

whose elements depend holomorphically on ζ. We warn the reader that Ω is not simply connected; however, one may define $E^\pm(\infty)$ by continuity. Then we apply Kato's construction in $\Omega \cup \{\infty\}$, which is simply connected.

An important point in the applications is that $E^\pm(\zeta)$ are "real" when $\zeta \in (1, \infty)$, in the sense that they are complexifications of real spaces. Then Kato's procedure yields bases formed of real vectors when $\zeta \in (1, \infty)$. Actually, one may extend this "reality" to $\zeta \in (-\infty, -1]$, remarking that $(1, -1] := (1, \infty] \cup [-\infty, -1]$ is a real "interval" in $\Omega \cup \{\infty\}$.

The Evans function is defined as the determinant of both bases:

$$D(\zeta) := \det(V_1^-(\zeta), \dots, V_n^-(\zeta), V_1^+(\zeta), \dots, V_n^+(\zeta)).$$

It is a holomorphic function on $\Omega \cup \{\infty\}$. It vanishes precisely at those numbers ζ such that (4.10) holds, that is at eigenvalues. One may prove that the order of a zero ζ_0 of D equals the order of the eigenvalue ζ_0 of L, but we shall not use this fact.

There is a lot of arbitrariness in the choice of the bases $\mathcal{B}^\pm(\zeta)$ and therefore D is not uniquely defined. This makes the computation somewhat difficult. In practice, noticing that D is real-valued along $(1, \infty)$, we concentrate on the *stability index*, defined as $+1$ if D takes values of the same sign near $\zeta = 1^+$ and $\zeta = \infty$, or -1 if it takes values of opposite signs[31]. Since complex eigenvalues come by conjugate pairs, the stability index $i(L)$ equals $+1$ if the number of eigenvalues in $\Omega \setminus (-\infty, 0)$ is even, and -1 if it is odd. It turns out that the stability index may be computed explicitly; we shall see below a few examples of such computations. Therefore we may state a *necessary condition for spectral stability* in the form $i(L) = +1$, or a sufficient condition of instability in the form $i(L) = -1$. We notice that although D is real on $(-\infty, -1]$, there does not seem to be anything worth to say on this side.

The Gap Lemma

The computation of the sign of D at $\zeta = 1^+$ is based on the remarkable fact that the Evans function admits an analytic continuation on a neighbourhood of $\zeta = 1$. This is by no mean obvious, since $\zeta = 1$ belongs to the boundary of Ω, but not to Ω itself. Actually, the spaces $E^\pm(\zeta)$ admit such analytic continuations and therefore the bases $\mathcal{B}^\pm(\zeta)$ extend analytically too. This is a consequence of the *Gap Lemma* stated below, due independently (in the context of ODEs), to Kapitula & Sanstede [KS] and to Gardner & Zumbrun [GZ]. We warn the reader that these extensions are purely of holomorphic nature, while the defining properties of $E^\pm(\zeta)$ are lost: These spaces are not any more the traces at $j = 0$ of the unstable/stable subspaces of the system (4.9).

---

[31] One often says "positive" or "even" for the index $+1$, and "negative" or "odd" for -1.

Lemma 4.1 (Gap Lemma) *Assume that the unstable/stable subspaces*

$$U^\pm(\zeta), \quad S^\pm(\zeta)$$

of[32] $A^\pm(\zeta)$ admit analytical extensions in some neighbourhood $\mathcal{V}^\pm$ of $\zeta = 1$, with

$$\mathbb{C}^{2n} = S^\pm(\zeta) \oplus U^\pm(\zeta). \tag{4.19}$$

(geometric separation.) Assume also that there is an $\omega \in (0,1)$ such that (exponential decay)

$$\|A_j(\zeta) - A^\pm(\zeta)\| \leq c\omega^{\pm j}, \quad j \in \pm\mathbb{N}. \tag{4.20}$$

Then the spaces $E^\pm(\zeta)$ extend analytically in some neighbourhood $\mathcal{V} \subset \mathcal{V}^\pm$ of $\zeta = 1$.

The exponential decay is crucial in the Gap Lemma, since the size of $\mathcal{V}$ depends in the decay rate $-\log\omega$. In particular, we need that the DSP $\bar{u}$ tend exponentially fast to its end values $u^{l,r}$. Usually, this decay holds when the steady shock $(u^l, u^r; s = 0)$ is not characteristic. This is a basic assumption of the theory, which is used in several places.

Thanks to the Gap Lemma, D is defined in $\Omega \cup \mathcal{V}$. Our next step is to compare the signs of D near $\zeta = 1^+$ and near infinity. The former could be computed as the sign of $D(1)$, provided this number is non-zero. This happens generically when the shock is under-compressive, namely when the shock has n incoming characteristics. However, Lax shocks have a completely different behaviour: We saw in Sect. 1.1 that they admit generally a continuous discrete profile $U(\tau)$ with $\bar{u}_j = U(j)$. Differentiating the profile equation with respect to τ on $\mathbb{Z}$, we obtain

$$L\bar{v} = \bar{v}, \tag{4.21}$$

where $\bar{v}_j := dU/d\tau(j)$. Since U and its derivatives decay exponentially fast, $\bar{v}$ lies in $\mathcal{X}$ and (4.21) tells that $\zeta = 1$ is an eigenvalue of L. This strongly suggests that $D(1)$ vanishes; however, that is not a proof of this fact, since the Evans function does not encode the spectral properties of L when ζ lies outside of Ω.

The Geometric Separation

We thus study the behaviour of $S^\pm(\zeta)$ and $U^\pm(\zeta)$. These are the stable/unstable subspaces of

$$A^\pm(\zeta) = \begin{pmatrix} 0_n & I_n \\ (\mathcal{D}_b F)^{-1} \mathcal{D}_a F & (\mathcal{D}_b F)^{-1} \left(\frac{1-\zeta}{\sigma} - \mathcal{D}_a F + \mathcal{D}_b F \right) \end{pmatrix},$$

---

[32] According to our notations, there holds $S^+(\zeta) = \mathcal{E}^+(\zeta)$ and $U^-(\zeta) = \mathcal{E}^-(\zeta)$.

where $\mathcal{D}_a F$ and $\mathcal{D}_b F$ are computed at (u^l, u^l) (subscript minus) or (u^r, u^r) (subscript plus). Recall that at such points, these differentials have the same eigenbases as $\mathcal{D} f^{l,r}$, while the spectrum of $\mathcal{D}_a F$ (respectively $\mathcal{D}_b F$) is positive (resp. negative.) Denote the eigenvectors of $\sigma \mathcal{D} f(u^{l,r})$ by $r_1^{\pm}, \ldots, r_n^{\pm}$ and the corresponding eigenvalues by $c_1^{\pm}, \ldots, c_n^{\pm}$. There follows that $A^{\pm}(\zeta)$ splits as a collection of 2×2 matrices that act on subspaces $\mathbb{C} r_k^{\pm} \times \mathbb{C} r_k^{\pm}$:

$$A_k^{\pm}(\zeta) = \begin{pmatrix} 0 & 1 \\ a_k^{\pm}/b_k^{\pm} & \left(1 - \zeta - a_k^{\pm} + b_k^{\pm}\right)/b_k^{\pm} \end{pmatrix},$$

where

$$b_k^{\pm} + a_k^{\pm} = c_k^{\pm}, \quad b_k^{\pm} < 0 < a_k^{\pm}, \quad 0 \neq |c_k^{\pm}| < 1.$$

Hereabove, we have use consistency and CFL condition. We are thus concerned with the geometric separation for matrices of the form

$$A(\zeta) = \begin{pmatrix} 0 & 1 \\ a/b & (1 - \zeta - a + b)/b \end{pmatrix},$$

where

$$b < 0 < a, \quad 0 \neq |a + b| < 1.$$

The matrix $A(\zeta)$ is analytic in ζ and has distinct eigenvalues 1 and a/b, the latter being negative. Hence the eigenvalues extend analytically, as well as the eigenspaces. Whence the geometric separation.

This analysis also provides a description of the stable/unstable eigenvalues of $A^{\pm}(\zeta)$ as $\zeta = 1^+$. The matrix $A_k^{\pm}(\zeta)$ has two eigenvalues of which one is stable and one is unstable, since $\zeta \in \Omega$. The nature of the one close to $a_k^{\pm}/b_k^{\pm}$ is given by the sign of $c_k^{\pm}$: It is stable if $c_k^{\pm} < 0$ and unstable if $c_k^{\pm} > 0$. Hence the eigenvalue close to the unit is stable if $c_k^{\pm} > 0$ and unstable if $c_k^{\pm} < 0$. We deduce a characterization of the spaces $\mathcal{E}^{\pm}(\zeta)$ at the critical value $\zeta = 1$. For instance, $\mathcal{E}^-(1)$ is spanned by the eigenvectors $(r_k^-, \mu_k^- r_k^-)$ of $A^-(1)$ associated to those eigenvalues μ_k^- that are limits of unstable ones. As discussed above, we have

$$\mu_k^- = \begin{cases} 1 & \text{if } c_k^- < 0, \\ -a_k^-/b_k^- & \text{if } c_k^- > 0. \end{cases} \tag{4.22}$$

Likewise, $\mathcal{E}^+(1)$ is spanned by the eigenvectors $(r_k^+, \mu_k^+ r_k^+)$ of $A^+(1)$ associated to the eigenvalues μ_k^+ that are limits of stable ones. Their values are

$$\mu_k^+ = \begin{cases} -a_k^+/b_k^+ & \text{if } c_k^+ < 0, \\ 1 & \text{if } c_k^+ > 0. \end{cases} \tag{4.23}$$

A complement to the Gap Lemma is that the space $E^-(1)$ is the set of initial data w_0 for which the solution of

$$w_{j+1} = A_j(1) w_j, \quad j \in \mathbb{Z}, \tag{4.24}$$

tends to a limit as $j \to -\infty$, this limit lying in the space spanned by the eigenvectors $(r_k^-, (a_k^-/b_k^-)r_k^-)$ with $c_k^- < 0$. Likewise, the space $E^+(1)$ is the set of initial data w_0 for which the solution of (4.24) tends to a limit as $j \to +\infty$, this limit lying in the space spanned by the eigenvectors $(r_k^+, (a_k^+/b_k^+)r_k^+)$ with $c_k^+ > 0$. In calculations, it is useful to work with bases $\mathcal{B}^\pm(\zeta)$ that generate solutions of (4.8) with the following asymptotics:

$$V_{k,j}^-(\zeta) \sim (\mu_k^-(\zeta))^j (r_k^-, \mu_k^-(\zeta)r_k^-), \quad \text{as } j \to -\infty \qquad (4.25)$$

$$V_{k,j}^+(\zeta) \sim (\mu_k^+(\zeta))^j (r_k^+, \mu_k^+(\zeta)r_k^+), \quad \text{as } j \to +\infty \qquad (4.26)$$

Recall that $(\bar{u}_0, \bar{u}_1)$ is an intersection point of the unstable manifold $W^u(u^l)$ with the stable manifold $W^s(u^r)$ of some dynamical system

$$(u_j, u_{j+1}) = H(u_{j-1}, u_j).$$

By differentiation, we see that $E^-(1)$ contains the tangent space to $W^u(u^l)$ at $(\bar{u}_0, \bar{u}_1)$, since these vectors generate solutions of (4.24) that tend to zero at $-\infty$. Likewise, $E^+(1)$ contains the tangent space to $W^s(u^r)$. For an extreme Lax shock, one equality occurs. For instance, for an n-Lax shock, we have

$$c_1^+, \ldots, c_n^+ < 0,$$

and therefore

$$E^+(1) = T_{(\bar{u}_0, \bar{u}_1)} W^s(u^r).$$

In the same spirit, if the DSP $\bar{u}$ is imbeded in a one-parameter[33] family, say $\bar{u}_j = U(j)$ where $U(y)$ is a continuous DSP, then differentiation with respect to y yields a non-trivial solution $\bar{v}$ of

$$L\bar{v} = \bar{v}, \quad \bar{v}_j := U'(j) \qquad (4.27)$$

that decay to zero at both ends. In particular, $(\bar{v}_0, \bar{v}_1)$ belongs to the intersection $E^-(1) \cap E^+(1)$. In this situation, the extension of the Evans function vanishes at $\zeta = 1$ since $\mathcal{B}^-(1) \cup \mathcal{B}^+(1)$ is not any more a basis of $\mathbb{C}^{2n}$.

5 Stability Analysis: Calculations

After the last paragraph, there remains to evaluate the sign of $D(\zeta)$ for $\zeta = 1^+$.

Since the results depend significantly on the nature of the steady shock wave (u^l, u^r), we need to fix some notations. The eigenvalues $c_k^\pm$ are labelled in the increasing order:

$$c_1^- \le \cdots \le c_n^-, \quad c_1^+ \le \cdots \le c_n^+.$$

---

[33] One could consider several parameters instead, especially if the shock is overcompressive.

Lax Shocks.

For a k-Lax shock, we have

$$c_k^+ < 0 < c_{k+1}^+, \quad c_{k-1}^- < 0 < c_k^-.$$

Hence there are $n+1$ incoming characteristics, corresponding to the velocities

$$c_1^+, \ldots, c_k^+, c_k^-, \ldots, c_n^-.$$

Under-Compressive Shocks.

These are shocks with exactly n incoming characteristics:

$$c_k^+ < 0 < c_{k+1}^+, \quad c_k^- < 0 < c_{k+1}^-.$$

Anti-Lax Shocks.

A shock (u^l, u^r) is anti-Lax if the opposite (u^r, u^l) is a Lax shock:

$$c_{k-1}^+ < 0 < c_k^+, \quad c_k^- < 0 < c_{k+1}^-.$$

Then there are only $n - 1$ incoming characteristics.

Of course, it is possible to treat shocks with $n - m$ incoming characteristics, $|m| > 1$, but they are pretty rare and we leave the reader making the computations in the interesting cases that (s)he could encounter.

For practical purpose, we need a few facts about the $V_m^\pm(1)$. Such a vector V has the form $(v_0, v_1)^T$ where v is a solution of $Lv = v$. The latter equation can be "integrated" once, in the form

$$\sigma \left(\mathcal{D}_a F_{j+1/2} v_j + \mathcal{D}_b F_{j+1/2} v_{j+1} \right) = \text{cst.}$$

The constant of integration can be computed by letting j going to either $-\infty$ (for $V_m^-(1)$), or $+\infty$ (for $V_m^+(1)$). For instance, with the minus sign, the constant is zero if $c_m^- > 0$, while it equals $c_k^- r_k^-$ if $c_m^- < 0$. With the plus sign, the constant is zero if $c_m^+ < 0$, while it equals $c_k^+ r_k^+$ if $c_m^+ > 0$. Therefore, we obtain

$$\sigma \left(\mathcal{D}_a F_{1/2} v_0 + \mathcal{D}_b F_{1/2} v_1 \right) = \begin{cases} c_m^- r_m^-, & \text{if } \pm = - \text{ and } c_m^- < 0, \\ 0, & \text{if } \pm = - \text{ and } c_m^- > 0, \\ 0, & \text{if } \pm = + \text{ and } c_m^+ < 0, \\ c_m^+ r_m^+, & \text{if } \pm = + \text{ and } c_m^+ > 0. \end{cases} \tag{5.1}$$

Exercise.

Use Formula (5.1) to prove that $D(1)$ vanishes as soon as there are at least $n + 1$ incoming characteristics. **Hint:** a suitable linear combination of lines that is used in the next paragraph.

5.1 Calculations with Lax Shocks

Consider a k-Lax shock. We place ourselves in one of the situations described in Sect. 1.1, where $\bar{u}$ is embeded into a continuous DSP U. As explained above, $D(1)$ vanishes because $\bar{v}$ belongs to both $E^{\pm}(1)$. Thus we have to compute the sign of $D'(1)$. Generically, $\bar{v}$ decays as slow as possible at both ends, meaning that we may choose[34]

$$V_k^-(1) = V_k^+(1) = \begin{pmatrix} \bar{v}_0 \\ \bar{v}_1 \end{pmatrix}. \tag{5.2}$$

Since D is a determinant of analytic functions, its derivative is a sum of $2n$ determinants, each one obtained by replacing one column vector $V(\zeta)$ by its derivative $V'(\zeta)$. But $2(n-1)$ of these determinants vanish because they still contain the equal columns V_k^- and V_k^+. Hence there remain two terms:

$$D'(1) = \det\left(V_1^-, \ldots, V_{k-1}^-, \frac{dV_k^-}{d\zeta}(1), V_{k+1}^-, \ldots, V_n^+\right)$$

$$+ \det\left(V_1^-, \ldots, V_{k-1}^+, \frac{dV_k^+}{d\zeta}(1), V_{k+1}^+, \ldots, V_n^+\right).$$

Using again the equality of V_1^- and V_n^+, this simplifies into

$$D'(1) = \det\left(V_1^-, \ldots, V_{k-1}^+, \frac{dV_k^+}{d\zeta}(1) - \frac{dV_k^-}{d\zeta}(1), V_{k+1}^+, \ldots, V_n^+\right). \tag{5.3}$$

We now establish an identity of the form (5.1) for the $(n+k)$-th column of $D'(1)$. Denote by $v^{\pm}(\zeta)$ the solutions of $Lv = \zeta v$ associated to $V_k^{\pm}(\zeta)$ and by $z^{\pm}$ their derivatives at $\zeta = 1$. Since $v^{\pm}(1) = \bar{v}$, a differentiation of the eigen-equation gives

$$Lz^{\pm} = z^{\pm} + \bar{v}.$$

This rewrites as

$$\sigma\left(\mathcal{D}_a F_{j-1/2} z_{j-1} + \mathcal{D}_b F_{j-1/2} z_j\right) - \sigma\left(\mathcal{D}_a F_{j+1/2} z_j + \mathcal{D}_b F_{j+1/2} z_{j+1}\right) = \bar{v}_j.$$

We "integrate" this equation, by summing either from $-\infty$ (for z^-), or up to $+\infty$ (for z^+); remark that $z_{-\infty}^- = z_{+\infty}^+ = 0$, since $c_k^+ < 0 < c_k^-$. We obtain

$$\sigma\left(\mathcal{D}_a F_{1/2} z_0^- + \mathcal{D}_b F_{1/2} z_1^-\right) = -\sum_{j=-\infty}^{0} \bar{v}_j, \tag{5.4}$$

and

---

[34] This is plenty justified when the shock is weak, since our construction of a continuous DSP in Sect. 1.1 provides a profile that is almost parallel to r_k.

$$\sigma\left(\mathcal{D}_a F_{1/2} z_0^+ + \mathcal{D}_b F_{1/2} z_1^+\right) = \sum_1^{+\infty} \bar{v}_j. \tag{5.5}$$

Summing (5.4) and (5.5), there comes

$$\sigma\left(\mathcal{D}_a F_{1/2}(z_0^+ - z_0^-) + \mathcal{D}_b F_{1/2}(z_1^+ - z_1^-)\right) = \sum_{j=-\infty}^{+\infty} \bar{v}_j \tag{5.6}$$

Thanks to (5.1) and (5.6), we may greatly simplify (5.3). To do so, we multiply at left the matrix

$$\left(V_1^-, \ldots, V_{k-1}^+, \frac{dV_k^+}{d\zeta}(1) - \frac{dV_k^-}{d\zeta}(1), V_{k+1}^+, \ldots, V_n^+\right)$$

by

$$\begin{pmatrix} I_n & 0_n \\ \sigma \mathcal{D}_a F_{1/2} & \sigma \mathcal{D}_b F_{1/2} \end{pmatrix}.$$

We receive

$$\det(\sigma \mathcal{D}_b F_{1/2}) D'(1) = \begin{vmatrix} \cdots & v_{k-1,0}^- & v_{k,0}^- & \cdots & v_{k-1,0}^+ & z_0^+ - z_0^- & v_{k+1,0}^+ & \cdots \\ \cdots & c_{k-1}^- r_{k-1}^- & 0 & \cdots & 0 & \sum_{j \in \mathbb{Z}} \bar{v}_j & c_{k+1}^+ r_{k+1}^+ & \cdots \end{vmatrix}. \tag{5.7}$$

We observe that there are n zeroes in the lower half of the right-hand side. Hence this determinant is blockwise triangular and indeed equals a product of two $n \times n$ determinants:

$$\det(\sigma \mathcal{D}_b F_{1/2}) D'(1) = (-1)^{n(k-1)} \det(v_{k,0}^-, \ldots, v_{k-1,0}^+) \times \tag{5.8}$$

$$\det\left(\ldots, c_{k-1}^- r_{k-1}^-, \sum_{j \in \mathbb{Z}} \bar{v}_j, c_{k+1}^+ r_{k+1}^+, \ldots\right).$$

The term

$$\gamma = \det(v_{k,0}^-, \ldots, v_{k-1,0}^+) \tag{5.9}$$

measures the transversality of $W^u(u^l)$ and $W^s(u^r)$ along their intersection, since their tangent spaces at $\bar{u}_0$ are spanned by $v_{k,0}^-, \ldots, v_{n,0}^-$ and by $v_{1,0}^+, \ldots, v_{k,0}^+$, respectively, and $v_{k,0}^+ = v_{k,0}^-$. Recall on the one hand that the spectrum of $\mathcal{D}_b F$ is always negative, and thus the sign of $\det(\mathcal{D}_b F)$ is $(-1)^n$. On the other hand, the c_j^-'s contribute for $(-1)^{k-1}$. Therefore

$$\operatorname{sgn} D'(1) = (-1)^{kn+k-1} \operatorname{sgn}(\gamma) \operatorname{sgn} \det\left(r_1^-, \ldots, r_{k-1}^-, \sum_{j \in \mathbb{Z}} \bar{v}_j, r_{k+1}^+, \ldots, r_n^+\right). \tag{5.10}$$

The Sign of γ:

Remark that, as long as $v_m^\pm$ enters in the definition of γ, it satisfies a *first* order induction

$$\mathcal{D}_a F_{j+1/2} v_{m,j} + \mathcal{D}_b F_{j+1/2} v_{m,j+1} = 0.$$

Therefore, we have a "Wronskian" formula:

$$\gamma \prod_{l=0}^{j-1} \det\left(-(\mathcal{D}_b F_{j+1/2})^{-1} \mathcal{D}_a F_{j+1/2}\right) = \det(v_{k,j}^-, \ldots, v_{k-1,j}^+).$$

For general values of k, this identity does not help, since we only know the behaviour of the $v_{m,j}^-$'s near $-\infty$ and that of the $v_{m,j}^+$'s near $+\infty$. Hence the sign of γ remains mysterious in general. However, it can be computed explicitly when the (u^l, u^r) is an *extreme shock*, meaning that either $k = 1$ or $k = n$. For instance, if $k = n$, we may write

$$\gamma = (-1)^{n-1} \det(v_{1,0}^+, \ldots, v_{n,0}^+),$$

since $v_n^- = v_n^+$ (they both equal $\bar{v}$.) Using the Wronskian formula, plus the fact that the spectra of the matrices $-\mathcal{D}_b F_{j+1/2}$ and $\mathcal{D}_a F_{j+1/2}$ are positive, we have

$$\mathrm{sgn}\gamma = (-1)^{n-1}\mathrm{sgn}\det(v_{1,j}^+, \ldots, v_{n,j}^+),$$

for every j in $\mathbb{Z}$. We let $j \to +\infty$ and use the fact that

$$v_{m,j}^+ \sim \left(-\frac{a_m^+}{b_m^+}\right)^j r_m^+,$$

where

$$-\frac{a_m^+}{b_m^+} \in (0,1),$$

to derive

$$\mathrm{sgn}\gamma = (-1)^{n-1}\mathrm{sgn}\det(r_1^+, \ldots, r_n^+). \tag{5.11}$$

The formula for a 1-shock is similar:

$$\mathrm{sgn}\gamma = \mathrm{sgn}\det(r_1^-, \ldots, r_n^-). \tag{5.12}$$

The Homotopy from $\zeta = 1$ to ∞

Remark that (5.10) is void as long as it remains isolated, since the vectors $v_{j,0}^\pm$ are defined only up to scalars, which could be either positive or negative. Hence the sign of $D'(1)$ is not intrinsic, contrary to that of the product $D'(1)D(\zeta)$ for $\zeta \gg 1$. The last delicate point is thus to relate the right-hand side of (5.10) to the behaviour of D near infinity. To do so, we shall use the continuity of the bases $\mathcal{B}^\pm(\zeta)$ and of the vectors $V_j^\pm(\zeta)$, with respect to ζ. The idea used below was introduced by Benzoni & al. in [BSZ] in the context of differential operators. We begin with an algebraic lemma that corresponds to Lemma 7.2 in [BSZ] (see also Lemma 3.5 in [GZ].)

Lemma 5.1 *Let $\zeta \in [1, +\infty)$ be given. Then the projections from the subpaces $\mathcal{E}^{\pm}(\zeta)$ of $\mathbb{C}^{2n} = \mathbb{C}^n \times \mathbb{C}^n$ on the first or second factors $\mathbb{C}^n$ are one-to-one.*

Proof. Let V be a solution of the homogeneous induction

$$V_{j+1} = A^+(\zeta)V_j.$$

We have $V_j = (v_j, v_{j+1})^T$ for some v that satisfies

$$(L^+ - \zeta)v = 0. \tag{5.13}$$

Our goal is to prove that if $v_{+\infty} = 0$ and $v_0 = 0$, then $v \equiv 0$.

Since $\mathcal{D}_a F$ and $-\mathcal{D}_b F$ are diagonalizable in a common basis (the eigenbasis of $\mathcal{D}f$), we choose a Euclidian structure in which both are symmetric. And since their eigenvalues are positive, we may assume that these matrices are symmetric positive definite. Let us take the scalar product of the j-th term of (5.13) with v_j. Then sum over j from 1 to $+\infty$. After some rearrangements, we obtain

$$v_0^T \mathcal{D}_a F v_0 + v_1^T \mathcal{D}_b F v_1 = \sum_1^{\infty} \left\{ 2\sigma^{-1}(\zeta - 1)|v_j|^2 + (v_j - v_{j-1})^T \mathcal{D}_a F(v_j - v_{j-1}) \right.$$
$$\left. -(v_{j+1} - v_j)^T \mathcal{D}_b F(v_{j+1} - v_j) \right\}.$$

We notice that the right-hand side is always non-negative. If v_0 vanishes, the left hand-side reduces to $v_1^T \mathcal{D}_b F v_1$, a non-positive term. Hence both sides vanish, implying that $v_1 = 0$ and therefore $V \equiv 0$.

This shows that the projection from $\mathcal{E}^+(\zeta)$ on the first $\mathbb{C}^n$-factor is one-to-one. A shift shows that the projection on the second factor is one-to-one as well. The case of $\mathcal{E}^-(\zeta)$ is similar.

□

The elements $V_m^-(\zeta)$ of the basis $\mathcal{B}^-(\zeta)$ correspond to solutions $v_m^-(\zeta)$ of the equation $(L - \zeta)v = 0$. We may choose $v_m^-(\zeta)$ real when ζ is real. Let us form the number

$$\Delta_j^-(\zeta) := \det(v_{1,j}^-(\zeta), \ldots, v_{n,j}^-(\zeta)).$$

For $j \ll -1$, Lemma 5.1 tells that this determinant does not vanish. And since the eigenvalues of $A^-(\zeta)$ are positive, its sign is actually independent of the large negative j; this sign is denoted by $\Delta^-(\zeta)$. Since it is a continuous function of its arguments, it is constant and we may simply write Δ^-. This is the core of the homotopy argument: We have proved that the behaviour of the bases $\mathcal{B}^-(\zeta)$ are correlated as ζ varies. Of course, the same is true for $\mathcal{B}^+(\zeta)$.

The Large Wave-Length Analysis

Since the matrices $A_j(\zeta)$ actually depend only on $\varrho = (\zeta - 1)/\sigma$, we specialize to large positive values of ϱ. From the two-step induction

$$(\varrho - \mathcal{D}_b F_{j-1/2}' + \mathcal{D}_a F_{j+1/2})v_j - \mathcal{D}_a F_{j-1/2}v_{j-1} + \mathcal{D}_b F_{j+1/2}v_{j+1} = 0,$$

we derive the following properties:

Lemma 5.2 *Assume that $\varrho \gg 1$. Then solutions of $(L - \zeta)v = 0$ that tend to zero as $j \to -\infty$ satisfy*

$$v_{j+1} = \left(\varrho(-\mathcal{D}_b F_{j+1/2})^{-1} + \mathcal{O}(1)\right)v_j, \tag{5.14}$$

uniformly in $j \in -\mathbb{N}$. Likewise, the solutions that tend to zero as $j \to +\infty$ satisfy

$$v_{j+1} \sim \frac{1}{\varrho}\mathcal{D}_a F_{j+1/2}v_j, \tag{5.15}$$

uniformly in $j \in \mathbb{N}$.

From Lemma 5.2, we deduce that

$$D(\zeta) \sim \begin{vmatrix} \cdots & v_{n,0}^-(\zeta) & v_{1,0}^+(\zeta) & \cdots \\ \cdots & \varrho(-\mathcal{D}_b F_{1/2})^{-1}v_{n,0}^- & \varrho^{-1}\mathcal{D}_a F_{1/2}v_{1,0}^+ & \cdots \end{vmatrix}. \tag{5.16}$$

Let us multiply he right-hand side of (5.16) by the determinant

$$1 = \begin{vmatrix} I_n & 0_n \\ \varrho(\mathcal{D}_b F_{1/2})^{-1} & I_n \end{vmatrix}.$$

We obtain

$$D(\zeta) \sim \begin{vmatrix} \cdots & v_{n,0}^- & v_{1,0}^+ & \cdots \\ \cdots & 0 & (\varrho^{-1}\mathcal{D}_a F_{1/2} + \varrho \mathcal{D}_b F_{1/2}^{-1})v_{1,0}^+ & \cdots \end{vmatrix},$$

from which we derive

$$D(\zeta) \sim \Delta_0^-(\zeta)\,\Delta_0^+(\zeta)\,\det(\varrho^{-1}\mathcal{D}_a F_{1/2} + \varrho \mathcal{D}_b F_{1/2}^{-1}).$$

The sign of the last determinant is $(-1)^n$ because $\mathcal{D}_b F$ have negative eigenvalues and ϱ is large. Finally, we see that

$$\operatorname{sgn}D(\zeta) = (-1)^n \Delta^- \Delta^+, \quad \zeta \gg 1. \tag{5.17}$$

Conclusions

We gather Formulae (5.9), (5.10) and (5.17). When ζ is large and real, we obtain

$$\operatorname{sgn}D'(1)D(\zeta) = (-1)^{kn+n+k-1}\Delta^-\Delta^+ \operatorname{sgn}\det(v_{k,0}^-(1), \ldots, v_{k-1,0}^+(1))$$

$$\operatorname{sgn}\det\left(r_1^-, \ldots, r_{k-1}^-, \sum_{j\in\mathbb{Z}}\bar{v}_j, r_{k+1}^+, \ldots, r_n^+\right). \tag{5.18}$$

This formula might not look of practical interest in general. However, it becomes explicit in the case of extreme shocks. For instance, for an n-shock ($k = n$), we have

$$\det(v_{n,0}^-(1), \ldots, v_{n-1,0}^+(1)) = (-1)^{n-1} \Delta_0^+(1),$$

whence

$$\operatorname{sgn} D'(1) D(\zeta) = \Delta^- \operatorname{sgn} \det \left(r_1^-, \ldots, r_{n-1}^-, \sum_{j \in \mathbb{Z}} \bar{v}_j \right).$$

To make this formula explicit, we remark that Δ^- is the sign of $\Delta_j^-(1)$ for large negative j's, for which we know the asymptotics (4.25). Hence

$$\operatorname{sgn} D'(1) D(\zeta) = \operatorname{sgn} \det(r_1^-, \ldots, r_n^-) \operatorname{sgn} \det \left(r_1^-, \ldots, r_{n-1}^-, \sum_{j \in \mathbb{Z}} \bar{v}_j \right). \quad (5.19)$$

Remarks.

- Formula (5.19) is well-defined. The global sign is not affected by a flip of r_m^- for $m < n$, since this eigenvector enters twice in the right-hand side. On the other-hand, the orientation of r_n^- is fully determined by the fact that $v_{n,j}^-(1) = U'(j)$.
- The sum in the last determinant has an obvious significance. It is precisely the derivative $Y'(0)$ of the function Y considered in Sect. 2.1. It replaces the vector $[u] := u^r - u^l$ that appears in the computation of the stability index of viscous shock profiles for extreme shocks. Notice that

$$\int_0^1 Y'(h) \, dh = \int_{-\infty}^{+\infty} U'(h) \, dh = [u]. \quad (5.20)$$

Therefore, the average value of Y other the DSPs of the same extreme shock equals $[u]$. This tells that the average value of

$$\det \left(r_1^-, \ldots, r_{n-1}^-, \sum_{j \in \mathbb{Z}} \bar{v}_j \right) \quad (5.21)$$

equals $\det(r_1^-, \ldots, r_{n-1}^-, [u])$. The latter term enters into the formula of the stability index of viscous shock profiles (see [20, 10].) Hence in reasonable situations, this family of DSPs contains elements whose stability index is the same as that of the corresponding viscous profile.
- When the term (5.21) keeps a constant sign along the continuous DSP, there follows that every DSP extracted from the continuous one has the same stability index than the corresponding viscous profile.

- It is expected that weak Lax shocks are non-linearly stable, and in particular that they are spectrally stable. This should be proved through energy methods, as in the viscous case (see [19, 24, 40, 57] for the latter, and [30, 49] for large viscous shocks.) In both cases, the stability index must be even. This confirms the observation made in the previous remark.
- For a 1-Lax shock (the other kind of extreme shock), we leave the reader proving the result

$$\mathrm{sgn} D'(1) D(\zeta) = \mathrm{sgn}\det(r_1^+, \dots, r_n^+)\,\mathrm{sgn}\det\left(\sum_{j\in\mathbb{Z}} \bar{v}_j, r_2^+, \dots, r_n^+\right).$$
(5.22)

Theorem 5.1 *Let (u^r, u^l) be a steady k-Lax shock, for which a three-points difference scheme admits a continuous DSP U. Assume that, along the diagonal, $\mathcal{D}_a F(u, u)$ and $-\mathcal{D}_b F(u, u)$ have the same eigenbasis as $Df(u)$, with positive eigenvalues. Assume also that U has the generic behaviour at infinity:*

$$U(j) - u^{r,l} \sim (\mu_k^\pm)^j r_k^\pm, \quad j \to \pm\infty,$$

where $r_k^\pm$ is a k-th eigenvector of $Df(u^{r,l})$.
Then the stability index of $\bar{u} := (U(j))_{j\in\mathbb{Z}}$ equals

$$(-1)^{(k+1)(n+1)} \Delta^- \Delta^+ (\mathrm{sgn}\gamma)\,\mathrm{sgn}\det\left(r_1^-, \dots, r_{k-1}^-, Y'(0), r_{k+1}^+, \dots, r_n^+\right).$$

A necessary condition for spectral stability is that the stability index be even. In other words, a sufficient condition for spectral instability is that this index be odd.
For a 1-shock, the stability index equals

$$\mathrm{sgn}\det(r_1^+, \dots, r_n^+)\,\mathrm{sgn}\det\left(Y'(0), r_2^+, \dots, r_n^+\right),$$

while for an n-shock, it equals

$$\mathrm{sgn}\det(r_1^-, \dots, r_n^-)\,\mathrm{sgn}\det\left(r_1^-, \dots, r_{n-1}^-, Y'(0)\right).$$

When the shock is weak, we proved in Sect. 1.1 that the continuous DSP does exists. A careful examination of the bifurcation analysis shows that the range of U is approximately a segment from u^l to u^r. In particular, $Y'(0)$ is approximately parallel to $[u]$, with the same sense. Since on the other hand, $[u]$ is approximately parallel to r_k^- with the same sense, we conclude that $Y'(0) \sim \kappa r_k^-$ for a positive constant. From Theorem 5.1, we thus deduce

Corollary 5.1 *For a weak k-Lax shock, with the same assumptions as in Theorem 5.1, the stability index of $\bar{u}$ equals*

$$(-1)^{(k+1)(n+1)} \Delta^- \Delta^+ (\mathrm{sgn}\gamma)\,\mathrm{sgn}\det(r_1, \dots, r_n).$$

In particular, this index equals $+1$ for a weak extreme Lax shock.

Open Problem 5.1 *It seems that the stability index of every kind of weak Lax shock equals $+1$. That requires an accurate description of the intersection of $W^u(u^l)$ with $W^s(u^r)$.*

Remarks:

- In full gas dynamics, where $n = 3$, the middle characteristic field is linearly degenerate. The corresponding discontinuities are contacts, for which we do not expect that DSPs exist. Hence Theorem 5.1 and Corollary 5.1 apply to every reasonable shock. However, this is not true any more for more complex systems as MHD (magneto-hydrodynamics.)
- The Lax–Friedrichs scheme behaves a bit differently, still because of the independency of the subgrids defined by the parity of $j + m$. Not only the essential spectrum of L contains $\zeta = -1$, but also $\zeta = -1$ is a genuine eigenvalue. This fact can be checked by computing $D(-1) = 0$, left as an exercise. As a matter of fact, as long as we do not distinguish the sub-grids, a stability criterion will govern both sub-profiles. Therefore, it will involve the signs of both quantities

$$\det\left(r_1^-, \ldots, r_{k-1}^-, \sum_{j \in 2\mathbb{Z}} \bar{v}_j, r_{k+1}^+, \ldots, r_n^+\right)$$

and

$$\det\left(r_1^-, \ldots, r_{k-1}^-, \sum_{j \in 1+2\mathbb{Z}} \bar{v}_j, r_{k+1}^+, \ldots, r_n^+\right).$$

5.2 Calculations with Under-Compressive Shocks

Since the sum of the dimensions of $W^u(u^l)$ and $W^s(u^r)$ equals n, their intersection is usually discrete. Hence there is no special reason why D should vanish at $\zeta = 1$. Therefore, the stability index will be given by the sign of the product $D(1)D(\zeta)$ for $\zeta \gg 1$. We begin with the computation of $D(1)$. The notations are similar to that of Sect. 5.1, but with $c_k^\pm < 0 < c_{k+1}^\pm$. Notice that, since $W^u(u^l)$ and $W^s(u^r)$ must be non-void, we have $1 \leq k \leq n - 1$. Hence there does not exist a notion of "extreme" under-compressive shock.

The same linear combination of lines that in (5.7) gives here

$$\det(\sigma \mathcal{D}_b F_{1/2}) D(1) = \begin{vmatrix} \cdots & v_{k,0}^- & v_{k+1,0}^- & \cdots & v_{k,0}^+ & v_{k+1,0}^+ & \cdots \\ \cdots & c_k^- r_k^- & 0 & \cdots & 0 & c_{k+1}^+ r_{k+1}^+ & \cdots \end{vmatrix}. \tag{5.23}$$

Again, this is a block-triangular determinant, and we obtain

$$\det(\sigma \mathcal{D}_b F_{1/2}) D(1) = (-1)^{kn} \gamma \det(\ldots, c_k^- r_k^-, c_{k+1}^+ r_{k+1}^+, \ldots),$$

with

$$\gamma := \det(v_{k+1}^-, \ldots, v_k^+). \tag{5.24}$$

As before, the number γ measures the transversality of the intersection of $W^u(u^l)$ with $W^s(u^r)$. Since $\mathcal{D}_b F$ has a negative spectrum and since $c_k^- < 0 < c_{k+1}^+$, we obtain

$$\operatorname{sgn} D(1) = (-1)^{n+k+kn} (\operatorname{sgn}\gamma) \operatorname{sgn} \det(r_1^-, \dots, r_k^-, r_{k+1}^+, \dots, r_n^+). \qquad (5.25)$$

When computing the sign of $D(\zeta)$ at infinity, we may use the calculations of Section 5.1 without change. We obtain the final result

$$\operatorname{sgn} D(1)D(\zeta) = (-1)^{kn+k} \Delta^- \Delta^+ (\operatorname{sgn}\gamma) \operatorname{sgn} \det(r_1^-, \dots, r_k^-, r_{k+1}^+, \dots, r_n^+). \tag{5.26}$$

Remark:

As mentionned in Section 1.2, there are usually an even number of distinct DSPs fo the under-compressive shock. The signs $\Delta^\pm$ usually do not depend on the choice of the DSP. However, the sign of γ changes when passing from a DSP to the next one, because the orientation of the intersection is reversed. Hence two consecutive DSPs have *opposite* stability indices. In particular, at least one of both is unstable! When the other one is stable, it would be interesting to search for heteroclinic orbits of the scheme that connect an unstable DSP to a stable one, as in the reaction-diffusion case.

5.3 Results for the Godunov Scheme

We treat in this paragraph the spectral stability of the DSPs obtained in Sect. 1.1 for the Godunov scheme when $\eta = 0$ (namely $s = 0$). We assume that the shock $(u^l, u^r; 0)$ is a k-Lax-shock, so that the Godunov scheme displays some amount of numerical viscosity at both end states. We warn the reader that the matrices $\mathcal{D}_a F$ and $-\mathcal{D}_b F$ are only non-negative but not strictly positive at the diagonal points (u^l, u^l) and (u^r, u^r) in general. However, a careful analysis allows us to build the Evans function along essentially the same lines as in Sect. 4. This was done in [11]. Historically, this paper gave the first construction of an Evans function in a discrete context. The lack of strong viscosity was balanced by the constancy of the profile outside a central mesh, so that the linearized operator is almost a constant coefficients operator.

We concentrate in this section on the search of the unstable eigenvalues λ of the linearized operator L around the profile, leaving the reader checking that the essential spectrum lies in the stable disk. As usual, *unstable* means that $|\lambda| \geq 1$ and, either $\lambda \neq 1$, or $\lambda = 1$ is a non-semi-simple eigenvalue in some appropriate sense.

We recall that the profile has the form

$$u_j = \begin{cases} u^l, & j \leq -1, \\ u^m, & j = 0, \\ u^r, & j \geq 1, \end{cases}$$

where u^m is some point on the arc $\gamma(u^l, u^r)$. To avoid irrelevant difficulties, we assume that $u^m \neq u^{l,r}$. Recall that the operator L reads

$$(Lv)_j = v_j + \sigma(\mathcal{D}_a F_{j-1/2} v_{j-1} + \mathcal{D}_b F_{j-1/2} v_j - \mathcal{D}_a F_{j+1/2} v_j - \mathcal{D}_b F_{j+1/2} v_{j+1}).$$

For $|j| > 1$, this formula has constant coefficients. For instance, if $j \leq -2$, we have

$$\sigma \mathcal{D}_a F_{j-1/2} = \sigma \mathcal{D}_a F^l = (A^l)^+, \quad \sigma \mathcal{D}_b F_{j-1/2} = \sigma \mathcal{D}_b F^l = -(A^l)^-,$$

with the obvious notations

$$A = A^+ - A^-, \quad |A| = A^+ + A^-,$$

where we use the functional calculus for matrices. Therefore

$$(Lv)_j = (I_n - |A^l|)v_j + (A^l)^+ v_{j-1} + (A^l)^- v_{j+1}, \quad j \leq -2.$$

The above formula splits on the eigenbasis of $A^l = \sigma \mathcal{D} f^l$. On the unstable subspace (positive spectrum, $a > 0$), the equation $(Lv)_j = \lambda v_j$ reads

$$(1 - \lambda - a)y_j + a y_{j-1} = 0, \tag{5.27}$$

while on the stable subspace (negative spectrum, $a < 0$), the equation $(Lv)_j = \lambda v_j$ reads

$$(1 - \lambda - |a|)y_j + |a|y_{j+1} = 0. \tag{5.28}$$

Since $|a| < 1$ (from the CFL condition) and $1 - |\lambda|$ lies outside the disk $D(1,1)$ and is non-zero, we see that $|a| < ||a| + \lambda - 1|$. Therefore, the equation (5.28) ensures that $y_j \to 0$ as $j \to -\infty$. For the same reason, the only solution of (5.27) that tends to zero at $-\infty$ is the null solution. Therefore, we see that the stable subspace consists in sequences such that $v_j \in S(A^l)$ for every $j \leq -2$, these vectors being determined from v_{-1} by

$$v_{j-1} = ((A^l)^- - (1 - \lambda)I_n)^{-1}(A^l)^- v_j.$$

Likewise, the unstable space consists in sequences such that $v_j \in U(A^r)$ for $j \geq 2$, these vectors being determined from v_1 by

$$v_{j+1} = ((A^r)^+ - (1 - \lambda)I_n)^{-1}(A^r)^+ v_j.$$

We now consider the equation $(Lv)_1 = \lambda v_1$. We point out that the only term that involves v_2, namely $\sigma \mathcal{D}_b F^{mr} v_2$, vanishes automatically because on the one hand $\mathcal{D}_b F^{mr}$ filters the space $U(A^r)$ and on the other hand v_2 belongs to this space. Similarly, the equation $(Lv)_{-1} = \lambda v_{-1}$ does not involve v_{-2}. This yields the important conclusion that, in order to have a solution of $(L - \lambda)v = 0$, decaying at both ends, it is enough to solve the equations in meshes $j = -1, 0, 1$, but keeping only the unknowns v_{-1}, v_0 and v_1:

$$((1-\lambda)I_n + \sigma \mathcal{D}_b F^l - \sigma \mathcal{D}_a F^{lm})v_{-1} - \sigma \mathcal{D}_b F^{lm} v_0 = 0, \tag{5.29}$$

$$((1-\lambda)I_n + \sigma \mathcal{D}_b F^{lm} - \sigma \mathcal{D}_a F^{mr})v_0 + \sigma \mathcal{D}_a F^{lm} v_{-1} - \sigma \mathcal{D}_b F^{mr} v_1 = 0, \tag{5.30}$$

$$((1-\lambda)I_n + \sigma \mathcal{D}_b F^{mr} - \sigma \mathcal{D}_a F^r)v_1 + \sigma \mathcal{D}_a F^{mr} v_0 = 0. \tag{5.31}$$

Since $R(u^l, u^m) = u^l$, we still have $\sigma \mathcal{D}_a F^{lm} = (A^l)^+$ and therefore (5.29) rewrites

$$((1 - \lambda)I_n - |A^l|)v_{-1} - \sigma \mathcal{D}_b F^{lm} v_0 = 0.$$

From the CFL condition, the parenthesis is non-singular and we may eliminate v_{-1} from the system. Similarly, (5.31) serves to eliminate v_1, so that we end with only one linear equation in the unknown v_0. The latter admits a non-trivial solution, yielding a non-trivial kernel of $L - \lambda$, if and only if

$$\det \left((1 - \lambda)I_n + \sigma \mathcal{D}_b F^{lm} - \sigma \mathcal{D}_a F^{mr}\right.$$
$$+\sigma \mathcal{D}_a F^{lm}((1 - \lambda)I_n - |A^l|)^{-1}\sigma \mathcal{D}_b F^{lm}$$
$$\left.+\sigma \mathcal{D}_b F^{mr}((1 - \lambda)I_n - |A^r|)^{-1}\sigma \mathcal{D}_a F^{mr}\right) = 0. \tag{5.32}$$

Equation (5.32) looks a bit complicated, although all terms in it may be computed explicitly. However, there are additional simplifications, for the following reasons

- Because $R(u^l; u^m) = u^l$, there holds $\sigma \mathcal{D}_a F^{lm} = (A^l)^+$,
- Hence $\sigma \mathcal{D}_a F^{lm}$ commutes with $(1 - \lambda)I_n - |A^l|$ and with its inverse,
- On an other hand, the range of $\mathcal{D}_b F^{lm}$ is $S(A^l)$,
- Hence the product

$$\sigma \mathcal{D}_a F^{lm}((1 - \lambda)I_n - |A^l|)^{-1}\sigma \mathcal{D}_b F^{lm}$$

vanishes.

Likewise, there holds

$$\sigma \mathcal{D}_b F^{mr}((1 - \lambda)I_n - |A^r|)^{-1}\sigma \mathcal{D}_a F^{mr} = 0_n.$$

Finally, we obtain from (5.32) the Evans function

$$D(\lambda) = \det((\lambda - 1)I_n - \sigma \mathcal{D}_b F^{lm} + \sigma \mathcal{D}_a F^{mr}).$$

We point out that D is a polynomial. This is a rather unusual feature, as well as the fact that this Evans function is explicit! Notice that when the shock strength is small, then

$$\sigma \mathcal{D}_b F^{lm} - \sigma \mathcal{D}_a F^{mr} \sim |A^r|,$$

and the CFL condition ensures that the roots of D are stable.

The Computation of the Stability Index.

Since there is a continuous DSP (parametrized by the points of the arc $\gamma(u^l, u^r)$), we must have $D(1) = 0$. As a matter of fact, we have $R(u^l, c) = u^l$ and $R(c, u^r) = u^r$ for every c on this arc. Thus differentiation gives $\mathcal{D}_b R^{lm}\tau = 0$ and $\mathcal{D}_a R^{mr}\tau = 0$, where τ is the tangent vector to $\gamma(u^l, u^r)$ at u^m. By chain rule, we have likewise

$$\mathcal{D}_b F^{lm} \tau = \mathcal{D}_a F^{mr} \tau = 0.$$

This shows that the matrix

$$\sigma \mathcal{D}_b F^{lm} - \sigma \mathcal{D}_a F^{mr}$$

is singular, thus $D(1) = 0$.

Since the range of $\sigma \mathcal{D}_b F^{lm}$ is $S(A^l)$ and that of $\sigma \mathcal{D}_a F^{mr}$ is $U(A^r)$, and these subspaces are supplementary, we have generically

$$\mathrm{Ran}\left(\sigma \mathcal{D}_b F^{lm} - \sigma \mathcal{D}_a F^{mr}\right) = S(A^l) \oplus U(A^r). \tag{5.33}$$

We deduce that $\lambda = 1$ is non semi-simple if, and only if, τ belongs to this range, that is

$$\det(r_1(u^l), \ldots, r_{k-1}(u^l), \tau, r_{k+1}(u^r), \ldots, r_n(u^r)) = 0. \tag{5.34}$$

This is equivalent to saying that $D'(1)$ vanishes.

We now let u^m varying along $\gamma(u^l, u^r)$. Since the stability index can change only when the sign of $D'(1)$ changes, we easily see that it is proportional to the left-hand side of (5.34). Using a continuity argument, we shall have a closed form of the stability index once we have computed it when u^m is close to u^l. When $u^m \to u^l$, we have the following limit:

$$\sigma \mathcal{D}_b F^{lm} \to -(A^l)^-.$$

The structure of $\sigma \mathcal{D}_a F^{mr}$ is more complicated and we shall not detail it here, except in the case of an n-Lax shock. In this situation, the fact that the spectrum of $\mathcal{D}f^r$ is negative implies that $R(a, u^r) \equiv u^r$ for a in a neighbourhood of u^m ; therefore $\sigma \mathcal{D}_a F^{mr} = 0_n$. This yields the conclusion that the Evans function $D(\lambda; u^m)$ associated to the choice of the mid-point u^m, tends to

$$D(\lambda; u^l) := \det((\lambda - 1)I_n + (A^l)^-).$$

From the CFL condition, we see that a DSP for an n-Lax-shock with u^m close to u^l is spectrally stable. In particular, it has a positive stability index. Since $\tau(u^l)$ is nothing but the right eigenvector $r_n(u^l)$, a continuity argument yields the following result.

Theorem 5.2 *Let (u^l, u^r) be a steady k-Lax shock of a hyperbolic system of conservation laws, the k-th eigenvalue being simple. We assume that the Riemann problem is uniquely solvable, and that the equations*

$$R(u^l, c) = u^l, \quad R(c, u^r) = u^r$$

define a smooth arc $\gamma(u^l, u^r)$ from u^l to u^r. Let u^m be a point on this arc, and U be the corresponding DSP for the Godunov scheme.

Then:

- *The Evans function of this DSP is well-defined.*
- *If the shock strength is moderate, the DSP is spectrally stable.*
- *There exists a constant sign $\epsilon(u^l, u^r) = \pm$ such that the stability index of the DSP be*

$$\epsilon(u^l, u^r) \operatorname{sgn} \det(r_1(u^l), \ldots, r_{k-1}(u^l), \tau(u^m), r_{k+1}(u^r), \ldots, r_n(u^r))$$

  for every u^m along $\gamma(u^l, u^r)$, where $\tau(u^m)$ is the oriented tangent vector to $\gamma(u^l, u^r)$ at u^m. The sign $\epsilon(u^l, u^r)$ depends on the orientation.
- *If $k = n$, the stability index of the DSP equals the sign of the product*

$$\det(r_1(u^l), \ldots, r_{n-1}(u^l), \tau(u^m)) \, \det(r_1(u^l), \ldots, r_n(u^l)),$$

  where the orientation is chosen so that $\tau(u^l) = r_n(u^l)$.
- *If $k = n$ and u^m is close to u^l, then the DSP is spectrally stable.*
- *Similarly, if $k = 1$, the stability index of the DSP equals the sign of the product*

$$\det(\tau(u^m), r_2(u^r), \ldots, r_n(u^r)) \, \det(r_1(u^r), \ldots, r_n(u^r)),$$

  where the orientation is chosen so that $\tau(u^r) = r_1(u^r)$.
- *If $k = 1$ and u^m is close to u^r, then the DSP is spectrally stable.*

Remarks.

- We point out that in the case of an extreme shock, say an n-shock, the spectral stability of all the DSPs from u^l to u^r implies the monotonicity of the function

$$c \mapsto \det(r_1(u^l), \ldots, r_{n-1}(u^l), c)$$

  along $\gamma(u^l, u^r)$. In other words, an instability must appear for some DSPs as soon as this arc experiences an overshoot, crossing the hyperplane

$$u^r + \operatorname{Span}\{r_1(u^l), \ldots, r_{n-1}(u^l)\}$$

  before reaching its end u^r. See Figure 4. We shall give an example of such an instability in the next paragraph.
- Let us consider again an n-shock. The integral of

$$\det(r_1(u^l), \ldots, r_{n-1}(u^l), \tau(u^m))$$

  along $\gamma(u^l, u^r)$ equals

$$\det(r_1(u^l), \ldots, r_{n-1}(u^l), [u]).$$

  Therefore, a necessary condition for all the DSPs having a positive stability index is that the product

$$\det(r_1(u^l), \ldots, r_{n-1}(u^l), [u]) \, \det(r_1(u^l), \ldots, r_n(u^l))$$

  be positive. This is exactly saying that the stability index of a viscous shock profile, associated to the viscous tensor I_n, be positive (see [GZ].)

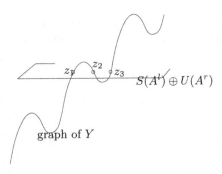

Fig. 4. A "non monotonous" continuous DSP for a k-Lax shock. The stability index for z_2 is opposite to that of z_1 and z_3.

The Case of Perfect Gases

Let us consider the Euler equations of a perfect gas

$$\partial_t \varrho + \partial_x(\varrho z) = 0,$$

$$\partial_t(\varrho z) + \partial_x(\varrho z^2 + (\gamma - 1)\varrho e) = 0,$$

$$\partial_t \left(\frac{1}{2}\varrho z^2 + \varrho e\right) + \partial_x \left((\frac{1}{2}\varrho z^2 + \gamma \varrho e)z\right) = 0,$$

where $\gamma > 1$ is the adiabatic constant. Physically relevant values are $\gamma = 7/5$ (di-atomic gases such as Air, Oxygen,...), $\gamma = 5/3$ (mono-atomic gases such as Argon, Helium,...) and $\gamma = 2$ (shallow water). Gases with more complex molecules yield lower values of γ.

All the shocks are extreme Lax shocks, since the intermediate characteristic field is linearly degenerate and the other ones are genuinely non-linear. According to Theorem 5.2, we only need to check whether the n-shock[35] curve $\mathcal{S}^3(u^l)$ is "monotonous" with respect to the plane spanned by $r_1(u^l)$ and $r_2(u^l)$. When the shock strength is moderate, we already know that this is the case. The interesting question is thus whether there exist large shocks for which this is not the case.

The complete analysis is made in [11]. It turns out that the answer to the question depends on γ. There exists a threshold γ^*, the unique root of the polynomial equation

$$X^4 + 3X^3 - 21X^2 + 17X + 8 = 0$$

in the interval $(1, 1 + 2/\sqrt{3})$. For γ larger than γ^* the monotonicity holds true, meaning that all DSPs of all steady shocks have a positive stability index. However, for γ less than γ^*, there exist large enough shocks for which the arc $\gamma(u^l, u^r)$ experiences an overshoot as in Figure 4. For such shocks, the DSPs that are associated to a mid-point u^m in the "returning part" of the arc, are spectrally unstable.

---

[35] The case of 1-shock is similar, by reversing the particle velocity.

Open Problem 5.2 *Prove that, for a perfect gas, every Godunov DSP with a positive stability index is actually spectrally stable. The numerical experiments in [11] support this conjecture.*

5.4 The Role of the Functional Y in the Nonlinear Stability

The theory of nonlinear stability of DSPs is a deep issue, which will not be described in details here. Let us only mention some results that we are aware of. We first recall that in the irrational case, the DSPs contructed by T.-P. Liu & H.-S. Yu [41] are asymptotically (orbitally) stable. In the rational case, J.-G. Liu & Z. Xin [37, 38] proved the nonlinear stability for DSPs of a modified Lax–Friedrichs scheme. The most important limitations of their work are

- The smallness of the underlying shock, which is usually required when performing direct energy estimates. It is also required in order that the existence Theorem of Majda & Ralston applies,
- The zero-mass of the initial disturbance, which avoids the development of diffusion waves. This assumption was dropped by L. Ying [58].

We restrict here to the rational case and assume as usual, for the sake of simplicity, that $\eta = 0$. Comparing with the situation of viscous shock profiles (see [24, 57, 19] and the works by K. Zumbrun & coll. from [30] to [25]) or semi-discrete shocks (see S. Benzoni, P. Huot & F. Rousset [9]), we expect that spectrally stable DSPs are nonlinearly orbitally stable, at least in the case where there exists a continuous DSP; typically, the shock is a Lax shock. Such a non-linear stability result would need an accurate estimate of a Green's function for a discrete operator, which would make use of the Evans function. The case of an under-compressive shock is less clear, since there is not any analogous object at the viscous level.

Open Problem 5.3 *Given a continuous DSP $\mathcal{U}$ for a steady Lax shock, assume that every DSP extracted from $\mathcal{U}$ is spectrally stable and that the scheme has some numerical viscosity. Prove that $\mathcal{U}$ is orbitally stable.*

In the sequel, we content ourselves computing a likely limit of the approximate solution $U^m = (u_j^m)_{j\in\mathbb{Z}}$, as $m \to +\infty$. To this end, we follow the strategy elaborated by T.-P. Liu in the viscous case. However, the situation is slightly richer in our discrete context.

Assuming that the initial data U^0 differs from some reference DSP U by a summable disturbance, we are looking for a likely limit V of U^m in the ℓ^p-norm for some $p > 1$:

$$\lim_{m\to+\infty} \|U^m - V\|_p = 0, \quad \|Z\|_p := \left(\sum_{j\in\mathbb{Z}} |z_j|^p\right)^{1/p}. \tag{5.35}$$

In practice, we do not expect that the limit hold in the ℓ^1 sense for the following reasons. On the one hand, V is a steady DSP (just pass to the limit in the scheme). On the other hand, there holds

$$\sum_{j\in\mathbb{Z}}(u_j^m - v_j) = \sum_{j\in\mathbb{Z}}(u_j^0 - v_j),$$

because of the conservativeness of the scheme. Hence a necessary condition for an ℓ^1-convergence is that the total mass

$$M^0 := \sum_{j\in\mathbb{Z}}(u_j^0 - v_j)$$

of the initial disturbance be zero. Actually, since there is in general a continuum of DSPs, and we do not know which one is expected to be the limit, it does not make sense to compare U^0 to the unknown limit V. However, it makes sense to compare U^0 to the reference DSP U. Then a necessary condition for the orbital stability in the ℓ^1 sense is that the total mass

$$m^0 := \sum_{j\in\mathbb{Z}}(u_j^0 - u_j)$$

of the initial disturbance be equal to one of the masses

$$\sum_{j\in\mathbb{Z}}(v_j - u_j)$$

between U and the other DSPs V. We point out that the latter vector is nothing but a value of the functional Y introduced in Section 2.1. Therefore, a necessary condition for nonlinear orbital stability in the ℓ^1-topology is that m^0 belong to the range of Y. This is unlikely when $n \geq 2$.

The way to overcome this difficulty and to make use of the mass conservation is similar to the viscous case: We must take in account the diffusion waves generated by the numerical viscosity. The diffusion waves are outgoing and therefore depend only on the linearized operator $L^{l,r}$ about the constant state $u^{l,r}$, and also on the amount of nonlinearity in each characteristic field. In particular, the numerical viscosity is only needed at the end states $u^{l,r}$ and therefore the theory applies to the Godunov scheme, as well as to the Lax–Friedrichs one.

For a conservative three-point scheme, the linearized operator about a constant state $\bar{u}$ reads

$$(LW)_j = Aw_{j-1} + Bw_j + Cw_{j+1},$$

where A, B, C are constant matrices with the property that

$$A + B + C = I_n.$$

For the sake of simplicity, we limit ourselves to the situation where A, B, C are diagonal in the same basis as $df(\bar{u})$. This assumption is fulfilled by the Godunov and Lax–Friedrichs schemes. Then each eigen-component is governed by a scalar operator

$$(Lz)_j = \alpha z_{j-1} + \beta z_j + \gamma z_{j+1},$$

where α, β, γ are constants such that

$$\alpha + \beta + \gamma = 1.$$

The scheme displays numerical viscosity if the three coefficients are non-negative, with at most one being zero. Under this assumption, the Central Limit Theorem tells that every solution Z of $Z^{m+1} = LZ^m$ with a summable initial data is asymptotic to a Gaussian function, traveling at constant grid velocity $\gamma - \alpha$, with a constant mass, and whose height decreases as $m^{-1/2}$. Such solutions are called "linear diffusion waves".

In practice, the diffusion waves are non-linear, since quadratic terms may not be negligible. Hence a distortion effect takes place in the Gaussian. However, what remains true is that to each outgoing mode $(\lambda_i^l < 0\,;r_i(u^l))$ or $(\lambda_i^r > 0\,;r_i(u^r))$, there exists a one-parameter family $W(i, m_i)$ of diffusion wave, with the property that it behaves like a function

$$\frac{1}{\sqrt{t}}\mathcal{W}_i\left(m_i\,;\frac{x - \lambda_i t}{\sqrt{t}}\right) r_i,$$

where $\mathcal{W}_i(m_i\,;\cdot)$ is of Schwartz class. Additionally, $m_i r_i$ is the mass of the wave:

$$\int_{\mathbb{R}} \mathcal{W}_i(m_i; y)\, dy = m_i. \tag{5.36}$$

The formulæ above show that the diffusion waves decay in every ℓ^p-norm but the ℓ^1. Hence they may enter as correctors in an ℓ^1-asymptotics, without polluting that in other ℓ^p's. Typically, the approximate solution U^m will be asymptotic in $\ell^1 \cap \ell^\infty$ to a sum

$$V + \sum_{\lambda_i^l < 0} W^l(i; m_i) + \sum_{\lambda_i^r > 0} W^r(i; m_i), \tag{5.37}$$

where V is a DSP for the same shock (u^l, u^r) and the other terms are diffusion waves.

There remains to determine V and the parameters m_i in the sum (5.37). To do so, we use the conservation of mass

$$\sum_{j\in\mathbb{Z}}(u_j^m - u_j) = \sum_{j\in\mathbb{Z}}(u_j^0 - u_j) =: m^0$$

and write that

$$U^m = V + \sum_{\lambda_i^l < 0} W^l(i; m_i) + \sum_{\lambda_i^r > 0} W^r(i; m_i) + Z^m,$$

where Z^m tends to zero in every norm, including that of ℓ^1. We decompose

$$\sum_{j \in \mathbb{Z}} (u_j^m - u_j) = \sum_{j \in \mathbb{Z}} (v_j - u_j) + \sum_{j \in \mathbb{Z}} z_j^m \qquad (5.38)$$

$$+ \sum_{j \in \mathbb{Z}} \sum_{\lambda_i^l < 0} W^l(i; m_i) + \sum_{j \in \mathbb{Z}} \sum_{\lambda_i^r > 0} W^r(i; m_i)$$

and pass to the limit as $m \to +\infty$. In the right-hand side of (5.38), the first term does not depend on m, the second one tends to zero and the other ones tend to vectors $m_i r_i$. We thus obtain:

$$m^0 = \sum_{j \in \mathbb{Z}} (v_j - u_j) + \sum_{\lambda_i^l < 0} m_i r_i^l + \sum_{\lambda_i^r > 0} m_i r_i^r. \qquad (5.39)$$

Equation (5.39) is the algebraic problem that we need to solve in order to guess the likely limit V of U^m. In its analysis, we must distinguish two cases:

Lax Shocks.

Let us assume that the shock is a k-Lax shock and that there is a continuous DSP $\mathcal{U}$. We may fix $u_j = \mathcal{U}(j)$. There are precisely $n - 1$ outgoing waves, $k - 1$ at left and $n - k$ at right. At last, every other DSP V has the form $v_j = \mathcal{U}(j + h)$ for some shift h. Hence (5.39) reads

$$m^0 = Y(h) + \sum_{i=1}^{k-1} m_i r_i^l + \sum_{i=k+1}^{n} m_i r_i^r. \qquad (5.40)$$

In other words, if we are only interested in the limit profile, that is in the shift h, we should write instead

$$Y(h) \in m^0 + S(\mathcal{D}f(u^l)) \oplus U(\mathcal{D}f(u^r)). \qquad (5.41)$$

Since $Y(h+1) = Y(h) + u^r - u^l$ and Y is continuous in h, a sufficient condition for the existence of a solution to (5.41) is that the Majda–Liu determinant

$$\Delta := \det(r_1^l, \ldots, r_{k-1}^l, [u], r_{k+1}^r, \ldots, r_n^r)$$

does not vanish. For if $\Delta \neq 0$, then the range $\mathcal{Y}$ of Y is a continuous curve of average direction $[u]$, and it must cross every affine hyperplane of direction $S(\mathcal{D}f(u^l)) \oplus U(df(u^r))$, thanks to the Intermediate Value Theorem.

On an other hand, it may happen that the curve $\mathcal{Y}$ has turning points with respect to the direction $S(\mathcal{D}f(u^l)) \oplus U(df(u^r))$. In other words, the sign of

$$\delta(h) := \det(r_1^l, \ldots, r_{k-1}^l, Y'(h), r_{k+1}^r, \ldots, r_n^r)$$

changes. In this case, there are masses m^0 such that the curve $\mathcal{Y}$ intersects $m^0 + S(\mathcal{D}f(u^l)) \oplus U(\mathcal{D}f(u^r))$ in more than one point (usually an odd number.) This is an obstacle to the unique determination of the limit V. This phenomenon is precisely the one that happened in Section 5.3. More generally, we found in Sects. 5.1 and 5.3 that the stability index of a DSP is proportional to the sign of $\delta(h)$. This explains *a posteriori* the major role played by this quantity for Lax shocks.

We point out a significant difference with the case of viscous shock profiles. For the latter, two profiles differ only by a shift, and therefore $Y(h) \equiv h[u]$. In particular, (5.40) is linear. Its unique solvability is equivalent to $\Delta \neq 0$. Whence a simpler picture.

Under-Compressive Shocks.

When the shock is under-compressive, the situation is much different since, on the one hand, there are n outgoing waves and on the other hand the set of DSPs is discrete. Typically, there are two DSPs, say U and U', with opposite "intersection signs", and all the other DSPs are deduced from U and U' by grid shifts. Hence the sum

$$\sum_{j \in \mathbb{Z}} (v_j - u_j)$$

in (5.39) may take only the values $\ell[u]$ and $\bar{m} + \ell[u]$, where ℓ is an integer and

$$\bar{m} := \sum_{j \in \mathbb{Z}} (u'_j - u_j).$$

In pratice, $\bar{m}$ does not belong to $\mathbb{Z}[u]$ and all the values described above are pairwise distinct. Our algebraic problem is therefore to find masses m_i and an integer ℓ such that

$$m^0 = \ell[u] + \sum_{i=1}^k m_i r_i^l + \sum_{i=k+1}^n m_i r_i^r + \begin{cases} \text{either } 0, \\ \text{or } \bar{m}. \end{cases} \tag{5.42}$$

Of course, if U is spectrally stable, U' is not (their stability indices are opposite) and we do not expect that it be the limit in general. Hence the choice $\bar{m}$ should be avoided in (5.42).

In the generic case where

$$\det(r_1^l, \ldots, r_k^l, r_{k+1}^r, \ldots, r_n^r) \neq 0,$$

this problem is underdetermined, since for every choice of ℓ in $\mathbb{Z}$ and of a vector in $\{0, \bar{m}\}$, there exists a unique set of masses m_i that satisfies (5.42). This is a cause of undeterminacy of the limit V.

Of course, if the initial disturbance is small enough (and then m^0 is small too), we expect that the limit be U itself, because of discreteness of the set of DSPs. However, this smallness will be extremely small in natural units, since the genuine L^1-norm is the product of the ℓ^1-norm by Δx. Hence, given a fixed non-trivial initial disturbance, and letting Δx tend to zero, this disturbance will be too large once Δx is small enough, and we cannot expect that the limit of U^m be U itself.

References

1. V. I. Arnold. *Geometrical methods in the theory of ordinary differential equations.* Grundlehren der mathematischen Wissenschaften **250**. Springer–Verlag, New York (1983).
2. P Baiti, A Bressan, H.-K Jenssen. Instability of Travelling Wave Profiles for the Lax-Friedrichs scheme. *Discrete Contin. Dynam. Systems*, 13 (2005), pp 877–899.
3. P. Bates, K. Lu, C. Zeng. Invariant foliations near normally hyperbolic invariant manifolds for semiflows. *Trans. Amer. Math. Soc.*, **352** (2000), pp 4641–4676.
4. S. Benzoni-Gavage. Contribution à l'étude des solutions régulières par morceaux de systèmes hyperboliques de lois de conservation. *Document de synthèse pour l'habilitation à diriger des recherches.*
5. S. Benzoni-Gavage. Semi-discrete shock profiles for hyperbolic systems of conservation laws. *Physica* **D115** (1998), pp 109–123.
6. S. Benzoni-Gavage. Sur la stabilité des profils semi-discrets au moyen d'une fonction d'Evans en dimension infinie. *Comptes Rendus Ac. Sci. Paris, I*, **329** (1999), pp 377–382.
7. S. Benzoni-Gavage. Stability of semi-discrete shock profiles by means of an Evans function in infinite dimension. *J. Dyn. Diff. Equations*, **14** (2002), pp 613–674.
8. S. Benzoni-Gavage, P. Huot. Existence of semi-discrete shocks. *Discrete Cont. Dyn. Systems*, **8** (2002), pp 163–190.
9. S. Benzoni-Gavage, P. Huot, F. Rousset. Nonlinear stability of semidiscrete shock waves. *SIAM J. Math. Anal.* **35** (2003), pp 639–707.
10. S. Benzoni-Gavage, D. Serre, K. Zumbrun. Alternate Evans functions and viscous shock waves. *SIAM J. Math. Anal.*, **32** (2001), pp 929–962.
11. M. Bultelle, M. Grassin, D. Serre. Unstable Godunov discrete profiles for steady shock waves. *SIAM J. Numer. Anal.*, **35** (1998), pp 2272–2297.
12. S.-N. Chow, J. Mallet-Paret, W. Shen. Traveling waves in lattice dynamical systems. *J. Diff. Equations*, **149** (1998), pp 248–291.
13. K. Chuey, C. Conley, J. Smoller, Positively invariant regions of nonlinear diffusion equations. *Indiana Univ. Math. J.* **26** (1977), pp 373–392.
14. W. Coppel. *Dichotomies in stability theory.* Lecture Notes in Mathematics, **629**. Springer–Verlag (1978).
15. C. Dafermos. *Hyperbolic conservation laws in continuum physics.* Grundlehren der mathematischen Wissenschaften, **325**. Springer–Verlag (2000), Heidelberg.
16. Fan, Haitao. *Existence and uniqueness of traveling waves and error estimates for Godunov schemes of conservation laws.* Math. Comp. **67** (1998), **no. 221**, 87–109.

17. B. Fiedler, J. Scheurle. *Discretization of homoclinic orbits, rapid forcing and "invisible chaos"*. Memoirs of the Amer. Math. Soc., **119** (1996), no 570.
18. M. Foy. Steady state solutions of hyperbolic systems of conservation laws with viscosity terms. *Comm. Pure Appl. Math.*, **17** (1964), pp 177–188.
19. C. Fries. Stability of viscous shock waves associated with non-convex modes. *Arch. Rational Mech. Anal.*, **152** (2000), pp 141–186.
20. R. Gardner, K. Zumbrun. The gap lemma and geometric criteria for instability of viscous shock profiles. *Comm. Pure Appl. Math*, **51** (1998), pp 797–855.
21. P. Godillon. *Stabilité des profils de chocs dans les systèmes de lois de conservation*. PhD Thesis. Ecole Normale Supérieure de Lyon. Lyon, France (2001).
22. E. Godlewski, P.-A. Raviart. *Numerical approximations of hyperbolic systems of conservation laws*. Springer–Verlag, New–York (1996).
23. S. Godunov. A difference scheme for numerical calculations of discontinuous solutions of the equations of hydrodynamics. *Math. Sb.*, **47** (1959), pp 271–306.
24. J. Goodman, Z. Xin. Viscous limits for piecewise smooth solutions to systems of conservation laws. *Arch. Rational Mech. Anal.*, **121** (1992), pp 235–265.
25. O. Guès, G. Métivier, M. Williams, K. Zumbrun. Multidimensional viscous shocks. I. Degenerate symmetrizers and long time stability, *J. Amer. Math. Soc.* **18** (2005), pp 61–120.
26. D. Hankerson, B. Zinner. Wavefronts for a cooperative tridiagonal system of differential equations. *J. Dyn. Differential Equations*, **5** (1993), pp 359–373.
27. A. Harten, J. M. Hyman, P. D. Lax. On finite-difference approximations and entropy conditions for shocks. With an appendix by B. Keyfitz. *Comm. Pure Appl. Math.*, **29** (1976), pp 297–322.
28. M. Hata. Rational approximations to π and some other numbers. *Acta Arith.*, **63** (1993), pp 335–349.
29. M. Hirsch, C. Pugh, M. Shub. *Invariant manifolds*. Lecture Notes in Mathematics, **583**. Springer–Verlag, Berlin (1977).
30. D. Howard, K. Zumbrun. Pointwise semi-group methods and stability of viscous shock waves. *Indiana Univ. Math. J.*, **47** (1998), pp 741–871.
31. G. Jennings. Discrete shocks. *Comm. Pure Appl. Math.*, **27** (1974), pp 25–37.
32. Shi Jin, J.-G. Liu. Oscillations induced by numerical viscosities. *Matemática Contemporânea*, **10** (1996), pp 169–180.
33. T. Kapitula, B. Sanstede. Stability of bright solitary-wave solutions to perturbed nonlinear Schrödinger equations. *Physica* **D124** (1998), pp 58–103.
34. T. Kato. *Perturbation theory for linear operators*. Springer-Verlag, Berlin (1995), Reprint of the 1980 edition.
35. P. D. Lax. Hyperbolic systems of conservation laws (II). *Comm. Pure Appl. Math.*, **10** (1957), pp 537–566.
36. R. J. Leveque. *Numerical methods for conservation laws*. Birkhäuser, Basel (1990).
37. J.-G. Liu, Z. Xin. L^1-stability of stationary discrete shocks. *Math. Comput.*, **60** (1993), pp 233–244.
38. J.-G. Liu, Z. Xin. Nonlinear stability of discrete shocks for systems of conservation laws. *Arch. Rational Mech. Anal.*, **125** (1994), pp 217–256.
39. T.-P. Liu. The entropy condition and the admissibility of shocks. *J. Math. Anal. Appl.*, **53** (1976), pp 78–88.
40. T.-P. Liu. *Nonlinear stability of shock waves for viscous conservation laws*, Memoirs of the Amer. Math. Soc., **328**. Providence (1985).

41. T.-P. Liu, H.-S. Yu. Continuum shock profiles for discrete conservation laws. *Comm. Pure Appl. Math.*, **52** (1999), I. Construction, pp 85–127 & II. Stability, pp 1047–73.
42. K. Mahler. On the approximation of π. *Indagationes Math.*, **15** (1953), pp 30–42.
43. A. Majda, J. Ralston. Discrete shock profiles for systems of conservation laws. *Comm. Pure Appl. Math.*, **32** (1979), pp 445–482.
44. J. Mallet–Paret. The Fredholm alternative for functional-difference equations of mixed type. *J. Dyn. Differential Equations*, **11** (1999), pp 1–47.
45. J. Mallet–Paret. The global structure of traveling waves in spatially discrete dynamical systems. *J. Dyn. Differential Equations*, **11** (1999), pp 49–127.
46. D. Michelson. Discrete shocks for difference approximations to systems of conservation laws. *Adv. Appl. Math.*, **5** (1984), pp 433–469.
47. J. Palis, W. de Melo. *Geometric theory of dynamical systems. An introduction.* Springer–Verlag, Berlin (1982).
48. G. Rhin, C. Viola. The group structure of $\zeta(3)$. *Acta Arith.*, **97** (2001), pp 269–293.
49. F. Rousset. Viscous approximation of strong shocks of systems of conservation laws. *SIAM J. Math. Anal.* **35** (2003), pp 492–519.
50. D. H. Sattinger. On the stability of waves of nonlinear parabolic systems. *Advances in Math.*, **22** (1976), pp 312–355.
51. D. Serre. Remarks about the discrete profiles of shock waves. *Matemática Contemporânea*, **11** (1996), pp 153–170.
52. D. Serre. Discrete shock profiles and their stability. *Hyperbolic problems: Theory, Numerics and Applications*, Zurich 1998. M. Fey, R. Jeltsch eds. ISNM **130**, Birkäuser (1999), pp 843–854.
53. D. Serre. *Systems of conservation laws, I.* Cambridge Univ. Press. Cambridge (1999).
54. D. Serre. *Systems of conservation laws, II.* Cambridge Univ. Press. Cambridge (2000).
55. D. Serre. L^1-stability of nonlinear waves in scalar conservation laws. Evolutionary equations, C. Dafermos, E. Feireisl eds. Vol. I, 473–553, *Handb. Differ. Equ.*, North-Holland, Amsterdam, 2004.
56. Y. Smyrlis. Existence and stability of stationary profiles of the LW scheme. *Comm. Pure Appl. Math.*, **42** (1990), pp 509–545.
57. A. Szepessy, Z. Xin. Nonlinear stability of viscous shock waves. *Arch. Rational Mech. Anal.*, **122** (1993), pp 53–103.
58. L. Ying. Asymptotic stability of discrete shock waves for the Lax–Friedrichs scheme to hyperbolic systems of conservation laws. *Japan J. Indus. Appl. Math.*, **14** (1997), pp 437–468.
59. B. Zinner. Existence of traveling wavefront solutions for the discrete Nagumo equation. *J. Differential Equations*, **96** (1992), pp 1–27.
60. K. Zumbrun, D. Serre. Viscous and inviscid stability of multidimensional shock fronts. *Indiana Univ. Math. J.*, **48** (1999), pp 937–992.

Stability of Multidimensional Viscous Shocks

Mark Williams [*]

Department of Mathematics, CB 3250, University of North Carolina
Chapel Hill, NC 27599, USA
williams@email.unc.edu

Summary. In the first four lectures we describe a recent proof of the short time existence of curved multidimensional viscous shocks, and the associated justification of the small viscosity limit for piecewise smooth curved inviscid shocks. Our goal has been to provide a detailed, readable, and widely accessible account of the main ideas, while avoiding most of the technical aspects connected with the use of pseudodifferential (or paradifferential) operators. The proof might be described as a combination of ODE/dynamical systems analysis with microlocal analysis, with the main new ideas coming in on the ODE side. In a sense the whole problem can be reduced to the study of certain linear systems of nonautonomous ODEs depending on frequencies as parameters. The frequency-dependent matrices we construct as conjugators or symmetrizers in the process of proving estimates for those ODEs serve as principal symbols of pseudodifferential operators used to prove estimates for the original PDEs.

The linearized problem one has to study in the multiD curved viscous shock problem is one for which there are no available constructive methods (in contrast to the 1D case). In other words we have no idea how to estimate solutions by first constructing them using tools like Fourier-Laplace transforms or Green's functions or even Fourier integral operators and their generalizations. Instead, we rely on energy estimates proved using Kreiss-type symmetrizers. Indeed, our main tool is a symmetrizer for hyperbolic-parabolic boundary problems which generalizes the kind of symmetrizer invented by Kreiss in the early 1970s to deal with hyperbolic boundary problems.

In the final lecture we describe how symmetrizers can be used to study the related (but nonequivalent) problem of long time stability for planar viscous shocks. For zero mass perturbations or nonzero mass perturbations in high space dimensions ($d \geq 5$), one can use symmetrizers just like those used for the first problem (*nondegenerate* symmetrizers). However, in order to get the strongest results by symmetrizer methods (nonzero mass perturbations for dimensions $d \geq 2$) we've had to introduce *degenerate* symmetrizers. In addition, we have to use them in a nonstandard way involving duality and interpolation arguments to get $L^1 - L^p$ estimates instead of L^2 estimates. We'll focus on the use of degenerate symmetrizers in lecture five.

[*] Research was supported in part by NSF grant DMS-0070684 (M.W.).

Preface.

These lectures are based on joint work with Olivier Guès, Guy Métivier, and Kevin Zumbrun contained in the papers [GMWZ1, GMWZ2, GMWZ3, GMWZ4] and also on the papers [GW, MZ]. I thank them all for an exciting collaboration that I hope will last well into the future.

There is a large literature dealing with small viscosity limits and long time stability questions for shocks in one space dimension. I'll mention now only some of the papers that deal with questions most closely analogous to the ones studied here. For the small viscosity problem the papers include [GX, MN, R, Y] and for long time stability [Go, KK, L1, L2, LZ, SX, ZH]. In 1D there are also remarkable small viscosity results by quite different methods where the inviscid limits are allowed to be much more general than the piecewise smooth shocks considered here. These results are discussed in the companion lectures by Alberto Bressan.

1 Lecture One: The Small Viscosity Limit: Introduction, Approximate Solution

In this lecture we set up the problem and construct an approximate solution. This construction is one of several places where the inviscid and viscous theories make close contact. In addition, it illuminates part of the role of our main stability hypothesis, and indicates the need to allow variation of the viscous front in the later stability analysis.

We begin with a simple case that still contains most of the main difficulties. Our regularity and hyperbolicity hypotheses can be weakened considerably, and more general viscosities can be treated by these methods (even the degenerate viscosity of compressible Navier-Stokes, we expect; we're checking NS as this is being typed). For those generalizations we refer to the GMWZ papers in the bibliography.

We work in two space dimensions just to make some things easier to write down. The same arguments work in any space dimension.

Consider the $m \times m$ hyperbolic system of conservation laws on $\mathbb{R}^3_{x,t,y}$

$$\partial_t u + \partial_x f(u) + \partial_y g(u) = 0, \qquad (1.1)$$

for which we are given a shock solution $(u^0_\pm, \psi_0)$. This means that u^0_+ (resp. u^0_-) satisfy (1.1) in the classical sense to the right (resp. left) of the shock surface $\mathcal{S}$ defined by $x = \psi_0(t, y)$ and in the distribution sense in a neighborhood of $\mathcal{S}$. The piecewise classical solution is a distribution solution near $\mathcal{S}$ if and only if the Rankine-Hugoniot jump condition holds:

$$\psi^0_t [u^0] + \psi^0_y [g(u^0)] - [f(u^0)] = 0 \text{ on } \mathcal{S}. \qquad (1.2)$$

Assumption 1.1 (Hypotheses on the inviscid shock) *(H1) For states u near $u_\pm^0$, the matrix $f'(u)\xi + g'(u)\eta$ has simple real eigenvalues for $(\xi, \eta) \in \mathbb{R}^2 \setminus 0$.*

(H2) The inviscid shock is piecewise smooth (C^∞), exists on the time interval $[0, T_0]$, is constant outside some compact set, and is a Lax shock. This means that the normal matrices

$$A_\nu(u_\pm^0, d\psi_0) \equiv f'(u_\pm^0) - \psi_t^0 I - \psi_y^0 g'(u_\pm^0) \tag{1.3}$$

are invertible, and that if we let k (resp. l) be the number of positive (resp. negative) eigenvalues of $A_\nu(u_+^0, d\psi^0)$ (resp. $A_\nu(u_-^0, d\psi^0)$), then $k + l = m - 1$.

Consider also a corresponding system of viscous conservation laws on $\mathbb{R}^3_{x,t,y}$

$$\partial_t u + \partial_x f(u) + \partial_y g(u) - \epsilon \triangle u = 0 \tag{1.4}$$

where

$$\triangle = \partial_x^2 + \partial_y^2.$$

A "weak" formulation of the problem we want to consider is: show that the given inviscid shock $(u_\pm^0)$ is the limit in some appropriate sense as $\epsilon \to 0$ of smooth solutions u^ϵ to the parabolic problem (1.4). An appropriate sense of convergence would be, for example, L^2_{loc} near the shock and pointwise away from the shock.

Imagine one had such a family of smooth u^ϵ. In order to have convergence to $u_\pm^0$, there must be a fast transition region located near the inviscid shock S. One might imagine the approximate "center" of that transition region as being defined by a surface S^ϵ that approaches S as $\epsilon \to 0$. If we knew S^ϵ we could treat it as an artificial boundary or transmission interface, and proceed to construct u^ϵ by constructing the boundary layers on each side of S^ϵ that describe the fast transition.

As a first guess one might take S^ϵ to be the known surface S itself, but that choice turns out to overdetermine the parabolic problem in a sense we'll clarify later. Thus, we are forced to introduce the unknown front S^ϵ given by

$$x = \psi^\epsilon(t, y) = \psi^0(t, y) + \epsilon \phi^\epsilon(t, y) \tag{1.5}$$

and to treat ψ^ϵ (or ϕ^ϵ) as an extra unknown (along with u^ϵ). We'll often refer to S^ϵ as *the viscous front*, although as we'll see it is not uniquely determined unless we add an extra condition.

To formulate the transmission problem more clearly, we flatten S^ϵ by the change of variables

$$(\tilde{x}, t, y) = (x - \psi^\epsilon(t, y), t, y). \tag{1.6}$$

If we set

$$\tilde{u}^\epsilon(\tilde{x}, t, y) = u^\epsilon(x, t, y) \tag{1.7}$$

and note that $\partial_x u = \partial_{\tilde{x}} \tilde{u}$, $\partial_t u = \partial_t \tilde{u} - \psi_t \partial_{\tilde{x}} \tilde{u}$, $\partial_y u = \partial_y \tilde{u} - \psi_y \partial_{\tilde{x}} \tilde{u}$, we find that the parabolic problem (1.4) in the new variables is (dropping tildes and epsilons)

$$\partial_t u + \partial_x f_\nu(u, d\psi) + \partial_y g(u) - \epsilon \triangle_\psi u = 0 \text{ on } \mathbb{R}^3_{x,t,y}, \tag{1.8}$$

where

$$\begin{aligned} f_\nu(u, d\psi) &= f(u) - \psi_t u - \psi_y g(u) \\ \triangle_\psi &= \partial_x^2 + (\partial_y - \partial_y \psi \partial_x)^2 = ((1 + \psi_y^2)\partial_x^2 - 2\psi_y \partial_{xy}^2 + \partial_y^2) - 2\psi_{yy}\partial_x. \end{aligned} \tag{1.9}$$

In the new coordinates the surface $\mathcal{S}^\epsilon$ is $x = 0$. Observe that solving (1.8) is equivalent to solving the transmission problem

$$\begin{aligned} &\partial_t u_\pm + \partial_x f_\nu(u_\pm, d\psi) + \partial_y g(u_\pm) - \epsilon \triangle_\psi u_\pm = 0 \text{ on } \pm x \geq 0, \\ &[u] = 0, [\partial_x u] = 0 \text{ on } x = 0. \end{aligned} \tag{1.10}$$

The transmission problem can easily be reformulated as a standard boundary problem on the half-space $x \geq 0$ by "doubling"; that is, for $x \geq 0$ one can define $\tilde{u}_+(x, t, y) = u_+(x, t, y)$ and $\tilde{u}_-(x, t, y) = u_-(-x, t, y)$. We won't do this yet though, to avoid having to write $\pm$ all the time. In fact, we'll usually write the transmission problem (1.10) without the $\pm$ on u. An important point is that with the transmission formulation we now have tools (like Kreiss-type symmetrizers) from the theory of boundary problems at our disposal to solve the original problem (1.8) on the full space.

Observe that with the extra unknown ψ we should expect the problem (1.10) to be underdetermined and to require an extra boundary condition involving ψ.

1.1 Approximate Solution

The first step in solving (1.10) is to construct a high order approximate solution, and the remaining step amounts to proving the stability of that solution. As we explain below uniform stability of the inviscid shock and transversality of the connection play a central part in the construction. The construction also illustrates the importance of allowing variation of the front in the viscous problem.

Since we expect solutions to (1.10) to undergo a fast transition near $x = 0$, it is natural to look for approximate solutions of the form (suppress $\pm$)

$$\begin{aligned} \tilde{u}^\epsilon(x, t, y) &= (\mathcal{U}^0(x, t, y, z) + \epsilon \mathcal{U}^1 + \cdots + \epsilon^M \mathcal{U}^M)|_{z=\frac{x}{\epsilon}} \\ \tilde{\psi}^\epsilon(t, y) &= \psi^0(t, y) + \epsilon \psi^1 + \cdots + \epsilon^M \psi^M, \end{aligned} \tag{1.11}$$

where each profile

$$\mathcal{U}^j(x,t,y,z) = U^j(x,t,y) + V^j(t,y,z) \tag{1.12}$$

is the sum of a slow U^j that describes behavior away from the viscous front and a fast boundary layer profile V^j which decays to 0 (exponentially, it turns out) as $z \to \pm\infty$.

Plug (1.11) into (1.10), collect coefficients of equal powers of ϵ, and separate slow from fast profiles to get

$$\sum_{-1}^{M} \epsilon^j \mathcal{F}^j(x,t,y,z)|_{z=\frac{x}{\epsilon}} + \epsilon^M R^{\epsilon,M}(x,t,y), \tag{1.13}$$

where

$$\mathcal{F}^j(x,t,y,z) = F^j(x,t,y) + G^j(t,y,z) \tag{1.14}$$

and (assuming smooth decaying profiles for the moment),

$$\begin{aligned}
|\partial_{t,y}^\alpha \partial_x^k R^{\epsilon,M}|_{L^\infty} &\leq C_{\alpha,k}\epsilon^{-k} \\
|\partial_{t,y}^\alpha \partial_x^k R^{\epsilon,M}|_{L^2} &\leq C_{\alpha,k}\epsilon^{\frac{1}{2}-k},
\end{aligned} \tag{1.15}$$

(caution: each estimate here is two estimates, one in $x \geq 0$ and one in $x \leq 0$). Observe that F^{-1} is automatically zero, and the equations obtained by setting F^0 and G^{-1} equal to zero are

$$\begin{aligned}
(a)\ & \partial_t U^0 + \partial_x f_\nu(U^0, d\psi^0) + \partial_y g(U^0) = 0, \\
(b)\ & -(1 + (\psi_y^0)^2)\partial_z^2 \mathcal{U}^0 + \partial_z f_\nu(\mathcal{U}^0, d\psi^0) = 0
\end{aligned} \tag{1.16}$$

respectively. Again, each equation here is really two equations; for example, in $\pm z \geq 0$ for (1.16)(b). The coefficient $(1 + (\psi_y^0)^2)$ will appear often in what follows; let's call it $B^0(t,y)$.

The equation (1.16)(a) is the inviscid shock problem (1.1) in the new coordinates, so a solution is given by $(U_\pm^0, \psi^0)$, where

$$U_\pm^0(x,t,y) = u_\pm^0(x + \psi^0(t,y), t, y). \tag{1.17}$$

In equation (1.16)(b) $\mathcal{U}^0$ is evaluated at $(0,t,y,z)$ instead of (x,t,y,z). The error of order $O(x)$ introduced by doing this is solved away at the stage of the next fast equation G^0 by writing

$$x = \epsilon\frac{x}{\epsilon} = \epsilon z. \tag{1.18}$$

The boundary conditions in (1.10) yield the boundary profile equations on $x = 0$, $z = 0$

$$\begin{aligned}
(a)\ & U_+^0 + V_+^0 = U_-^0 + V_-^0 \\
(b)\ & \partial_z V_+^0 = \partial_z V_-^0,
\end{aligned} \tag{1.19}$$

or equivalently,

$$[\mathcal{U}^0] = 0, \quad [\partial_z \mathcal{U}^0] = 0, \tag{1.20}$$

at the orders ϵ^0, ϵ^{-1} respectively.

Next integrate (1.16)(b) ($\int_{\pm\infty}^{z}$ in $\pm z \geq 0$) to obtain

$$B^0 \partial_z \mathcal{U}_{\pm}^0 = f_\nu(\mathcal{U}_{\pm}^0, d\psi^0) - f_\nu(U_{\pm}^0, d\psi^0), \tag{1.21}$$

where the unknowns are really $V_{\pm}^0(t, y, z)$, since $U_{\pm}^0$ are given.

The two boundary conditions in (1.20) clearly overdetermine this first order transmission problem, but note that the Rankine-Hugoniot condition on $(U_{\pm}^0, d\psi_0)$ is the necessary **compatibility condition**. More precisely, assume that $\mathcal{U}^0$ satisfies (1.21) and $[\mathcal{U}^0] = 0$. Then

$$[\partial_z \mathcal{U}^0] = 0 \text{ holds } \Leftrightarrow [f_\nu(U^0, d\psi^0)] = 0 \text{ on } x = 0. \tag{1.22}$$

In view of the transmission conditions (1.20), solving the problem (1.21) for unknowns $V_{\pm}^0 \to 0$ as $z \to \pm\infty$ is equivalent to solving the *connection* problem on $\mathbb{R}_z$ for $W(t, y, z)$

$$\begin{aligned}
&(a) B^0 \partial_z W = f_\nu(W, d\psi^0) - f_\nu(U_-^0, d\psi^0) \\
&(b) W(t, y, z) \to U_{\pm}^0(0, t, y) = u_{\pm}^0(\psi^0(t, y), t, y) \text{ as } z \to \pm\infty.
\end{aligned} \tag{1.23}$$

From this point of view the Rankine-Hugoniot condition is the statement that $U_{+}^0(t, y)$ is an equilibrium for the ODE (1.23)(a). We'll refer to the travelling wave equation (1.23)(a) as *the profile equation*. The solution $W(t, y, z) = \mathcal{U}^0(0, t, y, z)$ is variously referred to as a *connection*, a *profile*, and a *viscous shock*. Note that there is a lack of uniqueness due to translation invariance; that is, if $W(t, y, z)$ is a solution, so is $W(t, y, z + a)$ for any $a \in \mathbb{R}$.

Remark 1.1 *1. It is not hard to prove the existence of profiles $W(t, y, z)$ for sufficiently weak Lax shocks (see [MP], for example). To handle the case of strong shocks, we have to assume the existence of profiles.*

The fact that W decays exponentially to its endstates

$$|W(t, y, z) - U_{\pm}^0(0, t, y)| = O(e^{-\delta|z|}) \text{ as } z \to \pm\infty \tag{1.24}$$

is a consequence of the invertibility of the normal matrices $A_\nu(U_{\pm}^0, d\psi^0)$. The latter fact implies $U_{\pm}^0$ are hyperbolic equilibria for the ODE (1.23).

2. In view of (H2) we see that the range of $W(t, y, z)$ is contained in a compact subset of $\mathbb{R}^m$.

To see how the construction of the higher order profiles works, it will be enough just to consider the case of $(U_{\pm}^1(x, t, y), d\psi^1(t, y))$ and $V_{\pm}^1(t, y, z)$. The interior problems satisfied by $V_{\pm}^1$ are the fast problems at the order ϵ^0. As in the case of $V_{\pm}^0$, each problem is a second order ODE that can be integrated using the conservative structure to give a first order ODE. The equation for $V_\pm^1$ is a linearization (with respect to both $\mathcal{U}_\pm^0$ and ψ^0) of (1.21) with forcing $Q_\pm^0$ depending on previously determined functions:

$$
\begin{aligned}
B^0 \partial_z V_\pm^1 = {} & A_\nu(\mathcal{U}_\pm^0, d\psi^0)(V_\pm^1 + U_\pm^1) - \psi_t^1 \mathcal{U}_\pm^0 - \psi_y^1 g(\mathcal{U}_\pm^0) \\
& - \{A_\nu(U_\pm^0, d\psi^0)U_\pm^1 - \psi_t^1 U_\pm^0 - \psi_y^1 g(U_\pm^0)\} + Q_\pm^0(t, y, z),
\end{aligned}
\tag{1.25}
$$

where $Q_\pm^0 \to 0$ exponentially as $z \to \pm\infty$.

The interior equation for $U_\pm^1$ is a linearization (with respect to $U_\pm^0$) of (1.16)(a):

$$
H(U_\pm^0)\partial U_\pm^1 := \partial_t U_\pm^1 + A_\nu(U_\pm^0, d\psi^0)\partial_x U_\pm^1 + g'(U_\pm^0)\partial_y U_\pm^1 = P_\pm^0,
\tag{1.26}
$$

where again the forcing $P_\pm^0$ depends on previously determined functions.

Again, there are two boundary conditions at $x = 0$, $z = 0$:

$$
\begin{aligned}
&(a)\ U_+^1 + V_+^1 = U_-^1 + V_-^1 \\
&(b)\ \partial_x U_+^0 + \partial_z V_+^1 = \partial_x U_-^0 + \partial_z V_-^1,
\end{aligned}
\tag{1.27}
$$

so the first order problem for $V_\pm^1$ is overdetermined. Suppose for a moment that (1.25) and (1.27)(a) are satisfied. Then parallel to (1.22) we clearly have the compatibility condition

$$
\begin{aligned}
&[\partial_x U^0 + \partial_z V^1] = 0 \Leftrightarrow \text{ on } x = 0, z = 0 \text{ we have} \\
&[A_\nu(U^0, d\psi^0)U^1 - \psi_t^1 U^0 - \psi_y^1 g(U^0)] = B^0[\partial_x U^0] + [Q^0].
\end{aligned}
\tag{1.28}
$$

Thus, we may arrange the compatibility condition (1.28) by solving the following linearized shock problem for $(U_\pm^1, \psi^1)$:

$$
\begin{aligned}
&(a) H(U_\pm^0)\partial U_\pm^1 = P_\pm^0(x) \text{ on } \pm x \geq 0 \\
&(b) \psi_t^1[U^0] + \psi_y^1[g(U^0)] - [A_\nu(U^0, d\psi^0)U^1] = -B^0[\partial_x U^0] - [Q^0] \text{ on } x = 0,
\end{aligned}
\tag{1.29}
$$

The interior problem (1.29)(a) is the slow problem at the order ϵ^1, and the boundary operator in (b) is a linearization of the Rankine-Hugoniotconditions.

Linearized shock problems like (1.29) were first studied by Majda in [M2] as the first step in his proof of existence of curved multi-D inviscid shocks [M3]. It is a consequence of our main Evans assumption, Assumption (3.1), that the inviscid shock $(U_\pm^0, \psi^0)$ is *uniformly stable* in the sense of Majda [M2]. We'll discuss uniform stability more carefully later, but for now we just state informally that it is essentially equivalent to L^2 well-posedness of problems like (1.29).

So we now have the functions $(U_\pm^0, \psi^0)$, $V_\pm^0$, $(U_\pm^1, \psi^1)$, and the next step is to solve for $V_\pm^1$. We must choose initial data for $V_\pm^1$ at $z = 0$ so that both

(1.27)(a) holds and the solution $V_\pm^1$ to (1.25) decays exponentially to 0 as $z \to \pm\infty$. We explain next how a *transversality* condition implied by the same Evans assumption allows us to do this.

Consider again the travelling wave equation (1.23)(a) on $\mathbb{R}_z$, and recall that $U_\pm^0$ are both equilibrium points. Clearly, $W(t, y, z)|_{z=0}$ belongs to both the stable manifold of $U_+^0(0, t, y)$ and the unstable manifold of $U_-^0(0, t, y)$. The Evans assumption implies these manifolds intersect transversally at $W(t, y, 0)$.

For fixed (t, y) let $\mathbb{W}^s(t, y)$ and $\mathbb{W}^u(t, y)$, respectively, be the affine submanifolds of $\mathbb{R}^m$ consisting of initial data at $z = 0$ of solutions to (1.25) that decay as $z \to \pm\infty$. These submanifolds are translates of the tangent spaces (at $W(t, y, 0)$) to the above stable and unstable manifolds, so they too intersect transversally. Equivalently, the intersection of the affine submanifolds

$$(\mathbb{W}^s(t, y) \times \mathbb{W}^u(t, y)) \cap \{(v_1, v_2) \in \mathbb{R}^{2m} : v_1 - v_2 = U_-^1(0, t, y) - U_+^1(0, t, y)\} \tag{1.30}$$

is transversal, hence nonempty. In fact, since by assumption (H2) the dimension of $\mathbb{W}^s(t, y) \times \mathbb{W}^u(t, y)$ is $m + 1$, the intersection (1.30) is a line. Thus, we obtain a one-parameter family of choices of initial data for decaying solutions of (1.25) satisfying (1.27)(a).

We continue according to this pattern to solve for $(U_\pm^2, \psi^2)$, then $V_\pm^2$, then $(U_\pm^3, \psi^3)$, etc., always obtaining linearized Majda well-posed shock problems for $(U_\pm^j, \psi^j)$ whose boundary conditions are chosen as the compatibility conditions for the overdetermined problems satisfied by $V_\pm^j$.

Remark 1.2 *1. Later we'll add an extra boundary condition (2.10), and one effect of this will be to remove the nonuniqueness in the higher profiles.*

2. The boundary condition (1.27)(b) shows that in general $[\partial_z V^1] \neq 0$, so one can't solve for $V_\pm^1$ by solving a single ODE on $\mathbb{R}_z$ as we did for $V_\pm^0$.

3. The above construction doesn't work if one simply fixes $\tilde{\psi}^\epsilon = \psi^0$. If one does not allow the variation in the front given by ψ^1, for example, the problem (1.29) is overdetermined and generally unsolvable. A similar statement applies to ψ^j for $j > 1$.

1.2 Summary

Let's write the transmission problem (1.10) as

$$\begin{aligned} \mathcal{E}(u, \psi) &= 0 \\ [u] &= 0, \, [\partial_x u] = 0. \end{aligned} \tag{1.31}$$

Recalling (1.13) we now have an approximate solution $(\tilde{u}, \tilde{\psi})$ defined on a fixed time interval independent of ϵ (determined by the time of existence of the given inviscid shock) such that

$$\mathcal{E}(\tilde{u}, \tilde{\psi}) = \epsilon^M R^{\epsilon, M}$$
$$[\tilde{u}] = 0, [\partial_x \tilde{u}] = 0.$$

(1.32)

We proceed to look for an exact solution to (1.31) of the form

$$u = \tilde{u} + v, \psi = \tilde{\psi} + \phi.$$

(1.33)

The main difficulty is to obtain good L^2 estimates for the linearization of (1.31) about $(\tilde{u}, \tilde{\psi})$. Once these are in hand it is fairly routine to obtain higher derivative estimates by differentiating the equation, and to then solve the error equation for (v, ϕ) by Picard iteration (i.e., contraction).

Here is the main result:

Theorem 1.1 *Under assumptions (1.1) and (3.1) there exists an $\epsilon_0 > 0$ such that for $0 < \epsilon \leq \epsilon_0$ the parabolic transmission problem (1.10) has an exact solution on $[0, T_0] \times \mathbb{R}_{x,y}^2$ of the form*

$$u^\epsilon = \tilde{u} + v, \ \psi^\epsilon = \tilde{\psi} + \phi,$$

(1.34)

where $(\tilde{u}, \tilde{\psi})$ is an approximate solution satisfying (1.32). For arbitrary positive integers K and L, provided $M = M(K, L)$ in (1.32) is taken large enough, we have the estimates

$$|\partial^\alpha (v, \epsilon \partial_x v)|_{L^2(x,t,y)} + |\partial^\alpha (v, \epsilon \partial_x v)|_{L^\infty(x,t,y)} \leq \epsilon^L \ \ (\partial = \partial_{t,y})$$
$$|\partial^\alpha \phi|_{L^2(t,y)} + |\partial^\alpha \phi|_{L^\infty(t,y)} \leq \epsilon^L$$

(1.35)

for $|\alpha| \leq K$.

Remark 1.3 *The theorem asserts the stability of the boundary layer given by the approximate solution, and allows us to read off a precise sense in which the solution u^ϵ of (1.4) converges to the inviscid shock u^0. For example, we have convergence in L_{loc}^2 near the shock, and in C_{loc}^0 away from the shock.*

2 Lecture Two: Full Linearization, Reduction to ODEs, Conjugation to a Limiting Problem

2.1 Full Versus Partial Linearization

To find the error problem satisfied by (v, ϕ) we first rewrite (1.31)

$$\mathcal{E}(\tilde{u} + v, \tilde{\psi} + \phi) = \mathcal{E}(\tilde{u}, \tilde{\psi}) + \mathcal{E}'_u(\tilde{u}, \tilde{\psi})v + \mathcal{E}'_\psi(\tilde{u}, \tilde{\psi})\phi + Q(v, \phi) = 0, \quad (2.1)$$

where $\mathcal{E}'_u$ and $\mathcal{E}'_\psi$ are the linearizations of $\mathcal{E}$ with respect to u and ψ respectively, and Q is a sum of terms at least quadratic in $\partial^\alpha(v, \phi)$, $|\alpha| \leq 2$.

Thus, we must solve the transmission problem

$$\mathcal{E}'_u(\tilde{u}, \tilde{\psi})v + \mathcal{E}'_\psi(\tilde{u}, \tilde{\psi})\phi = -\epsilon^M R^{\epsilon,M} - Q(v, \phi)$$
$$[v] = [\partial_x v] = 0. \tag{2.2}$$

The explicit formulas for the linearizations are

$$\mathcal{E}'_u(\tilde{u}, \tilde{\psi})v = \partial_t v + \partial_x(A_\nu(\tilde{u}, \tilde{\psi})v) + \partial_y(g'(\tilde{u})v) - \epsilon \triangle_{\tilde{\psi}} v$$
$$\mathcal{E}'_\psi(\tilde{u}, \tilde{\psi})\phi = -\phi_t \partial_x \tilde{u} - \phi_y \left(g'(\tilde{u})\partial_x \tilde{u} - 2\epsilon(\partial_y - \tilde{\psi}_y \partial_x)\partial_x \tilde{u} \right) + \epsilon \phi_{yy} \partial_x \tilde{u}. \tag{2.3}$$

One should expect there to be some simple relationship between the two operators in (2.3). Much of what follows hinges on observing that

$$\mathcal{E}'_\psi(\tilde{u}, \tilde{\psi})\phi = -\mathcal{E}'_u(\tilde{u}, \tilde{\psi})(\phi \partial_x \tilde{u}) + \phi \partial_x(\mathcal{E}(\tilde{u}, \tilde{\psi})). \tag{2.4}$$

This can be proved by a direct verification; later we'll see that it becomes rather obvious after a few reductions.

This implies that the left side of (2.2) is the same as

$$\mathcal{E}'_u(\tilde{u}, \tilde{\psi})(v - \phi \partial_x \tilde{u}) + \phi \partial_x(\mathcal{E}(\tilde{u}, \tilde{\psi})),$$

so we reduce to solving

$$\mathcal{E}'_u(\tilde{u}, \tilde{\psi})(v - \phi \partial_x \tilde{u}) = -\epsilon^M R^{\epsilon,M} - Q(v, \phi) - \phi \partial_x(\epsilon^M R^{\epsilon,M}). \tag{2.5}$$

This suggests the strategy of reducing the study of the fully linearized operator given by the left side of (2.2) to that of the partially linearized operator $\mathcal{E}'_u$ by introducing the "good unknown"

$$v^{\#} = v - \phi \partial_x \tilde{u}. \tag{2.6}$$

Indeed this strategy turns out to work well in what we'll soon define as the medium and high frequency regimes, where $\mathcal{E}'_u$ is nonsingular. In the low frequency regime, we'll see that $\mathcal{E}'_u$ is singular, but that the singularity can be removed by the introduction of an extra boundary condition and a more subtle choice of "good unknown".

Remark 2.1 *The need to consider the full linearization is not obvious at this point. We have a high order approximate solution $(\tilde{u}, \tilde{\psi})$, so why not just fix $\tilde{\psi}$ once and for all, and solve (1.31) by solving*

$$\mathcal{E}(\tilde{u} + v, \tilde{\psi}) = 0$$
$$[v] = 0, [\partial_x v] = 0 \tag{2.7}$$

for v. Indeed, this was the strategy pursued in [GMWZ2]. What happens is that the singularity in $\mathcal{E}'_u$ in the low frequency regime leads to a slightly degenerate linearized L^2 estimate of the form

$$\sqrt{\epsilon}|v|_{L^2} \le |F|_{L^2}, \tag{2.8}$$

provided *we restrict how much the curved inviscid shock S can deviate from flatness. The estimate (2.8) somewhat surprisingly turns out to be good enough for Picard iteration, and this yields a solution of (1.31).*

But, of course, it is highly desirable to remove the restriction on how much S can curve, and we'll be able to do that by adding an extra boundary condition and working with the full linearization. That way we'll get an estimate without the $\sqrt{\epsilon}$ on the left as in (2.8).

2.2 The Extra Boundary Condition

Even though an extra boundary condition is not strictly needed except in the low frequency regime, where the unknown $v^{\#}$ in (2.6) is not used, we can use (2.5) to motivate our choice of boundary condition.

If we work with the full linearization we'll need estimates on ϕ in all frequency regimes, so the idea is to choose a boundary condition that will yield estimates for ϕ once we have an estimate on the trace of $v^{\#}$. ϕ is scalar so we choose a vector $l(t, y)$ and consider on $x = 0$

$$l \cdot v^{\#} = l \cdot v - \phi(l \cdot \partial_x \tilde{u}). \tag{2.9}$$

The leading part of $\partial_x \tilde{u}$ is $\frac{1}{\epsilon}\partial_z W$, so if we choose $l(t, y)$ so that $l(t, y) \cdot \partial_z W(t, y, 0) = -1$ and demand, for example, that

$$l \cdot v = \partial_t \phi - \epsilon \partial_y^2 \phi \text{ on } x = 0, \tag{2.10}$$

the leading part of the right side of (2.9) is

$$\partial_t \phi - \epsilon \partial_y^2 \phi + \frac{\phi}{\epsilon}.$$

Set $\tilde{\phi} = \frac{\phi}{\epsilon}$ and rewrite (2.9) to obtain

$$l \cdot v^{\#} = (\epsilon \partial_t - \epsilon^2 \partial_y^2 + 1)\tilde{\phi}, \tag{2.11}$$

which leads to an estimate for $\tilde{\phi}$ if one has an estimate for the trace of $v^{\#}$. Other similar choices are of course possible for the differential operator in (2.10), but we'll settle on (2.10) as the extra boundary condition. This condition also turns out to work well in the low frequency regime. So we must solve the nonlinear transmission problem

$$\begin{aligned} \mathcal{E}'_u(\tilde{u}, \tilde{\psi})v + \mathcal{E}'_\psi(\tilde{u}, \tilde{\psi})\phi &= -\epsilon^M R^{\epsilon, M} - Q(v, \phi) \\ [v] = [\partial_x v] = 0, \partial_t \phi - \epsilon \partial_y^2 \phi - l \cdot v &= 0 \text{ on } x = 0. \end{aligned} \tag{2.12}$$

Remark 2.2 *One can also impose the extra boundary condition in (1.31), before even constructing the approximate solution or defining (v, ϕ). Here a good choice is*

$$\partial_t \psi - \epsilon \partial_y^2 \psi - l \cdot u = \partial_t \psi^0 - \epsilon \partial_y^2 \psi^0 - l \cdot W(t, y, 0). \tag{2.13}$$

It's easy to check that imposing this extra condition in the construction of the approximate solution $(\tilde{u}, \tilde{\psi})$ removes the translational nonuniqueness that we observed for every fast profile $V_\pm^j(t, y, z)$. They are now pinned down. If one takes that uniquely determined solution $(\tilde{u}, \tilde{\psi})$ and defines (v, ϕ) as we did above, the extra condition on (v, ϕ) is exactly that in (2.12).

2.3 Corner Compatible Initial Data and Reduction to a Forward Problem

To determine a unique solution of (2.12) we must impose some initial conditions

$$v = \epsilon^{M_0} v_0, \phi = \epsilon^{M_0} \phi_0 \text{ at } t = 0. \tag{2.14}$$

We are free to choose these in many different ways, but our later work will be much easier if we choose the initial data to satisfy *corner compatibility* conditions to high order at the corner $x = 0, t = 0$ (recall, we can also formulate (2.12) as a doubled boundary problem, and then we really do have a corner). We show how to do that in [GW] or [GMWZ2], but those rather technical details seem like good ones to omit here. The key point is that such data will allow us to obtain solutions to the transmission problem that are piecewise smooth to high order in $\pm x \geq 0$ (recall the interior forcing term $R^{\epsilon, M}$ is piecewise smooth). The M_0 in (2.14) is about $M - k_0$, where k_0 is the order of compatibility. It's worth noting that the high order approximate solution plays an important role here. Arranging compatibility uses x derivatives, and each x derivative introduces a power of $\frac{1}{\epsilon}$. This leads to the reduction in M.

This choice of initial data allows us to reduce to a forward transmission problem, one where all data is zero in $t < 0$, and the boundary conditions are homogeneous. By a standard maneuver from linear PDE, we first transfer initial data to forcing at the price of introducing nonzero boundary forcing. But the above compatibility conditions are designed precisely so that the new boundary forcing vanishes to high order at $t = 0$. Next a similar maneuver transfers the nonzero boundary forcing to interior forcing, leaving us with the forward error problem (relabelling unknowns (v, ϕ) again) in $\mathbb{R}^3_{x,t,y}$

$$\begin{aligned} &\mathcal{E}'_u(\tilde{u}, \tilde{\psi})v + \mathcal{E}'_\psi(\tilde{u}, \tilde{\psi})\phi = -\epsilon^{M'} F - Q(v, \phi) \\ &[v] = [\partial_x v] = 0, \partial_t \phi - \epsilon \partial_y^2 \phi - l \cdot v = 0 \text{ on } x = 0 \\ &v = 0, \phi = 0 \text{ in } t < 0, \end{aligned} \tag{2.15}$$

where $F = 0$ in $t < 0$ and piecewise smooth to high order in $\pm x \geq 0$. Here $M' < M$, but M' is still large provided M was large compared to k_0. Q is not the same as before, but the earlier properties attributed to Q continue to hold.

Remark 2.3 *We're free to introduce a cutoff in time on the right in (2.15), so we can choose extensions of $\tilde{u}$ and $\tilde{\psi}$ to all time so that the coefficients in (2.15) are defined for all time.*

2.4 Principal Parts, Exponential Weights

The main task now is to prove a nondegenerate L^2 estimate for the fully linearized forward transmission problem on $\mathbb{R}^3_{x,t,y}$

$$\mathcal{E}'_u(\tilde{u}, \tilde{\psi})v + \mathcal{E}'_\psi(\tilde{u}, \tilde{\psi})\phi = f$$
$$[v] = 0, [\partial_x v] = 0, \partial_t \phi - \epsilon\partial_y^2 \phi - l \cdot v = 0 \text{ on } x = 0 \qquad (2.16)$$
$$v = 0, \phi = 0, f = 0 \text{ in } t < 0.$$

Recall that the leading parts of $\tilde{u}$ and $\tilde{\psi}$ are $W(t, y, z)|_{z=\frac{x}{\epsilon}}$ and $\psi^0(t, y)$ respectively. If we replace $(\tilde{u}, \tilde{\psi})$ by $(W(t, y, \frac{x}{\epsilon}), \psi^0(t, y))$ in (2.16), we introduce several error terms in the estimate, including some of size

$$O(|v|_{L^2}) + O(|\epsilon\partial_x v|_{L^2}) + O(|d\phi|_{L^2}) + O(\epsilon|d^2\phi|_{L^2}). \qquad (2.17)$$

Errors like (2.17) turn out to be absorbable by the left side of the estimate we obtain.

Remark 2.4 *The difference $U^0(x, t, y) - U^0(0, t, y)$ leads to another sort of error when $\tilde{u}$ is replaced by W that can't simply be absorbed as above. When $|x|$ is small though, the coefficients are perturbed only slightly by this difference, so the symmetrizer construction presented below for the slightly simplified case where this perturbation is ignored works in the same way for the case where the perturbation is included.*

Thus, the proof of the L^2 estimate for the original linearized PDE needs to be split in two parts; one where the solution is supported near the boundary, and another for solutions supported away from the boundary (errors introduced by cutoffs used to localize the estimate are again absorbable). In the second case there are no boundary conditions, no glancing modes, and no singular terms in the linearized operator, and the estimates can be proved by a much simpler argument (for details we refer to Propositions 5.6 and 5.7 of [MZ]). We focus on the symmetrizer construction needed for the estimates near the boundary in these notes.

There are other terms like $\partial_y(g'(W))v$ in (2.16) of size (2.17). Throwing away all such terms in (2.3) and replacing $(\tilde{u}, \tilde{\psi})$ by (W, ψ^0) we obtain the *principal parts*

$$\mathcal{L}_u(t, y, \frac{x}{\epsilon}, \partial_{x,t,y})v = \partial_t v + A_\nu(W, d\psi^0)\partial_x v + g'(W)\partial_y v +$$

$$+ \frac{(d_u A_\nu \cdot W_z)v}{\epsilon} - \epsilon \left(B^0 \partial_x^2 + \partial_y^2 - 2\psi_y^0 \partial_{xy}^2 \right) v,$$

$$\mathcal{L}_\psi(t, y, \frac{x}{\epsilon}, \partial_{t,y})\phi = -\frac{1}{\epsilon} \left(W_z \phi_t + g'(W)W_z\phi_y + 2\psi_y^0 W_{zz}\phi_y - \epsilon W_z \phi_{yy} \right).$$

$$(2.18)$$

At this point we introduce exponential weights in time $e^{-\gamma t}$, $\gamma \geq 1$ and define

$$v_\gamma = e^{-\gamma t}v, \phi_\gamma = e^{-\gamma t}\phi, f_\gamma = e^{-\gamma t}f. \qquad (2.19)$$

If (v, ϕ) satisfies

$$\mathcal{L}_u v + \mathcal{L}_\psi \phi = f \qquad (2.20)$$

with the above boundary conditions, then (v_γ, ϕ_γ) satisfies the same problem with ∂_t replaced by $\partial_t + \gamma$

$$\mathcal{L}_u(t, y, \frac{x}{\epsilon}, \partial_t + \gamma, \partial_{x,y})v_\gamma + \mathcal{L}_\psi(t, y, \frac{x}{\epsilon}, \partial_t + \gamma, \partial_y)\phi_\gamma = f_\gamma$$

$$[v_\gamma] = 0, [\partial_x v_\gamma] = 0, (\partial_t + \gamma)\phi_\gamma - \epsilon \partial_y^2 \phi_\gamma - l \cdot v_\gamma = 0 \text{ on } x = 0 \qquad (2.21)$$

$$v_\gamma = 0, \phi_\gamma = 0, f_\gamma = 0 \text{ in } t < 0.$$

Henceforth, we'll drop the subscripts γ, but the exponential weights are always there. Also, we'll often neglect to mention that all data is zero in the past.

Remark 2.5 *Why introduce exponential weights? One reason is that γ will be an important parameter in the later stability analysis, but that is hardly apparent now. Another reason is that we'll sometimes be able to absorb errors by taking γ large (not apparent now). Another is that later we'll freeze (t, y) in the coefficients, and it's common to take a Fourier-Laplace transform in time, thereby introducing an exponential weight, in the stability analysis of constant coefficient problems (not convincing yet, perhaps). Finally, a reason that seems clear even at this point is that it is technically much more convenient to estimate solutions to (2.21) on $\mathbb{R}^3_{x,t,y}$ instead of on a bounded time domain (e.g., we'll use pseudodifferential operators acting in (t, y)), and one can't expect solutions of (2.20) to have bounded L^2 norms for all time, but one can expect that to be true for (2.21) if γ is large enough.*

2.5 Some Difficulties

We have to find estimates for (2.21) that are uniform in ϵ as $\epsilon \to 0$, but a quick glance at $\mathcal{L}_u$ and $\mathcal{L}_\psi$ does not encourage optimism on that point. On the one hand the coefficients contain a mixture of powers of ϵ, including ϵ^{-1}. In addition, the crucial normal matrix A_ν, although nonsingular at the endstates

$U^0_\pm(0, t, y)$, is singular at some intermediate value of $z = \frac{x}{\epsilon}$ since one of its eigenvalues changes sign along the profile (the pth eigenvalue if the inviscid shock is a p shock). Indeed, the latter problem is one of the main difficulties in the entire analysis. Another difficulty we should expect from experience with hyperbolic boundary problems is that, since we are working in multiD we'll have to contend with *glancing points*. We'll define these later but for now we just point out that they correspond to characteristics that are tangent to the free boundary given by the curved viscous front $\mathcal{S}^\epsilon$. Except for special cases it is not known how to construct explicit solutions (using Green's functions, parametrices, Fourier integral operators, ...) for the linearized problem near glancing points. In the case of planar fronts the Fourier-Laplace transform can be used. In the case of first-order tangency the Fourier-Airy integral operators of Melrose and Taylor are available. For higher order tangency there are no constructive methods as far as we know. Fortunately, we will be able to construct Kreiss-type symmetrizers to deal with glancing points of any order. Indeed, one of the main points of our work is that, even in these singular hyperbolic-parabolic problems, one can avoid explicit constructions by using symmetrizers much as Kreiss did in the early 1970s for hyperbolic boundary problems.

2.6 Semiclassical Form

The equations become more balanced in ϵ if we simply multiply through by epsilon and write them in semiclassical form.

Let (τ, η) be dual variables to (t, y) (in the Fourier transform sense), and set $\zeta = (\tau, \gamma, \eta)$ and $\tilde{\zeta} = \epsilon\zeta$. Define semiclassical symbols

(a) $L_u(t, y, z, \tilde{\zeta}, \partial_z) =$

$\left((i\tilde{\tau} + \tilde{\gamma} + \tilde{\eta}^2)I + g'(W)i\tilde{\eta} + d_u A_\nu \cdot W_z\right) + \left(A_\nu(W, d\psi^0) + 2\psi^0_y i\tilde{\eta}\right)\partial_z - B^0\partial^2_z$

(b) $L_\psi(t, y, z, \tilde{\zeta}) = -\left(W_z(i\tilde{\tau} + \tilde{\gamma} + \tilde{\eta}^2) + g'(W)W_z i\tilde{\eta} + 2\psi^0_y W_{zz} i\tilde{\eta}\right)$

(c) $p(\tilde{\zeta}) = (i\tilde{\tau} + \tilde{\gamma} + \tilde{\eta}^2).$

$$(2.22)$$

Then if we set $\tilde{f} = \epsilon f$, $\tilde{\phi} = \frac{\phi}{\epsilon}$, $(D_t, D_y) = \frac{1}{i}(\partial_t, \partial_y)$, and change variables $z = \frac{x}{\epsilon}$, we can rewrite (2.21) in semiclassical form (dropping subscripts γ)

$$L_u(t, y, z, \epsilon D_t, \epsilon\gamma, \epsilon D_y, \partial_z)v + L_\psi(t, y, z, \epsilon D_t, \epsilon\gamma, \epsilon D_y)\tilde{\phi} = \tilde{f}$$
$$[v] = 0, \ [\partial_z v] = 0, \ p(\epsilon D_t, \epsilon\gamma, \epsilon D_y)\tilde{\phi} - l \cdot v = 0 \text{ on } z = 0.$$
$$(2.23)$$

Here we've used $\partial_z = \epsilon\partial_x$.

2.7 Frozen Coefficients; ODEs Depending on Frequencies as Parameters

Set $q = (t, y)$ and parallel to (2.23), consider on $\mathbb{R}_z$ the system of transmission ODEs depending on parameters $(q, \tilde{\zeta})$:

$$L_u(q, z, \tilde{\zeta}, \partial_z)v + L_\psi(q, z, \tilde{\zeta})\phi = f$$
$$[v] = 0, \ [\partial_z v] = 0, \ p(\tilde{\zeta})\phi - l \cdot v = 0 \text{ on } z = 0. \tag{2.24}$$

Here $v = v(z)$, $f = f(z)$, and ϕ is a scalar unknown.

Observe that if we freeze the variables $(t, y) = q$ in (2.23) and then take the Fourier transform in (t, y), we arrive at a problem just like (2.24) with

$$v(z) = \widehat{v}(\tau, \eta, z), \quad \phi = \widehat{\phi}(\tau, \eta), \quad f(z) = \widehat{f}(\tau, \eta, z). \tag{2.25}$$

Here the hat denotes Fourier transform and the absence of tildes on (τ, η) is correct in (2.25). Thus, there are two paths, equivalent of course, to the system of ODEs (2.24).

Consider the problem of proving estimates, uniform in the parameters $(q, \tilde{\zeta})$, for the system of transmission ODEs (2.24). Here we refer to estimates of

$$|v|_2 := |v|_{L^2(z)}, |\phi| := |\phi|_{\mathbb{C}}, \tag{2.26}$$

weighted by appropriate functions depending on the frequency, in terms of $|f|_2$. An immediate consequence of Plancherel's theorem and (2.25) is that such estimates imply corresponding estimates for the semiclassical system of PDEs (2.23) with coefficients frozen at $(t, y) = q$, and hence (after unravelling the changes of variables) estimates on the $|v|_{L^2(x,t,y)}$ and $|\partial_{t,y}\phi, \gamma\phi|_{L^2(t,y)}$ norms of solutions to the *frozen* version of (2.21).

Now it turns out, rather remarkably, that the same constructions needed (by us) to prove uniform estimates for the system of ODEs (2.24) are also the main steps in the proof of estimates even for the *variable coefficient* problem (2.21). To really see how this can be, one needs to use pseudodifferential (or paradifferential) operators and Garding inequalities, and we'll say a bit about those later. The point is that constructions for the ODEs (constructions of objects like conjugators, symmetrizers, ...) are exactly the same as the constructions of the *principal symbols* of the pseudodifferential operators that we use to prove estimates for the variable coefficient system of PDEs (2.21). So it's not too much of a stretch to say that if one takes pseudodifferential calculus as a given, the problem of proving estimates uniform with respect to frequency for the ODEs (2.24) is essentially equivalent to the problem of proving estimates uniform in epsilon for (2.21). We stress this because much of our effort from now on will be devoted to understanding the ODEs (2.24).

Remark 2.6 *The relationship we observed earlier between the linearizations $\mathcal{E}'_u$ and $\mathcal{E}'_\psi$ (2.4) is more obvious now at the ODE level. We have*

$$L_u(q, z, \tilde{\zeta}, \partial_z)W_z = -L_\psi(q, z, \tilde{\zeta}) \tag{2.27}$$

This is clear by inspection of (2.22) or can be deduced by looking at the leading $O(\frac{1}{\epsilon^2})$ term of (2.4). Observe that the leading part of $\partial_x(\mathcal{E}(\tilde{u}, \tilde{\psi}))$ in (2.4) is zero, since W satisfies the profile equation.

2.8 Three Frequency Regimes

Consider the frequency $\tilde{\zeta} = \epsilon(\tau, \gamma, \eta)$ that appears in (2.24). Estimates of solutions to (2.24) depend critically on the size of $|\tilde{\zeta}|$. The small, medium, and large frequency regimes (SF,MF,HF) are respectively

$$|\tilde{\zeta}| \le \delta,\ \delta \le |\tilde{\zeta}| \le R,\ |\tilde{\zeta}| \ge R \tag{2.28}$$

for small enough δ and large enough R to be determined.

The compact set of nonzero medium frequencies, where $L_u(q, z, \tilde{\zeta}, \partial_z)$ is nonsingular, is the easiest to handle. SF, where L_u is singular and hyperbolic and parabolic effects mix in a subtle way, is the hardest by far. In HF parabolic effects dominate and the hyperbolic part behaves like a perturbation, but care is needed because frequencies occur with mixed homogeneities and vary in an unbounded set.

Caution: When we speak of small, medium, or large frequencies, we are referring to the size of $\tilde{\zeta} = \epsilon\zeta$, not $|\zeta|$. For example, $|\zeta|$ can be extremely large in the small frequency regime, provided ϵ is small enough. *In our analysis of (2.24) we're going to drop the tilde on ζ from now on*, but it's important to remember the tilde is there before translating back to results for the PDE (2.21).

2.9 First-Order System

To prepare the way for conjugation and the construction of symmetrizers let's rewrite (2.24) as a $2m \times 2m$ first order system for the unknown (U, ϕ). Setting $U = (v, \partial_z v)$, we have

$$\partial_z U - G(q, z, \zeta)U = F + \begin{pmatrix} 0 \\ -(B^0(q))^{-1}L_\psi(q, z, \zeta)\phi \end{pmatrix} \tag{2.29}$$

$$[U] = 0,\ p(\zeta)\phi - l(q) \cdot v = 0 \text{ on } z = 0$$

where

$$F = \begin{pmatrix} 0 \\ -(B^0(q))^{-1}f \end{pmatrix} \text{ and}$$

$$G(q, z, \zeta) = \begin{pmatrix} 0 & I \\ \mathcal{M} & \mathcal{A} \end{pmatrix}, \tag{2.30}$$

with

$$\mathcal{M}(q, z, \zeta) = (B^0(q))^{-1} \left((i\tau + \gamma + \eta^2)I + g'(W)i\eta + d_u A_\nu \cdot W_z \right),$$
$$\mathcal{A} = (B^0(q))^{-1} \left(A_\nu(W, d\psi^0) + 2\psi_y^0 i\eta \right). \tag{2.31}$$

For now we'll ignore the L_ψ term in (2.29) and study the problem

$$\partial_z U - G(q, z, \zeta)U = F, \quad [U] = 0. \tag{2.32}$$

Since $W \to U_\pm^0(q)$ as $z \to \pm\infty$ we also have the limiting systems on $\pm z \geq 0$

$$\partial_z V - G_{\pm\infty}(q, \zeta)V = \tilde{F}, \tag{2.33}$$

where

$$G_{\pm\infty}(q, \zeta) = \begin{pmatrix} 0 & I \\ \mathcal{M}_{\pm\infty} & \mathcal{A}_{\pm\infty} \end{pmatrix} \tag{2.34}$$

with

$$\mathcal{M}_{\pm\infty}(q, \zeta) = (B^0(q))^{-1} \left((i\tau + \gamma + \eta^2)I + g'(U_\pm^0(q))i\eta \right)$$
$$\mathcal{A}_{\pm\infty} = (B^0(q))^{-1} \left(A_\nu(U_\pm^0(q), d\psi^0) + 2\psi_y^0 i\eta \right). \tag{2.35}$$

Here we've written $U_\pm^0(q)$ for $U_\pm^0(0, q)$.

2.10 Conjugation

It would be a great simplification to reduce the study of the variable coefficient problem (2.32) to that of the constant coefficient problem (2.33) (with an appropriately altered boundary condition), and we proceed to do that now in SF and MF but not HF. Since A_ν is nonsingular at the endstates $U_\pm^0(q)$ but not along $W(q, z)$, we'll then be in a much better position to construct symmetrizers and prove estimates.

There are classical results in the theory of ODEs (see, e.g. [Co]) that establish a correspondence between solutions to variable coefficient ODEs and solutions to corresponding limiting constant coefficient ODEs, when those limits exist and satisfy certain hypotheses. For such a correspondence to be useful here, we need it to be somehow uniform in the parameters (q, ζ). However, even if we restrict ζ to a compact set, the classical results fail to apply near $\zeta = 0$ because of the degeneracy

$$G_{\pm\infty}(q, 0) = \begin{pmatrix} 0 & I \\ 0 & \mathcal{A}_{\pm\infty}(q, 0) \end{pmatrix}. \tag{2.36}$$

Nevertheless, using the Gap Lemma of [GZ, KS] we can, locally near any basepoint $(\underline{q}, \underline{\zeta})$, reduce the study of (2.32) to (2.33) by constructing matrices $Z_\pm(q, z, \zeta)$ in $\pm z \geq 0$ depending smoothly on (q, z, ζ) such that

$(a)\partial_z Z = GZ - ZG_{\pm\infty}$ on $\pm z \geq 0$

$(b)|Z(q, z, \zeta) - I| = O(e^{-\theta|z|})$ for some $\theta > 0$ (2.37)

$(c)Z^{-1}(q, z, \zeta)$ is uniformly bounded for (q, ζ) near $(\underline{q}, \underline{\zeta})$, $\pm z \geq 0$.

Set $U = ZV$. Then expanding out $\partial_z(ZV)$ shows that U satisfies (2.32) if and only if V satisfies

$$\partial_z V - G_{\pm\infty}V = Z^{-1}F \text{ on } \pm z \geq 0,$$
$$[ZV] = 0 \text{ on } z = 0.$$
(2.38)

The acceptable price is a more complicated boundary condition. When we translate back to the PDE problem, the new boundary condition is pseudodifferential. The properties of Z in (2.37) show that L^2 estimates for (2.38) can be immediately transported to L^2 estimates for (2.32).

To construct Z_+ say, note that the matrix ODE (2.37)(a) can be written

$$\partial_z Z = \mathcal{L}Z + (\Delta G)Z$$
(2.39)

where $\mathcal{L}(q, \zeta)$ is the constant coefficient operator given by the commutator $[G_{+\infty}, \cdot]$ and ΔG is left multiplication by $G - G_{+\infty} = O(e^{-\delta z})$ (δ as in Remark 1.1).

The ODE (2.39) also has a limiting problem $\partial_z Y = \mathcal{L}Y$. The identity matrix I is an eigenvector of $\mathcal{L}$ associated to the eigenvalue 0, and hence $Y(z) = I$ solves the limiting problem. Suppose that the eigenvalues of $\mathcal{L}(q, \zeta)$, which are differences of eigenvalues of $G_{+\infty}(q, \zeta)$, avoid a line $\Re\mu = -\kappa$ for some $0 < \kappa < \delta$. This will always be true for (q, ζ) close enough to a fixed basepoint $(\underline{q}, \underline{\zeta})$.

A solution of (2.39) close to the solution I of $\partial_z Y = \mathcal{L}Y$ could then be found by solving the equation on $z \geq 0$

$$Z_+(q, z, \zeta) = I + \int_0^z e^{(z-s)\mathcal{L}}\pi_-(\Delta G)(s)Z_+(s)ds$$
$$- \int_z^{+\infty} e^{(z-s)\mathcal{L}}\pi_+(\Delta G)(s)Z_+(s)ds,$$
(2.40)

where $\pi_\pm(q, \zeta)$ are the spectral projectors on the generalized eigenspaces of $\mathcal{L}$ corresponding to eigenvalues with $\Re\mu > -\kappa$, $\Re\mu < -\kappa$ respectively. Note that the range of π_+ includes part of the *negative* eigenspace of $\mathcal{L}$, so we might expect the second integral in (2.40) to blow up since $z - s \leq 0$ there. But the integral is rescued by the exponential decay of ΔG and the fact that $0 < \kappa < \delta$ (this, essentially, is the Gap Lemma). The estimates of [GZ] show that we obtain a solution of (2.40) satisfying (2.37)(b) for $\theta < \kappa$.

Observe that if we set $D_+(z) = \det Z_+$, we have

$$\partial_z D_+ = tr(G(z) - G_{+\infty})D_+,$$
(2.41)

which implies D_+ is never 0 on $[0, \infty)$.

We'll sometimes refer to $Z_\pm$ as the MZ conjugator [MZ].

2.11 Conjugation to HP Form.

In SF we need another conjugation which separates $G_{\pm\infty}$ into one $m \times m$ block whose eigenvalues vanish as $|\zeta| \to 0$ and another whose eigenvalues have real parts bounded away from zero as $|\zeta| \to 0$. Inspection of (2.36) suggests there should be such a decomposition. Indeed, for $|\zeta|$ small there exist smooth matrices $Y_{\pm}(q, \zeta)$ with smooth inverses such that

$$Y_{\pm}^{-1} G_{\pm\infty} Y_{\pm} = \begin{pmatrix} H_{\pm}(q, \zeta) & 0 \\ 0 & P_{\pm}(q, \zeta) \end{pmatrix} := G_{HP\pm}, \qquad (2.42)$$

where

$$\begin{aligned} H_{\pm} &= -\mathcal{A}_{\pm\infty}^{-1} \mathcal{M}_{\pm\infty} + O(|\zeta|^2) = \\ &\quad - A_\nu (U_{\pm}^0, d\psi^0)^{-1} \left((i\tau + \gamma)I + g'(U_{\pm}^0)i\eta \right) + O(|\zeta|^2), \\ P_{\pm} &= \mathcal{A}_{\pm\infty} + O(|\zeta|) = B^0(q)^{-1} A_\nu (U_{\pm}^0, d\psi^0) + O(|\zeta|), \text{ and} \end{aligned} \qquad (2.43)$$

$$Y_{\pm}(q, 0) = \begin{pmatrix} I & \mathcal{A}_{\pm\infty}^{-1} \\ 0 & I \end{pmatrix}.$$

The forms of $H_{\pm}$ and $P_{\pm}$ can be motivated by a short computation which shows that for $|\zeta|$ small, small (resp. large) eigenvalues of $G_{\pm\infty}$ are close to eigenvalues of $-\mathcal{A}_{\pm\infty}^{-1}$ (resp. $\mathcal{M}_{\pm\infty} \mathcal{A}_{\pm\infty}$). One can then posit $Y_{\pm}$ of the given form and solve the equation

$$G_{\pm\infty} Y_{\pm} = Y_{\pm} G_{HP\pm} \qquad (2.44)$$

for the entries of $Y_{\pm}$.

3 Lecture Three: Evans Functions, Lopatinski Determinants, Removing the Translational Degeneracy

Before defining the Evans function we need to make a few observations about the spectrum of $G_{\pm\infty}(q, \zeta)$.

Proposition 3.1 *1. For $\zeta \neq 0$, $\gamma \geq 0$ $G_{\pm\infty}(q, \zeta)$ each have m eigenvalues counted with multiplicities in $\Re\mu > 0$ and m eigenvalues in $\Re\mu < 0$.*

2. $G_{\pm\infty}(q, 0)$ each have 0 as a semisimple eigenvalue of multiplicity m and nonvanishing eigenvalues equal to the eigenvalues of $\mathcal{A}_{\pm\infty}(q, 0)$.

$G_{+\infty}(q, 0)$ has $m - k$ eigenvalues in $\Re\mu < 0$ and $m - l$ eigenvalues in $\Re\mu > 0$, where $k + l = m - 1$.

Proof. **1.** Consider $G_{+\infty}(q, \zeta)$, where $\zeta \neq 0$, $\gamma \geq 0$. Then μ is an eigenvalue of $G_{+\infty} \Leftrightarrow$

$$\begin{pmatrix} 0 & I \\ \mathcal{M}_{+\infty} & \mathcal{A}_{+\infty} \end{pmatrix} \begin{pmatrix} u \\ v \end{pmatrix} = \mu \begin{pmatrix} u \\ v \end{pmatrix} \Leftrightarrow v = \mu u, \ \mathcal{M}_{+\infty} u + \mathcal{A}_{+\infty} v = \mu v, \qquad (3.1)$$

so $\mathcal{M}_{+\infty}u + \mathcal{A}_{+\infty}\mu u = \mu^2 u$. If $\mu = i\xi$, $\xi \in \mathbb{R}$, the latter equation is the same as

$$\left((i\tau+\gamma)I + A_\nu(U_+^0(q), d\psi^0)i\xi + g'(U_+^0(q))i\eta + (B^0(q)\xi^2 - 2\psi_y^0\eta\xi+\eta^2)\right)u = 0. \tag{3.2}$$

Thus, $i\tau + \gamma$ is an eigenvalue of

$$-\left(A_\nu(U_+^0(q), d\psi^0)i\xi + g'(U_+^0(q))i\eta\right) - \left(B^0(q)\xi^2 - 2\psi_y^0\eta\xi + \eta^2\right)$$

so hyperbolicity (H1) and positivity of the quadratic form given by the scalar second term imply

$$\gamma = -\left(B^0(q)\xi^2 - 2\psi_y^0\eta\xi + \eta^2\right) < 0 \text{ if } (\xi, \eta) \neq 0.$$

Now $\gamma \geq 0$ so we must have $(\xi, \eta) = 0$ which implies $i\tau + \gamma = 0$, contradicting $\zeta \neq 0$.

We conclude that the number of eigenvalues μ with positive (or negative) real part is constant for $\zeta \neq 0$, $\gamma \geq 0$. We can then take $(\tau, \eta) = 0$ and γ large to obtain an obvious count of m in each region.

2. The assertion follows immediately from (H2) and the explicit form of $G_{\pm\infty}(q, 0)$.

●

3.1 Evans Functions, Instabilities, the Zumbrun-Serre Result

The Evans function is a wronskian of solutions to

$$\partial_z U - G(q, z, \zeta)U = 0 \tag{3.3}$$

which contains information about the stability of the viscous profile and, less obviously, about the stability of the original inviscid shock. For $\zeta \neq 0$ let $E_\pm(q, \zeta)$ be the set of initial data $U(0)$ such that the solution of (3.3) with that data decays to zero as $z \to \pm\infty$. As one might expect from the degeneracy (2.36), these spaces are singular in ζ near $\zeta = 0$, and to resolve that singularity we blow up the origin using polar coordinates

$$\zeta = \rho\hat{\zeta}, \hat{\zeta} = (\hat{\tau}, \hat{\gamma}, \hat{\eta}), \text{ where}$$
$$\hat{\zeta} \in S_+^2 = \{\hat{\zeta} : |\hat{\zeta}| = 1, \hat{\gamma} \geq 0\}. \tag{3.4}$$

For $\zeta \neq 0$ we may just as well write $E_\pm(q, \hat{\zeta}, \rho)$ in place of $E^\pm(q, \zeta)$.

Proposition 3.2 *For $\zeta \neq 0$ the spaces $E_\pm(q, \hat{\zeta}, \rho)$ each have dimension m and are C^∞ in $(q, \hat{\zeta}, \rho)$. They extend continuously to $\rho = 0$.*

Proof (Partial proof). Suppose $\zeta \neq 0$, and let $F_\pm(q, \zeta)$ be the analogously defined decaying spaces for

$$\partial_z V - G_{\pm\infty}(q, \zeta)V = 0. \tag{3.5}$$

By Proposition 3.1 $F_\pm$ have dimension m and using the MZ conjugators we have

$$E_\pm(q, \zeta) = Z_\pm(q, 0, \zeta)F_\pm(q, \zeta). \tag{3.6}$$

In particular this shows that solutions of (3.3) with initial data in $E_\pm(q, \zeta)$ decay exponentially to zero.

The continuous extension to $\rho = 0$ is subtle because of glancing modes. A proof, which is best read after the reduction to *block structure* (4.24), is given in Appendix B. For now, note that when $\rho = 0$ (3.5) has nonzero constant solutions of the form $(r(q, \hat{\zeta}), 0)$. We'll see that solutions of (3.3) with data in $E_\pm(q, \hat{\zeta}, 0)$ decay as $z \to \pm\infty$ to limits that are constant solutions of (3.5).

$\bullet$

Definition 3.1 *For $\hat{\zeta} \in S_+^2$, $\rho \geq 0$ define the Evans function as the $2m \times 2m$ determinant*

$$D(q, \hat{\zeta}, \rho) = \det(E_+(q, \hat{\zeta}, \rho), E_-(q, \hat{\zeta}, \rho)). \tag{3.7}$$

Now suppose D vanishes for some (q_0, ζ_0) with $\zeta_0 = (\tau_0, \gamma_0, \eta_0)$ and $\gamma_0 > 0$. In this case we expect exponential instabilities of the boundary layers described by our approximate solution $(\tilde{u}, \tilde{\psi})$. Let us explain. Vanishing of $D(q_0, \zeta_0)$ means there is a smooth solution $w(z, \zeta_0)$ of

$$L_u(q_0, z, \zeta_0, \partial_z)w = 0 \quad (L_u \text{ as in (2.22)}) \tag{3.8}$$

on the whole line $\mathbb{R}_z$ which decays exponentially to zero as $z \to \pm\infty$. Direct computation shows that

$$w^\epsilon(x, t, y) = e^{\frac{(i\tau_0 + \gamma_0)t + iy\eta_0}{\epsilon}} w\left(\frac{x}{\epsilon}, \zeta_0\right) \tag{3.9}$$

is then a solution of the linearized transmission problem

$$\mathcal{L}_u(q_0, \frac{x}{\epsilon}, \partial_{x,t,y})w^\epsilon = 0 \quad (\mathcal{L}_u \text{ as in (2.18)}),$$
$$[w^\epsilon] = [\partial_x w^\epsilon] = 0. \tag{3.10}$$

Recalling that linearized equations describe evolution of small perturbations, we see from (3.9) that some small disturbances are amplified by the factor $e^{\frac{\gamma_0 t}{\epsilon}}$, so the boundary layers described by $(\tilde{u}, \tilde{\psi})$ should be completely destroyed on a time scale of $O(\epsilon)$. In this case there is no chance for L^2 estimates uniform in ϵ, and $(\tilde{u}, \tilde{\psi})$ is of no help in solving the small viscosity problem.

When $\zeta = 0$ note that (3.3) is the same as the linearized profile equation

$$L_u(q, z, 0, \partial_z)v = -B^0(q)\partial_z^2 v + \partial_z(A_\nu(W, d\psi^0)v) = 0. \qquad (3.11)$$

Since W satisfies the profile equation (1.23)(a), it follows by differentiating that equation twice that $W_z(q, z)$ satisfies (3.11). W_z decays to zero exponentially fast as $z \to \pm\infty$, so we conclude $D(q, \hat{\zeta}, 0) = 0$. This degeneracy reflects the translation-invariance of profiles, so we'll sometimes refer to it as the *translational degeneracy*.

The (q, z) dependence in $G(q, z, \zeta)$ enters through the viscous profile $W(q, z)$. Our main stability hypothesis is the following *Evans hypothesis* on $W(q, z)$:

Assumption 3.1 *(H3)* $D(q, \hat{\zeta}, \rho)$ *vanishes to exactly first order at $\rho = 0$ and has no other zeros for $(\hat{\zeta}, \rho) \in S_+^2 \times \{\rho \geq 0\}$.*

The preceding discussion shows that nonvanishing of D in $\gamma > 0$ is necessary for even for linearized stability.

In the construction of the approximate solution we had to use *transversality* of the connection (recall (1.30)) and *uniform stability* of the inviscid shock. An immediate corollary of the following theorem is that Assumption 3.1 implies both of these properties.

Theorem 3.1 ([ZS])

$$D(q, \hat{\zeta}, \rho) = \beta(q)\Delta(q, \hat{\zeta})\rho + o(\rho) \text{ as } \rho \to 0, \qquad (3.12)$$

where $\beta(q)$ is nonvanishing if and only if the connection is transverse at $W(q, 0)$. The second factor $\Delta(q, \hat{\zeta})$ is the Majda uniform stability determinant (6.5), which is nonvanishing if and only if $(U_\pm^0(0, q), d\psi^0(q))$ is uniformly stable.

A short proof is given in Appendix D.

Remark 3.1 *1. It follows from Assumption 1.3 that the stable/unstable manifolds of (1.23)(a) for the rest points $U_\pm^0(0, q)$ have dimensions $m-k$ and $m-l$ respectively, where $(m-k)+(m-l) = m+1$. Thus, the connection is transversal $\Leftrightarrow$ the intersection of these two manifolds is one dimensional $\Leftrightarrow$ the only $L^2(z)$ solution of $L_u(q, z, 0, \partial_z)w = 0$ on $\mathbb{R}_z$ is W_z.*

2. The singularity of $\mathcal{E}'_u$ in the low frequency regime that we referred to in Lecture 2 corresponds exactly to this one dimensional kernel of L_u when $\rho = 0$. The Evans hypothesis implies that for $\rho > 0$, the only L^2 solution of $L_u(q, z, \zeta, \partial_z)w = 0$ on $\mathbb{R}_z$ is $w = 0$.

3. The example of exponential blowup given above provides another (but belated) motivation for the introduction of exponential weights in Lecture 2.

4. The main Evans hypothesis (H3) is not easy to check, so we are glad to report that in recent work by Freistühler-Szmolyan [FS] and Plaza-Zumbrun [PZ], (H3) has been shown to hold for weak Lax shocks under mild structural assumptions satisfied by some of the important physical examples.

3.2 The Evans Function as a Lopatinski Determinant

Here is an equivalent definition of the Evans function that we'll use when constructing symmetrizers. With slight abuse of notation, write the transmission problem as

$$\partial_z U_\pm - G(q, z, \zeta)U_\pm = 0 \text{ on } \pm z \geq 0, \quad \Gamma U = 0 \qquad (3.13)$$

where $U = (U_+, U_-)$ and $\Gamma : \mathbb{C}^{4m} \to \mathbb{C}^{2m}$ is given by $\Gamma U = U_+ - U_-$.

Consider the $4m \times 4m$ determinant

$$\mathbb{D}(q, \hat{\zeta}, \rho) = \det \left(\ker \Gamma, E_+(q, \hat{\zeta}, \rho) \times E_-(q, \hat{\zeta}, \rho) \right). \qquad (3.14)$$

Performing a few row/column operations shows that $\mathbb{D} = cD$, for some $c \in \mathbb{C} \backslash 0$. Indeed, we should expect this since, clearly, having a nontrivial intersection of the subspaces on the right side of (3.14) is equivalent to have a nontrivial intersection of $E_\pm$.

In the theory of boundary problems determinants of this sort, which measure the degree of linear independence of a subspace giving the kernel of the boundary operator with the decaying eigenspace of the interior operator, are often called *Lopatinski* determinants. When these determinants are nonzero, we expect good L^2 estimates; when they vanish, we expect degenerate estimates or no estimates. Indeed, we saw above that vanishing of D in $\gamma > 0$ leads to exponential blowup. The vanishing of D at $\rho = 0$ (which implies $\gamma = \rho\hat{\gamma} = 0$) because of the translational degeneracy is a borderline case, and leads to degenerate L^2 estimates [GMWZ1, GMWZ2].

3.3 Doubling

For future use and to make the connection to boundary problems more explicit, let's double the problem (3.13) and write it as a $4m \times 4m$ system on $z \geq 0$. If $f(z)$ is any function (complex valued, matrix valued,...) defined on $z \leq 0$, we set

$$\tilde{f}(z) = f(-z) \text{ for } z \geq 0. \qquad (3.15)$$

Then the $2m \times 2m$ transmission problem on $\mathbb{R}_z$

$$\partial_z U_\pm - G(q, z, \zeta)U_\pm = F_\pm, \quad U_+ - U_- = 0 \qquad (3.16)$$

is equivalent to the $4m \times 4m$ boundary problem on $z \geq 0$

$$\partial_z U - \mathcal{G}(q, z, \zeta)U = \mathcal{F}, \quad \Gamma U = 0, \qquad (3.17)$$

where $U = (U_+, \tilde{U}_-)$, $\mathcal{F} = (F_+, -\tilde{F}_-)$, $\Gamma U = U_+ - \tilde{U}_-$, and

$$\mathcal{G} = \begin{pmatrix} G & 0 \\ 0 & -\tilde{G} \end{pmatrix}. \qquad (3.18)$$

Note that

$$\mathbb{E}(q, \hat{\zeta}, \rho) := E_+(q, \hat{\zeta}, \rho) \times E_-(q, \hat{\zeta}, \rho) \tag{3.19}$$

is the decaying generalized eigenspace for $U_z - \mathcal{G}U = 0$ on $z \geq 0$ and $\mathbb{D}$ is now exactly the Lopatinski determinant (in the classical sense) for the boundary problem (3.17).

It's easy to check that if we use the MZ conjugators $Z_\pm$ to define

$$Z(q, z, \zeta) = \begin{pmatrix} Z_+ & 0 \\ 0 & \tilde{Z}_- \end{pmatrix} \tag{3.20}$$

and define V by $U = ZV$, then U satisfies (3.17) $\Leftrightarrow$ V satisfies

$$\partial_z V - \mathcal{G}_\infty V = Z^{-1}\mathcal{F}, \quad \Gamma ZV = 0, \tag{3.21}$$

where

$$\mathcal{G}_\infty = \begin{pmatrix} G_{+\infty} & 0 \\ 0 & -G_{-\infty} \end{pmatrix}. \tag{3.22}$$

One advantage of doubling is that the two distinct limiting problems have become a single limiting problem at $z = +\infty$. Observe that each additional conjugation twists the boundary condition by an additional matrix factor to the right of Γ. We won't use the doubled form until the symmetrizer construction in Lecture 4.

3.4 Slow Modes and Fast Modes

Recall that the problems on $\pm z \geq 0$

$$U_z - GU = 0, V_z - G_{\pm\infty}V = 0, W_z - G_{HP\pm}W = 0 \tag{3.23}$$

are related by the conjugators: $U = ZV$, $V = YW$.

We have already defined the decaying spaces $E_\pm$, $F_\pm$ for the first two problems. For $\rho \neq 0$ set $K_\pm(q, \zeta)$ equal to the decaying generalized eigenspaces for

$$\partial_z W - G_{HP\pm}W = 0. \tag{3.24}$$

Clearly, $F_\pm = Y_\pm K_\pm$ and $E_\pm = (Z_\pm|_{z=0})Y_\pm K_\pm$.

Write $H_\pm(q, \hat{\zeta}, \rho) = \rho\hat{H}_\pm(q, \hat{\zeta}, \rho)$ where

$$\hat{H}(q, \hat{\zeta}, 0) = -A_\nu(U_\pm^0, d\psi^0)^{-1}\left((i\hat{\tau} + \hat{\gamma})I + g'(U_\pm^0)i\hat{\eta}\right). \tag{3.25}$$

Using obvious notation we can in turn decompose $K_\pm$ as

$$K_\pm(q, \hat{\zeta}, \rho) = K_{\hat{H}_\pm}(q, \hat{\zeta}, \rho) \oplus K_{P_\pm}(q, \hat{\zeta}). \tag{3.26}$$

By (H1) the dimensions of $K_{\hat{H}_\pm}$ are k (resp. l), while the dimensions of $K_{P_\pm}$ are $m-k$ (resp. $m-l$). By the result proved in Appendix B, $K_\pm(q,\hat{\zeta},\rho)$ extend continuously to $\rho = 0$. The main step is to prove continuous extendability of $K_{\hat{H}_\pm}$.

We may now define *slow* (resp. *fast*) *modes* as solutions of (3.3) of the form

$$U_\pm = Z_\pm Y_\pm W_\pm, \tag{3.27}$$

where $W_\pm(q,z,\hat{\zeta},\rho)$ is a solution of (3.24) with $W_\pm(q,0,\hat{\zeta},\rho) \in K_{\hat{H}_\pm}$ (resp. $K_{P_\pm}$).

Fast modes decay exponentially to zero as $z \to \pm\infty$ for $\rho \geq 0$, and slow modes decay exponentially to zero (but slowly) for $\rho > 0$ small. For $\rho = 0$ slow modes can decay to nonzero constant vectors. In Appendix C we identify those limits, a knowledge of which is needed for the Zumbrun-Serre result (Appendix D) and also for removal of the translational degeneracy. Here is the result on slow modes.

Proposition 3.3 *Let* $U_\pm(q,z,\hat{\zeta},\rho)$ *be a slow mode. Then*

$$\lim_{z\to\pm\infty} U_\pm(q,z,\hat{\zeta},0) \text{ exists and belongs to } K_{\hat{H}_\pm}(q,\hat{\zeta},0). \tag{3.28}$$

Let's rephrase this in a manner useful for the applications. The spaces $K_{\hat{H}_\pm}(q,\hat{\zeta},\rho)$ have bases

$$\{(r_+^s(q,\hat{\zeta},\rho),0)\}_{s=1}^k \text{ and } \{(r_-^t(q,\hat{\zeta},\rho),0)\}_{t=1}^l \tag{3.29}$$

respectively, where the $r_\pm^j$ are smooth in $\rho > 0$ and extend continuously to $\rho = 0$. Since

$$\hat{H}_\pm(q,\hat{\zeta},0) = \mathbb{H}_\pm(q,\hat{\zeta}) \text{ (for } \mathbb{H} \text{ as in Appendix A),} \tag{3.30}$$

we may choose the $r_\pm^j$ to agree at $\rho = 0$ with the vectors $r_\pm^j(q,\hat{\zeta})$ that appear in the definition of the uniform stability determinant $\Delta(q,\hat{\zeta})$ (6.5). Thus, for each choice of sign

$$\lim_{z\to\pm\infty} U_\pm(q,z,\hat{\zeta},0) \in \operatorname{span}\{(r_\pm^j(q,\hat{\zeta}),0)\}, \tag{3.31}$$

where $r_\pm^j(q,\hat{\zeta})$ span the decaying generalized eigenspaces of $\mathbb{H}_\pm$ as in Appendix A.

3.5 Removing the Translational Degeneracy

Our next main task is to show how the extra boundary condition can be used to remove the translational degeneracy. In lecture two we sketched a strategy for reducing study of the fully linearized problem to the partially linearized one $(\mathcal{E}'_u)$. Let's recall that here in the context of our ODE problem.

Returning to transmission notation, where $U = (v, v_z)$, we recall (2.29) the fully linearized transmission problem

$$\partial_z U - G(q, z, \zeta)U = F + \begin{pmatrix} 0 \\ -(B^0(q))^{-1}L_\psi(q, z, \zeta)\phi \end{pmatrix} = F + \mathcal{B}(q, z, \zeta)\phi$$
$$[U] = 0, \ p(\zeta)\phi - l(q) \cdot v = 0 \text{ on } z = 0$$
$$(3.32)$$

Set $P(q, z) = (W_z, W_{zz})$. The relationship (2.27) between the partial linearizations

$$L_u(q, z, \zeta, \partial_z)W_z = -L_\psi(q, z, \zeta) \tag{3.33}$$

is equivalently expressed as

$$(\partial_z - G)P = \mathcal{B}. \tag{3.34}$$

Thus, (U, ϕ) satisfies (3.32) $\Leftrightarrow (v^\sharp, v_z^\sharp) = U^\sharp := U - P\phi$ satisfies

$$\partial_z U^\sharp - G(q, z, \zeta)U^\sharp = F, \ [U^\sharp] = 0, \tag{3.35}$$

together with the extra boundary condition

$$l(q) \cdot v^\sharp = (p(\zeta) + 1)\phi = (1 + i\tau + \gamma + \eta^2)\phi \text{ on } z = 0. \tag{3.36}$$

Now as we've seen, the main Evans hypothesis implies the Lopatinski determinant $\mathbb{D}(q, \hat{\zeta}, \rho)$ for the problem (3.35) is nonvanishing in MF or HF, so we expect good estimates there, including an estimate for the trace $|v^\sharp(0)|$. The extra condition (3.36) will then allow us to estimate $|\phi|$ in terms of $|v^\sharp(0)|$.

On the other hand we can't use this approach to get estimates uniform with respect to ζ in SF since in the notation of (3.14)

$$\ker \Gamma \cap (E_+(q, \hat{\zeta}, 0) \times E_-(q, \hat{\zeta}, 0)) =$$
$$= \operatorname{span}(W_z(q, 0), W_{zz}(q, 0), W_z(q, 0), W_{zz}(q, 0)), \tag{3.37}$$

so (3.20) is singular in SF.

To handle SF, instead of $U^\sharp$ we define as our "good unknown"

$$U^\flat = U - R(q, z, \zeta)\phi, \tag{3.38}$$

where $R_\pm = (r_\pm, s_\pm)$ is smooth in $\pm z \geq 0$ and constructed to satisfy

$$(a)\partial_z R - G(q, z, \zeta)R = \mathcal{B} \text{ in } \pm z \geq 0, \ l(q) \cdot r_\pm(q, 0, \zeta) = p(\zeta)$$
$$(b)R(q, z, 0) = 0, \ |R(q, z, \zeta)| \leq Ce^{-\delta|z|}. \tag{3.39}$$

The construction is not hard and is given in Appendix E. For now we note that since $\mathcal{B}(q, z, \zeta)$ also satisfies (3.39)(b) and we can use MZ conjugators to reduce to a constant coefficient problem, we should expect such R to exist.

Since $R_\pm(q, z, 0) = 0$ we may write

$$R_\pm(q, z, \zeta) = \rho \hat{R}_\pm(q, z, \hat{\zeta}, \rho). \tag{3.40}$$

Now (U, ϕ) satisfies (3.32)$\Leftrightarrow U^\flat = (u^\flat, u_z^\flat)$ satisfies

$$
\begin{aligned}
&(a) \partial_z U^\flat - G(q, z, \zeta) U^\flat = F, \\
&(b) [U^\flat] = -\phi[R] = -\phi \rho[\hat{R}] = -\hat{\phi}[\hat{R}], \\
&(c) l(q) \cdot u_+^\flat = l \cdot v - \phi l \cdot r_+ = p(\zeta)\phi - p(\zeta)\phi = 0.
\end{aligned} \tag{3.41}
$$

The following fact, proved below, is a consequence of uniform stability of the inviscid shock.

Proposition 3.4 *For all $\hat{\zeta} \in S_+^2$, we have $[\hat{R}(q, 0, \hat{\zeta}, 0] \neq 0$.*

So for ρ small there is a well-defined orthogonal projection

$$\pi(q, \hat{\zeta}, \rho) : \mathbb{C}^{2m} \to \mathbb{C}^{2m} \tag{3.42}$$

onto $[\hat{R}(q, 0, \hat{\zeta}, \rho)]^\perp$, the subspace of $\mathbb{C}^{2m}$ orthogonal to $[\hat{R}(q, 0, \hat{\zeta}, \rho)]$.

Now we can apply $\pi(q, \hat{\zeta}, \rho)$ to (3.41)(b) to project out the front, giving us the following problem for $U^\flat$:

$$
\begin{aligned}
&\partial_z U^\flat - G(q, z, \zeta) U^\flat = F \\
&\pi(q, \hat{\zeta}, \rho)[U^\flat] = 0, \quad l(q) \cdot u_+^\flat = 0.
\end{aligned} \tag{3.43}
$$

Using a notation like the one we used in defining $\mathbb{D}(q, \hat{\zeta}, \rho)$, let

$$\tilde{\Gamma} : \mathbb{C}^{4m} \to \mathbb{C}^{2m+1} \tag{3.44}$$

denote the boundary condition in (3.43). It's easy to check that $\dim(\ker \tilde{\Gamma}) = 2m$.

The next Proposition, a key to the whole analysis, expresses the fact that the modified problem (3.43) *does* satisfy the Lopatinski condition in SF:

Proposition 3.5 *Assume (H1), (H2), and (H3). Then for all $\hat{\zeta} \in S_+^2$*

$$\tilde{D}(q, \hat{\zeta}, 0) = \det\left(\ker \tilde{\Gamma}, E_+(q, \hat{\zeta}, 0) \times E_-(q, \hat{\zeta}, 0)\right) \neq 0,$$

or equivalently,

$$\ker \tilde{\Gamma} \cap (E_+(q, \hat{\zeta}, 0) \times E_-(q, \hat{\zeta}, 0)) = \{0\}.$$

By continuity we have $\tilde{D}(q,\hat{\zeta},\rho) \neq 0$ for ρ small. In other words the translational degeneracy has been removed. In view of our earlier discussion of Lopatinski determinants, we may expect good L^2 estimates for (3.43), including an estimate for the trace $U^b(0)$. Since $[U^b] = -\hat{\phi}[\hat{R}]$ and $[\hat{R}]$ is nonvanishing for ρ small (Proposition 3.4), we can then estimate $|\hat{\phi}|$ in terms of $|U^b(0)|$.

Remark 3.2 *The proof of Proposition 3.5, given in Appendix E, is based on nonvanishing of the uniform stability determinant $\Delta(q,\hat{\zeta})$. It is easy to see that if $(U_+(0), U_-(0))$ is a nonvanishing element of*

$$E_+(q,\hat{\zeta},0) \times E_-(q,\hat{\zeta},0),$$

the boundary condition $l(q) \cdot u_+ = 0$ implies

$$(U_+(0), U_-(0)) \notin \ker \Gamma \tag{3.45}$$

(use (3.37) and the fact that $l(q) \cdot W_z(q,0) = -1$). However, to prove the Proposition we need to show that $(U_+(0), U_-(0))$ is not in the larger space $\ker(\pi(q,\hat{\zeta},0)\Gamma)$.

4 Lecture Four: Block Structure, Symmetrizers, Estimates

In the construction of symmetrizers it is most convenient to work with the $4m \times 4m$ doubled problem (3.17)

$$\partial_z U - \mathcal{G}(q,z,\zeta)U = \mathcal{F}, \quad \Gamma U = 0, \tag{4.1}$$

where $U = (U_+, \tilde{U}_-)$, $\Gamma U = U_+ - \tilde{U}_-$, and

$$\mathcal{G} = \begin{pmatrix} G & 0 \\ 0 & -\tilde{G} \end{pmatrix}. \tag{4.2}$$

This is because the problems in $\pm z \geq 0$ are coupled by the boundary condition which depends on both U_+ and U_-, and the boundary estimates can't be done for $\pm$ blocks separately. Symmetrizer construction is much easier after further conjugation of $\mathcal{G}$ to an appropriate block structure, which is different in each of the three frequency regimes.

4.1 The MF Regime.

We've already described the MZ conjugation to the doubled limiting problem (3.21) on $z \geq 0$

$$\partial_z V - \mathcal{G}_\infty V = F, \quad \Gamma Z V = 0, \text{ where}$$

$$\mathcal{G}_\infty = \begin{pmatrix} G_{+\infty} & 0 \\ 0 & -G_{-\infty} \end{pmatrix}. \tag{4.3}$$

This conjugation can't be done uniformly in HF, so we use it only in the SF and MF regimes. Here we'll carry out the full details in MF, the easiest of the three regimes. This will clarify how Lopatinski conditions yield estimates.

We work near a basepoint $(\underline{q}, \underline{\zeta})$ with $\underline{\zeta} \neq 0$. Proposition 3.1 implies that for (q, ζ) near $(\underline{q}, \underline{\zeta})$ the spectrum of $\mathcal{G}_\infty$ lies in a compact subset of the complement of the imaginary axis, with $2m$ eigenvalues (counted with multiplicity) on each side. Thus, there exists a smooth conjugator $T_M(q, \zeta)$ such that

$$T_M^{-1} \mathcal{G}_\infty T_M = \begin{pmatrix} P_g(q, \zeta) & 0 \\ 0 & P_d(q, \zeta) \end{pmatrix} := \mathcal{G}_{gd} \qquad (4.4)$$

where for some $C > 0$

$$\Re P_g := \frac{P_g + P_g^*}{2} > CI; \quad \Re P_d < -CI. \qquad (4.5)$$

Note that each block in (4.4) is of size $2m$ with, for example, m of the eigenvalues of P_g coming from $G_{+\infty}$ and m from $-G_{-\infty}$. The letters g and d stand for "growing" and "decaying" respectively.

Redefining F and setting $V = T_M U$ (so $U = (U^1, U^2)$ is not the same as before), we have reduced the study of (4.3) to the study of

$$\begin{aligned} \partial_z U - \mathcal{G}_{gd} U &= F \text{ in } z \geq 0 \\ \Gamma'(q, \zeta) U &:= \Gamma Z(q, 0, \zeta) T_M(q, \zeta) U = 0 \text{ on } z = 0. \end{aligned} \qquad (4.6)$$

In Lecture 3 we defined the decaying generalized eigenspaces $E_\pm(q, \zeta)$ for $G(q, z, \zeta)$ on $\pm z \geq 0$. Observe that

$$\mathbb{E}(q, \zeta) := E_+(q, \zeta) \times E_-(q, \zeta) \qquad (4.7)$$

is the $2m$ dimensional decaying space for $\mathcal{G}$. The decaying space for $\mathcal{G}_{gd}(q, \zeta)$

$$\mathbb{F}(q, \zeta) := \{(0, U^2) : U^2 \in \mathbb{C}^{2m}\} \qquad (4.8)$$

is independent of (q, ζ) and evidently satisfies $\mathbb{E}(q, \zeta) = Z(q, 0, \zeta) T_M(q, \zeta) \mathbb{F}$.

The main Evans hypothesis (H3) implies that for (q, ζ) near the basepoint

$$\begin{aligned} \ker \Gamma \cap \mathbb{E}(q, \zeta) &= \{0\}, \text{ or equivalently} \\ \ker \Gamma'(q, \zeta) \cap \mathbb{F} &= \{0\}. \end{aligned} \qquad (4.9)$$

We're ready now to construct the symmetrizer. In fact, with the above preparation the construction is practically trivial. We'll write

$$U = U_g + U_d, \text{ where } U_g = (U^1, 0) \text{ and } U_d = (0, U^2). \qquad (4.10)$$

We'll use the notation

$$|U|_2 = |U(z)|_{L^2(z)}, \ |U| = |U(0)|_{\mathbb{C}^{4m}}, \ (U, V) = \int_0^\infty \langle U(z), V(z) \rangle dz, \qquad (4.11)$$

where $\langle\,,\,\rangle$ is the inner product in $\mathbb{C}^{4m}$.

The symmetrizer for (4.6) in MF is a matrix $S(q, z, \zeta)$ with the following properties

$$
\begin{aligned}
&(a)\, S = S^*, \ |S(q, z, \zeta)| \le C \\
&(b)\, \Re(S\mathcal{G}_{gd}) \ge I \ \text{in} \ z \ge 0 \\
&(c)\, S + C(\Gamma')^*\Gamma' \ge I \ \text{on} \ z = 0
\end{aligned}
\tag{4.12}
$$

for some $C > 0$. We may take S to be simply

$$
\begin{pmatrix} cI & 0 \\ 0 & -I \end{pmatrix}
\tag{4.13}
$$

for some large enough $c > 0$ to be chosen. Clearly, properties (4.12)(a),(b) are then satisfied.

The Lopatinski condition (4.9) is then equivalent to

$$
|\Gamma'(q, \zeta)U_d| \ge C|U_d| \ \text{for some} \ C > 0
\tag{4.14}
$$

near the basepoint, and this implies

$$
|U_d|^2 \le C|\Gamma'U_d|^2 \le C_1(|\Gamma'U|^2 + |U_g|^2).
\tag{4.15}
$$

To arrange (4.12)(c) observe that on $z = 0$

$$
\begin{aligned}
\langle SU, U \rangle &= c|U_g|^2 - |U_d|^2 = c|U_g|^2 + |U_d|^2 - 2|U_d|^2 \\
&\ge c|U_g|^2 + |U_d|^2 - 2C_1(|\Gamma'U|^2 + |U_g|^2) \\
&= (c - 2C_1)|U_g|^2 + |U_d|^2 - 2C_1|\Gamma'U|^2,
\end{aligned}
\tag{4.16}
$$

which gives (4.12)(c) for c big enough.

Integration by parts, the equation, and the property $S = S^*$ yield the identity

$$
-\langle SU(0), U(0) \rangle = \int_0^\infty \partial_z \langle SU, U \rangle dz = 2\Re(S\mathcal{G}_{gd}U, U) + 2\Re(SF, U), \quad (4.17)
$$

which together with (4.12) easily implies the L^2 estimate

$$
|U|_2^2 + |U|^2 \le C(|F|_2^2 + |\Gamma'U|^2),
\tag{4.18}
$$

for a C independent of (q, ζ) near the basepoint. Here we've used

$$
|(SF, U)| \le \delta|U|_2^2 + C_\delta|F|_2^2
\tag{4.19}
$$

and absorbed the $\delta|U|_2^2$ from the right. Conjugation via ZT_M and compactness of MF then yield the uniform estimate for (4.1) in MF

$$
|U|_2^2 + |U|^2 \le C(|F|_2^2 + |\Gamma U|^2).
\tag{4.20}
$$

Recall that the q dependence enters only through $W(q, z)$.

4.2 The SF Regime.

Symmetrizer construction was rather easy in MF thanks mainly to three things: the MZ conjugator, the compactness of MF, and the fact that for ζ in MF the spectrum of $\mathcal{G}_\infty$ is contained in $K_+ \cup K_-$ for compact sets $K_\pm$ in $\pm \Re \mu > 0$ respectively.

SF is more subtle because, even though we have the MZ conjugator and compactness here, the third property does not hold. As $\rho \to 0$ the spectrum creeps right up to the imaginary axis $\Re \mu = 0$ from both sides (and in a rather singular way). Indeed, recall that at $\zeta = 0$ we have

$$G_{\pm\infty}(q,0) = \begin{pmatrix} 0 & I \\ 0 & \mathcal{A}_{\pm\infty}(q,0) \end{pmatrix}. \tag{4.21}$$

In SF we first conjugate (4.1) via Z to $\mathcal{G}_\infty$, then again via

$$Y(q,\zeta) = \begin{pmatrix} Y_+ & 0 \\ 0 & Y_- \end{pmatrix} \text{ to}$$

$$\tilde{\mathcal{G}}_{HP}(q,\zeta) = \begin{pmatrix} H_+ & 0 & 0 & 0 \\ 0 & P_+ & 0 & 0 \\ 0 & 0 & -H_- & 0 \\ 0 & 0 & 0 & -P_- \end{pmatrix} \tag{4.22}$$

(recall (2.42)), and next via a constant matrix T_c to

$$\mathcal{G}_{HP}(q,\zeta) = \begin{pmatrix} H_+ & 0 & 0 & 0 \\ 0 & -H_- & 0 & 0 \\ 0 & 0 & P_+ & 0 \\ 0 & 0 & 0 & -P_- \end{pmatrix}. \tag{4.23}$$

We may write $H_\pm(q,\zeta) = \rho \hat{H}_\pm(q,\hat{\zeta},\rho)$.

So far the conjugations work uniformly for ρ small. For the final conjugation to *block structure* $\mathcal{G}_B$, we fix a basepoint $(\underline{q},\hat{\underline{\zeta}},0)$ and construct a matrix $T_B(q,\hat{\zeta},\rho)$ such that:

$$T_B^{-1}\mathcal{G}_{HP}T_B = \begin{pmatrix} H_B(q,\hat{\zeta},\rho) & 0 & 0 \\ 0 & P_g(q,\zeta) & 0 \\ 0 & 0 & P_d(q,\zeta) \end{pmatrix} := \mathcal{G}_B(q,\hat{\zeta},\rho) \tag{4.24}$$

for $(q,\hat{\zeta},\rho)$ near $(\underline{q},\hat{\underline{\zeta}},0)$. Here T_B is of the form

$$T_B = \begin{pmatrix} T_{BH}(q,\hat{\zeta},\rho) & 0 \\ 0 & T_{BP}(q,\zeta) \end{pmatrix} \tag{4.25}$$

with blocks of size $2m$, while the blocks H_B, P_g, and P_d are of sizes $2m$, $m-1$, and $m+1$ respectively. All the blocks appearing in T_B and $\mathcal{G}_B$ are smooth functions of their arguments.

The eigenvalues of P_g and P_d lie in compact subsets $K_\pm$ of $\pm\Re\mu > 0$ respectively. Their dimensions can be read off from Proposition 3.1. In addition

$$\Re P_g = \frac{1}{2}(P_g + P_g^*) \geq cI \text{ and } \Re P_d \leq -cI \tag{4.26}$$

for some $c > 0$.

The eigenvalues of H_B are those of $\mathcal{G}_\infty$ that approach zero as $\rho \to 0$. We may write $H_B(q, \hat\zeta, \rho) = \rho\hat{H}_B(q, \hat\zeta, \rho)$ with

$$\hat{H}_B(q, \hat\zeta, \rho) = \begin{bmatrix} Q_1 & \cdots & 0 \\ \vdots & \ddots & \vdots \\ 0 & \cdots & Q_s \end{bmatrix}. \tag{4.27}$$

The blocks Q_k are $\nu_k \times \nu_k$ matrices which satisfy one of the following conditions for $(q, \hat\zeta, \rho)$ near $(\underline{q}, \underline{\hat\zeta}, 0)$:
   i) $\Re Q_k > 0$.
   ii) $\Re Q_k < 0$.
   iii) $\nu_k = 1$, $\Re Q_k = 0$ when $\hat\gamma = \rho = 0$, and $\partial_{\hat\gamma}(\Re Q_k)\partial_\rho(\Re Q_k) > 0$.
   iv) $\nu_k > 1$, $Q_k(q, \hat\zeta, \rho)$ has purely imaginary coefficients when $\hat\gamma = \rho = 0$, there is $\alpha_k \in \mathbb{R}$ such that

$$Q_k(\underline{q}, \underline{\hat\zeta}, 0) = i \begin{bmatrix} \alpha_k & 1 & 0 & \\ 0 & \alpha_k & \ddots & 0 \\ & \ddots & \ddots & 1 \\ & & \cdots & \alpha_k \end{bmatrix}, \tag{4.28}$$

and the lower left corner a of Q_k satisfies $\partial_{\hat\gamma}(\Re a)\partial_\rho(\Re a) > 0$.

Proof (Sketch of proof).
   1. Starting from $\mathcal{G}_{HP}$ we construct smooth projectors $\mathcal{P}_g(q, \zeta)$ and $\mathcal{P}_d(q, \zeta)$

$$\mathcal{P}_{g,d}(q, \zeta) = \frac{1}{2\pi i} \int_{\mathcal{C}_{g,d}} \left(\xi - \begin{pmatrix} P_+ & 0 \\ 0 & -P_- \end{pmatrix}\right)^{-1} d\xi \tag{4.29}$$

using contours $\mathcal{C}_g$ and $\mathcal{C}_d$ in the right and left half planes, respectively, which enclose the eigenvalues of the submatrix $(P_+, -P_-)$ in those half planes. Applying $\mathcal{P}_{g,d}(q, \zeta)$ to a basis of $\mathcal{P}_{g,d}(q, 0)$ yields a basis of

$$\text{range } \mathcal{P}_{g,d}(q, \zeta)$$

varying smoothly with ζ.

The blocks H_+ and $-H_-$ are conjugated separately to block structure. Thus, there is a k_0 such that the blocks $Q_1, \ldots, Q_{k_0}$ in $\hat{H}_B$ correspond to $\hat{H}_+$, while blocks $Q_{k_0+1}, \ldots, Q_p$ correspond to $-\hat{H}_-$. For example, suppose

$$\{i\alpha_1, \ldots, i\alpha_{k_0-2}\} \tag{4.30}$$

are the distinct pure imaginary eigenvalues of $\hat{H}_+(q, \hat{\zeta}, 0)$ with $i\alpha_j$ of multiplicity m_j, and let $m_\pm$ be the number of eigenvalues of $\hat{H}_+(q, \hat{\zeta}, 0)$ counted with multiplicity in $\pm\Re\mu > 0$. Using projectors defined by integration on suitable contours as above, one obtains a decomposition of $\hat{H}_+$ near the basepoint in which there is a block of size m_j satisfying either (iii) or (iv) corresponding to each $i\alpha_j$, a block of size m_+ satisfying (i), and a block of size m_- satisfying (ii). As the basepoint changes so does the decomposition.

A further change of basis in cases (iii),(iv) puts $Q_k(q, \hat{\zeta}, 0)$ in Jordan form. Changing basis again using Ralston's Lemma [Ra, CP] makes Q_k pure imaginary in cases (iii),(iv) when $\hat{\gamma} = \rho = 0$. Observe that by hyperbolicity blocks satisfying conditions (iii) or (iv) only arise when $\hat{\gamma} = 0$.

2. We discuss the crucial *sign condition* in cases (iii) and (iv) in the next subsection.   •

4.3 The Sign Condition

Suppose we start with $\hat{H}_+(q, \hat{\zeta}, 0)$, where we assume $\hat{\gamma} = 0$ as in cases (iii) and (iv) above, and then perturb to $\hat{\gamma} > 0$ while holding $\rho = 0$ (case A), or to $\rho > 0$ holding $\hat{\gamma} = 0$ (case B). In both cases the perturbed matrix $\hat{H}_+$ has $m - k$ eigenvalues with positive real part and k with negative real part.

In case A this follows from hyperbolicity (H1), (H2), and the explicit form of $\hat{H}_+$ (2.43). Hyperbolicity implies that the number of eigenvalues with positive (or negative) real part is constant for $\hat{\gamma} > 0$, $\rho = 0$. We can then take $(\hat{\tau}, \hat{\eta}) = 0$ to obtain a count. In case B the count follows directly from (H2) and Proposition (3.1).

To describe this situation we'll say that changes in $\hat{\gamma}$ or ρ lead to *identical splitting* for $\hat{H}_+(q, \hat{\zeta}, 0)$ (this should not be confused with "consistent splitting", which has a different meaning).

Identical splitting follows from the sign condition in (iii) and (iv), but does not quite imply it. The sign condition implies more; namely, that we have identical splitting for each block Q_k. This is because the behavior of the lower left entry governs the splitting of $i\alpha_k$ when a block $Q_k(q, \hat{\zeta}, 0)$ has the Jordan form (4.28). The sign condition implies changes in $\hat{\gamma}$ or ρ have the same effect.

Let's try to understand the role of the lower left entry in a simple case. For $\alpha \in \mathbb{R}$ set

$$Q(\gamma) = i \begin{pmatrix} \alpha & 1 & 0 \\ 0 & \alpha & 1 \\ 0 & 0 & \alpha \end{pmatrix} + \gamma(b_{ij}) \tag{4.31}$$

where (b_{ij}) is some 3×3 constant matrix with $b_{31} \neq 0$. Compute the characteristic polynomial of $Q(\gamma)$ by expanding down the first column to get

$$\det(Q(\gamma) - \xi I) = (i\alpha - \xi)^3 - \gamma b_{31} + O(\gamma^2) + O(\gamma(i\alpha - \xi)). \tag{4.32}$$

Eigenvalues $\xi(\gamma)$ satisfy

$$i\alpha - \xi(\gamma) = o(1) \text{ as } \gamma \to 0, \qquad (4.33)$$

so to determine the splitting we can solve

$$(i\alpha - \xi)^3 = \gamma b_{31} \qquad (4.34)$$

for ξ.

When $\Re b_{31} > 0$ we obtain two solutions with $\Re\xi > 0$, one with $\Re\xi < 0$. When $\Re b_{31} < 0$ we obtain one solution with $\Re\xi > 0$, two with $\Re\xi < 0$.

The sign condition allows one to construct symmetrizers by a modification of the ansatz used by Kreiss in [K]. In [K] there was only one perturbation parameter $\hat{\gamma}$ and one derivative to consider $\partial_{\hat{\gamma}}\Re a$. That derivative was nonzero as a consequence of his strict hyperbolicity assumption. Because of our sign condition, we can construct a symmetrizer in this two-parameter situation by adding an extra term, corresponding to the ρ parameter, to the kth block of S in the Kreiss ansatz. As explained in Appendix B the sign condition is also used in the proof that the decaying eigenspaces $E_{\pm}(q,\hat{\zeta},\rho)$ extend continuously to $\rho = 0$.

4.4 Glancing Blocks and Glancing Modes

Definition 4.1 *Blocks satisfying condition (iv) in the above theorem will be referred to as* glancing blocks. *From the above discussion we see these blocks correspond to coalescing eigenvalues of $\hat{H}_{\pm}$.*

In the remainder of this subsection we'll allow our 2D notation to represent any number of space dimensions. Thus, if $\eta \in \mathbb{R}^{d-1}$, when we write $g'(U_{\pm}^0)i\eta$ we mean

$$g'(U_{\pm}^0)i\eta := \sum_{j=1}^{d-1} g'_j(U_{\pm}^0)i\eta_j. \qquad (4.35)$$

To define glancing modes and relate them to glancing blocks we consider the linearized inviscid limiting operators

$$\mathbb{L}_{\pm}(q,\xi,\tau,\eta) = i\tau I + A_\nu(U_{\pm}^0(0,q), d\psi^0(q))i\xi + g'(U_{\pm}^0)i\eta \qquad (4.36)$$

and the corresponding scalar symbols

$$p_{\pm}(q,\xi,\tau,\eta) = \det \mathbb{L}_{\pm}(q,\xi,\tau,\eta). \qquad (4.37)$$

Definition 4.2 *Define the* glancing set $\mathcal{G}_q$ *to be the set of $(\tau,\eta) \in \mathbb{R}^d \setminus 0$ such that for at least one choice of sign the equation $p_{\pm}(q,\xi,\tau,\eta) = 0$ has a real root ξ of multiplicity ≥ 2. We'll refer to individual points $(\tau_0,\eta_0) \in \mathcal{G}$ as* glancing modes.

Clearly, at any point $(\tau_0, \eta_0) \in \mathcal{G}_q$ at least one real root ξ of $p_\pm(q, \xi, \tau, \eta) = 0$, has a branch singularity. (The degree of singularity with respect to τ (η held fixed) is equal to the integer s in (4.40) below.)

The hyperbolicity assumption (H1) implies there exist real functions

$$\tau_1^\pm(q, \xi, \eta), \ldots, \tau_m^\pm(q, \xi, \eta),$$

smooth and homogeneous of degree one in $(\xi, \eta) \neq 0$, such that

$$\tau_1^\pm < \cdots < \tau_m^\pm \text{ and} \tag{4.38}$$
$$p_\pm(q, \tau, \xi, \eta) = (\tau - \tau_1^\pm(q, \xi, \eta)) \cdots (\tau - \tau_m^\pm(q, \xi, \eta)).$$

If $(\tau_0, \eta_0) \in \mathcal{G}_q$, there exist ξ_0 and for at least one choice of sign a $\tau_j^\pm$ (with j uniquely determined by the choice of $\pm$ and (ξ_0, τ_0, η_0)) such that (dropping $\pm$)

$$\tau_0 = \tau_j(q, \xi_0, \eta_0), \text{ and} \tag{4.39}$$
$$\partial_\xi \tau_j(q, \xi_0, \eta_0) = 0.$$

Moreover, the multiplicity of ξ_0 as a root of $p(q, \xi_0, \tau_0, \eta_0) = 0$, and thus the degree of singularity (with respect to τ) of the associated branch point, is equal to s $(2 \leq s \leq m)$ if and only if

$$\partial_\xi^k \tau_j(q, \xi_0, \eta_0) = 0, \text{ for } k = 1, \ldots, s-1, \text{ but} \tag{4.40}$$
$$\partial_\xi^s \tau_j(q, \xi_0, \eta_0) \neq 0.$$

Note that this implies at the same time that $\partial_\xi \tau_j(\cdot, \eta_0)$ has no roots nearby ξ_0 other than ξ_0 itself.

To relate glancing modes to glancing blocks note, for example, that

$$p_+(q, \hat{\xi}, \hat{\tau}, \hat{\eta}) = \det\left(A_\nu(U_+^0(0, q), d\psi^0(q))(i\hat{\xi} - \hat{H}_+(q, \hat{\tau}, \hat{\gamma} = 0, \hat{\eta}, \rho = 0))\right). \tag{4.41}$$

So when we have a block of size ν associated to a multiple pure imaginary eigenvalue $i\alpha$ of $\hat{H}_+(q, \hat{\zeta}, 0)$, this means $(\hat{\tau}, \hat{\eta}) \in \mathcal{G}_q$ and that α is a root of multiplicity ν of

$$p_+(q, \xi, \hat{\tau}, \hat{\eta}) = 0.$$

The word *glancing* is used because characteristics $(x(t), t, y(t))$ associated to a glancing mode $(\tau_0, \eta_0) \in \mathcal{G}_q$ with $\tau_0 = \tau_j(q, \xi_0, \eta_0)$ satisfy

$$x'(t) = -\partial_\xi \tau_j(q, \xi_0, \eta_0) = 0, \tag{4.42}$$

and thus run parallel to the boundary $x = 0$.

4.5 Auxiliary Hypothesis for Lecture 5

This is a good place to state an auxiliary hypothesis that we'll need later. In Lecture 5 we deal with planar inviscid shocks so there is no q dependence anymore. In particular, we can remove the q dependence from all the functions appearing in the previous subsection and in place of (4.36) we have

$$\mathbb{L}_\pm(\xi, \tau, \eta) = i\tau I + A(W_\pm)i\xi + g'(W_\pm)i\eta \tag{4.43}$$

for constant states $W_\pm$.

Clearly, (4.40) and the implicit function theorem imply that for any such (τ_0, ξ_0, η_0) and function τ_j, there exists a function $\xi(\eta)$ such that locally near (ξ_0, η_0)

$$\partial_\xi^{s-1}\tau_j(\xi, \eta) = 0 \text{ precisely when } \xi = \xi(\eta). \tag{4.44}$$

Note that $\xi(\eta)$ is smooth and homogeneous of degree one away from $\eta = 0$.

We can now state the auxiliary assumption (H4):

Assumption 4.1 (H4) *For any* $(\tau_0, \eta_0) \in \mathcal{G}$, *corresponding root* ξ_0 *of multiplicity* s, *and functions* τ_j *and* $\xi(\eta)$ *as above, we have*

$$\partial_\xi^k \tau_j(\xi(\eta), \eta) = 0 \text{ for } k = 1, \ldots, s-1 \text{ and } \eta \text{ near } \eta_0. \tag{4.45}$$

In other words ξ_0 persists as a root $\xi(\eta)$ of multiplicity s of

$$p(\xi(\eta), \tau_j(\xi(\eta), \eta), \eta) = 0$$

for η near η_0, and (by the remark below (4.40)) there are no other nearby roots of multiplicity > 1.

A compactness argument using the fact that $\mathcal{G}$ is a closed conic set shows that under the assumption (H4) all such branch singularities are confined to a finite union of surfaces

$$\tau = \tau_{j,l}(\eta) \equiv \tau_j(\xi_l(\eta), \eta)$$

on which the singularity (with respect to τ) has order equal to s_l, the multiplicity of the root $\xi_l(\eta)$. We'll usually relabel and replace the double index j, l by a single index as in $\tau = \tau_k(\eta)$. Note that graphs τ_k may well intersect.

Remark 4.1 *1. The statements of this subsection (and the previous one) require only slight modification when the assumption of strict hyperbolicity (H1) is relaxed to the following more general hypothesis of [Z], [GMWZ3]:*

(H1'): $f'(u)\xi + g'(u)\eta$ has semisimple real eigenvalues of constant multiplicity for $(\xi, \eta) \in \mathbb{R}^d \setminus 0$ (nonstrict hyperbolicity with constant multiplicity).

In this case the multiplicity of ξ_0 as a root of $p(\xi_0, \tau_0, \eta_0) = 0$ is some integer multiple of s as in (4.40).

2. Condition (H4) is automatic in the cases $d = 1, 2$ and also in any dimension for rotationally invariant problems. In 1D the glancing set is empty. In the 2D case the homogeneity of τ_j and its derivatives implies that the ray through (ξ_0, η_0) is the graph of $\xi(\eta)$ and that (H4) holds there. (H4) also clearly holds if no real root ξ of $p(\xi, \tau, \eta) = 0$ has multiplicity > 2, in particular in the case that all eigenvalues $\tau_j(\xi, \eta)$ are linear or convex/concave in their dependence on ξ.

3. In the equations of gas dynamics all characteristics are linear combinations of (ξ, η) and $|\xi, \eta|$, hence the above results show that (H4) is valid whenever the constant multiplicity assumption (H1') applies. Thus, we see that (H4), though mathematically restrictive, nonetheless allows important physical applications.

4.6 The SF Estimate

Using the conjugator ZYT_cT_B we have reduced the study of

$$U_z - \mathcal{G}(q, z, \zeta)U = F, \quad \tilde{\Gamma}U = 0 \text{ to}$$
$$U_z - \mathcal{G}_B(q, \hat{\zeta}, \rho)U = F, \quad \tilde{\Gamma}ZYT_cT_BU := \Gamma'U = 0. \tag{4.46}$$

Here $\tilde{\Gamma}$ is the Lopatinski boundary condition from (3.44) obtained by removing the translational degeneracy.

The block form (4.24) of $\mathcal{G}_B$ determines a partition of $U \in \mathbb{C}^{4m}$ as $U = (u_1, \ldots, u_s, u_+, u_-)$, where $u_\pm$ correspond to the blocks P_g, P_d respectively. We caution that in the remainder of this subsection the meaning of $\pm$ is completely different from earlier usage.

Denote by β_j the number of eigenvalues of Q_j with $\Re\mu < 0$ for $\hat{\gamma} > 0$ (or $\rho > 0$), and write

$$u_j = (u_{j-}, u_{j+}) \tag{4.47}$$

where u_{j-} consists of the first β_j components of u_j.

Next set

$$\begin{aligned}
U_{P+} &= (0, \ldots, 0, u_+, 0) \\
U_{P-} &= (0, \ldots, 0, 0, u_-) \\
U_{H+} &= ((0, u_{1+}), \ldots, (0, u_{s+}), 0, 0) \\
U_{H-} &= ((u_{1-}, 0), \ldots, (u_{s-}, 0), 0, 0),
\end{aligned} \tag{4.48}$$

and write

$$U = U_{P_+} + U_{P_-} + U_{H_+} + U_{H_-}$$
$$U_\pm = U_{P_\pm} + U_{H_\pm}$$
$$U_P = U_{P_+} + U_{P_-} \tag{4.49}$$
$$U_H = U_{H_+} + U_{H_-}.$$

The definition of U_{H_-} is based on the observation, proved in Appendix B, that $\{(u_{j-}, 0) : u_{j-} \in \mathbb{C}^{\beta_j}\}$ is precisely the continuous extension *at the base-point* of the decaying space corresponding to the block Q_j. Note that the analogous statement is *not* true for U_{H_+}, because of the fact that continuous extensions of growing and decaying spaces have nontrivial intersection (and sometimes even coincide) at glancing basepoints. Rather, U_{H_+} is just a convenient choice of complementary vector.

The symmetrizer S has the form

$$S(q, \hat{\zeta}, \rho) = \begin{pmatrix} S_1(q, \hat{\zeta}, \rho) & & \\ & \ddots & \\ & & S_s(q, \hat{\zeta}, \rho) \\ & & & S_P(q, \zeta) \end{pmatrix}, \tag{4.50}$$

where the S_j, S_P are C^∞ functions of their arguments. We'll sometimes write

$$S = \begin{pmatrix} S_H & \\ & S_P \end{pmatrix}, \tag{4.51}$$

where each block is of size $2m$.

Let

$$U_{H_j} = U_{H_{j+}} + U_{H_{j-}} \tag{4.52}$$

where the terms on the right have obvious meanings in view of (4.48). The S_j are constructed so that $S = S^*$, with interior estimates

$$(\mathrm{Re}\, S\mathcal{G}_B U_P, U_P) \geq C|U_P|_2^2$$
$$(\mathrm{Re}\, S\mathcal{G}_B U_{H_j}, U_{H_j}) \geq (\gamma + \rho^2)|U_{H_j}|_2^2, \tag{4.53}$$

as well as boundary estimates

$$(a)\ (SU_P, U_P) \geq c|U_{P_+}|^2 - |U_{P_-}|^2$$
$$(b)\ (SU_{H_j}, U_{H_j}) \geq c|U_{H_{j+}}|^2 - |U_{H_{j-}}|^2 \tag{4.54}$$

both holding uniformly near the basepoint.

Note that S_P can be taken to be simply

$$S_P = \begin{pmatrix} cI & \\ & -I \end{pmatrix} \tag{4.55}$$

for some large $c > 0$, where the blocks are of sizes $m - 1$ and $m + 1$. Details of the construction of S_H are the same as for the case of Dirichlet boundary layers and are given in [MZ].

Since Γ' satisfies the uniform Lopatinski condition at $(\underline{q}, \hat{\zeta}, 0)$ we have

$$|U_-|^2 \leq C|\Gamma' U_-|^2 \leq C(|\Gamma' U|^2 + |U_+|^2) \tag{4.56}$$

at $(\underline{q}, \hat{\zeta}, 0)$ and in fact uniformly near the basepoint by continuity.

We are now in a position to argue precisely as we did in the MF regime. From (4.54) and (4.56) we find as before

$$S + C_1 (\Gamma')^* \Gamma' \geq I \text{ on } z = 0 \tag{4.57}$$

for some $C_1 > 0$ provided c was big enough. Continuing we obtain the SF estimate

$$(|U_P|_2^2 + (\gamma + \rho^2)|U_H|_2^2) + |U|^2 \leq C\left(|F_P|_2^2 + \frac{1}{(\gamma + \rho^2)}|F_H|_2^2\right) + C|\Gamma' U|^2, \tag{4.58}$$

uniformly near $(\underline{q}, \hat{\zeta}, 0)$. Here we've written $F = F_H + F_P$ in the obvious way and used

$$|(SF, U)| \leq (C_\delta |F_P|_2^2 + \delta |U_P|_2^2) + \left(\frac{C_\delta}{(\gamma + \rho^2)}|F_H|_2^2 + \delta(\gamma + \rho^2)|U_H|_2^2\right) \tag{4.59}$$

to absorb terms from the right.

4.7 The HF Regime

In HF the spectrum of $\mathcal{G}(q, z, \zeta)$ stays well away from the imaginary axis, so the argument here is similar to the one for MF, except for the extra difficulty that the set of frequencies is noncompact. However, since the parabolic part of the linearized operator is dominant in HF, we can use the natural parabolic homogeneity to reduce to a compact set of parameters. This argument is given in Appendix F.

4.8 Summary of Estimates

We recall that our goal, established in Lecture 2, has been to prove L^2 estimates for the fully linearized transmission problem (2.24)

$$\begin{aligned} L_u(q, z, \zeta, \partial_z)v + L_\psi(q, z, \zeta)\phi &= f \\ [v] = 0, \ [\partial_z v] = 0, \ p(\zeta)\phi - l \cdot v &= 0 \text{ on } z = 0, \end{aligned} \tag{4.60}$$

which was reformulated as a first order system for the unknowns $U = (v, v_z)$ and ϕ in (2.29):

$$\partial_z U - G(q, z, \zeta)U = F + \begin{pmatrix} 0 \\ -(B^0(q))^{-1} L_\psi(q, z, \zeta)\phi) \end{pmatrix}$$

$$[U] = 0, \; p(\zeta)\phi - l(q) \cdot v = 0 \text{ on } z = 0 \tag{4.61}$$

In Lecture 3 we defined a good unknown for the problem (4.61), namely $U^\sharp = U - P\phi$, which allows us to reduce the study of (4.61) in MF and HF to the study of (3.35):

$$\partial_z U^\sharp - G(q, z, \zeta)U^\sharp = F, \; [U^\sharp] = 0. \tag{4.62}$$

In SF the good unknown is $U^\flat = U - R\phi$, and we showed that this allows us to reduce the study of (4.61) to that of (3.43):

$$\partial_z U^\flat - G(q, z, \zeta)U^\flat = F$$

$$\pi(q, \hat\zeta, \rho)[U^\flat] = 0, \; l(q) \cdot u^\flat_+ = 0. \tag{4.63}$$

Tracing back through the conjugations, the estimates (4.58) in SF, (4.20) in MF, and (11.25) in HF imply the following estimates for $U^\sharp = (v^\sharp, v^\sharp_z)$ and $U^\flat = (u^\flat, u^\flat_z)$. We define

$$h(\zeta) = \begin{cases} (\gamma + \rho^2)^{1/2}, \; \rho \le 1 \\ \langle \zeta \rangle, \; \rho > 1 \end{cases}, \tag{4.64}$$

where

$$\langle \zeta \rangle := (\tau^2 + \gamma^2 + \eta^4)^{1/4} \sim |\tau, \gamma|^{1/2} + |\eta|. \tag{4.65}$$

Noting that the forcing term F in the above first order systems satisfies

$$|F|_2 \le C|f|_2, \tag{4.66}$$

for f as in (4.60)

We have

$$(a) h^2 |u^\flat|_2 + h|u^\flat_z|_2 + h|U^\flat| \le C|f|_2 \text{ in SF}$$
$$(b) |v^\sharp|_2 + |v^\sharp_z|_2 + |U^\sharp| \le C|f|_2 \text{ in MF} \tag{4.67}$$
$$(c) \langle \zeta \rangle^2 |v^\sharp|_2 + \langle \zeta \rangle |v^\sharp_z|_2 + \langle \zeta \rangle^{3/2} |v^\sharp| + \langle \zeta \rangle^{1/2} |v^\sharp_z| \le C|f|_2 \text{ in HF}.$$

From (3.36) we deduce

$$\langle \zeta \rangle^2 |\phi| \le |v^\sharp| \text{ in MF and HF} \tag{4.68}$$

and from (3.41)(b) and the nonvanishing of $[\hat{R}]$ we get

$$|\rho\phi| \leq C|U^\flat|. \tag{4.69}$$

In view of (4.68) and (4.69), the three estimates above immediately imply the following estimates for the solution (v, ϕ) to the fully linearized transmission problem (4.60):

$(a) h^2|v|_2 + h|v_z|_2 + h|v, v_z| + h\rho|\phi| \leq C|f|_2$ in SF

$(b) |v|_2 + |v_z|_2 + |v, v_z| + |\phi| \leq C|f|_2$ in MF

$(c) \langle\zeta\rangle^2|v|_2 + \langle\zeta\rangle|v_z|_2 + \langle\zeta\rangle^{3/2}|v| + \langle\zeta\rangle^{1/2}|v_z| + \langle\zeta\rangle^{7/2}|\phi| \leq C|f|_2$ in HF.

$$\tag{4.70}$$

Finally, let's summarize all three estimates in a single estimate. First define

$$h_1(\zeta) = \begin{cases} \rho, & \rho \leq 1 \\ \langle\zeta\rangle^2, & \rho > 1 \end{cases}. \tag{4.71}$$

Then for all ζ we have

$$h^2|v|_2 + h|v_z|_2 + h(1 + \langle\zeta\rangle)^{1/2}|v| + h(1 + \langle\zeta\rangle)^{-1/2}|v_z| + h(1 + \langle\zeta\rangle)^{1/2}h_1|\phi|$$
$$\leq C|f|_2. \tag{4.72}$$

5 Lecture Five: Long Time Stability via Degenerate Symmetrizers

In this lecture we focus mainly on obtaining linearized estimates via symmetrizers. Details of the nonlinear endgame are given in [GMWZ1]. There is much overlap with the earlier results of [Z] obtained by construction of Green's functions. However, as described in [GMWZ1] each approach seems to yield some results inaccessible to the other. The symmetrizer approach has the advantage, as we saw in the first four lectures, of applying to curved shocks as well as planar shocks. We note that the nonlinear endgame for Theorem 5.1 is inspired by that in [KK], while the one for Theorem 5.2 is essentially that of [Z]. A brief discussion of these arguments is given at the end of this Lecture.

Consider the $m \times m$ system of viscous conservation laws

$$u_t + f(u)_x + g(u)_y - \triangle u = 0, \tag{5.1}$$

where now $y \in \mathbb{R}^{d-1}$, $d \geq 2$, but we continue to use 2D notation as in (4.35).

We are given a stationary inviscid shock $x = 0$ with constant states $W_\pm$ and a stationary solution $W(x)$ (the profile) of (5.1) satisfying

$$W_x = f(W) - f(W_-)$$
$$\lim_{x \to \pm\infty} W(x) = W_\pm. \tag{5.2}$$

5.1 Nonlinear Stability

We wish to understand the stability of the profile $W(x)$ under multidimensional perturbations. Let $\mathcal{A}$ denote some set of admissible perturbations to be specified later.

Definition 5.1 (1) For $v_0 \in \mathcal{A}$ let $u(x,t,y)$ be the solution to the system (5.1) with initial data at $t = 0$ given by

$$u_0(x,y) = W(x) + \delta v_0(x,y). \tag{5.3}$$

We say that W is nonlinearly stable with respect to perturbations in $\mathcal{A}$ if there exists a $\delta_0 > 0$ (depending on $|v_0|_\mathcal{A}$) such that for $\delta \leq \delta_0$, the solution $u(x,t,y)$ exists for all time and

$$|u(x,t,y) - W(x)|_{L^\infty(x,y)} \to 0 \text{ as } t \to \infty. \tag{5.4}$$

(2) We refer to v_0 as a zero mass perturbation if it has the form $v_0 = \operatorname{div} V_0$ for some V_0. General perturbations not necessarily of this form are called nonzero mass perturbations.

We look for u of the form

$$u(x,t,y) = W(x) + \delta z(x,t,y). \tag{5.5}$$

To obtain a problem with zero initial data we take

$$z(x,t,y) = v(x,t,y) + e^{-t}v_0(x,y) \tag{5.6}$$

and after a short computation we obtain the following error problem for v, where we set $A = f'$:

$$v_t + (A(W(x))v)_x + g'(W(x))v_y + \delta\operatorname{div}_{x,y}(B(x,t,y)v) + \delta\operatorname{div}_{x,y}(h(x,t,y,v)) - \triangle v = f$$
$$v|_{t=0} = 0, \tag{5.7}$$

where $h(x,t,y,v) = O(|v|^2)$.

The main step is to prove appropriate estimates for the corresponding linearized problem

$$u_t + (A(W(x))u)_x + g'(W(x))u_y - \triangle u = f$$
$$u|_{t=0} = 0. \tag{5.8}$$

We'll derive these by studying instead the eigenvalue equation, obtained by Fourier-Laplace transform of (5.8) after extending f and u by zero in $t < 0$:

$$\hat{u}_{xx} - (A(W(x))\hat{u})_x - s(x,\zeta)\hat{u} = f \text{ on } \mathbb{R}_x, \text{ where}$$
$$s(x,\zeta) = g'(W(x))i\eta + (i\tau + \gamma + \eta^2)I. \tag{5.9}$$

Dropping the hat on u we set $U = (u, u_x)$ and

$$G(x, \zeta) = \begin{pmatrix} 0 & I \\ \mathcal{M} & A(W)) \end{pmatrix}, \text{ where} \tag{5.10}$$
$$\mathcal{M}(x, \zeta) = s(x, \zeta) + A'(W) \cdot W_x.$$

Next replace (5.9) by the equivalent $4m \times 4m$ doubled boundary problem on $x \geq 0$

$$U_x - \mathcal{G}(x, \zeta)U = F \text{ on } x \geq 0, \ \Gamma U = 0, \tag{5.11}$$

where now $U = (U_+, \tilde{U}_-)$ (recall (3.15)) and

$$\mathcal{G} = \begin{pmatrix} G & 0 \\ 0 & -\tilde{G} \end{pmatrix}, \quad \Gamma U = U_+(0) - \tilde{U}_-(0). \tag{5.12}$$

The final preparatory step is to conjugate to block structure via a conjugator ZYT_cT_B just as in Lecture 4:

$$U_x - \mathcal{G}_B(\hat{\zeta}, \rho)U = F, \ \Gamma ZYT_cT_B U := \Gamma^\# U = 0. \tag{5.13}$$

The only difference is that here we retain the degenerate boundary condition $\Gamma U = 0$ in the SF region.

5.2 $L^1 - L^2$ Estimates

The main step is to establish $L^1 - L^2$ estimates in SF for (5.13), assuming now the auxiliary Assumption 4.1 (or (H4)). That is, we assume that branch singularities of characteristic roots ξ (considered as functions of (τ, η)) are confined to a finite union of smooth surfaces $\tau = \tau_j(\eta)$ on which the singularity has constant order equal to s_j, the multiplicity of the corresponding root ξ.

In MF and HF we can use the estimates established in Lecture 4. The following Proposition is proved in the next few subsections.

Proposition 5.1 *Fix a basepoint $X_0 = (\hat{\underline{\zeta}}, 0)$. Assume (H1),(H2),(H3),(H4), and*

$$-\theta\rho(\hat{\tau}^2 + \hat{\eta}^2) \leq \hat{\gamma} \leq C\rho \tag{5.14}$$

for some $C > 0$ and small enough $\theta > 0$.
Then, for $F \in L^1$ and $\rho > 0$ sufficiently small, the solution of the conjugated doubled boundary problem (5.13) satisfies

$$|U|^2_{L^2(x)} \leq \frac{C\beta^2 |F|^2_{L^1(x)}}{\rho^2} \tag{5.15}$$

for some $C > 0$ uniformly near the basepoint X_0, where

$$\beta(\hat{\zeta}, \rho) := \max_{j \geq 0} \beta_j, \tag{5.16}$$

with $\beta_0 := 1$ and

$$\beta_j := (|\hat{\tau} - \tau_j(\hat{\eta})| + \rho + \hat{\gamma})^{1/s_j - 1}. \tag{5.17}$$

(Note that $\beta = 1$ if the glancing set $\mathcal{G}$ is empty, in particular for $d = 1$.)

From (5.15) together with the MF and HF estimates from Lecture 4, we obtain readily the following linear estimate. This estimate leads directly to a proof of long time stability for nonzero mass perturbations in space dimensions $d \geq 3$.

Corollary 5.1 *Assume (H1), (H2), (H3), and (H4). Then, for $d \geq 3$, the solution of the linear problem (5.8) (nonzero mass) satisfies*

$$|u, u_y, u_t|_{L^2(x,t,y)} + |u_x|_{L^2(x,t,y)} \leq C(|f|_{L^1(x,t,y)} + |f|_{L^2(x,t,y)}). \tag{5.18}$$

Proof. We'll write $|F|_{L^1(x)} = |F|_1$.

Define V and H by $U = \hat{V}(x, \tau, \gamma, \eta)$, $F = \hat{H}$, where $0 < \gamma \leq C\rho^2$, and suppose now that U and F are supported in $\rho < \delta$.

(5.15) gives

$$|U|_2^2 \leq \frac{C\beta^2}{|\tau, \eta|^2} |F|_1^2. \tag{5.19}$$

Integrate (5.19) $d\tau d\eta$ (dimension of (τ, η) space is ≥ 3) to get

$$|e^{-\gamma t} V|_{L^2(x,t,y)}^2 \leq \int \frac{C\beta^2}{|\tau, \eta|^2} |\hat{H}(x, \tau, \gamma, \eta)|_{L^1(x)}^2 d\tau d\eta. \tag{5.20}$$

But

$$|\hat{H}(x, \tau, \gamma, \eta)| \leq C|H(x, t, y)|_{L^1(t,y)}, \tag{5.21}$$

so

$$|\hat{H}(x, \tau, \gamma, \eta)|_{L^1(x)} \leq C|H(x, t, y)|_{L^1(x,t,y)}. \tag{5.22}$$

Plug this into (5.20) to get

$$|e^{-\gamma t} V|_{L^2(x,t,y)}^2 \leq \int_{|\tau,\eta|<\delta} \frac{C\beta^2}{|\tau, \eta|^2} |H|_{L^1(x,t,y)}^2 d\tau d\eta \leq C|H|_{L^1(x,t,y)}^2. \tag{5.23}$$

Here we used the fact that for $d \geq 3$

$$\int_{|\tau,\eta|<\delta} \frac{\beta^2}{|\tau, \eta|^2} d\tau d\eta < \infty. \tag{5.24}$$

We note that a little care is needed in showing this since β is singular.

Finally, let $\gamma \to 0$ to get

$$|V|^2_{L^2(x,t,y)} \leq C|H|^2_{L^1(x,t,y)}. \tag{5.25}$$

For U and F supported in MF or HF, the results of Lecture 4 yield estimates with only the L^2 norm of H on the right. The corollary follows.

•

Note that (5.24) fails for $d = 1, 2$ and some different ideas are needed. For $d = 1$ we refer to [ZH]. For $d = 2$ we need $L^1 - L^p$ estimates for (5.13).

In what follows we'll occasionally interpolate between L^2 and L^∞ using the following elementary inequalities:

$$|u|_{L^p} \leq |u|^{1-\frac{2}{p}}_{L^\infty} |u|^{\frac{2}{p}}_{L^2} \leq |u|_{L^\infty} + |u|_{L^2}. \tag{5.26}$$

From (5.15) we obtain readily the following $L^1 \to L^p$ bounds.

Corollary 5.2 *Assume (H1), (H2), (H3), (H4), and (5.14). Then, for $F \in L^1$ and $\rho > 0$ sufficiently small, the solution of the conjugated doubled boundary problem (5.13) satisfies*

$$|u|_{L^p} \leq \frac{C\beta|F|_{L^1}}{\rho} \tag{5.27}$$

for all $2 \leq p \leq \infty$, for some $C > 0$ uniformly near the basepoint X_0, where β is defined as in Proposition 5.1.

Proof. Recall that $|U|$ bounds both $|u|$ and $|u_x|$. Thus, the result for $p = \infty$ follows from the standard one-dimensional Sobolev inequality

$$|f|_\infty \leq |f|^{1/2}_2 |f_x|^{1/2}_2, \tag{5.28}$$

and the general result $2 \leq p \leq \infty$ by interpolation between L^2 and L^∞ norms.

•

5.3 Proof of Proposition 5.1.

Our strategy in proving Proposition 5.1 will be to establish an $L^2 \to L^\infty$ bound for the adjoint problem, then appeal to duality. In deriving adjoint $L^2 \to L^\infty$ bounds, we use duality in a second way, to first conclude adjoint $L^2 \to L^2$ bounds from the $L^2 \to L^2$ bounds of the forward equation (slightly refined). From the adjoint L^2 bounds, $L^2 \to L^\infty$ bounds are then readily obtained by a standard energy estimate/integration by parts.

Remark 5.1 *It is worth noting that we do not in this argument apply degenerate symmetrizers to the adjoint equation. Indeed, because of an asymmetry between forward vs. dual equations, our standard degenerate symmetrizer estimate would not recover the sharp bound available by duality. (Specifically, the degeneracy in the boundary condition for the dual equation occurs in hyperbolic modes, though we shall not show it here.)*

5.4 The Dual Problem

Consider a general boundary problem

$$LU := U_x - G(x, \zeta)U = F$$
$$\Gamma U = 0 \text{ on } x = 0.$$
$$(5.29)$$

The dual problem is then defined via L^2 inner product on $\mathbb{R}^+$ as

$$L^*V := -V_x - G^*(x, \zeta)U = G$$
$$\Gamma^*V = 0 \text{ on } x = 0,$$
$$(5.30)$$

where the kernel of Γ^* is the orthogonal complement of the kernel of Γ, i.e., by the property that

$$\langle LU, V \rangle = \langle U, L^*V \rangle \qquad (5.31)$$

for $\Gamma U(0) = \Gamma^* V(0) = 0$.

A formality is to first establish well-posedness of both problems.

Proposition 5.2 *For $\rho > 0$, both forward and dual problems have a unique H^1 solution for any data in L^2.*

Proof. It is sufficient to prove uniqueness, which follows in both cases from the *standard* (nondegenerate) symmetrizer construction carried out for fixed $\rho \neq 0$. The interior estimates thereby obtained feature constants that may blow up arbitrarily fast in ρ as $\rho \to 0$; however, this is of no consequence for the present purpose. ●

Corollary 5.3 *The bound of Proposition 5.1 is equivalent to the dual bound*

$$|V|_{L^\infty}^2 \leq \frac{C\beta^2}{\rho^2} |G|_2^2 \qquad (5.32)$$

for solutions of the dual conjugated boundary problem, for $G \in L^2$.

Proof. We have

$$|U|_{L^2} = \sup_{|G|_{L^2}=1} \langle U, G \rangle = \langle U, L^*V \rangle = \langle LU, V \rangle = \langle F, V \rangle \leq |F|_{L^1}|V|_{L^\infty}, \quad (5.33)$$

from which we obtain the forward direction

$$|U|_{L^2}/|F|_{L^1} \leq |V|_{L^\infty}/|G|_{L^2}. \qquad (5.34)$$

A reverse calculation yields the backward direction. ●

5.5 Decomposition of $U_{H\pm}$

To establish (5.32), we will need to sharpen the basic $L^2 \to L^2$ estimate for the forward equation. To do this, we shall need to decompose the hyperbolic modes U_H in decomposition (4.49) as the sum $U_H = U_{H_+} + U_{H_-}$, where

$$U_{H\pm} = U_{H_h\pm} + U_{H_e\pm} + U_{H_g\pm}. \tag{5.35}$$

Each vector appearing in (5.35) has $4m$ components, and the decomposition depends on $(\hat{\zeta}, \rho)$. While U_H here is the same as the vector U_H appearing in (4.49), to avoid confusion it is important to note that the definitions of $U_{H\pm}$ are different now as we explain below.

We shall write

$$U_{H_h} = U_{H_h+} + U_{H_h-}$$

and do similarly for e and g. The hyperbolic mode $U_{H_h\pm}$ has nonvanishing components corresponding (only) to the blocks Q_k in (4.27) which are 1×1 with real part vanishing at the base point, but with real part > 0 (resp.< 0) when $\rho > 0$. The elliptic mode $U_{H_e\pm}$ has nonvanishing components corresponding to blocks with $\Re Q_k$ positive or negative definite at the base point. Finally, the glancing mode U_{H_g} has nonvanishing components corresponding to blocks of size larger than 1×1 which are purely imaginary at the base point (glancing blocks).

Further, we shall diagonalize the glancing blocks by a $4m \times 4m$ matrix $T_{H_g}(\hat{\zeta}, \rho)$:

$$U'_{H_g} := T_{H_g}^{-1} U_{H_g}, \tag{5.36}$$

where $U_{H_g} := U_{H_g+} + U_{H_g-}$. Here $U_{H_g\pm}$ are defined as the projections of U_{H_g} onto the growing (resp. decaying) eigenspaces of $\hat{H}_B$ in (4.27) corresponding to glancing blocks. Call these subspaces $H_{g\pm}$. Clearly, T_{H_g} also has a block structure and we may construct it so that in any given block corresponding to a glancing block Q_j, the first columns are eigenvectors of Q_j associated (for $\rho > 0$) to eigenvalues with $\Re \mu < 0$. The remaining blocks of T_{H_g} are identity matrices.

We denote by

$$U' := T_{H_g}^{-1} U \tag{5.37}$$

the full variable with U_{H_g} diagonalized, and all other components unchanged. By calculations similar to those in [Z], we obtain the following estimates.

Lemma 5.1 *The diagonalizing transformation T_{H_g} may be chosen so that*

$$|T_{H_g}| \leq C, \tag{5.38}$$

$$|T_{H_g}^{-1}| \leq C\beta, \tag{5.39}$$

and

$$|T^{-1}_{H_g|H_{g-}}| \le C\alpha, \tag{5.40}$$

where $\beta := \max_j \beta_j$, $\alpha := \max_j \alpha_j$, *with*

$$\beta_j := \theta_j^{1-s_j}, \qquad \alpha_j := \theta_j^{1-[(s_j+1)/2]}, \tag{5.41}$$

$$\theta_j := (|\hat{\tau} - \tau_j(\hat{\eta})| + \hat{\gamma} + \rho)^{1/s_j}, \tag{5.42}$$

and $T^{-1}_{H_g|H_{g-}}$ *denotes the restriction of* $T^{-1}_{H_g}$ *to subspace* H_{g-}. *In particular,*

$$\beta\alpha^{-2} \ge 1. \tag{5.43}$$

Remark 5.2 *The quantities* β *and* α, *and their sharp estimation, we regard as a key to the analysis of long-time stability in multidimensions.*

Proof. Clearly, it is sufficient to establish for a single block Q_j of size s_j that there exist diagonalizing matrices whose inverses are bounded by β_j, α_j, respectively. Let μ denote the multiple pure imaginary eigenvalue appearing in Q_j evaluated at the basepoint $(\hat{\tau}, \hat{\eta})$. From here on, we drop the j subscript.

Set $\sigma = |\hat{\tau} - \tau(\hat{\eta})| + \hat{\gamma}$ so $\theta = (\sigma + \rho)^{1/s}$. By a classic matrix perturbation argument (e.g., [Z], Lemma 4.8) the eigenvalue μ splits for $\sigma + \rho > 0$ small into s eigenvalues.

$$\mu_k = \mu + \pi_k + o(|\sigma, \rho|^{1/s}), \quad k = 1, \dots, s \tag{5.44}$$

Here

$$\pi_k = \epsilon^k i(p\sigma - iq\rho)^{1/s} \text{ with}$$
$$\epsilon = 1^{1/s}, \tag{5.45}$$
$$p(\hat{\eta}) \text{ and } q(\hat{\eta}) \text{ are real and } \sim 1, \text{ and sgn } p = \text{sgn } q.$$

Moreover, correponding eigenvectors are given in appropriate coordinates by

$$(1, \pi_k, \pi_k^2, \dots, \pi_k^{s-1}) + o(|\sigma, \rho|^{1/s}). \tag{5.46}$$

Thus, there exists a matrix T_{H_g} of eigenvectors of the $s \times s$ block Q that is approximately given by a vandermonde matrix with generators distance at least θ apart related by s roots of unity.

By Kramer's rule, we may therefore estimate β as the quotient of two vandermonde determinants, the numerator of size $s - 1$ and the denominator of size s, taken from the same set of equally spaced generators. The standard formula for vandermonde determinants gives then

$$\beta \sim \theta^{\binom{s-1}{2} - \binom{s}{2}} = \theta^{1-s} \tag{5.47}$$

as claimed.

Denoting by

$$\binom{t_1}{t_2} \tag{5.48}$$

the matrix consisting of the $k \leq [(s+1)/2]$ stable eigenvectors of Q, i.e., the first k columns of T_{H_g}, and noting that t_1 as a vandermonde matrix is invertible, we find that H_{g-} consists of vectors of form

$$\binom{w}{t_2 t_1^{-1} w} = \binom{t_1}{t_2} t_1^{-1} w, \tag{5.49}$$

where $w \in \mathbb{C}^k$ is arbitrary.

From $|(w, t_2 t_1^{-1} w)| \geq |w|$ and the computation

$$
\begin{aligned}
\left| T_{H_g}^{-1} \binom{w}{t_2 t_1^{-1} w} \right| &= \left| \binom{t_1 \ *}{t_2 \ *}^{-1} \binom{t_1}{t_2} t_1^{-1} w \right| \\
&= \left| \binom{I_k}{0} t_1^{-1} w \right| \\
&= |t_1^{-1} w|
\end{aligned}
\tag{5.50}
$$

we thus obtain that $|T_{H_g \,|H_{g-}}^{-1}| \leq |t_1^{-1}|$.

Observing that t_1 is a $k \times k$ vandermonde matrix with generators taken from the same equally spaced set, and applying Kramer's rule similarly as before, we obtain

$$|t_1^{-1}| \leq C\theta^{1-[(s+1)/2]}, \tag{5.51}$$

and thus $\alpha = \theta^{1-[(s+1)/2]}$ as claimed.   ●

We define similar decompositions on the dual variable V, and also the forcing terms F and G.

5.6 Interior Estimates

We begin by carrying out a basic degenerate symmetrizer estimate for the diagonalized forward problem. Note that the treatment of glancing modes is considerably simpler in diagonalized coordinates, and indeed has nothing to do with that of the original Kreiss construction.

Lemma 5.2 *For the forward diagonalized problem, we have the interior bound*

$$|U'|_{L^2}^2 \leq C \frac{|F'|_{L^2}^2}{\rho^2(\gamma + \rho^2)}. \tag{5.52}$$

Proof. In diagonalized coordinates, we must deal with a new degeneracy of order α^{-1} in the glancing modes of the diagonalized boundary condition $\Gamma' :=$ ΓT_{H_g} for the forward problem, as may be seen by the calculation

$$|\Gamma' U'_{H_{g-}}| = |\Gamma U_{H_{g-}}| \geq C^{-1}|U_{H_{g-}}| \geq \frac{C^{-1}|U'_{H_{g-}}|}{|T^{-1}_{H_g}|_{H_{g-}}|}. \tag{5.53}$$

On the other hand, there are no coalescing modes, and so we may dispense with the usual Kreiss construction, treating glancing modes in the same way as hyperbolic and elliptic modes. Precisely, in each S_H block except for those corresponding to glancing modes, we make the same choice of nondegenerate symmetrizer as in Lecture 4, while for each glancing blocks we choose a degenerate symmetrizer

$$S_{H_g} = diag(S_{H_{g+}}, S_{H_{g-}}) := diag(C, -\alpha^{-2}) \tag{5.54}$$

(recall, $\alpha^{-1} \to 0$ as $\sigma + \rho \to 0$).

In view of the glancing degeneracy (5.53) and the translational degeneracy (which we have not removed in the long time problem), there holds

$$|\Gamma' U'_{-}| \geq C(\delta|U'_{H_{h-}}| + \delta|U'_{H_{e-}}| + \alpha^{-1}|U'_{H_{g-}}| + \rho|U'_{P-}|). \tag{5.55}$$

So if we take the S_P also to be degenerate

$$S_P = \begin{pmatrix} cI & 0 \\ 0 & -\rho^2 \end{pmatrix}, \tag{5.56}$$

we again obtain good trace terms in the resulting symmetrizer estimate.

It remains to check that we retain good interior (L^2) bounds. Let $\mu_{k\pm}$ denote the eigenvalue associated with the kth mode of U'_{H_g}. Taylor expanding the expression (5.45) for π_k about $\rho/\sigma = 0$ yields,

$$|\Re\mu_{k\pm}| \geq C^{-1}\rho^2\beta, \tag{5.57}$$

whence we obtain from the fact that $\beta\alpha^{-2} \geq 1$ the lower bound

$$|\Re\mu_{k\pm}| \geq C^{-1}\alpha^2\rho^2, \tag{5.58}$$

and thereby the key interior estimate

$$(\text{Re } SG'_B(\infty)U'_{H_g}, U'_{H_g}) \geq \alpha^2\rho^2|U'_{H_{g+}}|^2_2 + \rho^2|U'_{H_{g-}}|^2_2. \tag{5.59}$$

That is, we still find that $\Re SG'_B(\infty) \geq \rho^2$, and therefore the rest of the symmetrizer argument goes through as before to give the claimed estimate.

●

Remark 5.3 *Since T_{H_g} diagonalizes the forward problem, $T_{H_g}^{-1*}$ diagonalizes the dual problem.*

By duality, this yields

Corollary 5.4 *For the dual diagonalized problem, we have the interior bound*

$$|V'|_{L^2}^2 \le \frac{C|G'|_{L^2}^2}{(\gamma + \rho^2)\rho^2}. \tag{5.60}$$

In fact, the above estimates can be somewhat refined. Let $U'_{H_{g\pm,j}}$ denotes the jth growing/decaying glancing mode, and $\mu_{j\pm}$ the associated growth/decay rate (eigenvalue of G_B).

Lemma 5.3 *For the forward diagonalized problem, we have the refined interior bounds*

$$|U'|_2^2 \le C \frac{|F_P|_2^2 + (\gamma + \rho^2)^{-1}|F_{H_h}|_2^2 + \rho^{-1}|F_{H_e}|_2^2 + \sum_{j,\pm} |\Re\mu_{j\pm}|^{-1}|F_{H_{g\pm,j}}|_2^2}{\rho^2}. \tag{5.61}$$

Proof. Parabolic modes have growth/decay rates with real part bounded in absolute value above and below by order one; elliptic modes have growth/decay rates bounded above and below by order ρ; hyperbolic modes have growth/decay rates bounded above and below by order $(\gamma + \rho^2)$. Glancing modes are treated individually in the diagonalized coordinates, and have growth/decay rates with absolute value of real part $|\Re\mu_{j\pm}|$. Using this sharp information in the degenerate symmetrizer estimate described just above, specifically in the application of Young's inequality to estimate $|(SF', U')|$ we obtain the claimed bound. ●

Corollary 5.5 *For the dual diagonalized problem, we have the interior bounds*

$$|V'_P|_{L^2}^2 + (\gamma + \rho^2)|V'_{H_h}|_{L^2}^2 + \rho|V'_{H_e}|_{L^2}^2 + \sum_{j\pm} |\Re\nu_{j\pm}||V'_{H_{g\pm,j}}|_{L^2}^2) \le \frac{C|G'|_{L^2}^2}{\rho^2}, \tag{5.62}$$

where $\nu_{j\pm} = -\mu_{j\mp}^*$ denote growth/decay rates for the dual problem (eigenvalues of $-G_B^*$).

Proof. Integration by parts, exactly as in the proof of Corollary 5.3, but mode by mode. For example, to obtain the bound

$$\rho|V'_{H_e}|_{L^2}^2 \le \frac{C|G'|_{L^2}^2}{\rho^2}, \tag{5.63}$$

we begin with bound

$$\rho|U'|_{L^2}^2 \le C\rho^{-2}|F'_{H_e}|_{L^2}^2 \tag{5.64}$$

for the forward problem $L'U' = F'_{H_e}$, and calculate

$$|V'_{H_e}|_{L^2} = \sup_{|F'_{H_e}|=1} \langle V'_{H_e}, F'_{H_e} \rangle = \sup\langle V', L'U' \rangle = \sup\langle L'^*V', U' \rangle \tag{5.65}$$

$$= \sup |G'|_{L^2}|U'|_{L^2} \le |G'|_{L^2}C\rho^{-3/2}|F'_{H_e}|_{L^2} = C\rho^{-3/2}|G'|_{L^2}. \tag{5.66}$$

●

5.7 L^∞ Estimates

With these preparations, L^∞ estimates are now easily obtained.

Lemma 5.4 *For the dual problem, we have the bounds*

$$|V'|_\infty^2 \le \frac{C|G'|_{L^2}^2}{\rho^2}, \qquad |V|_\infty^2 \le \frac{C\beta^2|G|_{L^2}^2}{\rho^2}. \tag{5.67}$$

Proof. Working in diagonalized coordinates, we may take the real part of the L^2 inner product of V' with equation $(L')^*V' = G'$ from $x_0 \ge 0$ to plus infinity to obtain after integration by parts the estimate

$$|V'(x_0)|^2 \le C(|V'_P|_2^2 + (\gamma + \rho^2)|V'_{H_h}|_2^2 + \rho|V'_{H_e}|_2^2 + \sum_{j\pm}|\Re\nu_j||V'_{H_{g\pm,j}}|_2^2) \tag{5.68}$$
$$+ C|V'|_2|G'|_2.$$

Bounding the first term on the righthand side using Corollary 5.5 and the second term using Corollary 5.4, we obtain the first asserted bound. The second asserted bound then follows by change of coordinates and the Jacobian bounds of Lemma 5.1.   ●

This completes the proof of Proposition 5.1.

5.8 Nonlinear Stability Results

Recall from Definition 5.1 the definition of nonlinear stability with respect to a given family of perturbations $\mathcal{A}$. Define

$$\mathcal{A}_p = \{v_0(x,y) : v_0 \in W^{p+2,2} \cap W^{2,1}\}$$
$$\mathcal{A}_\infty = \{v_0(x,y) : v_0 \in L^\infty \cap L^1\}, \tag{5.69}$$

where $W^{k,s}$ is the standard Sobolev space (k is order of differentiation; s is the L^s exponent).

Theorem 5.1 (nonzero mass, $d \geq 3$)
Assume (H1),(H2),(H3),(H4) and $p > \frac{d}{2}$, where the number of space dimensions is $d \geq 3$. Then the viscous profile $W(x)$ is nonlinearly stable with respect to $\mathcal{A}_p$.

This theorem follows from the linear estimate of Corollary 5.1 and the nonlinear endgame of [KK]. The idea is to use the linear estimate together with standard Sobolev and Moser inequalites to show that the perturbation v as in (5.7) satisfies

$$|v|_{W^{p+1,2}(T)} + |v_t|_{W^{p,2}(T)} \leq E \qquad (5.70)$$

for a fixed E independent of $[0, T]$. This implies that $|v|_{L^\infty(x,y)}$ decays to zero as $t \to \infty$.

The proof of Corollary 5.1 does not work when $d = 2$ since β^2/ρ^2 is not integrable then. This reflects the underlying fact that the linearized response to nonzero mass L^1 initial data in general decays in L^p, $p \geq 2$ no faster than a d-dimensional heat kernel.

The endgame in dimension 2 seems to require a special argument similar to the one in [Z]. The corresponding nonlinear stability result is Theorem 5.2. Here, the inverse Laplace transform of the solution to the linearized error problem is estimated via an integral on a parabolic contour $\gamma = -\theta |\tau, \eta|^2$ rather than the flat contour $\gamma = 0$, to take into account the additional decay due to diffusion in the parabolic case. The main ingredient for this argument is the $L^1 - L^p$ estimate of Corollary 5.2.

Theorem 5.2 (nonzero mass, $d \geq 2$)
Assume (H1),(H2),(H3), and (H4), where the number of space dimensions is $d \geq 2$. Then the viscous profile $W(x)$ is nonlinearly stable with respect to $\mathcal{A}_\infty$. Moreover, the perturbation v decays in L^p, $p \geq 2$ at the rate $|v|_p(t) \leq C(p,d)(1+t)^{-\frac{d-1}{2}(1-\frac{1}{p})}$ of a $(d-1)$-dimensional heat kernel, where $C(p,d)$ is monotone increasing in p, finite for $p < \infty$, and uniformly bounded for $d \geq 3$.

6 Appendix A: The Uniform Stability Determinant

Consider the homogeneous version of the linearized inviscid shock problem

$$\partial_t v + A_\nu(U^0, d\psi^0)\partial_x v + g'(U^0)\partial_y v = 0 \text{ in } \pm x \geq 0$$
$$\phi_t[U^0] + \phi_y[g(U^0)] - [A_\nu(U^0, d\psi^0)v] = 0 \text{ on } x = 0. \qquad (6.1)$$

We've already encountered this problem (with nonzero forcing) in the construction of higher profiles (1.29). To obtain a stability condition we freeze $q = (t, y)$ in $(U^0(0, t, y), \psi^0(t, y))$, Fourier-Laplace transform in t and Fourier transform in y to get (with $\zeta = (\tau, \gamma, \eta)$, $\gamma \geq 0$)

$$(i\tau + \gamma)\hat{v} + A_\nu \partial_x \hat{v} + g'(U^0)i\eta\hat{v} = 0$$
$$(i\tau + \gamma)\hat{\phi}[U^0] + i\eta\hat{\phi}[g(U^0)] - [A_\nu\hat{v}] = 0, \tag{6.2}$$

or, rearranging a little,

$$\partial_x \hat{v} - \mathbb{H}(q, \zeta)\hat{v} = 0 \text{ in } \pm x \geq 0$$
$$\hat{\phi}\left((i\tau + \gamma)[U^0] + i\eta[g(U^0)]\right) - [A_\nu\hat{v}] = 0 \text{ on } x = 0, \text{ where} \tag{6.3}$$
$$\mathbb{H}_\pm(q, \zeta) := -A_\nu(U^0_\pm, d\psi^0)^{-1}\left((i\tau + \gamma)I + i\eta g'(U^0_\pm)\right).$$

For $\gamma > 0$ the negative (resp. positive) generalized eigenspace of $\mathbb{H}_\pm$ has dimension k (resp. l), varies smoothly with (q, ζ), and extends *continuously* to $\gamma \geq 0$ in $\{\zeta \neq 0\}$. Here negative/positive refers to $\Re\mu$. To see the dimensions are correct use (H1), set $(\tau, \eta) = 0$, and note the minus sign in the definition of $\mathbb{H}$. Continuous extensions of decaying eigenspaces are discussed in Appendix B.

Thus, we may choose bases $\{r^1_+(q, \zeta), \ldots, r^k_+\}$ and $\{r^1_-(q, \zeta), \ldots, r^l_-\}$ for these spaces, where the $r^j_\pm$ are homogeneous of degree zero in ζ and have the same regularity (as the spaces) in (q, ζ) for $\zeta \neq 0$. Clearly, there will be unstable modes growing exponentially with time if for some $\gamma > 0$ the m vectors

$$(i\tau + \gamma)[U^0] + i\eta[g(U^0)], A_\nu(U^0_+, d\psi^0)r^s_+, A_\nu(U^0_-, d\psi^0)r^t_-$$
$$(s = 1, \ldots, k; t = 1, \ldots, l) \tag{6.4}$$

are linearly dependent.

As before we let $S^2_+ = \{\hat{\zeta} : |\hat{\zeta}| = 1, \hat{\gamma} \geq 0\}$. The inviscid shock $(U^0(0, q), \psi^0(q))$ is *uniformly stable* if for all q the $m \times m$ determinant

$$\Delta(q, \hat{\zeta}) :=$$
$$\det\left((i\hat{\tau} + \hat{\gamma})[U^0] + i\hat{\eta}[g(U^0)], A_\nu(U^0_+, d\psi^0)r^{(s)}_+(q, \hat{\zeta}), A_\nu(U^0_-, d\psi^0)r^{(t)}_-(q, \hat{\zeta})\right) \tag{6.5}$$

is nonvanishing (here (s) indicates k columns).

In [M2] Majda showed that uniform stability implies optimal L^2 estimates for the linearized problem. In [M3] he used those estimates to construct curved multiD shocks.

7 Appendix B: Continuity of Decaying Eigenspaces

In Lecture 3 we defined

$$\mathbb{E}(q, \hat{\zeta}, \rho) := E_+(q, \hat{\zeta}, \rho) \times E_-(q, \hat{\zeta}, \rho), \tag{7.1}$$

the decaying generalized eigenspace for $U_z - \mathcal{G}(q, z, \zeta)U = 0$ on $z \geq 0$. We know that $E_\pm$ are C^∞ functions of their arguments in $\hat{\gamma} + \rho > 0$. Here we show

Proposition 7.1 $E_\pm(q,\hat{\zeta},\rho)$ *extend continuously to the corner* $\hat{\gamma} + \rho = 0$.

Proof. **1.** We work near a basepoint $\underline{X} = (\underline{q},\hat{\underline{\zeta}},0)$. In Lecture 4 we saw that the change of variables $U = ZYT_cT_BV$ reduces the study of $U_z - \mathcal{G}(q,z,\zeta)U = 0$ near $\underline{X}$ to that of $V - \mathcal{G}_B(q,\hat{\zeta},\rho)V = 0$, where $\mathcal{G}_B(q,\hat{\zeta},\rho)$ is the block structure matrix:

$$\mathcal{G}_B(q,\hat{\zeta},\rho) = \begin{pmatrix} H_B(q,\hat{\zeta},\rho) & 0 & 0 \\ 0 & P_g(q,\zeta) & 0 \\ 0 & 0 & P_d(q,\zeta) \end{pmatrix} \text{ with}$$

$$\hat{H}_B(q,\hat{\zeta},\rho) = \begin{bmatrix} Q_1 & \cdots & 0 \\ \vdots & \ddots & \vdots \\ 0 & \cdots & Q_s \end{bmatrix}$$

(7.2)

for Q_j as in (4.27). We'll show that $\mathbb{E}$ extends continuously to the corner, but the same block by block argument shows the individual factors $E_\pm$ extend continuously as well.

2. Let $\mathbb{F}(q,\hat{\zeta},\rho)$ be the decaying generalized eigenspace for $\mathcal{G}_B$ in $\hat{\gamma}+\rho > 0$. It suffices to obtain a continuous extension of $\mathbb{F}$ to the corner. In the obvious way we write

$$\mathbb{F}(q,\hat{\zeta},\rho) = (\oplus_{j=1}^s \mathbb{F}_j) \oplus \mathbb{F}_g \oplus \mathbb{F}_d, \tag{7.3}$$

where, for example, $\mathbb{F}_g = \{0\}$ and $\mathbb{F}_d = \mathbb{C}^{m+1}$ in $\hat{\gamma} + \rho > 0$, each having a smooth extension to the corner.

If Q_j is a block of size ν_j satisfying $\pm\Re Q_j > 0$, we have $\mathbb{F}_j = \{0\}$ (resp. $\mathbb{C}^{\nu_j}$) in $\hat{\gamma} + \rho > 0$, so again there is a smooth extension to the corner.

If $\nu_j = 1$ we use the sign condition to see that $\mathbb{F}_j = \{0\}$ or $\mathbb{C}$ in $\hat{\gamma} + \rho > 0$, depending on whether the common sign of $\partial_{\hat{\gamma}}\Re Q_j$ and $\partial_\rho\Re Q_j$ is positive or negative. Here too we have a smooth extension.

3. There remains the case of a glancing block $Q_j(X)$ of size $\nu_j > 1$, where $X = (q,\hat{\zeta},\rho)$ and $Q_j(\underline{X})$ has the Jordan form (4.28). Here we follow an argument in Chapter 7 of [CP].

Setting $Q_j = Q$, $\mathbb{F}_j = \mathbb{F}_-$, $i\alpha_j = i\alpha$, and $\nu_j = \nu$, for $\hat{\gamma} + \rho > 0$ we may write

$$\det(\xi - Q(X)) = \prod_{\Re\mu_k > 0} (\xi - \mu_k)^{\beta_{k+}} \prod_{\Re\mu_k < 0} (\xi - \mu_k)^{\beta_{k-}} := p_+(\xi, X)p_-(\xi, X),$$

(7.4)

where again the sign condition implies the numbers

$$\beta_\pm = \sum_k \beta_{k\pm} \tag{7.5}$$

are independent of X for X close to $\underline{X}$ and $\hat{\gamma} + \rho > 0$. Note that

$$\mathbb{F}_-(X) = \oplus_{\Re\mu_k<0} \ker(Q(X) - \mu_k)^{\beta_{k-}} \tag{7.6}$$

and define

$$\mathbb{F}_+(X) = \oplus_{\Re\mu_k>0} \ker(Q(X) - \mu_k)^{\beta_{k+}}, \tag{7.7}$$

so

$$\mathbb{F}_+(X) \oplus \mathbb{F}_-(X) = \mathbb{C}^\nu \text{ for } \hat\gamma + \rho > 0. \tag{7.8}$$

If we define matrices $\mathcal{P}_\pm(X) = p_\pm(Q(X), X)$, we have in view of the decomposition (7.8)

$$\mathbb{F}_\pm = \ker \mathcal{P}_\pm, \quad \text{rank } \mathcal{P}_\pm = \dim \mathbb{F}_\mp. \tag{7.9}$$

The Cayley-Hamilton theorem implies $\mathcal{P}_-\mathcal{P}_+ = 0$, so

$$\text{range } \mathcal{P}_+(X) \subset \ker \mathcal{P}_-(X) = \mathbb{F}_-(X). \tag{7.10}$$

Thus, $\mathbb{F}_-(X) = \text{image } \mathcal{P}_+(X)$ for $\hat\gamma + \rho > 0$.

4. To define a continuous extension of $\mathbb{F}_-$ we first use continuity of eigenvalues to extend $\mathcal{P}_+(X)$ continuously to $\mathcal{P}_+(\underline{X})$ by defining

$$\mathcal{P}_+(\underline{X}) = (Q(\underline{X}) - i\alpha)^{\beta_+} = (Q(\underline{X}) - i\alpha)^{\nu-\beta_-}. \tag{7.11}$$

We then define $\mathbb{F}_-(\underline{X}) = \text{image } \mathcal{P}_+(\underline{X})$, and to see this extension is continuous we just need to check that the rank doesn't drop. But the image of $(Q(\underline{X}) - i\alpha)^{\nu-\beta_-}$ is clearly spanned by the first β_- standard basis vectors of $\mathbb{C}^\nu$, so we are done.   •

Remark 7.1 *1. The same argument shows* $\mathbb{F}_+(X) = \text{image } \mathcal{P}_-(X)$ *extends continuously to $\underline{X}$ with $\mathbb{F}_+(\underline{X})$ the span of the first β_+ standard basis vectors of $\mathbb{C}^\nu$.*

2. The result of this appendix extends to much more general viscosities by a different argument. See [MZ2].

8 Appendix C: Limits as $z \to \pm\infty$ of Slow Modes at Zero Frequency

We've allotted this separate short appendix to the proof of Proposition 3.3 because of its central importance for understanding the connection between viscous and inviscid stability. The result is used in the proof of the Zumbrun-Serre theorem, and also in the propositions of Lecture 3 that remove the translational degeneracy in the SF regime.

Proof (Proof of Proposition 3.3.). We may write elements of $K_{\hat{H}}(q, \hat{\zeta}, \rho)$ as $(w_\pm, 0)$, so slow modes can be expressed as

$$U_\pm = Z_\pm Y_\pm \begin{pmatrix} e^{zH_\pm} w_\pm \\ 0 \end{pmatrix}. \tag{8.1}$$

Recalling properties of $Z_\pm(q, z, \zeta)$ (2.37) and $Y_\pm(q, \zeta)$ (2.43) we compute

$$\lim_{z \to \pm\infty} Z_\pm(q, z, 0) Y_\pm(q, 0) = \begin{pmatrix} I & 0 \\ 0 & I \end{pmatrix} \begin{pmatrix} I & \mathcal{A}_{\pm\infty}^{-1}(q, 0) \\ 0 & I \end{pmatrix}. \tag{8.2}$$

Since $H_\pm(q, 0) = 0$ we obtain

$$\lim_{z \to \pm\infty} U_\pm(q, z, \hat{\zeta}, 0) = \begin{pmatrix} w_\pm(q, \hat{\zeta}, 0) \\ 0 \end{pmatrix} \in K_{\hat{H}}(q, \hat{\zeta}, 0.) \tag{8.3}$$

•

9 Appendix D: Evans $\Rightarrow$ Transversality + Uniform Stability

In this section we prove the Zumbrun-Serre result, Theorem 3.1. We'll make use of the discussion of slow and fast modes in Lecture 3.

The curvature of the inviscid shock plays no role here, so it's enough to consider an $m \times m$ system of viscous conservation laws

$$u_t + f_1(u)_x + f_2(u)_y - \Delta u = u_t + A_1(u)u_x + A_2(u)u_y - \Delta u = 0, \tag{9.1}$$

a planar inviscid shock $(x = 0, U_\pm)$, and a stationary solution $W(x)$ (the profile) of (9.1) satisfying the integrated profile equation

$$W' = f_1(W) - f_1(U_-) \text{ and} \tag{9.2}$$
$$W(x) \to U_\pm \text{ as } x \to \pm\infty.$$

Linearize (9.1) about W, Fourier-Laplace transform in t, and Fourier transform in y to get the eigenvalue equation

$$(i\tau + \gamma)w + (A_1(W)w)' + i\eta A_2(W)w - w'' + \eta^2 w = 0. \tag{9.3}$$

Let $\zeta = (\tau, \gamma, \eta) = \rho\hat{\zeta}$, where $\gamma \geq 0$, $0 \leq \rho \leq \rho_0$. For $\rho > 0$ let $\{w_1^\pm(x, \hat{\zeta}, \rho), \ldots, w_m^\pm\}$ be a basis for the decaying solutions of (9.3) in $\pm x \geq 0$. By the result of Appendix B these functions can be chosen to be smooth in $\rho > 0$ with continuous extensions to $\rho = 0$. The Evans function is the wronskian

$$D(\hat{\zeta}, \rho) = \det \begin{pmatrix} w_+^1 & \cdots & w_+^m & w_-^1 & \cdots & w_-^m \\ w_+^{1'} & \cdots & w_+^{m'} & w_-^{1'} & \cdots & w_-^{m'} \end{pmatrix} |_{x=0}. \tag{9.4}$$

Note that when $\rho = 0$ equation (9.3) becomes the linearized profile equation

$$w'' - (A_1(W)w)' = 0. \tag{9.5}$$

Since W' satisfies (9.5) on the whole line, we have $D(\hat{\zeta}, 0) = 0$. We want to show

$$D(\hat{\zeta}, \rho) = \rho\beta\Delta(\hat{\zeta}) + o(\rho) \text{ as } \rho \to 0, \tag{9.6}$$

where β is a transversality constant defined below and $\Delta(\hat{\zeta})$ is the uniform stability determinant (6.5).

Suppose $A_1(U_+)$ has k positive eigenvalues and $A_1(U_-)$ has l negative eigenvalues. The inviscid shock is a Lax shock, so $k + l = m - 1$, and hence $(m-k) + (m-l) = m+1$. There are $m - k$ exponentially decaying solutions of (9.5) in $x \geq 0$, and $m - l$ in $x \leq 0$. Calling these $w_+^1(x), \ldots, w_+^{m-k}$ and $w_-^1(x), \ldots, w_-^{m-l}$ respectively, we may arrange so that the similarly labeled elements $w_{\pm}^j(x, \hat{\zeta}, \rho)$ in the bases chosen above are smooth extensions of the $w_{\pm}^j(x)$ to small nonzero frequencies; so $w_{\pm}^j(x, \hat{\zeta}, 0) = w_{\pm}^j(x)$ (see (10.5) for this kind of extension). These extensions decay exponentially to zero in x for $\rho \geq 0$. We call these the *fast* modes. The vectors given by the $w_{\pm}^j(0)$ (a total of $m+1$ vectors) span the tangent spaces to the stable/unstable manifolds of (9.2) at $W(0)$ for the rest points $U_{\pm}$. Moreover, we may suppose the $w_{\pm}^j$ are chosen so that $w_+^1(x) = w_-^1(x) = W'(x)$. The connection is *transversal* $\Leftrightarrow$ the $m \times m$ determinant

$$\det(w_+^1(0), \ldots, w_+^{m-k}(0), w_-^2(0), \ldots, w_-^{m-l}(0)) = \beta \tag{9.7}$$

is not zero.

Note that the limiting versions of (9.5), namely

$$w'' - A_1(U_{\pm})w' = 0 \text{ in } \pm x \geq 0 \tag{9.8}$$

also have constant solutions. The remaining $k + l = m - 1$ basis elements used in defining D (called the *slow* modes)

$$w_{\pm}^j(x, \hat{\zeta}, \rho), j = m - k + 1, \ldots, m \ (+case); \ j = m - l + 1, \ldots, m \ (-case) \tag{9.9}$$

have the property that $w_{\pm}^j(x, \hat{\zeta}, 0)$ decay to nonzero constant vectors as $x \to \pm\infty$. Recalling (3.31) we see that we can choose the slow modes so that

$$\lim_{x \to \pm\infty} w_{\pm}^j(x, \hat{\zeta}, 0) = r_{\pm}^j(\hat{\zeta}), \tag{9.10}$$

where the $r_{\pm}^j(\hat{\zeta})$ are the vectors appearing in the definition of $\Delta(\hat{\zeta})$ (6.5) with indices relabeled.

Now we begin the computation of $D(\hat{\zeta}, \rho)$. Set $z_\pm(x) = \partial_\rho w_\pm^1(x, \hat{\zeta}, 0)$. Using the special property of $w_\pm^1$ we may write

$$\begin{pmatrix} w_+^1 \\ w_+^{1'} \end{pmatrix} - \begin{pmatrix} w_-^1 \\ w_-^{1'} \end{pmatrix} = \rho \begin{pmatrix} z_+ - z_- \\ z_+' - z_-' \end{pmatrix} + o(\rho). \tag{9.11}$$

Use the column operation given by the left side of (9.11) to replace the w_-^1 column in (9.4) by the right side of (9.11).

Since (9.5) implies that $w' - A_1(W)w$ is constant, this suggests using the row operation $w' - A_1 w$ to simplify the determinant, provided we can identify the constants.

Integrate (9.5) from $\pm\infty$ to x in $\pm x \geq 0$ to get

$$w' - A_1(W)w = -(A_1(W)w)|_{x=\pm\infty} = -A_1(U_\pm)w(\pm\infty, \hat{\zeta}, 0). \tag{9.12}$$

There are three cases. First, for fast modes the right side of (9.12) is clearly 0.

Second, by (9.10) for slow modes $w_\pm^j$ the right side of (9.12) is $-A_1(U_\pm)r_\pm^j(\hat{\zeta})$ for $r_\pm^j$ as above.

The third case is $z_+ - z_-$. Recall that $w_\pm^1$ satisfy (9.3) on $\pm x \geq 0$. Write the frequencies in (9.3) in polar coordinates, substitute in $w_\pm^1(x, \hat{\zeta}, \rho)$, differentiate with respect to ρ, and evaluate at $\rho = 0$ to get

$$(i\hat{\tau} + \hat{\gamma})W' + (A_1(W)z_\pm)' + i\hat{\eta}A_2(W)W' - z_\pm'' = 0 \tag{9.13}$$

(recall $w_\pm^1(x, \hat{\zeta}, 0) = W'(x)$). Integrate from $\pm\infty$ to x in $\pm x \geq 0$ to get

$$(i\hat{\tau} + \hat{\gamma})W + A_1(W)z_\pm + i\hat{\eta}f_2(W) - z_\pm' - \{(i\hat{\tau} + \hat{\gamma})U_\pm + i\hat{\eta}f_2(U_\pm)\} = 0 \tag{9.14}$$

(we used $z_\pm(\pm\infty, \hat{\zeta}, 0) = 0$). Finally, subtract the $(-)$ equation from the $(+)$ equation to get

$$(z_- - z_+)' - A_1(z_- - z_+) = (i\hat{\tau} + \hat{\gamma})[U] + i\hat{\eta}[f(U)]. \tag{9.15}$$

So in cases 1,2,3 the row operation $w' - A_1 w$ produces the results 0, $-A_1(U_\pm)r_\pm^j(\hat{\zeta})$, and $(i\hat{\tau} + \hat{\gamma})[U] + i\hat{\eta}[f(U)]$ respectively. Apply the row operation to get (up to a sign)

$$\rho \det \begin{pmatrix} w_\pm^j(0) \; (m \; fast) & * & * \\ 0 & A_1(U_\pm)r_\pm^j(\hat{\zeta}) \; (m-1 \; slow) & (i\hat{\tau} + \hat{\gamma})[U] + i\hat{\eta}[f(U)] \end{pmatrix} + o(\rho). \tag{9.16}$$

The upper left $m \times m$ determinant is β, the transversality constant. The lower right $m \times m$ determinant is $\Delta(\hat{\zeta})$, the Majda uniform stability determinant, so up to a sign we have (9.6). We can redefine β to correct the sign if necessary.

10 Appendix E: Proofs in Lecture 3

10.1 Construction of R

To construct R as in (3.39) we must find (for ρ small) an exponentially decaying function, vanishing at $\rho = 0$ which satisfies

$$
\begin{aligned}
&R_z - GR = \mathcal{B} \text{ in } \pm z \geq 0, \; l(q) \cdot r_\pm = -p(\zeta) \text{ where} \\
&\mathcal{B} = (B^0)^{-1} \left(0, (i\tau + \gamma + \eta^2)W_z + i\eta g'(W)W_z + 2i\eta\psi_y^0 W_{zz} \right).
\end{aligned}
\tag{10.1}
$$

First find an exponentially decaying function $R^1 = (r^1, r^2)$ such that $R_z^1 - GR^1 = \mathcal{B}$ with

$$
R^1 = (B^0)^{-1} \left((i\tau + \gamma + \eta^2)R^{11} + i\eta R^{12} + 2i\eta\psi_y^0 R^{13} \right),
\tag{10.2}
$$

where R^{1j}, $j = 1, 2, 3$ satisfy $R_z^{1j} - GR^{1j} = F$, with

$$
F = (0, W_z), (0, g'(W)W_z), (0, W_{zz})
\tag{10.3}
$$

respectively. This is easy after MZ conjugation which replaces G by $G_{\pm\infty}$, whose spectrum is described at the beginning of Lecture 3.

Next we must add a correction to arrange the boundary condition. Recall that $\mathcal{P}(q, z) = (W_z, W_{zz})$ is a fast decaying solution of $\mathcal{P}_z - G(q, z, 0)\mathcal{P} = 0$. Using the correspondence with $G_{HP\pm}$ form, this means that

$$
\mathcal{P}(q, z) = Z_\pm(q, z, 0)Y_\pm(q, 0) \begin{pmatrix} 0 \\ e^{zP_\pm(q,0)}c_\pm(q) \end{pmatrix},
\tag{10.4}
$$

for some $c_\pm(q)$ in the negative (resp., positive) eigenspace of $P_\pm(q, 0)$. Thus, we can construct smooth extensions to nonzero frequency $\mathcal{P}_\pm(q, z, \zeta)$ satisfying $\mathcal{P}_z - G\mathcal{P} = 0$ in $\pm z \geq 0$ by taking

$$
\mathcal{P}(q, z, \zeta) = Z_\pm(q, z, \zeta)Y_\pm(q, \zeta) \begin{pmatrix} 0 \\ e^{zP_\pm(q,0)}\pi_\pm(q, \zeta)c_\pm(q) \end{pmatrix}
\tag{10.5}
$$

where $\pi_\pm(q, \zeta)$ projects onto the negative (resp., positive) eigenspace of $P_\pm$. Writing $\mathcal{P} = (p^1, p^2)$ we have

$$
l(q) \cdot p_\pm^1(q, 0, \zeta) = d_\pm(q, \zeta) \sim 1 \text{ for } \rho \text{ small.}
\tag{10.6}
$$

So if we define

$$
\begin{aligned}
&R = R^1 - \alpha(q, \zeta)\mathcal{P}, \text{ where} \\
&\alpha_\pm = \left(l(q) \cdot r_\pm^1(q, 0, \zeta) + p(\zeta) \right) \cdot \frac{1}{d_\pm(q, \zeta)},
\end{aligned}
\tag{10.7}
$$

then R has all the required properties.

10.2 Propositions 3.4 and 3.5

Proof (Proof of Proposition 3.4). From (3.39) we obtain

$\partial_z \hat{R} - G(q, z, 0)\hat{R} = \hat{B}$ or equivalently,

$$L_u(q, z, \hat{\zeta}, 0, \partial_z)\hat{r} = -L_\psi(q, z, \hat{\zeta}, 0) = (i\hat{\tau} + \hat{\gamma})W_z + i\hat{\eta}g'(W)W_z + i\hat{\eta}2\psi_y^0 W_{zz}. \tag{10.8}$$

In turn this is the same as

$$-B^0 \partial_z^2 \hat{r}_\pm + \partial_z(A_\nu(W, d\psi^0)\hat{r}) = \partial_z \hat{\beta}, \tag{10.9}$$

where $\hat{\beta}$ is a primitive of $-L_\psi$. Integrating $\int_{\pm\infty}^z$ in $\pm z \geq 0$ gives

$$-B^0 \partial_z \hat{r}_\pm + A_\nu \hat{r}_\pm = \hat{\beta}(z) - \left((i\hat{\tau} + \hat{\gamma})U_\pm^0 + i\hat{\eta}g(U_\pm^0)\right). \tag{10.10}$$

Finally, subtract the $+$ equation from the $-$ equation and evaluate at $z = 0$ to find

$$[B^0 \partial_z \hat{r} - A_\nu \hat{r}] = (i\hat{\tau} + \hat{\gamma})[U^0] + i\hat{\eta}[g(U^0)] \neq 0, \tag{10.11}$$

since the uniform stability determinant $\Delta(q, \hat{\zeta}) \neq 0$ (see (6.5)). The Proposition follows immediately from (10.11).

●

Proof (Proof of Proposition 3.5). Suppose the intersection

$$\ker \tilde{\Gamma} \cap (E_+(q, \hat{\zeta}, 0) \times E_-(q, \hat{\zeta}, 0))$$

is nontrivial. Then there exist $U_\pm(q, z, \hat{\zeta}, 0)$ satisfying $U_z - G(q, \zeta, 0)U = 0$ in $\pm z \geq 0$ with initial data in $E_+(q, \hat{\zeta}, 0) \times E_-(q, \hat{\zeta}, 0)$ such that (with $U = (u, u_z)$)

$$[U](q, 0, \hat{\zeta}, 0) = c[\hat{R}](q, 0, \hat{\zeta}, 0) \text{ for some } c \text{ and } l(q) \cdot u_+ = 0. \tag{10.12}$$

In Remark 3.2 we ruled out the possibility $c = 0$. So suppose $c \neq 0$ and define $\mathcal{U} = (\mu, \mu_z)$ by

$$\mathcal{U}_\pm = c\hat{R}_\pm(q, z, \hat{\zeta}, 0) - U_\pm(q, z, \hat{\zeta}, 0). \tag{10.13}$$

Then

$$\partial_z \mathcal{U} - G(q, z, 0)\mathcal{U} = c\hat{B}(q, 0, \hat{\zeta}, 0) \text{ and } [\mathcal{U}] = 0 \tag{10.14}$$

or equivalently

$$L_u(q, z, \hat\zeta, 0, \partial_z)\mu_\pm = -cL_\psi(q, z, \hat\zeta, 0) =$$
$$= c\left((i\hat\tau + \hat\gamma)W_z + i\hat\eta g'(W)W_z + i\hat\eta 2\psi_y^0 W_{zz}\right), \quad (10.15)$$
$$[\mu] = [\mu_z] = 0.$$

Note that by (8.3)

$$\lim_{z\to\pm\infty} \mu_\pm(q, z, \hat\zeta, 0) = -\lim_{z\to\pm\infty} u_\pm \in \text{span}\{r_\pm^j(q, \hat\zeta)\}$$
$$\lim_{z\to\pm\infty} \partial_z\mu_\pm = 0 \quad (10.16)$$

for $r_\pm^j(q, \hat\zeta)$ as in (6.5) (or (3.31)).

Next integrate (10.15) $\int_{\pm\infty}^z$ in $\pm z \geq 0$ to get with β as in (10.9)

$$-B^0\partial_z\mu_\pm + A_\nu\mu_\pm - A_\nu\mu_\pm(\pm\infty) = c\hat\beta(z) - c\left((i\hat\tau + \gamma)U_\pm^0 + i\hat\eta g(U_\pm^0)\right). \quad (10.17)$$

Because of (10.16), if we set $z = 0$ and subtract the $+$ equation from the $-$ equation, we can find a nontrivial linear combination (since $c \neq 0$) of the vectors

$$A_\nu r_\pm^j(q, \hat\zeta) \text{ and } (i\hat\tau + \hat\gamma)[U^0] + i\hat\eta[g(U^0)] \quad (10.18)$$

that vanishes. This contradicts uniform stability of the inviscid shock.

●

11 Appendix F: The HF Estimate

The first step is to understand the spectrum of $\mathcal{G}(q, z, \zeta)$ for large $|\zeta|$. We'll set

$$\langle\zeta\rangle = (\tau^2 + \gamma^2 + \eta^4)^{1/4}, \quad (11.1)$$

reflecting the parabolic quasihomogeneity where τ and γ have weight two, and η has weight one.

Proposition 11.1 *For C large enough and $|\zeta| \geq C$, $\mathcal{G}(q, z, \zeta)$ has $2m$ eigenvalues in $\Re\mu > 0$ and $2m$ eigenvalues in $\Re\mu < 0$ with*

$$|\Re\mu| > C\langle\zeta\rangle. \quad (11.2)$$

Proof. Consider the block

$$G_+(q, z, \zeta) = \begin{pmatrix} 0 & I \\ \mathcal{M} & \mathcal{A} \end{pmatrix} \quad (11.3)$$

(recall (2.31)).

Define

$$\lambda = \frac{1}{\langle \zeta \rangle}, \quad \check{\zeta} = (\check{\tau}, \check{\gamma}, \check{\eta}) = \left(\frac{\tau}{\langle \zeta \rangle^2}, \frac{\gamma}{\langle \zeta \rangle^2}, \frac{\eta}{\langle \zeta \rangle} \right), \tag{11.4}$$

and note that $\langle \check{\zeta} \rangle = 1$, so $\check{\zeta} \in \check{S}_+^2$, the parabolic unit half sphere in $\check{\gamma} \geq 0$. One might call these "parabolic polar coordinates at ∞".

Write

$$\mathcal{M}(q, z, \zeta) = \langle \zeta \rangle^2 \check{\mathcal{M}}(q, z, \check{\zeta}, \lambda)$$
$$\mathcal{A}(q, z, \zeta) = \langle \zeta \rangle \check{\mathcal{A}}(q, z, \check{\zeta}, \lambda), \tag{11.5}$$

where

$$\check{M}(q, z, \check{\zeta}, \lambda) = (B^0)^{-1} \left((i\check{\tau} + \check{\gamma} + \check{\eta}^2)I \right) + O(\lambda) = \check{\mathcal{M}}_0(q, z, \check{\zeta}) + O(\lambda)$$
$$\check{A}(q, z, \check{\zeta}, \lambda) = (B^0)^{-1} 2\psi_y^0 i\check{\eta} I + O(\lambda) = \check{\mathcal{A}}_0(q, z, \check{\zeta}) + O(\lambda). \tag{11.6}$$

We have

$$\begin{pmatrix} \langle \zeta \rangle I & 0 \\ 0 & I \end{pmatrix} \begin{pmatrix} 0 & I \\ \mathcal{M} & \mathcal{A} \end{pmatrix} \begin{pmatrix} \langle \zeta \rangle^{-1} I & 0 \\ 0 & I \end{pmatrix} = \langle \zeta \rangle \begin{pmatrix} 0 & I \\ \check{\mathcal{M}} & \check{\mathcal{A}} \end{pmatrix} = \langle \zeta \rangle \begin{pmatrix} 0 & I \\ \check{\mathcal{M}}_0 & \check{\mathcal{A}}_0 \end{pmatrix} + O(1). \tag{11.7}$$

Since (q, z) dependence enters only through $W(q, z)$, we see that

$$\check{G}_0 = \begin{pmatrix} 0 & I \\ \check{\mathcal{M}}_0 & \check{\mathcal{A}}_0 \end{pmatrix} \tag{11.8}$$

depends on a compact set of parameters. We claim $\check{G}_0$ has no eigenvalues on the imaginary axis. Then, setting $(\check{\tau}, \check{\eta}) = 0$ easily yields a count of m eigenvalues in each of the regions $\pm \Re \mu > 0$, thus completing the proof.

To prove the claim, suppose $i\check{\xi}$ is a pure imaginary eigenvalue of $\check{G}_0$ associated to the eigenvector (u, v). A short computation shows

$$(B^0 \check{\xi}^2 + \check{\eta}^2 - 2\psi_y^0 \check{\eta}\check{\xi} + (i\check{\tau} + \check{\gamma}))u = 0. \tag{11.9}$$

Since

$$B^0 \check{\xi}^2 + \check{\eta}^2 - 2\psi_y^0 \check{\eta}\check{\xi} \geq C|\check{\xi}, \check{\eta}|^2 \tag{11.10}$$

for some $C > 0$, we conclude $\check{\gamma} \leq -C|\check{\xi}, \check{\eta}|^2$. Hence $|\check{\xi}, \check{\eta}| = 0$ which implies $|\check{\tau}, \check{\gamma}| = 0$, and this contradicts $\langle \check{\zeta} \rangle = 1$.

•

Remark 11.1 *This proof makes precise the sense in which the parabolic part of the operator is "dominant" in the HF regime.*

11.1 Block Stucture

First we rewrite the $2m \times 2m$ transmission problem

$$U_z - G(q, z, \zeta)U = F \text{ in } \pm z \geq 0, [U] = 0. \tag{11.11}$$

With $U = (u, v)$ set

$$U_1 = (u_1, v_1), \text{ with } u_1 = \langle \zeta \rangle u, \ v_1 = v. \tag{11.12}$$

Then (11.11) is the same as

$$\partial_z U_1 - \langle \zeta \rangle \check{G}(q, z, \check{\zeta}, \lambda)U_1 = F \text{ in } \pm z \geq 0, [U_1] = 0, \tag{11.13}$$

where

$$\check{G} = \begin{pmatrix} 0 & I \\ \check{M} & \check{A} \end{pmatrix}. \tag{11.14}$$

Next, with doubling notation as in (3.17) we rewrite the problem as the $4m \times 4m$ system on $z \geq 0$

$$\partial_z U_1 - \langle \zeta \rangle \check{\mathcal{G}} U_1 = \mathcal{F}, \ \Gamma U_1 = 0 \tag{11.15}$$

where

$$\check{\mathcal{G}}(q, z, \check{\zeta}, \lambda) = \begin{pmatrix} \check{G}(z) & 0 \\ 0 & -\check{G}(-z) \end{pmatrix}. \tag{11.16}$$

The spectral separation proved in Proposition 11.1 implies for $|\zeta|$ large that there exists a smooth conjugator $T(q, z, \check{\zeta}, \lambda)$ such that

$$T^{-1} \check{\mathcal{G}} T = \begin{pmatrix} P_g(q, z, \check{\zeta}, \lambda) & 0 \\ 0 & P_d \end{pmatrix} := \check{\mathcal{G}}_{gd}, \tag{11.17}$$

where $\Re P_g > CI$, $\Re P_d < -CI$ with the spectrum of P_g, P_d contained in a compact subset of $\pm \Re \mu > 0$ respectively.

11.2 Symmetrizer and Estimate

Setting $U_1 = TU_2$ and noting that T has z dependence, we reduce (11.15) to

$$\partial_z U_2 - \langle \zeta \rangle \check{\mathcal{G}}_{gd} U_2 = T^{-1}\mathcal{F} - T^{-1}T_z U_2 := \mathcal{F}',$$
$$\Gamma T U_2 := \Gamma'(q, \check{\zeta}, \lambda)U_2 = 0. \tag{11.18}$$

Let $\mathbb{F}^\infty(q, \check{\zeta}, \lambda)$ be the $2m$ dimensional generalized eigenspace of $\check{\mathcal{G}}(q, 0, \check{\zeta}, \lambda)$ corresponding to eigenvalues with negative real part. Clearly,

$$\mathbb{F}^\infty(q, \check{\zeta}, \lambda) = T(q, 0, \check{\zeta}, \lambda)\mathbb{F} \tag{11.19}$$

where $\mathbb{F} = \{(0, a) : a \in \mathbb{C}^{2m}\}$ as before (see (4.8)), and

$$\ker \Gamma'(q, \check{\zeta}, \lambda) \cap \mathbb{F} = \{0\} \Leftrightarrow \ker \Gamma \cap \mathbb{F}^\infty(q, \check{\zeta}, \lambda) = \{0\}. \tag{11.20}$$

In view of Proposition 11.1 and (11.16), the second equality in (11.20) holds for λ small, since a vector w is in the negative invariant space of $\check{G}(0)$ if and only if it is in the positive invariant space of $-\check{G}(0)$.

We can now finish by arguing just as in the MF case (4.10)–(4.20). In view of the block structure of (11.18) we set $U = U_g + U_d$ as before and take S of the form

$$S = \begin{pmatrix} cI & 0 \\ 0 & -I \end{pmatrix}. \tag{11.21}$$

For c large enough we have

$$\begin{aligned} \Re(S\check{\mathcal{G}}_{gd}) &\geq I \text{ in } z \geq 0 \\ S + C(\Gamma')^*\Gamma' &\geq I \text{ on } z = 0 \end{aligned} \tag{11.22}$$

for some $C > 0$. This implies

$$\begin{aligned} \langle\zeta\rangle|U_2|_2^2 + |U_2|^2 &\leq C\frac{|\mathcal{F}'|_2^2}{\langle\zeta\rangle}, \text{ or} \\ \langle\zeta\rangle|U_2|_2 + \langle\zeta\rangle^{1/2}|U_2| &\leq C|\mathcal{F}'|_2. \end{aligned} \tag{11.23}$$

Recall the definition of $\mathcal{F}'$, take $|\zeta|$ large to absorb the $T^{-1}T_z U_2$ term, and replace U_2 by U_1 to get

$$\langle\zeta\rangle|U_1|_2 + \langle\zeta\rangle^{1/2}|U_1| \leq C|\mathcal{F}|_2. \tag{11.24}$$

Since $u_1 = \langle\zeta\rangle u$, $v_1 = v$ for $U = (u, v)$ as in (11.11), when the forcing F has the form $F = (0, f)$ we obtain

$$\langle\zeta\rangle^2|u|_2 + \langle\zeta\rangle|u_z|_2 + \langle\zeta\rangle^{3/2}|u| + \langle\zeta\rangle^{1/2}|u_z| \leq C|f|_2. \tag{11.25}$$

This finishes the proof of the estimate (4.67)(c) quoted in Lecture 4.

Remark 11.2 *Note that we have not used the Evans assumption in HF at all. Instead we were able to use the positive definiteness of the viscosity to deduce that the Evans condition (11.20) holds uniformly for large $|\zeta|$.*

12 Appendix G: Transition to PDE Estimates

In this appendix we try to give a brief indication of how the same matrix symbols we've constructed here in the course of proving uniform estimates for systems of ODEs depending on frequencies as parameters can be used to prove estimates for the original linearized system of PDEs. For details we refer to the appendix of [GMWZ2].

For purposes of illustration it is simplest to work with smooth matrix symbols $a(q, \zeta)$ supported in MF, the midfrequency region. First, remember that ζ here is really $\tilde{\zeta} = \epsilon\zeta = \epsilon(\tau, \gamma, \eta)$ (recall (2.22)). Starting with $a(q, \tilde{\zeta})$ we unfreeze $q = (t, y)$ and replace $\tilde{\zeta}$ by $\epsilon\zeta$:

$$a(q, \tilde{\zeta}) \to a(q, \epsilon\zeta) \to a(t, y, \epsilon\zeta). \tag{12.1}$$

Next we associate a semiclassical pseudodifferential operator a_D to $a(t, y, \epsilon\zeta)$ whose action on a function $u(t, y)$ is given by

$$(a_D u)(t, y) = \int e^{it\tau + iy\eta} a(t, y, \epsilon\tau, \epsilon\gamma, \epsilon\eta)\hat{u}(\tau, \eta) d\tau d\eta. \tag{12.2}$$

Note that the Fourier inversion formula implies that semiclassical *differential* operators are special cases of the operators defined by (12.2).

It is not hard using basic properties of the Fourier transform to show that

$$a_D : L^2 \to L^2. \tag{12.3}$$

If $b(q, \tilde{\zeta})$ is another such symbol, consider the composite operator $a_D b_D$. It is not immediately obvious that the composition is an operator of the same type. Ideally, one would have a relationship like $a_D b_D = (ab)_D$, where ab is the ordinary matrix product of the symbols a and b. Instead, one has

$$a_D b_D = (ab)_D + \epsilon R_D, \text{ where } R_D : L^2 \to L^2. \tag{12.4}$$

In other words the ideal relationship does hold, modulo an error with L^2 norm that can be taken arbitrarily small. Such errors are often negligible in energy estimates proved using the pseudodifferential calculus.

There is a similar relationship for adjoints

$$(a_D)^* = (a^*)_D + \epsilon R_D \tag{12.5}$$

for R_D as above. The main point here is that pseudodifferential operators behave just like their symbols under the operations of composition and taking adjoints, except for errors that are often negligible.

Suppose next that we have symbols $S(q, \tilde{\zeta})$, $G(q, \tilde{\zeta})$, and $\chi(\tilde{\zeta})$ such that

$$\text{Re } S(q, \tilde{\zeta})G(q, \tilde{\zeta}) \geq CI \text{ for } \tilde{\zeta} \in \text{supp } \chi \tag{12.6}$$

(recall for example the MF estimate in Lecture 4). Then *Garding's inequality*, which can easily be proved using the calculus outlined above, implies

$$\mathrm{Re}\ (S_D G_D \chi_D u, \chi_D u) \geq C(|\chi_D u|^2_{L^2} - \epsilon^2 |u|^2_{L^2}).\qquad (12.7)$$

Of course, this is the PDE analogue of the estimate on $\Re(S\mathcal{G}_{gd}U, U)$ that we used in (4.17), (4.18).

References

[Co]   Coppel, W. A., *Stability and asymptotic behavior of differential equations*, D.C. Heath, Boston.

[CP]   Chazarain J. and Piriou, A., *Introduction to the Theory of Linear Partial Differential Equations*, North Holland, Amsterdam, 1982.

[FS]   Freistühler, H. and Szmolyan,P., *Spectral stability of small shock waves, I*, Preprint, 2002.

[GZ]   Gardner, R. and Zumbrun, K., *The gap lemma and geometric criteria instability of viscous shock profiles*, CPAM 51. 1998, 797–855.

[Go]   Goodman, J., *Nonlinear asymptotic stability of viscous shock profiles for conservation laws*, Arch. Rational Mech. Analysis 95. 1986, 325–344.

[GX]   Goodman, J. and Xin, Z. *Viscous limits for piecewise smooth solutions to systems of conservation laws*, Arch. Rational Mech. Analysis 121. 1992, 235–265.

[GG]   Grenier, E. and Guès, O., *Boundary layers for viscous perturbations of noncharacteristic quasilinear hyperbolic problems*, J. Differential Eqns. 143. 1998, 110–146.

[GMWZ1] Gues, O., Metivier, G., Williams, M., and Zumbrun, K., *Multidimensional viscous shocks I: degenerate symmetrizers and long time stability*, Journal of the A.M.S. 18 (2005), 61–120.

[GMWZ2] Gues, O., Metivier, G., Williams, M., and Zumbrun, K., *Multidimensional viscous shocks II: the small viscosity limit*, to appear in Comm. Pure Appl. Math., available at http://www.math.unc.edu/Faculty/williams/

[GMWZ3] Gues, O., Metivier, G., Williams, M., and Zumbrun, K., Existence and stability of multidimensional shock fronts in the vanishing viscosity limit, Arch. for Rational Mechanics and Analysis, 175, 2004, 151–244.

[GMWZ4] Gues, O., Metivier, G., Williams, M., and Zumbrun, K., *Navier-Stokes regularization of multidimensional Euler shocks*, Ann.Sci. Ecole Norm. Sup. 39 (2006), 75–175.

[GR]   Grenier, E. and Rousset F., *Stability of one dimensional boundary layers by using Green's functions*, Comm. Pure Appl. Math. 54 (2001), 1343–1385.

[GW]   Gues, O. and Williams M., *Curved shocks as viscous limits: a boundary problem approach*, Indiana Univ. Math. J. 51. 2002, 421–450.

[K]   Kreiss, H.-O., *Initial boundary value problems for hyperbolic systems*, Comm. Pure Appl. Math. 23. 1970, pp. 277–298.

[KK]   Kreiss, G. and Kreiss, H.-O., *Stability of systems of viscous conservation laws*, CPAM. 50. 1998, 1397–1424.

[KS]   Kapitula, T. and Sandstede, B., *Stability of bright solitary-wave solutions to perturbed nonlinear Schrödinger equations*, Phys. D. 124. 1998, 58–103.

[L1]   Liu, T.-P., *Nonlinear stability of shock waves for viscous conservation laws*, Mem. Amer. Math. Soc. 56., no. 328, 1985.

[L2]   Liu, T.-P., *Pointwise convergence to shock waves for viscous conservation laws*, Comm. Pure Appl. Math. 50. 1997, 1113–1182.

[LZ]   Liu, T.-P. and Zeng, Y., *Large time behavior of solutions for general quasilinear hyperbolic-parabolic systems of conservation laws*, AMS Memoirs, 599, (1997).

[M2]   Majda, A., *The stability of multidimensional shock fronts*, Mem. Amer. Math. Soc. No. 275, AMS, Providence, 1983.

[M3]   Majda, A., *The existence of multidimensional shock fronts*, Mem. Amer. Math. Soc. No. 281, AMS, Providence, 1983.

[MP]   Majda, A. and Pego, R., *Stable viscosity matrices for systems of conservation laws*, J. Diff. Eqns. 56, 1985, 229–262.

[MN]   Matsumura, A. and Nishihara, K., *On the stability of travelling wave solutions of a one-dimensional model system for compressible viscous gas*, Japan J. Appl. Math. 2. 1985, 17–25.

[Met1] Metivier, G., *Stability of multidimensional shocks*, Advances in the theory of shock waves, Progress in Nonlinear PDE, 47, Birkhäuser, Boston, 2001.

[MZ]   Metivier, G. and Zumbrun, K., *Large viscous boundary layers for noncharacteristic nonlinear hyperbolic problems*, preprint, April 2002, available at http://www.maths.univ-rennes1.fr/ metivier/preprints.html

[MZ2]   Metivier, G. and Zumbrun, K., *Symmetrizers and continuity of stable subspaces for parabolic-hyperbolic boundary value problems*, preprint, Sept. 2002.

[PZ]   Plaza, R. and Zumbrun, K., *An Evans function approach to spectral stability of small-amplitude shock profiles*, Preprint, 2002.

[Ra]   Ralston, J., *Note on a paper of Kreiss*, Comm. Pure Appl. Math. 24. 1971, 759–762.

[R]   Rousset, F., *Viscous limits for strong shocks of systems of conservation laws*, Preprint, 2001, available at http://www.umpa.ens-lyon.fr/ frousset/

[SX]   Szepessy, A. and Xin, Z., *Nonlinear stability of viscous shock waves*, Arch. Rat. Mech. Anal. 122. 1993, 53–103.

[Y]   Yu, Shih-Hsien, *Zero-dissipation limit of solutions with shocks for systems of hyperbolic conservation laws*, Arch. Rational Mech. Analysis. 146. 1999, 275–370.

[Z]   Zumbrun, K., *Multidimensional stability of planar viscous shock waves*, Advances in the theory of shock waves, 304–516. Progress in Nonlinear PDE, 47, Birkhäuser, Boston, 2001.

[ZH]   Zumbrun, K. and Howard, P., *Pointwise semigroup methods and stability of viscous shock waves*, Indiana Univ. Math. J. 47. 1998, 741–871.

[ZS]   Zumbrun, K. and Serre, D. *Viscous and inviscid stability of multidimensional planar shock fronts*, Indiana Univ. Math. J. 48. 1999, 937–992.

Planar Stability Criteria for Viscous Shock Waves of Systems with Real Viscosity

Kevin Zumbrun*

Indiana University, Bloomington, IN 47405
kzumbrun@indiana.edu

Summary. We present a streamlined account of recent developments in the stability theory for planar viscous shock waves, with an emphasis on applications to physical models with "real," or partial viscosity. The main result is the establishment of necessary, or "weak", and sufficient, or "strong", conditions for nonlinear stability analogous to those established by Majda [M.1, M.2, M.3] in the inviscid case but (generically) separated by a codimension-one set in parameter space rather than an open set as in the inviscid case. The importance of codimension one is that transition between nonlinear stability and instability is thereby determined, lying on the boundary set between the open regions of strong stability and strong instability (the latter defined as failure of weak stability). Strong stability holds always for small-amplitude shocks of classical "Lax" type [PZ, FreS]; for large-amplitude shocks, however, strong instability may occur [ZS, Z.3].

* K.Z. thanks CIME and especially organizers C.M. Dafermos and P. Marcati for the opportunity to participate in the summer school at which this material was originally presented. Each section corresponds to a single 90-minute lecture. Thanks to G. Métivier, and M. Williams for their interest in the work and for many helpful conversations, and to O. Gues, G. Métivier, and M. Williams for their indirect contribution through our concurrent joint investigations of the closely related small-viscosity problem for real viscosity systems [GMWZ.4]. We note in particular that the simplified approach of obtaining high-frequency resolvent bounds entirely through Kawashima-type energy estimates was suggested to us by some small-viscosity investigations of M. Williams; indeed, the argument given here is a large-amplitude version of a small-amplitude argument developed by him in an earlier, since discarded version of [GMWZ.4] (see also related constant-coefficient and one-dimensional analyses in [KSh] and [HuZ], respectively). The novelty of the current presentation lies rather in the development of a nonlinear iteration scheme depending only on such bounds (standard for the small-viscosity problem, new for long-time stability). Thanks also to B. Texier for his careful reading and many helpful suggestions. Research of K.Z. was partially supported under NSF grants number DMS-0070765 and DMS-0300487.

1 Introduction: Structure of Physical Equations

Many equations of physics take the form of *hyperbolic conservation laws*

$$U_t + \sum_j F^j(U)_{x_j} = 0, \tag{1.1}$$

with associated *viscous conservation laws*

$$U_t + \sum_j F^j(U)_{x_j} = \nu \sum_{j,k} (B^{jk}(U)U_{x_k})_{x_j} \tag{1.2}$$

incorporating neglected transport effects of viscosity, heat conduction, etc. (more generally, hyperbolic and viscous *balance laws*[2] including also zero-order derivative terms $C(U)$, as especially in relaxation and combustion equations). Here, U, $F^j \in \mathbb{R}^n$, $B^{jk} \in \mathbb{R}^{n \times n}$, $x \in \mathbb{R}^d$, and $t \in \mathbb{R}$.

Examples 1.1 Euler and Navier–Stokes equations, respectively, of gas- or magnetohydrodynamics (MHD); see (1.44) below.

A fundamental feature of (1.1) is the appearance of *shock waves*

$$U(x,t) = \bar{U}(x - st) = \begin{cases} U_- & x_1 < st, \\ U_+ & x_1 \geq st, \end{cases} \tag{1.3}$$

discontinuous weak, or distributional solutions of (1.1) determined by the Rankine–Hugoniot conditions

$$RHs[U] = [F(U)];$$

see, e.g., [La, La.2, Sm]. Here and elsewhere, $[h(U)] := h(U_+) - h(U_-)$. Solution (1.3) may be uniquely identified by the "shock triple" (U_-, U_+, s).

Such waves in fact occur in applications, and do well-approximate experimentally observed behavior. However, in general they occur in only one direction, despite the apparent symmetry $(x, t, s) \to (-x, -t, s)$ in equations (1.1). That is, only one of the shock triples (U_-, U_+, s) (U_+, U_-, s) is typically observed, though they are indistinguishable from the point of view of (RH). The question of when and why a particular shock triple is physically realizable, known as the *shock admissibility problem*, is one of the oldest and most central problems in the theory of shock waves. For an interesting discussion of this issue from a general and surprisingly modern point of view, see the 1944 roundtable discussion of [vN].

Two basic approaches to admissibility are:

1. Hyperbolic stability in the Hadamard sense, i.e., short-time bounded stability, or well-posedness of (1.3) as a solution of (1.1), also known as *dynamical stability* [BE]: that is, internal consistency of the hyperbolic model.

---

[2] Outside the scope of these lectures, but accessible to the same techniques; see, e.g., [God, Z.3, MaZ.1, MaZ.5, Ly, LyZ.1, LyZ.2, JLy.1, JLy.2].

Here, there exists a well-developed theory; see, e.g., [M.1, M.2, M.3, Mé.1, Mé.2, Mé.3, Mé.4, FMé] and references therein.

2. Consistency with viscous or other regularization, in this case the viscous conservation law (1.2). (a) A simple version is the "viscous profile condition", requiring existence of an associated family of nearby traveling-wave solutions

$$U(x,t) = \bar{U}\left(\frac{x - st}{\nu}\right), \qquad \lim_{z \to \pm\infty} \bar{U}(z) = U_\pm \qquad (1.4)$$

of the viscous conservation law (1.2); see, e.g., [Ra, Ge, CF, BE, Gi, MP, P, MeP], and references therein. This is the planar version of the prepared-data "vanishing-" or "small-viscosity" problem (SV) treated for general, curved shocks in the parallel article of Mark Williams in this volume [W]. The viscous profile condition, augmented with the requirement that $\bar{U}$ be a transverse connection with respect to the associated traveling-wave ODE, is sometimes known as *structural stability* [BE, ZS, Z.3, Z.4]. (b) A more stringent version is the "stable viscous profile condition", requiring stability under perturbation of individual profiles (1.4) with viscosity coefficient ν held fixed. We denote by (LT) the associated problem of determining long-time viscous stability.

Definition 1.2 *Long-time viscous stability is defined as the property that, for some appropriately chosen norms $|\cdot|_X$ and $|\cdot|_Y$, for initial data U_0 sufficiently close to profile $\bar{U}$ in $|\cdot|_X$, the viscous problem (1.2), $\nu = 1$, has a (unique) global solution $U(\cdot,t)$ that converges to $\bar{U}$ as $t \to \infty$ in $|\cdot|_Y$. We refer to the latter property as asymptotic $|\cdot|_X \to |\cdot|_Y$ viscous stability.*

Remark 1.3 One may also consider the question whether solutions of (1.2) converge on a bounded time interval for fixed initial data as $\nu \to 0$ to a solution of (1.1), that is, the unprepared-data (SV) problem. This was considered for small-amplitude shock waves in one dimension by Yu [Yu], and, more recently, for general small-variation solutions in one dimension in the fundamental work of Bianchini and Bressan [BiB.1, BiB.2]; in multiple dimensions the problem remains completely open. This is a more stringent requirement than either of 2(a) or 2(b); indeed, it appears to be overly restrictive as an admissibility condition. In particular, as discussed in [Fre.1, Fre.2, FreL, L.4, Z.3, Z.4], there arise in (MHD) certain nonclassical "overcompressive" shocks that are both stable for fixed ν and play an important role in solution structure, yet which do not persist as $\nu \to 0$. The (LT) and (SV) problems are related by the scaling

$$(x,t,\nu) \to (x/T, t/T, \nu/T), \qquad (1.5)$$

with $T \to \infty$, $0 \leq t \leq T$, the difference lying in the prescription of initial data.

Conditions 1 and 2(a) may be formally derived by matched asymptotic expansion using the rescaling (1.5) and taking the zero-viscosity limit, as described, e.g., in [Z.3], Section 1.3. They have the advantage of simplicity, and for this reason have received the bulk of the attention in the classical mathematical physics literature; see, e.g., the excellent surveys [BE] and [MeP]. However, rigor (and also rectitude; see Section 1.4, [Z.3]) of the theory demands the study of the more complicated, but physically correct condition 2(b), motivating the study of the long-time viscous stability problem (LT). It is this problem that we shall consider here.

In contrast to the hyperbolic stability theory, progress in the multidimensional viscous stability theory has come only quite recently. The purpose of this article is to present an account of these recent developments, with an emphasis on (i) connections with, and refinement of the hyperbolic stability theory, and (ii) applications to situations of physical interest, i.e., *real viscosity, large amplitude*, and *real* (e.g., van der Waals-type) *gas equation of state*. Our modest goal is to present sharp and (at least numerically) computable planar stability criteria analogous to the Lopatinski condition obtained by Majda [K, M.1, M.2, M.3] in the hyperbolic case.

This is only the first step toward a complete theory; in particular, evaluation of the stability criteria/classification of stability remain important open problems. Preliminary results in this direction include stability of general small-amplitude shock profiles [Go.1, Go.2, HuZ, PZ, FreS]; geometric conditions for stability, yielding instability of certain large-amplitude shock profiles [GZ, BSZ, FreZ, God, Z.2, Z.3, Z.4, Ly, LyZ.1, LyZ.2]; and the development of efficient algorithms for numerical testing of stability [Br.1, Br.2, BrZ, BDG, KL]. On the other hand, the techniques we use here are completely general, applying also to relaxation, combustion, etc.; see, e.g., [Z.3] and references therein.

We begin in this section with some background discussion of a mainly historical nature concerning the common structural properties relevant to our investigations of various equations arising in mathematical physics, at the same time introducing some basic energy estimates of which we shall later make important use.

1.1 Symmetry and normal forms. We may write (1.1) and (1.2) in quasilinear form as

$$U_t + \sum_j A^j U_{x_j} = 0 \qquad (1.6)$$

and

$$U_t + \sum_j A^j U_{x_j} = \nu \sum_{j,k} (B^{jk} U_{x_k})_{x_j}, \qquad (1.7)$$

where $A^j := dF^j(U)$ and $B^{jk} := B^{jk}(U)$. We assume the further structure

$$U = \begin{pmatrix} u^I \\ u^{II} \end{pmatrix}, \qquad A^j = \begin{pmatrix} A^j_{11} & A^j_{12} \\ A^j_{21} & A^j_{22} \end{pmatrix}, \qquad B^{jk} = \begin{pmatrix} 0 & 0 \\ b^{jk}_I & b^{jk}_{II} \end{pmatrix}, \qquad (1.8)$$

$u^I \in \mathbb{R}^{n-r}$, $u^{II} \in \mathbb{R}^r$ typical in physical applications, identifying a distinguished, "inviscid" variable u^I, with

$$\Re\sigma \sum \xi_j \xi_k b_{II}^{jk} \geq \theta|\xi|^2, \tag{1.9}$$

$\theta > 0$, for all $\xi \in \mathbb{R}^d$. Here and below, σM denotes spectrum of a matrix or linear operator M. In the case of an unbounded operator, we use the simplest definition of spectrum as the complement of the resolvent set $\rho(M)$, defined as the set of $\lambda \in \mathbb{C}$ for which $\lambda - M$ possesses a bounded inverse with respect to a specified norm and function space; see, e.g., [Kat], or Appendix A.

1.1.1. Inviscid equations. Local stability of constant solutions (well-posedness) requires hyperbolicity of (1.6), defined as the property that $A(\xi) := \sum_j A^j \xi_j$ have real, semisimple eigenvalues for all $\xi \in \mathbb{R}^d$. As pointed out by Godunov and Friedrichs [G, F.1, F.2], this may be guaranteed by *symmetriz-ability*, defined as existence of a "symmetrizer" $\tilde{A}^0$ such that $\tilde{A}^0$ is symmetric positive definite and $\tilde{A}^j := \tilde{A}^0 A^j$ are symmetric. Left-multiplication by $\tilde{A}^0$ converts (1.6) to (quasilinear) *symmetric hyperbolic form*

$$\tilde{A}^0 U_t + \sum_j \tilde{A}^j U_{x_j} = 0. \tag{1.10}$$

A nonlinear version of this procedure is an invertible coordinate change $U \to W$ such that (1.1) considered as an equation in W takes the form

$$U(W)_t + \sum_j F^j(U(W))_{x_j} = 0, \tag{1.11}$$

where $\tilde{A}^0 := \partial U / \partial W$ is symmetric positive definite and $\tilde{A}^j := dF^j(\partial U / \partial W)$ are symmetric. This is more restrictive, but has the advantage of preserving divergence form; see Remarks 1.12 1-2 below.

Symmetric form yields hyperbolic properties directly through elementary energy estimates/integration by parts, using the *Friedrichs symmetrizer relation*

$$\Re\langle U, SU_{x_j}\rangle = -\frac{1}{2}\langle U, S_{x_j}^j U\rangle \tag{1.12}$$

for self-adjoint operators $S \in \mathbb{C}^{n \times n}$, $U \in \mathbb{C}^n$ (exercise). Here and below, $\langle \cdot, \cdot \rangle$ denotes the standard, (complex) L^2 inner product with respect to variable x. We shall require also the following elementary bounds.

Lemma 1.4 (Strong Sobolev embedding principle) *For $s > d/2$,*

$$|f|_{L^\infty(\mathbb{R}^d)} \leq |\hat{f}|_{L^1(\mathbb{R}^d)} \leq C|f|_{H^s(\mathbb{R}^d)}, \tag{1.13}$$

where $\hat{f}$ denotes Fourier transform of f.

Proof. The first inequality follows by Hausdorff–Young's inequality, the second by

$$\begin{aligned}
|\hat{f}(\xi)|_{L^1} &= |\hat{f}(\xi)(1 + |\xi|^s)(1 + |\xi|^s)^{-1}|_{L^1} \\
&\leq |(1 + |\xi|^s)^{-1}|_{L^2}|\hat{f}(\xi)(1 + |\xi|^s)|_{L^2} \leq C|f|_{H^s}.
\end{aligned} \tag{1.14}$$

●

Lemma 1.5 (Weak Moser inequality) *For $s = \sum |\alpha_j|$ and $k \geq d/2$,*

$$
\begin{aligned}
|(\partial^{\alpha_1} v_1) \cdots (\partial^{\alpha_r} v_r)|_{L^2(\mathbb{R}^d)} &\leq \sum_{i=1}^{r} |v_i|_{H^s(\mathbb{R}^d)} \Big(\prod_{j \neq i} |\hat{v}_j|_{L^1(\mathbb{R}^d)} \Big) \\
&\leq C \sum_{i=1}^{r} |v_i|_{H^s(\mathbb{R}^d)} \Big(\prod_{j \neq i} |v_j|_{H^k(\mathbb{R}^d)} \Big).
\end{aligned}
\tag{1.15}
$$

Proof. By repeated application of the Hausdorff–Young inequality $|f * g|_{L^p} \leq |f|_{L^2} |g|_{L^p}$, where $*$ denotes convolution, we obtain the first inequality,

$$
\begin{aligned}
|(\partial^{\alpha_1} v_1) \cdots (\partial^{\alpha_r} v_r)|_{L^2} &= |(\xi^{\alpha_1} \hat{v}_1) * \cdots * (\xi^{\alpha_r} \hat{v}_r)|_{L^2} \\
&\leq \Big\| |\xi^s \hat{v}_1| * \cdots * |\hat{v}_r| \Big\|_{L^2} + \cdots + \Big\| |\hat{v}_1| * \cdots * |\xi^s \hat{v}_r| \Big\|_{L^2} \\
&\leq \sum_{i=1}^{r} |v_i|_{H^s} \Big(\prod_{j \neq i} |\hat{v}_j|_{L^1} \Big).
\end{aligned}
\tag{1.16}
$$

The second follows by (1.13). $\bullet$

Proposition 1.6 ([Fr, G]) *Symmetric form (1.10) implies local well - posedness in H^s, and bounded local stability $|U(t)|_{H^s} \leq C|U_0|_{H^s}$, provided $\tilde{A}^j(\cdot) \in C^s$ and $s \geq [d/2] + 2$.*

Proof. Integration by parts together with (1.10) yields the basic L^2 estimate

$$
\begin{aligned}
\frac{1}{2} \langle U, \tilde{A}^0 U \rangle_t &= \frac{1}{2} \langle U, \tilde{A}^0_t U \rangle + \langle U, \tilde{A}^0 U_t \rangle \\
&= \frac{1}{2} \langle U, \tilde{A}^0_t U \rangle - \langle U, \sum_j \tilde{A}^j U_{x_j} \rangle \\
&= \frac{1}{2} \langle U, (\tilde{A}^0_t + \sum_j \tilde{A}^j_{x_j}) U \rangle \\
&\leq C|U|^2_{L^2} |U|_{W^{1,\infty}} \leq C|U|^2_{L^2} |U|_{H^s}.
\end{aligned}
\tag{1.17}
$$

Here, we have used (1.12) in equating $\langle U, \sum_j \tilde{A}^j U_{x_j} \rangle = -\frac{1}{2} \langle U, \sum_j \tilde{A}^j_{x_j} U \rangle$, original equation $U_t = -\sum_j A^j U_{x_j}$ in estimating $\tilde{A}^0_t \leq C|U_t| \leq C_2|U_x|$ in the second-to-last inequality, and (1.13) in the final inequality.

A similar, higher-derivative calculation yields the H^s estimate

$$\frac{1}{2}\Big(\sum_{r=0}^{s}\langle\partial_x^r,\tilde{A}^0\partial_x^r U\rangle\Big)_t = \frac{1}{2}\sum_{r=0}^{s}\langle\partial_x^r U,\tilde{A}_t^0\partial_x^r U\rangle + \sum_{r=0}^{s}\langle\partial_x^r U,\tilde{A}^0\partial_x^r U_t\rangle$$

$$= \sum_{r=0}^{s}\frac{1}{2}\langle\partial_x^r U,\tilde{A}_t^0\partial_x^r U\rangle - \sum_{r=0}^{s}\langle\partial_x^r U,\sum_j \tilde{A}^0\partial_x^r A^j U_{x_j}\rangle$$

$$= \frac{1}{2}\sum_{r=0}^{s}\langle\partial_x^r U,(\tilde{A}_t^0 + \sum_j \tilde{A}_{x_j}^j)\partial_x^r U\rangle$$

$$+ \sum_{r=0}^{s}\sum_{\ell=1}^{r}\langle\partial_x^r U,\partial_x^\ell\tilde{A}^0\partial_x^{r-\ell}A^j U_{x_j}\rangle$$

$$\leq C|U|_{H^s}^2\big(|U|_{H^s} + (|U|_{H^s}^s\big),$$

$$(1.18)$$

where the final inequality follows by (1.15); see Exercise 1.8 below.
    Defining

$$\zeta(t) := \frac{1}{2}\Big(\sum_{r=0}^{\ell}\langle\partial_x^r,\tilde{A}^0\partial_x^r U\rangle\Big), \qquad (1.19)$$

we have, therefore, the Ricatti-type inequality

$$\zeta_t \leq C(\zeta + \zeta^s), \qquad (1.20)$$

yielding $\zeta(t) \leq C\zeta(0)$ for small t, provided $\zeta(0)$ is sufficiently small. Observing that $\zeta^{1/2}$ is a norm equivalent to $|U|_{H^\ell}$, we obtain bounded local stability, provided a solution exists.

Essentially the same a priori estimate can be used to show existence and uniqueness of solutions. Define the standard nonlinear iteration scheme (see, e.g., [Fr, M.3])

$$\tilde{A}^0(U^n)U_t^{n+1} + \sum_j \tilde{A}^j(U^n)U_{x_j}^{n+1} = 0, \qquad U(0) = U_0. \qquad (1.21)$$

For $U^n \in H^s$, an H^s solution $\mathcal{T}U^{n+1}$ of (1.21) may be obtained by linear theory; see Remark 3.8, Section 3.2. By the estimate already obtained, we find, that $\mathcal{T}$ takes the ball $\mathcal{B} := \{U : |U|_{L^\infty([0,\tau];H^s(x))} \leq 2|U_0|_{H^s(x)}\}$ to itself, for $\tau > 0$ sufficiently small. A similar energy estimate on the variation $e := \mathcal{T}(U_1) - \mathcal{T}(U_2)$, for $U_j \in \mathcal{B}$ yields that $\mathcal{T}$ is contractive in $L^\infty([0,\tau]; L^2(x))$ on the invariant set $\mathcal{B}$, and stable in $L^\infty[0,\tau]; H^s(x)$, yielding existence of a fixed-point solution $U \in L^\infty([0,T]; H^s(x))$ (Exercise 1.9). Likewise, uniqueness of solutions may be obtained by a stability estimate on the nonlinear variation $e := U_1 - U_2$, where U_1 and U_2 denote solutions of (1.1).   •

Remarks 1.7   1. Clearly, we do not obtain global well-posedness by this argument, since solutions of Ricatti-type equations in general blow up in finite time. Indeed, it is well-known that shock-type discontinuities may form in finite time even for arbitrarily smooth initial data, corresponding to blow-up in H^1; see, e.g., [La.1, La.2, J, KlM, Si].

2. Using the *strong Moser inequality*

$$|(\partial^{\alpha_1} v_1) \cdots (\partial^{\alpha_r} v_r)|_{L^2(\mathbb{R}^d)} \leq C \sum_{i=1}^{r} |v_i|_{H^s(\mathbb{R}^d)} \left(\prod_{j \neq i} |v_j|_{L^\infty(\mathbb{R}^d)} \right) \qquad (1.22)$$

for $s = \sum |\alpha_j|$ (proved using Gagliardo–Nirenberg inequalities [T]), the same argument may be used to show that smooth continuation of the solution is possible so long as $|U|_{W^{1,\infty}}$ remains bounded; see [M.3], Chapter 2.

Exercise 1.8 *Verify the final inequality in (1.18) by showing that*

$$|\partial_x^{r-s} A(U) U_x| \leq C \sum_{\sum |\alpha_j| \leq r+1-s} \Pi_{1 \leq j \leq r+1-s} |\partial_x^{\alpha_j} U|. \qquad (1.23)$$

Exercise 1.9 *If $u_n \in H^s(x)$ are uniformly bounded in H^s, and convergent in $L^2(x)$, show that $\lim_{n \to \infty} u_n \in H^s$, using the definition of H^s as the set of $v \in L^2$ such that, for all $1 \leq r \leq s$, $\langle v, \partial_x^r \phi \rangle \leq C|\phi|_{L^2}$ for all test functions $\phi \in C_0^\infty$, together with the fact that limits and distributional derivatives commute.*

Symmetrizability is at first sight a rather restrictive requirement in more than one spatial dimension. However, it turns out to be satisfied in many physically interesting situations, in particular for gas dynamics and MHD. Indeed, a fundamental observation of Godunov is that symmetrizability is closely related with existence of an associated convex entropy.

Definition 1.10 *A hyperbolic entropy, entropy flux ensemble is a set of scalar functions (η, q^j) such that*

$$d\eta dF^j = dq^j, \qquad (1.24)$$

or equivalently

$$\eta(U)_t + \sum_j q^j(U)_{x_j} = 0 \qquad (1.25)$$

for any smooth solution U of (1.1).

Proposition 1.11 *([God, Mo, B, KSh]) For U lying in a convex set $\mathcal{U}$, existence of a convex entropy η is equivalent to symmetrizability of (1.1) by an invertible coordinate change $U \to W := d\eta$ (known as an "entropy variable"), i.e., writing*

$$U(W)_t + \sum_j F^j(W)_{x_j} = 0, \qquad (1.26)$$

we have $\tilde{A}^0 := (\partial U / \partial W) = (d^2 \eta)^{-1}$ symmetric positive definite and $\tilde{A}^j := (\partial F^j / \partial W) = A^j \tilde{A}^0$ symmetric.

Proof. (Exercise) ($\Rightarrow$) Differentiate (1.24) and use symmetry of $d^2 \eta$. ($\Leftarrow$) Reverse the calculation to obtain (1.24) with $d\eta := W$, then note that $d\eta$ is exact, due to symmetry of $dW := \tilde{A}^0$. •

Remarks 1.12   1. Symmetrizability by coordinate change implies symmetrizability in the usual quasilinear sense (exercise), but not the converse. In particular, (nonlinear) symmetrization by coordinate change preserves divergence form, whereas (quasilinear) symmetrization by a left-multiplier $\tilde{A}^0$ does not, cf. (1.26) and (1.8).

2. Symmetrizability by coordinate change implies also (1.25), which yields the additional information that $\int \eta dx$, without loss of generality equivalent to the L^2 norm of U, is conserved for smooth solutions. Thus, we find in the discussion of Remark 1.7 that blowup occurs in a derivative of U and not in U itself.

3. For gas dynamics and MHD, there exists a convex entropy in the neighborhood of any thermodynamically stable state, namely the negative of the thermodynamical entropy s; see, e.g. [Kaw, MaZ.4, Z.4]. In particular, for an ideal gas, there exists a global convex entropy.

4. For the hyperbolic shock stability problem, hypotheses of hyperbolicity, symmetrizability, etc., are relevant only in neighborhoods of $U_\pm$ and not between. Thus, a shock may be stable even if $U_\pm$ are entirely separated by unstable constant states, as, e.g., for phase-transitional shocks in van der Waals gas dynamics; see [Fre.3, B–G.2, B–G.2, B–G.3].

1.1.2. Viscous equations. Analogous to (1.10) in the setting of the viscous equations (1.2) is the *symmetric hyperbolic–parabolic form*

$$\tilde{A}^0 W_t + \sum_j \tilde{A}^j W_{x_j} = \sum_{j,k} (\tilde{B}^{jk} W_{x_k})_{x_j} + \begin{pmatrix} 0 \\ \tilde{g} \end{pmatrix}, \tag{1.27}$$

$\tilde{A}^0$ symmetric positive definite, $\tilde{A}^j_{11}$ symmetric, $\tilde{B}^{jk} = \text{block-diag} \{0, \tilde{b}^{jk}\}$ with

$$\sum \xi_j \xi_k \tilde{b}^{jk} \geq \theta |\xi|^2, \quad \theta > 0, \tag{1.28}$$

for all $\xi \in \mathbb{R}^d$, and

$$\tilde{G} = \begin{pmatrix} 0 \\ \tilde{g}(\partial_x W) \end{pmatrix} \tag{1.29}$$

with $\tilde{g} = \mathcal{O}(|W_x|^2)$, to be achieved by an invertible coordinate change $U \to W$ combined with left-multiplication by an invertible lower block-triangular matrix $S(U)$; see [Kaw, KSh] and references therein, or Appendix A1, [Z.4].

Remark 1.13   As pointed out in the references, A^0 may be taken without loss of generality to be block-diagonal, thus identifying "hyperbolic" and "parabolic" variables w^I and w^{II}. Indeed, w^I may be taken without loss of generality as u^I and w^{II} as any variable satisfying the (clearly necessary) integrability condition $B^{jk} U_{x_k} = \beta^{jk}(U) w^{II}_{x_k}$ for all j, k [GMWZ.4, Z.4].

Similarly as in the hyperbolic case, we may deduce local well-posedness of (1.27) directly from the structure of the equations, using energy estimates/integration by parts. Here, we shall require also a standard but essential tool for multidimensional parabolic systems, the *Gårding inequality*

$$\sum_{j,k}\langle \partial_{x_j}f, \tilde{b}^{jk}\partial_{x_k}f\rangle \geq \tilde{\theta}|\partial_x f|^2_{L^2} - C|f|^2_{L^2} \tag{1.30}$$

for Lipshitz $\tilde{b}^{jk} \in \mathbb{R}^{n\times n}$ satisfying uniform ellipticity condition (1.28), and $0 < \tilde{\theta} < \theta$, with $C = C(\tilde{\theta}, |\partial_x \tilde{b}^{jk}|_{L^\infty}) \leq C_2|\partial_x \tilde{b}^{jk}|_{L^\infty}/|\theta - \tilde{\theta}|$ for some uniform $C_2 > 0$.

Exercise 1.14 *(i) Prove (1.30) with $C = 0$ in the case $\tilde{b}^{jk} \equiv$ constant, using the Fourier transform and Parseval's identity. (ii) Prove (1.30) with $C = 0$ in the case that $\tilde{b}^{jk}$ varies by less that $|\theta - \tilde{\theta}|/d^2$ from some constant value, i.e, oscillation $\tilde{b}^{jk} \leq 2|\theta - \tilde{\theta}|/d^2$. (iii) Prove the general case using a partition of unity $\{\chi_r\}$, $\chi_r^{1/2} \in C_0^\infty$, such that oscillation $\tilde{b}^{jk} \leq 2|\theta - \tilde{\theta}|/d^2$ on the support of each χ_r, and the estimates*

$$\sum_{j,k}\langle \partial_{x_j}f, \tilde{b}^{jk}\partial_{x_k}f\rangle = \sum_r\sum_{j,k}\Big(\langle \partial_{x_j}\chi_r^{1/2}f, \tilde{b}^{jk}\partial_x\chi_r^{1/2}f\rangle $$
$$ + \mathcal{O}(|\partial_x f|_{L^2}|(\partial_x \chi_r^{1/2})f|_{L^2} + |(\partial_x \chi_r^{1/2})f|^2_{L^2})\Big) \tag{1.31}$$

and

$$|\partial_x f|^2_{L^2} = \sum_r |\chi_r^{1/2}\partial_x f|^2_{L^2} = \sum_r\Big(|\partial_x\chi_r^{1/2}f|_{L^2} $$
$$ + \mathcal{O}(|\partial_x f|_{L^2}|(\partial_x \chi_r^{1/2})f|_{L^2} + |(\partial_x \chi_r^{1/2})f|_{L^2}))\Big). \tag{1.32}$$

Remark 1.15 The Gårding inequality (1.30) is an elementary example of a pseudodifferential estimate. Pseudodifferential techniques play a fundamental role in the analysis of the curved shock problem; see [M.1, M.2, M.3, Me.4], [GMWZ.1, GMWZ.3, GMWZ.4, W], and references therein.

Proposition 1.16 ([Kaw]) *Symmetric form (1.27) implies local well-posedness in H^s, and bounded local stability $|U(t)|_{H^s} \leq C|U_0|_{H^s}$, provided $\tilde{A}^j(\cdot)$, $\tilde{B}^{jk}(\cdot) \in C^s$ and $s \geq [d/2] + 3$.*

Proof. For simplicity, take $\tilde{g} \equiv 0$; the general case is similar. Similarly as in (1.17), we have

$$\frac{1}{2}\langle W, \tilde{A}^0 W\rangle_t = \frac{1}{2}\langle W, \tilde{A}^0_t W\rangle + \langle W, \tilde{A}^0 W_t\rangle$$

$$= \frac{1}{2}\langle W, \tilde{A}^0_t W\rangle - \langle W, \sum_j \tilde{A}^j W_{x_j} - \sum_{j,k}(\tilde{B}^{jk} W_{x_k})_{x_j}\rangle$$

$$= \frac{1}{2}\langle W, \tilde{A}^0_t W\rangle + \langle w^I, \sum_j \tilde{A}^j_{11,x_j} w^I\rangle + \mathcal{O}(|\partial_x w^{II}||W|) \qquad (1.33)$$

$$- \sum_{j,k}\langle w^{II}_{x_j}, \tilde{B}^{jk} w^{II}_{x_k}\rangle$$

$$\leq C|W|^2_{L^2}(1 + |W|_{W^{2,\infty}})$$

$$\leq C|W|^2_{L^2}(1 + |W|_{H^{[d/2]+3}}),$$

where in the second-to-last inequality we have used (1.30) with $\tilde{\theta} = \theta/2$, together with Young's inequality $|\partial_x w^{II}||W| \leq (1/2C)|\partial_x w^{II}|^2 + (C/2)|W|^2$ with $C > 0$ sufficiently large, and used the original equation to bound $|W_t| \leq C(|W_x| + |\partial_x^2 w^{II}|)$, and in the final inequality we have used the Sobolev inequality (1.13).

Likewise, we obtain by a similar calculation

$$\frac{1}{2}\Big(\sum_{r=0}^s \langle \partial_x^r, \tilde{A}^0 \partial_x^r W\rangle\Big)_t \leq C|W|^2_{H^s}\big(1 + |W|^{s-1}_{H^{[d/2]+2}}\big) \qquad (1.34)$$

$$\leq C\big(|W|^2_{H^s} + |W|^{s+1}_{H^s}\big),$$

by the Moser inequality (1.15) and $s \geq [d/2]+3$, from which the result follows by the same argument as in the proof of Proposition 1.6.   •

The following beautiful results of Kawashima et al (see [Kaw, KSh] and references therein) generalize to the viscous case the observations of Godunov et al regarding entropy and the structure of the inviscid equations.

Definition 1.17 ([Kaw, Sm]) *A viscosity-compatible convex entropy, entropy flux ensemble (η, q^j) for (1.7) is a convex hyperbolic entropy, entropy flux ensemble such that*

$$d^2\eta \sum_{j,k} \tilde{B}^{jk}\xi_j\xi_k \geq 0 \qquad (1.35)$$

and

$$\text{rank } \Re d^2\eta \sum_{j,k} \tilde{B}^{jk}\xi_j\xi_k = \text{rank } \sum_{j,k} \tilde{B}^{jk}\xi_j\xi_k \equiv r \qquad (1.36)$$

for all nonzero $\xi \in \mathbb{R}^d$.

Exercise 1.18 *Using (1.8)–(1.9), and the fact that*

$$(d^2\eta)^{1/2}B^{jk}(d^2\eta)^{-1/2} = (d^2\eta)^{-1/2}(d^2\eta B^{jk})(d^2\eta)^{-1/2} \qquad (1.37)$$

is similar to B^{jk}, show that $d^2\eta B^{jk}$ symmetric is sufficient for (1.35)–(1.36).

Proposition 1.19 ([KSh, Yo][3]) *A necessary and sufficient condition that a system (1.2), (1.9) can be put in symmetric hyperbolic–parabolic form (1.27) with $\tilde{G} \equiv 0$ and $\tilde{A}^j$ symmetric ($\tilde{A}^0$ not necessarily block-diagonal) by a change of coordinates $U \to W(U)$ for U in a convex set $\mathcal{U}$, is existence of a convex viscosity-compatible entropy, entropy flux ensemble η, q^j defined on $\mathcal{U}$, with $W = d\eta(U)$.*

Proof. Identical with that of Proposition 1.11 as concerns $\tilde{A}^0$ and $\tilde{A}^j$. Regarding $\tilde{B}^{jk} = B^{jk}(d^2\eta)^{-1} = (d^2\eta)^{-1}(d^2\eta B^{jk})(d^2\eta)^{-1}$, we have that conditions $\sum_{j,k} \tilde{B}^{jk}\xi_j\xi_k \geq 0$ and rank $\Re \tilde{B}^{jk}\xi_j\xi_k \equiv r$ for all nonzero $\xi \in \mathbb{R}^d$ are equivalent both to (1.35)–(1.36) and to (1.28) and $\tilde{B}^{jk} = $ block-diag $\{0, \tilde{b}^{jk}\}$ (exercise). The latter assertion depends on the observation that nonnegativity of $\sum_{j,k} \tilde{B}^{jk}\xi_j\xi_k$ together with structure (1.8) implies that $\Re \sum_{j,k} \tilde{B}^{jk}\xi_j\xi_k$ is block-diagonal and vanishing in the first diagonal block. This completes the argument.   ●

Definition 1.20 ([Kaw]) *Augmenting the local conditions (1.27)–(1.28), we identify the time-asymptotic stability conditions[4] of "first-order symmetry," $\tilde{A}^j$ symmetric, and "genuine coupling:"*

$$\text{No eigenvector of } \sum_j \xi_j dF^j \text{ lies in ker} \sum \xi_j \xi_k B^{jk}, \text{ for } \xi \neq 0 \in \mathbb{R}^d. \quad \text{(GC)}$$

Proposition 1.21 ([Kaw][5]) *Symmetric hyperbolic–parabolic form (1.27) together with the time-asymptotic stability conditions of first-order symmetry, $\tilde{A}^j$ symmetric, and genuine coupling, (GC), implies $L^1 \cap H^s \to H^s$ time-asymptotic stability of constant solutions, $\bar{U} \equiv$ constant, provided that $\tilde{A}^j$, $\tilde{B}^{jk} \in C^s$ and $s \geq [d/2] + 3$, with rate of decay*

$$|U - \bar{U}|_{H^s}(t) \leq C(1+t)^{-\frac{d}{4}}|U - \bar{U}|_{H^s \cap L^1}(0) \quad (1.38)$$

for $|U - \bar{U}|_{H^s \cap L^1}(0)$ sufficiently small, equal to that of a d-dimensional heat kernel.

Proposition 1.22 ([Kaw][6]) *Existence of a convex viscosity-compatible entropy, together with genuine coupling, (GC), implies global well-posedness of (1.7)–(1.9) in H^s, and bounded stability of constant solutions,*

$$|U - \bar{U}|_{H^s}(t) \leq C|U - \bar{U}|_{H^s}(0), \quad (1.39)$$

provided $\tilde{A}^j$, $\tilde{B}^{jk} \in C^s$ and $s \geq [d/2] + 3$.

Propositions 1.21 and 1.22 depend on a circle of ideas associated with the phenomenon of "hyperbolic–parabolic smoothing", as indicated by the existence of parabolic-type energy estimates

---

[4] This refers to stability of constant solutions; see Proposition 1.21 below.

$$|W(t)|^2_{H^s} + \int_0^t \left(|\partial_x W|^2_{H^{s-1}} + |\partial_x w^{II}|^2_{H^s} \right) ds \leq C \left(|W(0)|^2_{H^s} + \int_0^t |W(s)|^2_{L^2} ds \right)$$
(1.40)

for s sufficiently large. We defer discussion of these important concepts to the detailed treatment of Sections 3 and 4.

Exercise 1.23 *Result (1.39) was established for gas dynamics in the seminal work of Matsumura and Nishida [MNi]. The key point is existence of an L^2 energy estimate*

$$|W(t)|^2_{L^2} + \int_0^t |\partial_x w^{II}|^2_{L^2} \leq C|W(0)|^2_{L^2},$$
(1.41)

which, coupled with the general machinery used to obtain (1.40), yields an improved version of (1.40) in which the $\int |U(s)|^2_{L^2} ds$ term on the righthand side does not appear. This in turn implies (1.39). Prove (1.41) in the general case using the viscous version

$$\eta(U)_t + \sum_j q^j(U)_{x_j} = \sum_{j,k} (d\eta B^{jk} U_{x_k})_{x_j} - \sum_{j,k} U^t_{x_j} d^2 \eta B^{jk} U_{x_k}$$
(1.42)

of entropy equation (1.25) and Exercise 1.14(ii), under the assumption that $|W|_{H^s}$ (hence $|W|_{L^\infty}$) remains sufficiently small, for s sufficiently large.

The Gårding inequality is not required for gas dynamics or MHD, for which there hold the strengthened ellipticity condition

$$\sum_{j,k} \langle f_{x_j}, d^2 \eta B^{jk} f_{x_k} \rangle \geq \theta |\partial x f|^2.$$
(1.43)

Under this assumption, show using (1.42) that (1.39) holds for $s = 0$, independent of the size of $|W(0)|_{L^2}$.

Remark 1.24 Similarly as in the inviscid case, the negative of the thermodynamical entropy serves as a convex viscosity-compatible entropy for gas and MHD in the neighborhood of any thermodynamically stable state; see [Kaw, MaZ.4, Z.4]. In particular, for an ideal gas, there exists a global convex viscosity-compatible entropy.

Example 1.25 The Navier–Stokes equations of compressible gas dynamics, may be written as

$$\rho_t + \text{div}(\rho u) = 0,$$

$$(\rho u)_t + \text{div}(\rho u \otimes u) + \nabla p = \overbrace{\mu \Delta u + (\lambda + \mu)\nabla \text{div} u}^{\text{div}\tau},$$
(1.44)

$$(\rho(e + \tfrac{1}{2}u^2))_t + \text{div}(\rho(e + \tfrac{1}{2}u^2)u + pu) = \text{div}(\tau \cdot u) + \kappa \Delta T,$$

where $\rho > 0$ denotes density, $u \in \mathbb{R}^d$ fluid velocity, $T > 0$ temperature, $e = e(\rho, T)$ internal energy, and $p = p(\rho, T)$ pressure. Here, $\tau := \lambda \operatorname{div}(u)I + 2\mu Du$, where $Du_{jk} = \frac{1}{2}(u_{x_j}^j + u_{x_k}^j)$ is the deformation tensor, $\lambda(\rho, T) > 0$ and $\mu(\rho, T) > 0$ are viscosity coeffients, and $\kappa(\rho, T) > 0$ is the coefficient of thermal conductivity. The thermodynamic entropy s is defined implicitly by the underlying thermodynamic relations $e = \hat{e}(v, s)$, $T = \hat{e}_s$, $p = -\hat{e}_v$, where $v = \rho^{-1}$ is specific volume. (These may or may not be integrable for arbitrary choices of $p(\cdot, \cdot)$, $e(\cdot, \cdot)$; on the other hand, each choice of $\hat{e}$ satisfying $\hat{e}_s = T > 0$ uniquely determines functions p and e.) Thermodynamic stability is the condition that $\hat{e}$ be a convex function of (v, s), or equivalently (exercise) $e_T, p_\rho > 0$. The Euler equations of compressible gas dynamics are (1.44) with the righthand side set to zero.

Evidently, (1.44) is of form (1.2), (1.8), with "conservative variables" $U = (\rho, \rho u, \rho(e + \frac{1}{2}u^2))$. By Remark 1.13, symmetric hyperbolic–parabolic form (1.27), if it exists, may be expressed in the "natural variables" $W = (\rho, u, T)$. Indeed, this can be done [KSh], with

$$\tilde{A}^0 = \begin{pmatrix} p_\rho/\rho & 0 & 0 \\ 0 & \rho I_d & 0 \\ 0 & 0 & \rho e_T/T \end{pmatrix}, \tag{1.45}$$

$$\sum_j \tilde{A}^j \xi_j = \begin{pmatrix} (p_\rho)u \cdot \xi & p_\rho \xi & 0 \\ p_\rho \xi^t & \rho(u \cdot \xi)I_d & p_T \xi^t \\ 0 & p_T \xi & (\rho e_T/T)u \cdot \xi \end{pmatrix}, \tag{1.46}$$

and

$$\sum_{j,k} \tilde{B}^{jk} \xi_j \xi_k = \begin{pmatrix} 0 & 0 & 0 \\ 0 & \mu|\xi|^2 I_d + (\mu + \lambda)\xi^t \xi & 0 \\ 0 & 0 & T^{-1}|\xi|^2 \end{pmatrix}, \tag{1.47}$$

whenever $p_\rho, e_T > 0$ (thermodynamic stability), in which case we also have the time-asymptotic stability conditions of symmetry of $\tilde{A}^j$ and genuine coupling (GC) (by inspection, equivalent to $p_\rho \neq 0$). If only $e_T > 0$, as for example for a van der Waals-type equation of state, we may still achieve symmetric hyperbolic–parabolic form, but without symmetry of $\tilde{A}^j$ or genuine coupling, by dividing through p_ρ from the first equation ($\sim$ first row of each matrix $\tilde{A}^0$, $\tilde{A}^j$, $\tilde{B}^{jk}$). That is, we obtain in this case local well-posedness, but not asymptotic stability of constant states. Similar considerations hold in the case of MHD; see [Kaw, KSh].

2 Description of Results

We now turn to the long-time stability of viscous shock waves. Fixing the viscosity coefficient ν, consider a viscous shock solution (1.4) of a system of viscous conservation laws (1.2) of form (1.8)–(1.9). Changing coordinates if necessary to a rest frame moving with the speed of the shock, we may without loss of generality arrange that shock speed s vanish, so that (1.4) becomes a stationary, or standing-wave solution. Hereafter, we take $\nu \equiv 1$ and $s \equiv 0$, and suppress the parameters ν and s.

2.1. Assumptions. The classical Propositions 1.21 and 1.22 concern long-time stability of a single, thermodynamically stable equilibrium, whereas a viscous shock solution typically consists of two different thermodynamically stable equilibria connected by a profile that in general may pass through regions of thermodynamical instability; see Remark 1.12.4, and examples, [MaZ.4, Z.4]. Accordingly, we make the following structural assumptions, imposing stability at the endstates, but only local well-posedness along the profile.

Assumptions 2.1

(A1) *For U in a neighborhood of profile $\bar{U}(\cdot)$, there is an invertible change of coordinates $U \to W$ such that (1.2) may be expressed in W coordinates in symmetric hyperbolic–parabolic form, i.e, in form (1.27) with $\tilde{A}^0$ symmetric positive definite and (without loss of generality) block-diagonal and $\tilde{A}^j_{11}$ symmetric, $\tilde{B}^{jk} =$ block-diag $\{0, \tilde{b}^{jk}\}$ with $\tilde{b}^{jk}$ satisfying uniform ellipticity condition (1.28), and $\tilde{G} = \begin{pmatrix} 0 \\ \tilde{g} \end{pmatrix}$ with $\tilde{g} = \mathcal{O}(|\partial_x W|^2)$.*

(A2) *At endstates $U_\pm$, the coefficients $\tilde{A}^j$, $\tilde{B}^{jk}$ defined in (A1) satisfy the constant-coefficient stability conditions of first-order symmetry, $\tilde{A}^j_\pm$ symmetric, and genuine coupling, (GC).*

Remark 2.2 Conditions (A1)–(A2) are satisfied for gas- and magnetohydrodynamical shock profiles connecting thermodynamically stable endstates, under the mild assumption $e_T > 0$ on the equation of state $e = e(\rho, T)$ relating internal energy e to density ρ and temperature T: in particular for both ideal and van der Waals-type equation of state. (Exercise: verify this assertion for gas dynamics, starting from form (1.45)–(1.47) and modifying all coeffients M by the transformation

$$M \to \chi M + (1 - \chi)\text{block-diag}\ \{p_\rho^{-1}, I_d, 1\}M, \tag{2.1}$$

where $\chi = \chi(W)$ is a smooth cutoff function supported on the region of thermodynamic stability $p_\rho > 0$.)

To Assumptions 2.1, we add the following technical hypotheses.

Assumptions 2.3

(H0) F^j, B^{jk}, $W(\cdot)$, $\tilde{A}^0 \in C^{q+1}$, $q \geq q(d) := [d/2] + 3$.

(H1) $\tilde{A}^1_{11}(\bar{U})$ *is uniformly definite, without loss of generality $\tilde{A}^1_{11} \geq \theta > 0$. (For necessary conditions, $\det \tilde{A}^1_{11} \neq 0$ is sufficient.)*

(H2) $\det \tilde{A}^1_\pm \neq 0$, *or equivalently $\det A^1_\pm \neq 0$.*

(H3) *Local to $\bar{U}$, the solutions of (1.4) form a smooth manifold $\{\bar{U}^\delta\}$, $\delta \in \mathcal{U} \subset \mathbb{R}^\ell$, with $\bar{U}^0 = \bar{U}$. See ODE discussion, Section 3.1.*

Condition (H0) gives the regularity needed for our analysis. Condition (H2) is the standard inviscid requirement that the hyperbolic shock triple $(U_-, U_+, 0)$ be noncharacteristic. Condition (H1) states that convection in the reduced, hyperbolic part of (1.27) governing coordinate w^I is either up- or downwind, in particular uniformly noncharacteristic, everywhere along the profile; as discussed in [MaZ.4, Z.4], this is satisfied for all gas-dynamical shocks and at least generically for magnetohydrodynamical shocks. Condition (H3) is a weak form of (implied by but not implying) transversality of the connection $\bar{U}$ as a solution of the associated traveling-wave ODE.

These hypotheses suffice for the investigation of necessary conditions for stability. In our investigation of sufficient conditions, we shall require two further hypotheses at the level of the inviscid stability problem.

Assumption 2.4

(H4) *The eigenvalues of $\sum_j A_\pm^j \xi_j$ have constant multiplicity with respect to $\xi \in \mathbb{R}^d$, $\xi \neq 0$.*

Denote by $a_j^\pm(\xi)$, $j = 1, \dots, n$ the eigenvalues of $\sum_j A_\pm^j \xi_j$, necessarily real by (A2), indexed by increasing order. These are positive homogeneous degree one and, by (H4), locally analytic on $\xi \in \mathbb{R}^d \setminus \{0\}$.[7] Here, and elsewhere, real homogeneity refers to homogeneity with respect to positive reals. Let

$$P_\pm(\xi, \tau) := i\tau + \sum_j i\xi_j A_\pm^j \qquad (2.2)$$

denote the frozen-coefficient symbols associated with (1.6) at $U = U_\pm$. Then, $\det P_\pm(\xi, \tau) = 0$ has n locally analytic, positive homogeneous degree one roots

$$i\tau = -ia_r^\pm(\xi), \qquad r = 1, \dots, n, \qquad (2.3)$$

describing dispersion relations for the frozen-coefficient initial-value problem (IVP). The corresponding objects for the initial–boundary-value problem (IBVP) are relations

$$i\xi_1 = \mu_r(\tilde{\xi}, \tau), \qquad r = 1, \dots, n, \qquad (2.4)$$

$\xi = (\xi_1, \dots, \xi_d) =: (\xi_1, \tilde{\xi})$, describing roots of $\det Q_\pm(\tau) = 0$, where

$$Q_\pm(\xi, \tau) := i\xi_1 + (A_\pm^1)^{-1}\left(i\tau + \sum_{j=2}^d i\xi_j A_\pm^j\right) \qquad (2.5)$$

---

[7] In fact, they are globally analytic, since they maintain fixed order; however, this is unimportant in the analysis.

denote the frozen-coefficient initial–boundary-value symbols associated with
(1.6) at $U = U_\pm$. Evidently, graphs (2.3) and (2.4) describe the same sets, since
$\det A^1 \det Q_\pm(\tau) = \det P_\pm$ and $\det A^1_\pm \neq 0$; the roots $i\tau$ describe characteristic
rates of temporal decay, whereas $\mu = i\xi_1$ describe characteristic rates of spatial
decay in the x_1 direction.

Definition 2.5 *Setting $\xi = (\xi_1, \ldots, \xi_d) =: (\xi_1, \tilde{\xi})$, we define the glancing sets*
$\mathcal{G}(P_\pm)$ *as the set of all $(\tilde{\xi}, \tau)$ such that, for some real ξ_1 and $1 \leq r \leq n$,*
$\tau = -a^\pm_r(\xi_1, \tilde{\xi})$ *and* $(\partial a^\pm_r/\partial \xi_1)(\xi_1, \tilde{\xi}) = 0$*: that is, the projection onto $(\tilde{\xi}, \tau)$*
of the set of real roots (ξ, τ) of $\det P_\pm = 0$ at which (2.3) is not analytically
invertible as a function (2.4). The roots (ξ, τ) are called glancing points.

Assumption 2.6

(H5) *Each glancing set $\mathcal{G}(P_\pm)$ is the (possibly intersecting) union of fi-*
nitely many smooth curves $\tau = \eta^\pm_q(\tilde{\xi})$, on which the root ξ_1 of $i\tau + a^\pm_r(\cdot, \tilde{\xi}) = 0$
has constant multiplicity s_q (by definition ≥ 2).

Among other useful properties (see, e.g., [Me.3, Z.3, MeZ.2]), condition
(H4) implies the standard block structure condition [M.1, M.2, M.3, Me.2]
of the inviscid stability theory [Me.3]. Condition (H5), introduced in [Z.3],
imposes a further, laminar structure on the glancing set that is convenient
for the viscous stability analysis. In either context, glancing is the fundamental obstacle in obtaining resolvent estimates; see, e.g., [K, M.1, M.2, M.3,
Z.3, MeZ.1, GMWZ.1, GMWZ.2, GMWZ.3, GMWZ.4]. The term "glancing"
derives from the fact that, at glancing points (ξ, τ), null bicharacteristics of
$\det P_\pm$ lie parallel to the shock front $x_1 \equiv 0$.

Remarks 2.7 1. Condition (H5) holds automatically in dimensions one (vacuous) and two (trivial), and also for the case that all characteristics $a_r(\xi)$ are
either linear or convex/concave in ξ_1. (Exercise [GMWZ.2]: prove the latter
assertion using the Implicit Function Theorem.) In particular, (H5) holds for
both gas dynamics and MHD.

2. Condition (H4) holds always for gas dynamics in any dimension, and
generically for MHD in one dimension, but fails always for MHD in dimension greater than or equal to two. For a refined treatment applying also to
multidimensional MHD, see [MeZ.3].

Assumption 2.8 *For definiteness, consider a classical, "pure" Lax p-shock,
satisfying*

$$a^-_p > s = 0 > a^+_p, \tag{L}$$

where $a^\pm_j := a^\pm_j(1, 0, \ldots, 0)$ denote the eigenvalues of $A^1_\pm := dF^1(U_\pm)$, and
$\ell = 1$ *in (H3). See, e.g., [ZS, Z.3, Z.4] for discussions of the general case,*
including the interesting situation of nonclassical over- or undercompressive
shocks.

2.2. Classical stability conditions. Recall the classical admissibility conditions described in the introduction of:

1. *Structural stability*, defined as existence of a transverse viscous profile. For gas dynamics, this is equivalent to the (hyperbolic) Liu–Oleinik admissibility condition; see [L.2, Gi, MeP]. In general, it may be a complicated ODE problem involving both (1.1) and the specific form of the viscosity B^{jk}.

2. *Dynamical stability*, defined as local hyperbolic well-posedness, or bounded, bounded-time stability of ideal shock (1.3). Following Majda [M.1, M.2, M.3], we define *weak dynamical stability* (Majda's Lopatinski condition) as the absence of unstable spectrum $\Re \lambda > 0$ for the linearized operator (appropriately defined) about the shock, an evident necessary condition for stability. (Recall, equations (1.1) are positive homogeneous degree one, hence spectrum lies on rays through the origin and instabilities, should they occur, occur to all orders.)

In the present context of a Lax p-shock, under assumptions (A1)–(A2) and (H2), the necessary condition of weak dynamical stability may be expressed in terms of the *Lopatinski determinant*

$$\Delta(\tilde{\xi}, \lambda) := \det\left(r_1^-, \cdots, r_p^-, r_{p+1}^+, \cdots, r_n^+, \lambda[U] + i[F^{\tilde{\xi}}]\right) \qquad (2.6)$$

as

$$\Delta(\tilde{\xi}, \lambda) \neq 0 \quad \text{for } \tilde{\xi} \in \mathbb{R}^{d-1} \text{ and } \Re \lambda > 0, \qquad (2.7)$$

where $r_j^\pm = r_j^\pm(\tilde{\xi}, \lambda)$ denote bases for the unstable (resp. stable) subspace of the coefficient matrix

$$\mathcal{A}_\pm := (A^1)^{-1}(\lambda + iA^{\tilde{\xi}})_\pm \qquad (2.8)$$

arising in the IBVP symbol $Q_\pm$ with $\lambda := i\tau$, $F^{\tilde{\xi}} := \sum_{j \neq 1} F^j \xi_j$, and $A^{\tilde{\xi}} := \sum_{j \neq 1} A^j \xi_j$. Functions $r_j^\pm$, and thus Δ, are well-defined on $\Re \lambda > 0$ and may be chosen analytically in $(\tilde{\xi}, \lambda)$, by a standard lemma of Hersch [H] asserting that $\mathcal{A}_\pm$ has no center subspace on this domain; see exercise 2.10 below. Zeroes of Δ correspond to "normal modes", or solutions $(U, X) = e^{\lambda t} e^{i\tilde{\xi} \cdot \tilde{x}} (\hat{U}(x_1), \hat{X})$ of the linearized perturbation equations, where $x_1 = X(\tilde{x}, t)$ denotes location of the shock surface. We define *strong dynamical instability* as failure of (2.7).

Under assumptions (A1)–(A2), (H2), and (H4), $r_j^\pm$, and thus Δ, may be extended continuously to the boundary $\Re \lambda = 0$; see [Me.3, CP]. *Strong dynamical stability* (Majda's uniform Lopatinski condition) is then defined as

$$\Delta(\tilde{\xi}, \lambda) \neq 0 \quad \text{for } \tilde{\xi} \in \mathbb{R}^{d-1}, \ \Re \lambda \geq 0, \text{ and } (\tilde{\xi}, \lambda) \neq (0, 0). \qquad (2.9)$$

Strong dynamical stability under our assumptions is sufficient for dynamical stability, by the celebrated result of Majda [M.1, M.2, M.3] together with the result of Métivier [Me.3] that (H4) implies Majda's block structure condition.

Remarks 2.9 1. In the one-dimensional case, $\tilde{\xi} = 0$, both (2.7) and (2.9) reduce to nonvanishing of the Liu–Majda determinant

$$\delta := \det\left(r_1^-, \cdots, r_p^-, r_{p+1}^+, \cdots, r_n^+, [U]\right), \qquad (2.10)$$

where $r_j^\pm$ denote eigenvectors of $A_\pm^1$ associated with characteristic modes outgoing from the shock. As a single condition, this is generically satisfied, whence we find that shocks of Lax type (L) are generically stable in one dimension from the hyperbolic point of view.

2. As pointed out by Majda [M.1, M.2, M.3], for gas dynamics the region of strong dynamical stability is typically separated from the region of strong dynamical instability by an open set in parameter space $\alpha := (U_-, U_+, s)^8$ of indeterminate stability, on which $\Delta(\tilde\xi, \lambda)$ has roots λ lying precisely on the imaginary axis, but none with $\Re\lambda > 0$; *thus, the point of transition from stability to instability is not determined in the hyperbolic stability theory.* A similar situation holds in MHD [Bl, BT.1, BT.2, BT.3, BT.4, BTM.1, BTM.2]. The reason for this at first puzzling phenomenon, as described in [BRSZ, Z.4], is that, under appropriate normalization, $\Delta(\tilde\xi, \cdot)$ takes imaginary $\lambda = i\tau$ to imaginary Δ for $|\tau|$ sufficiently large relative to $|\tilde\xi|$: more precisely, for $(\tilde\xi, \tau)$ lying in the "hyperbolic regions" $\mathcal{H}_\pm$ bounded by glancing sets $\mathcal{G}(P_\pm)$. Thus, fixing $\tilde\xi$, and explicitly noting the dependence on parameters α, we find that imaginary zeroes $\lambda = i\tau$ of $\Delta^\alpha(\tilde\xi, \cdot)$ for which $(\tilde\xi, \tau) \in \mathcal{H}_\pm$, such as occur for gas dynamics and MHD, generically persist under perturbation in α, rather than moving into the stable or unstable complex half-planes $\Re\lambda < 0$ and $\Re\lambda > 0$ as would otherwise be expected; see exercise 2.11 below.

Exercise 2.10 *Under (A1)–(A2), (H2), show that* $\det\left(A_\pm(\tilde\xi, \lambda) - i\xi_1\right) = 0$ *implies* $\det(\sum_j i\xi_j A^j + \lambda) = 0$, $A_\pm$ *defined as in (2.8), violating hyperbolicity of (1.1) if* $\Re\lambda \neq 0$, $\Im\xi_1 = 0$. *Thus,* $A_\pm$ *has no center subspace on the domain* $\Re\lambda > 0$. *It follows by standard matrix perturbation theory [Kat] that the stable and unstable subspaces of* $A_\pm$ *have constant dimension on* $\Re\lambda > 0$, *and the associated* $A_\pm$-*invariant projections are analytic in* $(\tilde\xi, \lambda)$. *This implies the existence of analytic bases* $r_j^\pm$, *by a lemma of Kato asserting that analytic projections induce analytic bases on simply connected domains; see [Kat], pp. 99–102.*

Exercise 2.11 *Let* $f(\alpha, \lambda)$ *be continuous in* α *and* $\lambda \in \mathbb{C}$, *with the additional property that* $f(\alpha, \cdot) : \mathbb{R} \to \mathbb{R}$.
(a) Show that in the vicinity of any root $(0, \lambda_0)$, λ_0 *real, such that* $f_\lambda(0, \lambda_0)$ *exists and is nonzero, there is a family of roots* $\lambda(\alpha)$, *continuous at* $\alpha = 0$, *with* $\lambda(\alpha)$ *real (Intermediate Value Theorem). More generally, odd topological degree of* $f(0, \cdot)$ *at* $\lambda = \lambda_0$ *is sufficient; for even degrees* $f(\alpha, \lambda) = (\alpha + \lambda^2)^m$, $\alpha \in \mathbb{R}$, $\lambda_0 = 0$ *is a counterexample.*
(b) If f *is analytic in* λ, *show that* $\bar{f}(\alpha, \bar\lambda) = f(\alpha, \lambda)$, *where* $\bar z$ *denotes complex conjugate of* z, *hence nonreal roots occur in conjugate pairs.*

---

[8] More precisely, the subset of α corresponding to Lax p-shocks on which Δ was defined, or an open neighborhood with Δ extended by (2.6).

2.3. Viscous stability conditions. We'll both augment and refine the classical stability conditions through a viscous stability analysis, at the same time providing their rigorous justification.

2.3.1. Spectral stability. We begin by adding to the conditions of structural and dynamical stability a third condition of (viscous) spectral stability. Let L denote the linearized operator about the wave, i.e., the generator of the linear equations $U_t = LU$ obtained by linearizing (1.2) about the profile $\bar{U}$, and $L_{\tilde{\xi}}$ the family of operators obtained from L by Fourier transform in the directions $\tilde{x}$ parallel to the front, indexed by frequency $\tilde{\xi}$. Explicit representations of L and $L_{\tilde{\xi}}$ are given in Section 3.2.

Definition 2.12 Similarly as in the hyperbolic stability theory, we define *weak spectral stability* (clearly necessary for viscous stability) as the absence of unstable L^2 spectrum $\Re\lambda > 0$ for the linearized operator L about the wave, or equivalently

$$\lambda \notin \sigma(L_{\tilde{\xi}}) \quad \text{for } \tilde{\xi} \in \mathbb{R}^{d-1} \text{ and } \Re\lambda > 0. \tag{2.11}$$

We define *strong spectral stability* (neither necessary nor sufficient for viscous stability) as

$$\lambda \notin \sigma(L_{\tilde{\xi}}) \quad \text{for } \tilde{\xi} \in \mathbb{R}^{d-1},\ \Re\lambda > 0, \text{ and } (\tilde{\xi}, \lambda) \neq (0, 0). \tag{2.12}$$

We define *strong spectral instability* as failure of (2.11).

Conditions (2.11) and (2.12) may equivalently be expressed in terms of the *Evans function* $D(\tilde{\xi}, \lambda)$ (defined Section 5), a spectral determinant analogous to the Lopatinski determinant of the hyperbolic case, as

$$D(\tilde{\xi}, \lambda) \neq 0 \quad \text{for } \tilde{\xi} \in \mathbb{R}^{d-1} \text{ and } \Re\lambda > 0 \tag{2.13}$$

and

$$D(\tilde{\xi}, \lambda) \neq 0 \quad \text{for } \tilde{\xi} \in \mathbb{R}^{d-1},\ \Re\lambda \geq 0, \text{ and } (\tilde{\xi}, \lambda) \neq (0, 0), \tag{2.14}$$

respectively, where zeroes of D correspond to normal modes, or solutions $U = e^{\lambda t} e^{i\tilde{\xi} \cdot \tilde{x}} \hat{U}(x_1)$ of the linearized perturbation equations: equivalently, eigenvalues λ of the operator $L_{\tilde{\xi}}$ obtained from L by Fourier transform in the directions $\tilde{x}$ parallel to the front. Under assumptions (A1)–(A2), D may be chosen analytically in $(\tilde{\xi}, \lambda)$ on $\{\tilde{\xi}, \lambda : \Re\lambda \geq 0\} \setminus \{(0, 0)\}$; see [ZS, Z.3, Z.4] or Section 5 below.

Remarks 2.13 1. It is readily verified that $L\bar{U}' = L_0\bar{U}' = 0$, a consequence of translational invariance of the original equations (1.2), from which we obtain $0 \in \sigma_{\text{ess}}(L)$ and in particular $D(0, 0) = 0$. The exclusion of the origin in definition 2.14 is therefore necessary for the application to stability of viscous shock profiles. If there held the stronger condition of uniform spectral stability,

$$\Re\sigma(L) < 0, \tag{2.15}$$

we could conclude exponential stability of the linearized solution operator e^{LT} by the generalized Hille–Yosida theorem (Proposition 6.11, Appendix A) together with the high-frequency resolvent bounds we obtain later. However, in the absence of a spectral gap between 0 and $\Re\sigma(L)$, even bounded linearized stability is a delicate question. Moreover, we here discuss nonlinear asymptotic stability, for which we require a rate of linearized decay sufficient to carry out a nonlinear iteration. This is connected with the rate at which the spectrum of $L_{\tilde{\xi}}$ moves into the stable complex half-plane $\Re\lambda < 0$ as $\tilde{\xi}$ is varied about the origin, information that is encoded in the conditions of structural and refined dynamical stability defined just below.

2. In the one-dimensional case $\tilde{\xi} \equiv 0$, strong spectral stability is roughly equivalent to linearized stability with respect to zero-mass initial perturbations; see [ZH, HuZ, MaZ.2, MaZ.3, MaZ.4]. Zero-mass stability is exactly the property that was used by Goodman and Xin [GoX] to rigorously justify the small-viscosity matched asymptotic expansion about small-amplitude shock waves in one dimension; indeed, their argument implies validity of matched asymptotic expansion in one dimension for any structurally and dynamically stable wave about which the associated profiles exhibit zero-mass stability.

2.3.2. Refined dynamical stability conditions. A rigorous connection between viscous stability and the formal conditions of structural and dynamical stability is given by the following fundamental result (proved in the Lax case in Section 5).

Proposition 2.14 ([ZS, Z.3, MeZ.2][9]) *Given (A1) – (A2) and (H0) – (H4), and under appropriate normalizations of D and Δ,*

$$D(\tilde{\xi},\lambda) = \overbrace{\gamma\Delta(\tilde{\xi},\lambda)}^{\mathcal{O}(|\tilde{\xi},\lambda|^{\ell})} + o(|\tilde{\xi},\lambda|^{\ell}), \tag{2.16}$$

uniformly on $\tilde{\xi} \in \mathbb{R}^{d-1}$, $\Re\lambda \geq 0$, for $\rho := |\tilde{\xi},\lambda|$ sufficiently small, where γ is a constant measuring transversality of connection $\bar{U}$ as a solution of the associated traveling-wave ODE, Δ is an appropriate Lopatinski condition, and ℓ is as in (H3). In the present, Lax case, Δ is as in (2.6) and $\ell = 1$.

Remark 2.15 In the absence of (H4), (2.16) still holds along individual rays through the origin, provided that they lie in directions of analyticity for Δ, in particular for $\Re\lambda > 0$, by the original argument of [ZS]. This is sufficient for the necessary stability conditions derived below; however, uniformity is essential for our sufficient stability conditions.[10]

---

[10] This point was not clearly stated in [Z.3]; indeed, (2.16) is hidden in the detailed estimates of the technical Section 4 of that reference, which depend on the additional hypothesis (H5). For a more satisfactory treatment, see [MeZ.2, GMWZ.3, GMWZ.4, Z.4].

Definition 2.16 We define *refined weak dynamical stability* as (2.7) augmented with the second-order condition $\Re\beta(\tilde{\xi}, i\tau) \geq 0$ for any real $\tilde{\xi}, \tau$ such that Δ and $D^{\tilde{\xi}, \lambda}(\rho) := D(\rho\tilde{\xi}, \rho\lambda)$ are analytic at $(\tilde{\xi}, i\tau)$ and $(\tilde{\xi}, i\tau, 0)$, respectively, with $\Delta(\tilde{\xi}, i\tau) = 0$ and $\Delta_\lambda(\tilde{\xi}, i\tau) \neq 0$, where

$$\beta(\tilde{\xi}, i\tau) := \frac{(\partial/\partial\rho)^{\ell+1} D(\rho\tilde{\xi}, \rho i\tau)|_{\rho=0}}{(\partial/\partial\lambda)\bar{\Delta}(\tilde{\xi}, i\tau)} = \frac{(\partial/\partial\rho)^{\ell+1} D(\rho\tilde{\xi}, \rho\lambda)}{(\partial/\partial\lambda)(\partial/\partial\rho)^{\ell} D(\rho\tilde{\xi}, \rho\lambda)}\Big|_{\rho=0, \lambda=i\tau}. \tag{2.17}$$

Under (A1)–(A2), (H2), (H4), analyticity of Δ and $D^{\tilde{\xi}, \lambda}(\rho)$ are equivalent, and hold for all real $(\tilde{\xi}, \tau)$ lying outside the glancing sets $\mathcal{G}(P_\pm)$ [ZS, Z.3, Z.4]. In this case, we define *strong refined dynamical stability* as (2.7) augmented with the conditions that: (i) $\Re\beta(\tilde{\xi}, i\tau) > 0$, $\Delta_\lambda(\tilde{\xi}, i\tau) \neq 0$, and $\{r_1^-, \ldots, r_{p-1}^-, r_{p+1}^+, \ldots, r_n^+\}(\tilde{\xi}, i\tau)$ are independent, where $r_j^\pm$ are defined as in (2.6) (automatic for "extreme" shocks $p = 1$ or n), for any real $(\tilde{\xi}, \tau) \notin \mathcal{G}(P_\pm)$ such that $\Delta(\tilde{\xi}, i\tau) = 0$ and Δ is analytic, and (ii) the zero-level set in $(\tilde{\xi}, \tau)$ of $\Delta(\tilde{\xi}, i\tau) = 0$ intersects the glancing sets $\mathcal{G}_\pm$ transversely in $\mathbb{R}^{d-1} \times \mathbb{R}$ (in particular, their intersections are trivial in dimension $d = 2$).

Remark 2.17 The coefficient β may be recognized as a viscous correction measuring departure from homogeneity: note that β would vanish if D were homogeneous degree ℓ. It has a physical interpretation as an "effective viscosity" governing transverse propagation of deformations in the front [Go.2, GM, ZS, Z.3, Z.4, HoZ.1, HoZ.2].

2.3.3. Main results. With these definitions, we can now state our main results, to be established throughout the remainder of the article.

Theorem 2.1 ([ZS, Z.3]) *Given (A1)–(A2), (H0)–(H3), structural stability $\gamma \neq 0$, and one-dimensional inviscid stability $\Delta(0, 1) \neq 0$, weak spectral stability and weak refined dynamical stability are necessary for bounded $C_0^\infty \to L^p$, viscous stability in all dimensions $d \geq 2$, for any $1 \leq p \leq \infty$, and for profiles of any type.*

Theorem 2.2 ([Z.3, Z.4]) *Given (A1) - (A2), (H0) - (H5), strong spectral stability plus structural and strong refined dynamical stability are sufficient for asymptotic $L^1 \cap H^s \to H^s \cap L^p$ viscous stability of pure, Lax-type profiles in dimensions $d \geq 3$, for $q(d) \leq s \leq q$, q and $q(d)$ as defined in (H0), and any $2 \leq p \leq \infty$, or linearized viscous stability in dimensions $d \geq 2$, for $1 \leq s \leq q$ and $2 \leq p \leq \infty$, with rate of decay (in either case)*

$$|U(t) - \bar{U}|_{H^s} \leq C(1+t)^{-(d-1)/4+\epsilon}|U(0) - \bar{U}|_{L^1 \cap H^s} \tag{2.18}$$

approximately equal to that of a $(d-1)$-dimensional heat kernel, for any fixed $\epsilon > 0$ and $|U(0) - \bar{U}|_{L^1 \cap H^s}$ sufficiently small. Strong spectral plus structural and strong (inviscid, not refined) dynamical stability are sufficient for nonlinear stability in all dimensions $d \geq 2$, with rate of decay (2.18), $\epsilon = 0$, exactly *equal to that of a $(d-1)$-dimensional heat kernel. Similar results hold for profiles of nonclassical, over- or undercompressive type, as described in [Z.3, Z.4].*

Remarks 2.18 1. Rate (2.18) with $\epsilon = 0$ is sharp for weakly inviscid stable shocks, as shown by the scalar case [Go.3, GM, HoZ.1, HoZ.2]. (As pointed out by Majda, scalar shocks are always weakly inviscid stable [M.1, M.2, M.3, ZS, Z.3].)

2. Given (A1)–(A2), (H0)–(H3), $\gamma\Delta(0,1) \neq 0$ is necessary for one-dimensional viscous stability [ZH, MaZ.3] with respect to C_0^∞ perturbations. (Note: $C_0^\infty(x_1) \not\subset C_0^\infty(x)$.)

3. Under (A1)–(A2), (H0)–(H3) plus the mild additional assumptions that the principal characteristic speed $a_p(\xi)$ be simple, genuinely nonlinear, and strictly convex (resp. concave) with respect to $\tilde\xi$ at $\xi = (\xi_1, \tilde\xi) = (1, 0_{d-1})$, the sufficient stability conditions of Theorem 2.2 are satisfied for sufficiently small-amplitude shock profiles [FreS, PZ]. On the other hand, the necessary conditions of Theorem 2.1 are known to fail for certain large-amplitude shock profiles under (A1)–(A2), (H0)–(H3) [GZ, FreZ, ZS, Z.3, Z.4].

4. The boundary between the necessary conditions of Theorem 2.1 and the sufficient conditions of Theorem 2.2 is generically of codimension one (exercise; see [Z.4], Remark 1.22.2). Thus, transition from viscous stability to instability is in principle determined, in contrast to the situation of the inviscid case (Remark 2.9).

5. It is an important open problem which of the conditions of refined dynamical and structural stability in practice determines the transition from viscous stability to instability as shock strength is varied; see [MaZ.4, Z.3, Z.4] for further discussion.

3 Analytical Preliminaries

The rest of this article is devoted to the proof of Theorems 2.1 and 2.2. We begin in this section by assembling some needed background results.

3.1. Profile facts. Consider the standing wave ODE

$$
\begin{aligned}
(F^1)^I(U)' &= 0, \\
(F^1)^{II}(U)' &= ((B^1)^{II}U')',
\end{aligned}
\tag{3.1}
$$

$U^I, (F^1)^I \in \mathbb{R}^{n-r}, U^{II}, (F^1)^{II} \in \mathbb{R}^r$, where "'" denotes d/dx and superscripts I and II refer respectively to first- and second-block rows. Integrating from $-\infty$ to x gives an implicit first-order system

$$
(B^1)^{II}U' = (F^1)^{II}(U) - (F^1)^{II}(U_-)
\tag{3.2}
$$

on the r-dimensional level set

$$
(F^1)^I(U) \equiv (F^1)^I(U_-).
\tag{3.3}
$$

Lemma 3.1 ([MaZ.3]) *Given (A1), the weak version* $\det A_{11}^1 \neq 0$ *of (H1) is equivalent to the property that (3.3) determines a nondegenerate r-dimensional manifold on which (3.2) determines a nondegenerate first-order ODE. Moreover, assuming (A1)–(A2) and $\det A_{11}^1 \neq 0$, (H2) is equivalent to hyperbolicity of rest states $U_\pm$.*

Remark 3.2 This result motivated the introduction of the weak form $\det A_{11}^1 \neq 0$ of (H1) in [Z.3, MaZ.3]; the importance of the strong form $A_{11}^1 > 0$ (< 0) was first pointed out in [MaZ.4].

Corollary 3.3 ([MaZ.3]) *Given (A1)–(A2), and (H0)–(H2)*

$$|\partial_x^k (\bar{U} - U_\pm)| \leq C e^{-\theta|x|}, \qquad x \gtrless 0, \tag{3.4}$$

for $0 \leq k \leq q+2$, where q is as defined in (H0).

Proof of Lemma 3.1. This result was proved in somewhat greater generality in [MaZ.3]. Here, we give a simpler proof making use of structure (A1)–(A2). From (A1), it can be shown (exercise; see [Z.4], appendix A1) that $\partial U / \partial W$ is lower block-triangular and $(B^1)^{II}(\partial U / \partial W) = \text{block-diag}\,\{0, \hat{b}\}$, $\hat{b}$ nonsingular, with $\partial(F^1)^I / \partial w^I \neq 0$ if and only if $\tilde{A}_{11}^1 \neq 0$, with $\tilde{A}^1$ as in (A1). Considering (3.3)–(3.2) with respect to coordinate W, we readily obtain the first assertion, with resulting ODE

$$\hat{b}(w^{II})' = (F^1)^{II}(U(w^I(w^{II}), w^{II}) - (F^1)(U_-), \tag{3.5}$$

where $w^I(w^{II})$ is determined by the Implicit Function Theorem using relation $(F^1)^I(U(W)) \equiv (F^1)^I(U_-)$.

Linearizing (3.5) about rest point $W = W_\pm$ and using the Implicit Function Theorem to calculate $\partial w^I / \partial w^{II}$, we obtain

$$(w^{II})' = M_\pm w^{II} := (\hat{b})^{-1}(-\alpha_{21}\alpha_{11}^{-1}\alpha_{12} + \alpha_{22})_\pm w^{II}, \tag{3.6}$$

where $\alpha_\pm := (\partial F^1 / \partial W)_\pm$, and $\hat{b}_\pm$ are evaluated at $W_\pm$.

Noting that $\det(-\alpha_{21}\alpha_{11}^{-1}\alpha_{12} + \alpha_{22}) = \det \alpha / \det \alpha_{11}$, with $\alpha_{11} \neq 0$ as a consequence of $\det \tilde{A}_{11}^1 \neq 0$, we find that the coefficient matrix $M_\pm$ of the linearized ODE has zero eigenvalues if and only if $\det \alpha_\pm$, or equivalently $\det A_\pm^1$ vanishes. On the other hand, existence of nonzero pure imaginary eigenvalues $i\xi_1$ would imply, going back to the original equation (1.27) linearized about $W_\pm$, that $\det(i\xi_1 \tilde{A}^1 - \xi_1^2 \tilde{B}^{11})_\pm = 0$ for some real $\xi_1 \neq 0$. But, this is precluded by (A1)–(A2) (nontrivial exercise; see (K3), Lemma 3.18, below). Thus, the coefficient matrix M has a center subspace if and only if (H2) fails, verifying the second assertion.

Proof of Corollary 3.3. Standard ODE estimates.

For general interest, and to show that our assumptions on the profile are not vacuous, we include without proof the following generalization of a result of [MP] in the strictly parabolic case, relating structure of a viscous profile to its hyperbolic type.

Lemma 3.4 ([MaZ.3]) *Let d_+ denote the dimension of the stable subspace of M_+ and d_- the dimension of the unstable subspace of M_-, $M_\pm$ defined as in (3.6). Then, existence of a connecting profile $\bar{U}$ together with (A1)–(A2) and (H1)–(H2), implies that*

$$\hat{\ell} := d_+ + d_- - r = i_+ + i_- - n, \qquad (3.7)$$

where i_+ denotes the dimension of the stable subspace of A_+^1 and i_- denotes the dimension of the unstable subspace of A_-^1; in the Lax case, $\hat{\ell} = 1$.

Remark 3.5 $\hat{\ell} = \ell$ in the case of a transverse profile, ℓ as in (H3), whereas $i_+ + i_- - n = 1$ may be recognized as the Lax characteristic condition (L). Assumption 2.8 could be rephrased in this case as $\hat{\ell} = 1 = i_+ + i_- - n$.

Proof. See [MaZ.3], Appendix A1, or [Z.4], Appendix A2.   ●

3.2. Spectral resolution formulae.

Consider the linearized equations

$$U_t = LU := -\sum_j (A^j U)_{x_j} + \sum_{jk} (B^{jk} U_{x_k})_{x_j}, \qquad U(0) = U_0, \qquad (3.8)$$

where

$$B^{jk} := B^{jk}(\bar{U}(x_1)), \qquad A^j v := dF^j(\bar{U}(x_1))v - (dB^{j1}(\bar{U}(x_1))v)\bar{U}'(x_1). \quad (3.9)$$

Since the coefficients depend only on x_1, it is natural to Fourier-transform in $\tilde{x}$, $x = (x_1, \ldots, x_d) =: (x_1, \tilde{x})$, to reduce to a family of partial differential equations (PDE)

$$\hat{U}_t = L_{\tilde{\xi}}\hat{U} := \overbrace{(B^{11}\hat{U}')' - (A^1\hat{U})'}^{L_0\hat{U}} - i\sum_{j\neq 1} A^j \xi_j \hat{U} + i\sum_{j\neq 1} B^{j1}\xi_j \hat{U}'$$

$$+ i\sum_{k\neq 1}(B^{1k}\xi_k \hat{U})' - \sum_{j,k\neq 1} B^{jk}\xi_j \xi_k \hat{U}, \qquad \hat{U}(0) = \hat{U}_0 \qquad (3.10)$$

in (x_1, t) indexed by frequency $\tilde{\xi} \in \mathbb{R}^{d-1}$, where "$'$" denotes $\partial/\partial x_1$ and $\hat{U} = \hat{U}(x_1, \tilde{\xi}, t)$ denotes the Fourier transform of $U = U(x, t)$.

Finally, taking advantage of autonomy of the equations, we may take the Laplace transform in t to reduce, formally, to the resolvent equation

$$(\lambda - L_{\tilde{\xi}})\hat{\hat{U}} = \hat{U}_0, \qquad (3.11)$$

where $\hat{\hat{U}}(x_1, \tilde{\xi}, \lambda)$ denotes the Laplace–Fourier transform of $U = U(x, t)$ and $\hat{U}_0(x_1)$ denotes the Fourier transform of initial data $U_0(x)$. A system of ODE in x_1 indexed by $(\tilde{\xi}, \lambda)$, (3.11) may be estimated sharply using either explicit representation formulae/variation of constants for systems of ODE or Kreiss symmetrizer techniques; see, e.g., [Z.3, Z.4] and [GMWZ.2], respectively. The following proposition makes rigorous sense of this procedure.

Proposition 3.6 ([MaZ.3, Z.4]) *Given (A1), L generates a C^0 semigroup $|e^{Lt}| \le Ce^{\gamma_0 t}$ on L^2 with domain $\mathcal{D}(L) := \{U : U, LU \in L^2\}$ (Lu defined in the distributional sense), satisfying the generalized spectral resolution (inverse Laplace–Fourier transform) formula*

$$e^{Lt} f(x) = \text{P.V.} \int_{\gamma - i\infty}^{\gamma + i\infty} \int_{\mathbb{R}^{d-1}} e^{i\tilde{\xi} \cdot \tilde{x} + \lambda t} (\lambda - L_{\tilde{\xi}})^{-1} \hat{f}(x_1, \tilde{\xi}) \, d\tilde{\xi} d\lambda \qquad (3.12)$$

for $\gamma > \gamma_0$, $t \ge 0$, and $f \in \mathcal{D}(L)$, where $\hat{f}$ denotes Fourier transform of f. Likewise,

$$\int_0^t e^{L(t-s)} f(s) \, ds = \text{P.V.} \int_{\gamma - i\infty}^{\gamma + i\infty} \int_{\mathbb{R}^{d-1}} e^{i\tilde{\xi} \cdot \tilde{x} + \lambda t} (\lambda - L_{\tilde{\xi}})^{-1} \widehat{\widehat{f^T}}(x_1, \tilde{\xi}, \lambda) \, d\tilde{\xi} d\lambda$$
$$(3.13)$$

for $\gamma > \gamma_0$, $0 \le t \le T$, and $f \in L^1([0,T]; \mathcal{D}(L))$, where $\hat{g}$ denotes Laplace–Fourier transform of g and

$$f^T(x, s) := \begin{cases} f(x, s) & \text{for } 0 \le s \le T \\ 0 & \text{otherwise.} \end{cases} \qquad (3.14)$$

Remark 3.7 Bound (3.13) concerns the inhomogeneous, zero-initial-data problem $U_t - LU = f$, $U(0) = 0$. Provided that L generates a C^0 semigroup e^{LT}, the "mild", or semigroup solution of this equation is defined (see, e.g., [Pa], pp. 105–110) by Duhamel formula

$$U(t) = \int_0^t e^{L(t-s)} f(s) \, ds. \qquad (3.15)$$

This corresponds to a solution in the usual weak, or distributional sense, with somewhat stronger regularity in t. Formally, $\hat{V} := (\lambda - L)^{-1} \widehat{\widehat{f^T}}$ describes the Laplace–Fourier transform of the solution of the truncated-source problem $V_t - LV = f^T$, $V(0) = 0$. Evidently, $U(s) = V(s)$ for $0 \le s \le T$.

Proof. Let $u_n \to u$ and $Lu_n \to f$, $u_n \in \mathcal{D}(L)$, $u, f \in L^2$, "$\to$" denoting convergence in L^2 norm. Then, $Lu_n \rightharpoonup Lu$ by $u_n \to u$, where Lu is defined in the distributional sense, but also $Lu_n \rightharpoonup f$ by $Lu_n \to f$, whence $Lu = f$ by uniqueness of weak limits, and therefore $u \in \mathcal{D}(L)$. Moreover, $\mathcal{D}(L)$ is dense in L^2, since $C_0^\infty \subset \mathcal{D}(L)$. Thus, L is a closed, densely defined operator on L^2 with domain $\mathcal{D}(L)$; see Definition 6.1, Appendix A.

By standard semigroup theory, therefore (Proposition 6.11, Appendix A), L generates a C^0-semigroup $|e^{LT}| \le Ce^{\gamma_0 t}$ if and only if it satisfies resolvent estimate

$$|(\lambda - L)^{-k}|_{L^2} \le \frac{C}{|\lambda - \gamma_0|^k} \qquad (3.16)$$

for some uniform $C > 0$, for all real $\lambda > \gamma_0$ and all $k \ge 1$, or equivalently

$$|U|_{L^2} \leq \frac{C|(\lambda - L)^k U|_{L^2}}{|\lambda - \gamma_0|^k} \tag{3.17}$$

for all $k \geq 1$ and

$$\text{range}(\lambda - L) = L^2. \tag{3.18}$$

In the present case, the latter condition is superfluous, since it is easily verified by results of Henry [He, BSZ] that $\lambda - L$ is Fredholm for sufficiently large real λ. More generally, (3.18) may be verified by a corresponding estimate

$$|U|_{L^2} \leq \frac{C|(\lambda - L^*)U|_{L^2}}{|\lambda - \gamma_0|} \tag{3.19}$$

on the adjoint operator L^*, equivalent to (3.18) given (3.17) (or, given (3.18), to (3.17), $k = 1$) and often available (as here) by the same techniques; see Corollary 6.17, Appendix A.

A sufficient condition for (3.17), and the standard means by which it is proved, is

$$|W|_{L^2} \leq \frac{|S(\lambda - L)S^{-1}W|_{L^2}}{|\lambda - \gamma_0|} \tag{3.20}$$

for all real $\lambda > \gamma_0$, for some uniformly invertible transformation S: in this case the change of coordinates $S : U \to W$ defined by

$$W := (\tilde{A}^0)^{1/2}(\partial W/\partial U)(\bar{U}(x_1))U. \tag{3.21}$$

Once (3.17)–(3.18) have been established, we obtain by standard properties of C^0 semigroups the bounds $|Se^{Lt}S^{-1}| \leq e^{\gamma_0 t}$ and $|e^{Lt}| \leq Ce^{\gamma_0 t}$ and also the inverse Laplace transform formulae

$$e^{Lt}f(x) = \text{P.V.} \int_{\gamma - i\infty}^{\gamma + i\infty} e^{\lambda t}(\lambda - L)^{-1}f \, d\lambda \tag{3.22}$$

for $f \in \mathcal{D}(L)$ and

$$\int_0^T e^{L(T-t)}f(x) = \text{P.V.} \int_{\gamma - i\infty}^{\gamma + i\infty} e^{\lambda t}(\lambda - L)^{-1}\widehat{f^T}(\lambda) \, d\lambda \tag{3.23}$$

for $f \in L^1([0, T]; \mathcal{D}(L))$, for any $\gamma > \gamma_0$, where $\hat{f}$ denotes Laplace transform of f; see Propositions 6.11, 6.24, and 6.25, Appendix A. Formulae (3.12)–(3.13) then follow from (3.22)–(3.23) by Fourier transform/distribution theory. Thus, it remains only to verify (3.20) in order to complete the proof.

To establish (3.20), rewrite resolvent equation $(\lambda - L)U = f$ in W-coordinates, as

$$\lambda \tilde{A}^0 W + \sum_j \tilde{A}^j W_{x_j} + \sum_{jk}(\tilde{B}^{jk}W_{x_k})_{x_j} + \tilde{C}W + \sum_j \tilde{D}^j w_{x_j}^{II} = \tilde{A}^0 \tilde{f}, \tag{3.24}$$

where $\tilde{f} := (\partial W / \partial U)(\bar{U}(x_1)) f$, C is uniformly bounded, and coefficients $\tilde{A}^j$, $\tilde{B}^{jk}$ satisfy (A1) with $\tilde{G} \equiv 0$. (Exercise: check that properties (A1) are preserved up to lower-order terms $\tilde{C} W + \sum_j \tilde{D}^j w^{II}_{x_j}$, using commutation of coordinate-change and linearization.) Using the block-diagonal form of $\tilde{A}^0$, we arrange by coordinate change $W \to (\tilde{A}^0)^{1/2} W$ if necessary that $\tilde{A}^0 = I$ without loss of generality; note that lower-order commutator terms arising through change of coordinates are again of the form $\tilde{C} W + \sum_j \tilde{D}^j w^{II}_{x_j}$, hence do not change the structure asserted in (3.24).

Taking the real part of the complex inner product of W against (3.24), and using (1.28), (1.30) together with symmetry of $\tilde{A}^j_{11}$ terms and Young's inequality, similarly as in the proof of Proposition 1.16, we obtain

$$\Re \lambda |W|_{L^2} + \theta |w^{II}|^2_{H^1} \leq \gamma_0 |W|^2_{L^2} + |\tilde{f}|_{L^2} |W|_{L^2} \qquad (3.25)$$

for $\gamma_0 > 0$ sufficiently large. Dropping the favorable w^{II} term and dividing both sides by $|W|_{L^2}$, we obtain

$$(\Re \lambda - \gamma_0) |W|_{L^2} \leq |\tilde{f}|_{L^2} \qquad (3.26)$$

Noting that $S(\lambda - L) S^{-1} W = \tilde{f}$, we are done. $\qquad \bullet$

Remark 3.8 The properties and techniques used to establish Proposition 3.6 are linearized versions of the ones used to establish Proposition 1.16. For a still more direct link, see Remark 6.18 and Exercise 6.22, Appendix A, concerning the Lumer–Phillips Theorem, which assert that, in the favorable W-coordinates, existence of a C^0 contraction semigroup $|e^{Lt}| \leq e^{\gamma_0 t}$ is equivalent to the a priori estimate

$$(d/dt)(1/2)|W|^2_{L^2} = \Re \langle u, Lu \rangle \leq \gamma_0 |W|^2, \qquad (3.27)$$

the autonomous version of (1.17). That is, Proposition 3.6 essentially concerns local linearized well-posedness, with no global stability properties either asserted or required so far.

The semigroup framework, besides validating the spectral resolution formulae, gives also a convenient means to generate solutions of the nonautonomous linear equations arising in the nonlinear iteration schemes described in Section 1, based only on the already-verified instantaneous version of (3.27); specifically, Proposition 6.31, Appendix A, assuming (3.27), guarantees a solution satisfying $|u(t)|_{L^2} \leq e^{\gamma_0 t} |u(0)|_{L^2}$. A standard way of constructing the autonomous semigroup is by semidiscrete approximation using the first-order implicit Euler scheme (see Remark 6.12), the nonautonomous solution operator then being approximated by a concatenation of frozen-coefficient semigroup solutions on each mesh block; see [Pa]. It is interesting to compare this approach to Friedrichs' original construction of solutions by finite difference approximation [Fr].

3.3. Asymptotic ODE theory: the gap and conjugation lemmas.
Consider a general family of first-order ODE

$$\mathbb{W}' - \mathbb{A}(x, \Lambda)\mathbb{W} = \mathbb{F} \tag{3.28}$$

indexed by a spectral parameter $\Lambda \in \Omega \subset \mathbb{C}^m$, where $W \in \mathbb{C}^N$, $x \in \mathbb{R}$ and "'" denotes d/dx.

Examples 3.9 1. Eigenvalue equation $(L_{\tilde{\xi}} - \lambda)U = 0$, written in phase coordinates $\mathbb{W} := (W, (w^{II})')$, $W := (\partial W/\partial U)(\bar{U}(x))U$, with $\Lambda := (\tilde{\xi}, \lambda)$ and $\mathbb{F} := 0$.

2. Resolvent equation $(L_{\tilde{\xi}} - \lambda)U = f$, written in phase coordinates $\mathbb{W} := (W, (w^{II})')$, $W := (\partial W/\partial U)(\bar{U}(x))U$, and $|\mathbb{F}| = \mathcal{O}(|f|)$.

Assumption 3.10

(h0) *Coefficient $\mathbb{A}(\cdot, \Lambda)$, considered as a function from Ω into $C^0(x)$ is analytic in Λ. Moreover, $\mathbb{A}(\cdot, \Lambda)$ approaches exponentially to limits $\mathbb{A}_{\pm}$ as $x \to \pm\infty$, with uniform exponential decay estimates*

$$|(\partial/\partial x)^k (\mathbb{A} - \mathbb{A}_{\pm})| \leq C_1 e^{-\theta|x|/C_2}, \quad \text{for } x \gtrless 0, \, 0 \leq k \leq K, \tag{3.29}$$

C_j, $\theta > 0$, *on compact subsets of Ω.*

Lemma 3.11 (The gap lemma [KS, GZ, ZH])
Consider the homogeneous version $\mathbb{F} \equiv 0$ of (3.28), under assumption (h0). If $V^-(\Lambda)$ is an eigenvector of $\mathbb{A}_-$ with eigenvalue $\mu(\Lambda)$, both analytic in Λ, then there exists a solution of (3.28) of form

$$\mathbb{W}(\Lambda, x) = V(x, \Lambda)e^{\mu(\Lambda)x}, \tag{3.30}$$

where V is C^1 in x and locally analytic in Λ and, for any fixed $\bar{\theta} < \theta$, satisfies

$$V(x, \Lambda) = V^-(\Lambda) + \mathbf{O}(e^{-\bar{\theta}|x|}|V^-(\Lambda)|), \quad x < 0. \tag{3.31}$$

Proof. Setting $\mathbb{W}(x) = e^{\mu x}V(x)$, we may rewrite $\mathbb{W}' = \mathbb{A}\mathbb{W}$ as

$$V' = (\mathbb{A}_- - \mu I)V + \theta V, \qquad \theta := (\mathbb{A} - \mathbb{A}_-) = \mathbf{O}(e^{-\theta|x|}), \tag{3.32}$$

and seek a solution $V(x, \Lambda) \to V^-(x)$ as $x \to \infty$. Choose $\bar{\theta} < \theta_1 < \theta$ such that there is a spectral gap $|\Re(\sigma\mathbb{A}_- - (\mu + \theta_1))| > 0$ between $\sigma\mathbb{A}_-$ and $\mu + \theta_1$. Then, fixing a base point Λ_0, we can define on some neighborhood of Λ_0 to the complementary $\mathbb{A}_-$-invariant projections $P(\Lambda)$ and $Q(\Lambda)$ where P projects onto the direct sum of all eigenspaces of $\mathbb{A}_-$ with eigenvalues $\tilde{\mu}$ satisfying $\Re(\tilde{\mu}) < \Re(\mu) + \theta_1$, and Q projects onto the direct sum of the remaining eigenspaces, with eigenvalues satisfying $\Re(\tilde{\mu}) > \Re(\mu) + \theta_1$. By basic matrix perturbation theory (eg. [Kat]) it follows that P and Q are analytic in a neighborhood of Λ_0, with

$$\left| e^{(\mathbb{A}_- - \mu I)x} P \right| \leq C(e^{\theta_1 x}), \quad x > 0, \qquad \left| e^{(\mathbb{A}_- - \mu I)x} Q \right| \leq C(e^{\theta_1 x}), \quad x < 0. \tag{3.33}$$

It follows that, for $M > 0$ sufficiently large, the map $\mathcal{T}$ defined by

$$\mathcal{T}V(x) = V^- + \int_{-\infty}^{x} e^{(\mathbb{A}_- - \mu I)(x-y)} P\theta(y)V(y)dy \tag{3.34}$$
$$- \int_{x}^{-M} e^{(\mathbb{A}_- - \mu I)(x-y)} Q\theta(y)V(y)dy$$

is a contraction on $L^\infty(-\infty, -M]$. For, applying 3.33, we have

$$|\mathcal{T}V_1 - \mathcal{T}V_2|_{(x)} \leq C|V_1 - V_2|_\infty \left(\int_{-\infty}^{x} e^{\theta_1(x-y)} e^{\theta y} dy + \int_{x}^{-M} e^{\theta_1(x-y)} e^{\theta y} dy \right)$$
$$\leq C_1|V_1 - V_2|_\infty \left(e^{\theta_1 x} e^{(\theta - \theta_1)y} |_{-\infty}^{x} + e^{\theta_1 x} e^{(\theta - \theta_1)y} |_{x}^{-M} \right)$$
$$\leq C_2|V_1 - V_2|_\infty e^{-\bar\theta M} < \frac{1}{2}|V_1 - V_2|_\infty. \tag{3.35}$$

By iteration, we thus obtain a solution $V \in L^\infty(-\infty, -M]$ of $V = \mathcal{T}V$ with $V \leq C_3|V^-|$; since $\mathcal{T}$ clearly preserves analyticity $V(\Lambda, x)$ is analytic in Λ as the uniform limit of analytic iterates (starting with $V_0 = 0$). Differentiation shows that V is a bounded solution of $V = \mathcal{T}V$ if and only if it is a bounded solution of 3.32 (exercise). Further, taking $V_1 = V$, $V_2 = 0$ in 3.35, we obtain from the second to last inequality that

$$|V - V^-| = |\mathcal{T}(V) - \mathcal{T}(0)| \leq C_2 e^{\bar\theta x}|V| \leq C_4 e^{\bar\theta x}|V^-|, \tag{3.36}$$

giving 3.31. Analyticity, and the bounds 3.31, extend to $x < 0$ by standard analytic dependence for the initial value problem at $x = -M$.   •

Remark 3.12 The title "gap lemma" alludes to the fact that we do not make the usual assumption of a spectral gap between $\mu(\Lambda)$ and the remaining eigenvalues of $\mathbb{A}_-$, as in standard results on asymptotic behavior of ODE [Co]; that is, the lemma asserts that exponential decay of $\mathbb{A}$ can substitute for a spectral gap. Note also that we require only analyticity of μ and not its associated eigenprojection Π_μ, allowing crossing eigenvalues of arbitrary type (recall, Π_μ is analytic only if μ is semisimple; indeed, Π_μ blows up at a nontrivial Jordan block [Kat]). This is important in the following application; see Exercise 3.16 below.

Corollary 3.13 (The conjugation lemma [MeZ.1])
Given (h0), there exist locally to any given $\Lambda_0 \in \Omega$ invertible linear transformations $P_+(x, \Lambda) = I + \Theta_+(x, \Lambda)$ and $P_-(x, \Lambda) = I + \Theta_-(x, \Lambda)$ defined on $x \geq 0$ and $x \leq 0$, respectively, $\Phi_\pm$ analytic in Λ as functions from Ω to $C^0[0, \pm\infty)$, such that:

(i) *For any fixed $0 < \bar{\theta} < \theta$ and $0 \le k \le K+1$, $j \ge 0$,*

$$|(\partial/\partial\Lambda)^j (\partial/\partial x)^k \Theta_\pm| \le C(j) C_1 C_2 e^{-\theta|x|/C_2} \quad \text{for } x \gtrless 0. \tag{3.37}$$

(ii) *The change of coordinates $\mathbb{W} =: P_\pm \mathbb{Z}$, $\mathbb{F} =: P_\pm \mathbb{G}$ reduces (3.28) to*

$$\mathbb{Z}' - \mathbb{A}_\pm \mathbb{Z} = \mathbb{G} \quad \text{for } x \gtrless 0. \tag{3.38}$$

Equivalently, solutions of (3.28) may be factored as

$$\mathbb{W} = (I + \Theta_\pm) \mathbb{Z}_\pm, \tag{3.39}$$

where $\mathbb{Z}_\pm$ satisfy the limiting, constant-coefficient equations (3.38) and $\Theta_\pm$ satisfy bounds (3.37).

Proof. Substituting $\mathbb{W} = P_- Z$ into (3.28), equating to (3.38), and rearranging, we obtain the defining equation

$$P'_- = \mathbb{A}_- P_- - P_- \mathbb{A}, \qquad P_- \to I \quad \text{as} \quad x \to -\infty. \tag{3.40}$$

Viewed as a vector equation, this has the form $P'_- = \mathcal{A} P_-$, where $\mathcal{A}$ approaches exponentially as $x \to -\infty$ to its limit $\mathcal{A}_-$, defined by

$$\mathcal{A}_- P := \mathbb{A}_- P - P \mathbb{A}_-. \tag{3.41}$$

The limiting operator $\mathcal{A}_-$ evidently has analytic eigenvalue, eigenvector pair $\mu \equiv 0$, $P_- \equiv I$, whence the result follows by Lemma 3.11 for $j = k = 0$. The x-derivative bounds $0 < k \le K+1$ then follow from the ODE and its first K derivatives, and the Λ-derivative bounds from standard interior estimates for analytic functions. A symmetric argument gives the result for P_+.   ●

Remark 3.14 Equation (3.39) gives an explicit connection to the inviscid, bi-constant-coefficient case, for bounded frequencies $|(\tilde{\xi}, \lambda)| \le R$ (low frequency $\sim$ inviscid regime). For high-frequencies, the proper analogy is rather to the frozen-coefficient case of local existence theory.

Exercise 3.15 ([B]) *Use Duhamel's formula to show that*

$$PV(x) = V^- + \int_{-\infty}^{x} e^{(\mathbb{A}_- - \mu I)(x-y)} P\theta(y) V(y) dy \tag{3.42}$$

and

$$QV(x) = QV(-M) + \int_{-M}^{x} e^{(\mathbb{A}_- - \mu I)(x-y)} Q\theta(y) V(y) dy, \tag{3.43}$$

hence $V(x) = \mathcal{T}V(x)$ for the unique solution V of 3.32 determined by conditions $V(-\infty) = V_- = PV_-$ and $QV(-M) = 0$.

Exercise 3.16 *(i) If $\{r_j\}$ and $\{l_k\}$ are dual bases of (possibly generalized) right and left eigenvectors of $\mathbb{A}_-$, with associated eigenvalues μ_j and μ_k, show that $\{r_j l_k^*\}$ is a basis of (possibly generalized) right eigenvectors of the operator $\mathcal{A}_-$ defined in (3.41), with associated eigenvalues $\mu_j - \mu_k$. (ii) Show that $\mu = 0$ is a semisimple eigenvalue of $\mathcal{A}_-$ if and only if each eigenvalue of $\mathbb{A}_-$ is semisimple.*

3.4. Hyperbolic–parabolic smoothing. We next introduce the circle of ideas associated with hyperbolic–parabolic smoothing and estimate (1.40).

Lemma 3.17 ([Hu, MaZ.5]) *Let $A = $ block-diag $\{a_j I_{m_j}\}$ be diagonal, with real entries a_j appearing with prescribed multiplicities m_j in order of increasing size, and let B be arbitrary. Then, there exists a smooth skew-symmetric matrix-valued function $K(A, B)$ such that*

$$\text{Re } (B - KA) = \Re\text{block-diag } B, \tag{3.44}$$

where block-diag B denotes the block-diagonal part of B, with blocks of dimension m_j equal to the multiplicity of the corresponding eigenvalues of A.

Proof. It is straightforward to check that the symmetric matrix $\Re KA = (1/2)(KA - A^t K)$ may be prescribed arbitrarily on off-diagonal blocks, by setting $K_{ij} := (a_i - a_j)^{-1} M_{ij}$, where M_{ij} is the desired block, $i \neq j$. Choosing $M = \Re B$, we obtain $\Re(B - KA) = \Re\text{block-diag } (B)$ as claimed.   •

Lemma 3.18 ([KSh]) *Let $\tilde{A}^0$, A, and B denote real-valued matrices such that $\tilde{A}^0$ is symmetric positive definite and $\tilde{A} := \tilde{A}^0 A$ and $\tilde{B} := \tilde{A}^0 B$ are symmetric, $\tilde{B} \geq 0$. Then, the following are equivalent:*

(K0) (Genuine coupling) No eigenvector of A lies in $\ker B$ (equivalently, in $\ker \tilde{B}$).

(K1) block-diag $LBR = R^t \tilde{B} R > 0$, where $L := \tilde{O}^t (\tilde{A}^0)^{1/2}$ and $R := (\tilde{A}^0)^{-1/2} \tilde{O}$ are matrices of left and right eigenvectors of A block-diagonalizing LAR, with O orthonormal. Here, as in Lemma 3.17, block-diag M denotes the matrix formed from the diagonal blocks of M, with blocks of dimension equal to the multiplicity of corresponding eigenvalues of LAR.

(K2) (hyperbolic compensation) There exists a smooth skew-symmetric matrix-valued function $K(\tilde{A}, \tilde{B}, \tilde{A}^0)$ such that

$$\text{Re } \left(\tilde{B} - KA \right) > 0. \tag{3.45}$$

(K3) (Strict dissipativity)

$$\Re\sigma(-i\xi A - |\xi|^2 B) \leq -\theta|\xi|^2/(1 + |\xi|^2), \quad \theta > 0. \tag{3.46}$$

Proof. (K0) $\Leftrightarrow$ (K1) by the property of symmetric nonnegative matrices M that $v^t M v = 0$ if and only if $M v = 0$ (exercise), which implies that block-diag $\{R_j^t \Re \tilde{B} R_j\}$ has a kernel if and only if $\alpha^t R_j^t \Re \tilde{B} R_j \alpha = 0$, if and only if $\Re \tilde{B} R_j \alpha = 0$, where R_j denotes a block of eigenvectors with common eigenvalue.

(K1) $\Rightarrow$ (K2) follows readily from Lemma 3.18, by first converting to the case of symmetric A and $B \leq 0$ by the transformations

$$A \to (\tilde{A}^0)^{1/2} A \tilde{A}^0)^{-1/2}, \quad B \to (\tilde{A}^0)^{1/2} B \tilde{A}^0)^{-1/2}, \tag{3.47}$$

from which the original result follows by observing that

$$M > 0 \Leftrightarrow (\tilde{A}^0)^{1/2} M (\tilde{A}^0)^{1/2} > 0, \tag{3.48}$$

then converting by an orthonormal change of coordinates to the case that A is diagonal and $B \leq 0$. Variable multiplicity eigenvalues may be handled by partition of unity/interpolation, noting that $\Re(B - KA) < 0$ persists under perturbation.

(K2) $\Rightarrow$ (K3) follows upon rearrangement of energy estimate

$$
\begin{aligned}
0 &= \Re \langle (C(1 + |\xi|^2) \tilde{A}^0 + i\xi K) w, (\lambda + i\xi A + |\xi|^2 B) w \rangle \\
&= \Re \lambda \langle w, \left(C(1 + |\xi|^2) \tilde{A}^0 + i\xi K \right) w \rangle + |\xi|^2 \langle w, \Re(\tilde{C} B - KA) w \rangle \\
&\quad + \Re \langle w, -i|\xi|^3 K (\tilde{A}^0)^{-1} \tilde{B} w \rangle + C |\xi|^4 \langle w, \tilde{B} w \rangle,
\end{aligned}
\tag{3.49}
$$

which yields

$$
\begin{aligned}
\Re \lambda \langle w, & \left(C(1 + |\xi|^2) \tilde{A}^0 + i\xi K \right) w \rangle \\
&\leq -\left(\theta |\xi|^2 |w|^2 + M|\xi|^3 (|w||\tilde{B} w| + (C/M)|\xi|^4 |\tilde{B} w|^2 \right) \\
&\leq -\theta |\xi|^2 |w|^2
\end{aligned}
\tag{3.50}
$$

and thereby

$$\Re \lambda (1 + |\xi|^2) |w|^2 \leq -\theta_1 |\xi|^2 |w|^2, \tag{3.51}$$

for $M, C > 0$ sufficiently large and $\theta, \theta_1 > 0$ sufficiently small, by positivity of $\tilde{A}^0$ and $\Re(\tilde{B} - KA)$.

Finally, (K3) $\Rightarrow$ (K1) follows by by first-order Taylor expansion at $\xi = 0$ of the spectrum of $LAR - i\xi LBR$ (well-defined, by symmetry of LAR), together with symmetry of $R^t \tilde{B} R$. $\bullet$

Corollary 3.19 *Under (A1)–(A2), there holds the uniform dissipativity condition*

$$\Re \sigma \left(\sum_j i\xi_j A^j - \sum_{j,k} \xi_j \xi_k B^{jk} \right)_{\pm} \leq -\theta |\xi|^2 / (1 + |\xi|^2). \tag{3.52}$$

Moreover, there exist skew-symmetric "compensating matrices" $K_\pm(\xi)$ *that are smooth and homogeneous degree one in* ξ, *such that*

$$\Re\Big(\sum_{j,k}\xi_j\xi_k\tilde{B}^{jk} - K(\xi)(\tilde{A}^0)^{-1}\sum_k\xi_k\tilde{A}^k\Big)_\pm \geq \theta > 0 \tag{3.53}$$

for all $\xi \in \mathbb{R}^d \setminus \{0\}$.

Proof. By the block–diagonal structure of $\tilde{B}^{jk}$ (GC) holds also for $A_\pm^j$ and $\hat{B}^{jk} := (\tilde{A}^0)^{-1}\Re\tilde{B}^{jk}$, since

$$\ker\sum_{j,k}\xi_j\xi_k\hat{B}^{jk} = \ker\sum_{j,k}\xi_j\xi_k\Re\tilde{B}^{jk} = \ker\sum\xi_j\xi_k\tilde{B}^{jk} = \ker\sum\xi_j\xi_k B^{jk}. \tag{3.54}$$

Applying Lemma 3.18 to

$$\tilde{A}^0 := \tilde{A}_\pm^0, \quad A := \Big((\tilde{A}^0)^{-1}\sum_k\xi_k\tilde{A}^k\Big)_\pm, \quad B := \Big((\tilde{A}^0)^{-1}\sum_{j,k}\xi_j\xi_k\Re\tilde{B}^{jk}\Big)_\pm, \tag{3.55}$$

we thus obtain (3.53) and

$$\Re\sigma\Big[(\tilde{A}^0)^{-1}\Big(-\sum_j i\xi_j\tilde{A}^j - \sum_{j,k}\xi_j\xi_k\Re\tilde{B}^{jk}\Big)\Big]_\pm \leq -\theta_1|\xi|^2/(1+|\xi|^2), \tag{3.56}$$

$\theta_1 > 0$, from which we readily obtain

$$\Big(-\sum_j i\xi_j\tilde{A}^j - \sum_{j,k}\xi_j\xi_k\tilde{B}^{jk}\Big)_\pm \leq -\theta_2|\xi|^2/(1+|\xi|^2) \tag{3.57}$$

and thus (3.52) (exercise, using $M > \theta_1 \Leftrightarrow (\tilde{A}^0)_\pm^{-1/2}M(\tilde{A}^0)_\pm^{-1/2} > \theta$ and $\sigma(\tilde{A}^0)_\pm^{-1/2}M(\tilde{A}^0)_\pm^{-1/2} > \theta \Leftrightarrow \sigma(\tilde{A}^0)_\pm^{-1}M > \theta$, together with $S > \theta \Leftrightarrow \sigma S > \theta$ for S symmetric). Because all terms other than K in the lefthand side of (3.53) are homogeneous, it is evident that we may choose $K(\cdot)$ homogeneous as well (restrict to the unit sphere, then take homogeneous extension). $\bullet$

3.4.1. Basic estimate. Energy estimate (3.49)–(3.51) may be recognized as the Laplace–Fourier transformed version of a corresponding time-evolutionary estimate

$$\begin{aligned}
(d/dt)(1/2)\Big(&C\langle\tilde{A}^0 W, W\rangle + C\langle\tilde{A}^0 W_x, W_x\rangle + \langle K\partial_x W, W\rangle\Big) \\
&= -\langle W_x, \Re(C\tilde{B} - K(\tilde{A}_0)^{-1}\tilde{A})W_x\rangle - \langle W_x, K(\tilde{A}^0)^{-1}\tilde{B}W_{xx}\rangle \\
&\quad - C\langle W_{xx}, \tilde{B}W_{xx}\rangle, \\
&\leq -\theta(|W_x|_{L^2}^2 + |\tilde{B}W_x|_{H^1}^2)
\end{aligned} \tag{3.58}$$

for the one-dimensional, linear constant-coefficient equation

$$\tilde{A}^0 W_t + \tilde{A} W_x = \tilde{B} W_{xx}, \tag{3.59}$$

which in the block-diagonal case $\tilde{B} = \text{block-diag}\ \{0, \tilde{b}\}$, $\tilde{b} > 0$, yields (1.40) for $s = 1$ by the observation that

$$\mathcal{E}(W) := (1/2)\Big(C\langle \tilde{A}^0 W, W\rangle + \langle K\partial_x W, W\rangle + C\langle \tilde{A}^0 W_x, W_x\rangle\Big) \tag{3.60}$$

for $C > 0$ sufficiently large determines a norm $\mathcal{E}^{1/2}(\cdot)$ equivalent to $|\cdot|_{H^1}$. This readily generalizes to

$$(d/dt)\mathcal{E}(t) \leq -\theta(|\partial_x W|^2_{H^{s-1}} + |\tilde{B}\partial_x W|^2_{H^s}), \tag{3.61}$$

$$\mathcal{E}(W) := (1/2)\sum_{r=0}^{s} c^{-r}\langle \partial_x^r W, \tilde{A}^0 \partial_x^r W\rangle + (1/2)\sum_{r=0}^{s-1} c^{-r-1}\langle K\partial_x^{r+1} W, \partial_x^r W\rangle,$$
$$\tag{3.62}$$

$c > 0$ sufficiently large, where $\mathcal{E}^{1/2}(\cdot)$ is a norm equivalent to $|\cdot|_{H^s}$, yielding (1.40) for $s \geq 1$. We refer to energy estimates of the general type (K3), (3.58)–(3.62) as "Kawashima-type" estimates.

Proposition 3.20 *Assuming (A1)–(A2), (H0) for $W_- = 0$, energy estimate (1.40) is valid for all $q(d) \leq s \leq q$, $q(d)$ and q as defined in (H0), so long as $|W|_{H^s}$ remains sufficiently small.*

Proof. In the linear, constant-coefficient case, (3.53) together with a calculation analogous to that of (3.58) and (3.61) yields

$$(d/dt)\mathcal{E}(t) \leq -\theta(|\partial_x W|^2_{H^{s-1}} + |\partial_x w^{II}|^2_{H^s}) \tag{3.63}$$

for

$$\mathcal{E}(W) := (1/2)\sum_{r=0}^{s} c^{-r}\langle \partial_x^r W, \tilde{A}^0 \partial_x^r W\rangle + (1/2)\sum_{r=0}^{s-1} c^{-r-1}\langle K_-(\partial_x)\partial_x^r W, \partial_x^r W\rangle,$$
$$\tag{3.64}$$

$c > 0$ sufficiently large, where operator $K_-(\partial_x)$ is defined by

$$\widehat{K_-(\partial x f)}(\xi) := iK_-(\xi)\hat{f}(\xi) \tag{3.65}$$

with $K_-(\cdot)$ as in (3.53), where $\hat{g}$ denotes Fourier transform of g, for all $s \geq 1$.

In the general (nonlinear, variable-coefficient) case, we obtain by a similar calculation

$$(d/dt)\mathcal{E}(t) \leq -\theta(|\partial_x W|^2_{H^{s-1}} + |\partial_x w^{II}|^2_{H^s}) + C|W|^2_{L^2} \tag{3.66}$$

for $C > 0$ sufficiently large, for

$$\mathcal{E}(W) := (1/2)\sum_{r=0}^{s} c^{-r}\langle \partial_x^r W, \tilde{A}^0(W)\partial_x^r W\rangle$$
$$+ (1/2)\sum_{r=0}^{s-1} c^{-r-1}\langle K_-(\partial_x)\partial_x^r W, \partial_x^r W\rangle, \tag{3.67}$$

$c > 0$ sufficiently large, provided $s \geq q(d) := [d/2] + 3$ and the coefficient functions possess sufficient regularity C^s, i.e., $s \leq q$.

Here, we estimate time-derivatives of $\tilde{A}^0$ terms in straightforward fashion, using integration by parts as in the proof of Proposition 1.6. We estimate terms involving the nondifferential operator $K_-(\partial_x)$ in the frequency domain as

$$(d/dt)\frac{1}{2}\langle K(\partial_x)\partial_x^r W, \partial_x^r W\rangle = (d/dt)\frac{1}{2}\langle iK(\xi)(i\xi)^r \hat{W}, (i\xi)^r \hat{W}\rangle$$
$$= \langle iK(\xi)(i\xi)^r \hat{W}, (i\xi)^r \hat{W}_t\rangle$$
$$= \langle (i\xi)^r \hat{W}, -K(\xi)(\tilde{A}^0_-)^{-1}\big(\sum_j \xi_j \tilde{A}^j_-\big)(i\xi)^r \hat{W}\rangle$$
$$+ \langle (iK(\xi)i\xi)^r \hat{W}, (i\xi)^r \hat{H}\rangle \tag{3.68}$$

using Plancherel's identity together with the equation, written in the frequency domain as

$$\hat{W}_t = -\sum_j i\xi_j(\tilde{A}^0_-)^{-1}\tilde{A}^j_-\hat{W} + \hat{H}, \tag{3.69}$$

where

$$H := \sum_j \big((\tilde{A}^0_-)^{-1}\tilde{A}^j_- - (\tilde{A}^0)^{-1}\tilde{A}^j(W)\big)W_{x_j}\big)$$
$$+ \sum_{j,k}(\tilde{A}^0)^{-1}(\tilde{B}^{jk}W_{x_k})_{x_j} + (\tilde{A}^0)^{-1}\tilde{G}(W_x, W_x)\big). \tag{3.70}$$

By a calculation similar to those in the proofs of Propositions 1.6 and 1.16 (exercise, using smallness of $|\tilde{A}^j_- - \tilde{A}^j(W)| \sim |W|$ and the Moser inequality (1.15)), we obtain

$$|\partial_x^r H|_{L^2} \leq C|\partial_x^{r+2}w^{II}|_{L^2} + C|W|_{H^{r+1}}\big(|W|_{H^{[d/2]+2}} + |W|^r_{H^{[d/2]+2}}\big). \tag{3.71}$$

Thus, using homogeneity, $|K(\xi)| \leq C|\xi|$, together with the Cauchy–Schwartz inequality and Plancherel's identity, we may estimate the final term on the righthand side of (3.68) as

$$\langle (iK(\xi)i\xi)^r \hat{W}, (i\xi)^r \hat{H}\rangle \leq C|\partial_x^{r+1}W|_{L^2}|\partial_x^r H|_{L^2}$$
$$\leq C|\partial_x^{r+1}W|_{L^2}|\partial_x^{r+2}w^{II}|_{L^2} + \epsilon|W|^2_{H^{r+1}}, \tag{3.72}$$

any $\epsilon > 0$, for $|W|_{H^{[d/2]+2}}$ sufficiently small: that is, a term of the same form arising in the constant-coefficient case plus an absorbable error.

Combining $\tilde{A}^0$- and $K(\xi)_-$-term estimates, and using (3.53) together with

$$\sum_{j,k}\langle\partial_x^r W_{x_j}, \tilde{B}^{jk}(W)\partial_x^r W_{x_k}\rangle = \langle(i\xi)^r\hat{W}, \sum_{j,k}\xi_j\xi_k\tilde{B}_-^{jk}(i\xi)^r\hat{W}\rangle$$

$$+ \mathcal{O}(|\partial_x^{r+1}W|_{L^2}^2|W|_{L^\infty})$$

$$\geq \langle(i\xi)^r\hat{W}, \sum_{j,k}\xi_j\xi_k\tilde{B}_-^{jk}(i\xi)^r\hat{W}\rangle - \epsilon|\partial_x^{r+1}W|_{L^2}^2$$

$$(3.73)$$

for any $\epsilon > 0$, for $|W|_{L^\infty}$ sufficiently small, we obtain the result, similarly as in the constant-coefficient case.   •

Exercise 3.21 *(Alternative proof) Show using (1.34) together with (3.68)–(3.73) that*

$$(d/dt)\mathcal{E} \leq C|W|_{H^{s-1}}^2 - \theta_1(|\partial_x^s W|_{L^2}^2 + |\partial_x^{s+1}w^{II}|_{L^2}) \tag{3.74}$$

for $C > 0$ sufficiently large and $\theta_1 > 0$ sufficiently small, for

$$\mathcal{E}(W) := (1/2)\sum_{r=0}^s\langle\partial_x^r W, \tilde{A}^0\partial_x^r W\rangle + (1/2C)\langle K_-(\partial_x)\partial_x^{s-1}W, \partial_x^{s-1}W\rangle. \tag{3.75}$$

Using the H^s interpolation formula

$$|f|_{H^{s_*}}^2 \leq \beta C^{1/\beta}|f|_{H^{s_1}}^2 + (1-\beta)C^{-1/(1-\beta)}|f|_{H^{s_2}}^2, \quad \beta = (s_2 - s_*)/(s_2 - s_1), \tag{3.76}$$

valid for $s_1 \leq s_* \leq s_2$, taking $s_1 = 0$, $s_2 = s$, and C sufficiently large, show that (3.74) implies (3.66).

3.4.2. Alternative formulation. A basic but apparently new observation is that energy estimate (3.67) implies not only (1.40) but also the following considerably stronger estimate.

Proposition 3.22 *Under the assumptions of Proposition 3.20,*

$$|W(t)|_{H^s}^2 \leq C\left(e^{-\theta_1 t}|W(0)|_{H^s}^2 + \int_0^t e^{-\theta_1(t-s)}|W(s)|_{L^2}^2 \, ds\right), \qquad \theta_1 > 0, \tag{3.77}$$

so long as $|W|_{H^s}$ remains sufficiently small.

Proof. Rewriting (3.66) using equivalence of $\mathcal{E}^{1/2}$ and H^s as

$$(d/dt)\mathcal{E}(t) \leq -\theta_1\mathcal{E}(t) + C|W(t)|_{L^2}^2, \tag{3.78}$$

we obtain by Gronwall's inequality

$$e^{\theta_1 t}\mathcal{E}|_0^t \leq C\int_0^t e^{\theta_1 s}|W(s)|_{L^2}^2 \, ds$$

and thereby the result (exercise).   •

Proof of Proposition 1.22. Exercise, using (3.77) together with the fact that $|U(t)|_{L^2} \leq C|U(0)|_{L^2}$ for $|W|_{L^\infty}$ sufficiently small (Exercise 1.23).   •

In Appendix B, we sketch also a simple proof of Proposition 1.21 using (3.77) together with linearized stability estimates. This may be helpful for the reader in motivating the nonlinear stability argument carried out in Section 4 for the more complicated variable-coefficient situation of a viscous shock profile.

4 Reduction to Low Frequency

We now begin our main analysis, carrying out high- and mid-frequency estimates reducing the nonlinear long-time stability problem (LT) to the study of the linear resolvent equation $(\lambda - L_{\tilde{\xi}})U = f$ in the low-frequency regime $|(\tilde{\xi}, \lambda)|$ small described in Proposition 2.14. The novelty of the present treatment lies in the simplified argument structure based entirely on energy estimates. More detailed linearized estimates were obtained in [Z.4] using energy estimates together with the pointwise machinery of [MaZ.3].

4.1. Nonlinear estimate. Define the nonlinear perturbation

$$U := \tilde{U} - \bar{U}, \qquad (4.1)$$

where $\tilde{U}$ denotes a solution of (1.2) with initial data $\tilde{U}_0$ close to $\bar{U}$. Our first step is to establish the following large-variation version of Proposition 3.22.

Proposition 4.1 ([MaZ.4, Z.4]) *Given (A1) – (A2), (H0) – (H3), and Assumption 2.8,*

$$|U(t)|^2_{H^s} \leq C\left(e^{-\theta_2 t}|U(0)|^2_{H^s} + \int_0^t e^{-\theta_2(t-s)}|U(s)|^2_{L^2}\,ds\right), \quad \theta_2 > 0, \qquad (4.2)$$

so long as $|U|_{H^s}$ remains sufficiently small, for all $q(d) \leq s \leq q$, $q(d)$ and q as defined in (H0).

4.1.1. "Goodman-type" weighted energy estimate. The obvious difficulty in proving Proposition 4.1 is that (A2) is assumed only at endpoints $W_\pm$, hence we cannot hope to establish smoothing in the key hyperbolic modes w^I through the circle of ideas discussed in Section 3.4, except in the far field $x_1 \to \pm\infty$. The complementary idea needed to treat the "near field" or "interior" region $x_1 \in [-M, M]$ is that propagation in the hyperbolic modes thanks to assumption (H1) is uniformly transverse to the shock profile, so that signals essentially spend only finite time in the near field, at all other times experiencing smoothing properties of the far field.

The above observation may be conveniently quantified using a type of weighted-norm energy estimate introduced by Goodman [Go.1, Go.2] in the study of one-dimensional stability of small-amplitude shock waves. Consider a linear hyperbolic equation $\tilde{A}^0 W_t + \sum_{j=1}^d \tilde{A}^j W_{x_j} = 0$ with large- but finite-variation coefficients depending only on x_1,

$$|A^j_{x_1}| \leq \Theta, \quad \int \Theta(y)\, dy < \infty, \quad j = 0, 1, \ldots d, \tag{4.3}$$

$\tilde{A}^j$ symmetric, $\tilde{A}^0 > 0$, under the "upwind" assumption

$$\tilde{A}^1 \geq \theta > 0. \tag{4.4}$$

Define the scalar weight $\alpha(x_1) := e^{\int_0^{x_1} -(2C\Theta/\theta)(y)\, dy}$, positive and bounded above and below by (4.3), where $C > 0$ is sufficiently large.

Then, the zero-order Goodman's estimate, in this simple setting, is just

$$
\begin{aligned}
(d/dt)(1/2)\langle W, \alpha \tilde{A}^0 W \rangle &= \langle W, \alpha \tilde{A}^0 W_t \rangle \\
&= -\langle W, \alpha \sum_j \tilde{A}^j W_{x_j} \rangle \\
&= (1/2)\langle W, (\alpha \tilde{A}^1)_{x_1} W \rangle \\
&= (1/2)\langle W, \alpha(-(2C\Theta/\theta)\tilde{A}^1 + \tilde{A}^1_{x_1})W \rangle \\
&\leq -(C/2)\langle W, \alpha \Theta W \rangle,
\end{aligned}
\tag{4.5}
$$

yielding time-exponential decay in $|W|_{L^2}$ on any finite x_1-interval $[-M, M]$.

Exercise 4.2 *Verify the corresponding s-order estimate*

$$(d/dt)(1/2)\sum_{r=0}^s \langle \partial_x^r W, \alpha \tilde{A}^0 \partial_x^r W \rangle \leq -(C/2)\sum_{r=0}^s \langle \partial_x^r W, \alpha \Theta \partial_x^r W \rangle \tag{4.6}$$

for $C > 0$ sufficiently large, yielding time-exponential decay in $|W|_{H^s}$ on any finite interval $x_1 \in [-M, M]$.

Remark 4.3 Proper accounting of the favorable effects of transverse propagation is a recurring theme in the analysis of stability of viscous waves; see, e.g., [Go.1, Go.2, L.4, LZ.1, LZ.2, LZe.1, SzZ, L.1, ZH, BiB.1, BiB.2, GrR, MeZ.1, GMWZ.1, GMWZ.2, GMWZ.3, GMWZ.4].

4.1.2. Large-amplitude hyperbolic–parabolic smoothing. Combining the weighted energy estimate just described above with the techniques of Section 3.4, we are now ready to establish Proposition 4.1.

Proof of Proposition 4.1. Equivalently, we establish

$$|W(t)|^2_{H^s} \leq C\big(e^{-\theta_1 t}|W(0)|^2_{H^s} + \int_0^t e^{-\theta_1(t-s)}|W(s)|^2_{L^2}\, ds\big), \quad \theta_1 > 0, \tag{4.7}$$

where

$$W(x, t) := \tilde{W}(x, t) - \bar{W}(x_1) := W(\tilde{U}(x, t)) - W(\bar{U}(x_1)). \tag{4.8}$$

(Exercise: Show $|U|_{H^r} \sim |W|_{H^r}$ for all $0 \leq r \leq s$, in particular 0 and s, using $s \geq [d/2] + 3$ and same estimates used to close the energy estimate in the proof of Proposition 1.16.)

Noting that both $\tilde{W}$ and $\bar{W}$ satisfy (1.27), we obtain by straightforward calculation [MaZ.4, Z.4] the nonlinear perturbation equation

$$
\begin{aligned}
\tilde{A}^0 W_t + \sum_j \tilde{A}^j W_{x_j} = \sum_{j,k} (\tilde{B}^{jk} W_{x_k})_{x_j} \\
+ M_1 \bar{W}_{x_1} + \sum_j (M_2^j \bar{W}_{x_1})_{x_j} + \sum_j M_3^j W_{x_j} + G(x, \partial_x W),
\end{aligned}
\tag{4.9}
$$

where

$$
\tilde{A}^j(x,t) := \tilde{A}^j(\tilde{W}(x,t)), \quad \tilde{B}^{jk}(x,t) := \tilde{B}^{jk}(\tilde{W}(x,t)), \tag{4.10}
$$

$$
M_1 = M_1(W, \bar{W}) := \tilde{A}^1(\tilde{W}) - \tilde{A}^1(\bar{W}) = \left(\int_0^1 d\tilde{A}^1(\bar{W} + \theta W) \, d\theta \right) W \tag{4.11}
$$

$$
M_2^j = M_2^j(W, \bar{W}) := \tilde{B}^{j1} - \bar{B}^{j1} = \begin{pmatrix} 0 & 0 \\ 0 & (\int_0^1 db^{j1}(\bar{W} + \theta W) \, d\theta) W \end{pmatrix}. \tag{4.12}
$$

$$
M_3^j = M_3^j(x_1) := (\partial \tilde{G}/\partial W_{x_j})|_{\partial_x \bar{W}} = \begin{pmatrix} 0 & 0 \\ \mathcal{O}(|\bar{W}_{x_1}|) & \mathcal{O}(|\bar{W}_{x_1}|) \end{pmatrix}. \tag{4.13}
$$

and

$$
\begin{aligned}
G = \begin{pmatrix} 0 \\ g \end{pmatrix} &:= \begin{pmatrix} 0 \\ g(\partial_x \tilde{W}, \partial_x \tilde{W}) - g(\partial_x \bar{W}, \partial_x \bar{W}) - dg(\bar{W}) \partial_x W \end{pmatrix} \\
&= \begin{pmatrix} 0 \\ \mathcal{O}(|\partial_x W|^2) \end{pmatrix}.
\end{aligned}
\tag{4.14}
$$

(In the case that there exists a global convex viscosity-compatible entropy, we may further arrange that $G \equiv 0$.)

To clarify the argument, we drop terms M and G in (4.9). The reader may verify that these generate harmless, absorbable error terms in our energy estimates (exercise; see [MaZ.4, Z.4]). With this simplification, we reduce to the familiar situation

$$
\tilde{A}^0 W_t + \sum_j \tilde{A}^j W_{x_j} = \sum_{j,k} (\tilde{B}^{jk} W_{x_k})_{x_j}, \tag{4.15}
$$

with the difference that $\tilde{A}^j = \tilde{A}^j(\bar{W} + W)$ and $\tilde{B}^{jk} = \tilde{B}^{jk}(\bar{W} + W)$ are no longer approximately constant, but for small $|W|_{H^s}$ approach the possibly rapidly varying values $\tilde{A}^j(\bar{W})$ and $\tilde{B}^{jk}(\bar{W})$ along the background profile.

By Corollary 3.3 and Assumption (H1), respectively, we have

$$
|(d/d_{x_1})^k \tilde{A}^j(\bar{W})|, \ |(d/d_{x_1})^k \tilde{B}^{jk}(\bar{W})| \leq \Theta(x_1) := C e^{-\theta|x|} \tag{4.16}
$$

and

$$A^1 \geq \theta > 0, \tag{4.17}$$

for some $C, \theta > 0$. Define

$$\alpha(x_1) := e^{\int_0^{x_1} -(2C_*\Theta/\theta)(y)\,dy}, \tag{4.18}$$

where $C_* > 0$ is a sufficiently large constant to be determined later.

By assumption (A1), we have $\tilde{A}_{11}^j$ symmetric, and also $\bar{A}^j$ symmetric at $x_1 = \pm\infty$. Defining smooth linear interpolants $\hat{A}_{12}^j$ and $\hat{A}_{12}^j$ between the values of $\bar{A}_{12}^j$ and $\bar{A}_{12}^j$ at $\pm\infty$, we thus obtain symmetry,

$$\hat{A}_{12}^j = (\hat{A}_{21}^j)^*, \tag{4.19}$$

at the cost of an $\mathcal{O}(\Theta)$ error

$$\partial_x^q(\bar{A}_{12}^j - \hat{A}_{12}^j) = \mathcal{O}(\Theta), \qquad \partial_x^q(\bar{A}_{12}^j - \hat{A}_{12}^j) = \mathcal{O}(\Theta) \tag{4.20}$$

for $0 \leq q \leq q(d)$.

Combining calculations (1.33) and (4.6), we obtain

$$\begin{aligned}
\frac{1}{2}\langle W, \alpha\tilde{A}^0 W\rangle_t &= \frac{1}{2}\langle W, \alpha\tilde{A}_t^0 W\rangle + \langle W, \alpha\tilde{A}^0 (W)_t\rangle \\
&= \frac{1}{2}\langle W, \alpha\tilde{A}_t^0 W\rangle - \langle W, \sum_j \alpha\tilde{A}^j(W)_{x_j} + \sum_{j,k}\alpha(\tilde{B}^{jk}(W)_{x_k})_{x_j}\rangle \\
&= \Big\{\frac{1}{2}\langle w^I, (\alpha\tilde{A}_{11}^1(\bar{W}))_{x_1} w^I\rangle - \sum_{j,k}\langle (w^{II})_{x_j}, \alpha\tilde{B}^{jk}(w^{II})_{x_k}\rangle\Big\} \\
&\quad + \frac{1}{2}\langle w^I, (\alpha\hat{A}_{12}^1)_{x_1} w^{II}\rangle + \frac{1}{2}\langle w^{II}, (\alpha\hat{A}_{21}^1)_{x_1} w^I\rangle \\
&\quad + \frac{1}{2}\langle W, \alpha\tilde{A}_t^0 W\rangle + \langle w^I, \sum_j \big(\alpha(\tilde{A}_{11}^j - \tilde{A}_{11}^j(\bar{W}))\big)_{x_j} w^I\rangle \\
&\quad + \frac{1}{2}\langle w^I, \sum_j \alpha(\tilde{A}_{12}^j - \hat{A}_{12}^j)(w^{II})_{x_j}\rangle - \frac{1}{2}\langle (\alpha w^{II})_{x_j}, \sum_j (\tilde{A}_{21}^j - \hat{A}_{21}^j)w^I\rangle \\
&\quad + \frac{1}{2}\langle w^{II}, \sum_j \alpha\tilde{A}_{22}^j(w^{II})_{x_j}\rangle - \sum_k \langle w^{II}, \alpha_{x_1}\tilde{B}^{1k}(w^{II})_{x_k}\rangle\Big\} \\
&\leq -\Big\{C_*\langle w^I, \alpha\Theta w^I\rangle + \theta\langle\partial_x w^{II}, \alpha\partial_x w^{II}\rangle\Big\} \\
&\quad + C(|W_t|_\infty + |W_x|_\infty + C_*|W|_\infty)\int \alpha|W|^2 \\
&\quad + CC_*\int \Theta\alpha|w^{II}|(|w^I| + |\partial_x w^{II}|),
\end{aligned} \tag{4.21}$$

for some $C > 0$ independent of C_*, M, for C_* sufficiently large. Estimating the unbracketed terms in the final line of (4.21) using Young's inequality, we find that

$$\frac{1}{2}\langle W, \alpha\tilde{A}^0 W\rangle_t \leq -\frac{C_*}{2}\langle w^I, \alpha\Theta w^I\rangle - \frac{\theta}{2}\langle\partial_x w^{II}, \alpha\partial_x w^{II}\rangle$$
$$+ C(C_*)\zeta\langle w^I, \alpha w^I\rangle. + C(C_*)\langle w^{II}, \alpha w^{II}\rangle, \qquad (4.22)$$

where $\zeta := |W|_{H^s}$ is arbitrarily small.

More generally, defining

$$\mathcal{E}_1(W) := (1/2)\sum_{r=0}^{s} c^{-r}\langle\partial_x^r W, \alpha\tilde{A}^0\partial_x^r W\rangle, \qquad (4.23)$$

for $c = c(C_*) > 0$ sufficiently large, we obtain by telescoping sum that

$$(d/dt)\mathcal{E}_1(W(t)) \leq \sum_{r=0}^{s} c^{-r}\Big(-\frac{C_*}{2}\langle\partial_x^r w^I, \alpha\Theta\partial_x^r w^I\rangle\Big)$$
$$+ \sum_{r=0}^{s} c^{-r}\Big(-\frac{\theta}{2}\langle\partial_x^{r+1}w^{II}, \alpha\partial_x^{r+1}w^{II}\rangle + C(C_*)\langle\partial_x^r w^{II}, \alpha\partial_x^r w^{II}\rangle\Big)$$
$$+ \sum_{r=0}^{s} c^{-r}\Big(CC_*\zeta\langle\partial_x^r w^I, \alpha\partial_x^r w^I\rangle\Big)$$
$$\leq \sum_{r=0}^{s} c^{-r}\Big(-\frac{C_*}{2}\langle\partial_x^r w^I, \alpha\Theta\partial_x^r w^I\rangle - \frac{\theta}{2}\langle\partial_x^{r+1}w^{II}, \alpha\partial_x^{r+1}w^{II}\rangle\Big)$$
$$+ \sum_{r=1}^{s} c^{-r}\Big(CC_*\zeta\langle\partial_x^r w^I, \alpha\partial_x^r w^I\rangle\Big) + C(C_*)|W|_{L^2}^2. \qquad (4.24)$$

Now, introduce C^∞ cutoff functions $\chi^\pm$ supported on $x_1 \geq M$ and $x_1 \leq -M$ and one on $x_1 \geq 2M$ and $x_1 \leq -2M$, respectively, for $M > 0$ sufficiently large, with

$$|d^r\chi^\pm(x_1)| \leq \epsilon := C/M. \qquad (4.25)$$

On the respective supports $x_1 \geq M$, $x_1 \leq -M$ of $W_F^\pm$, coefficients $\tilde{A}^j$ and $\tilde{B}^{jk}$ remain arbitrarily close in H^s to $\tilde{A}_\pm^j$ and $\tilde{B}_\pm^{jk}$ provided $|W|_{H^s}$ is taken sufficiently small and M sufficiently large.

Thus, defining $\mathcal{E} := \mathcal{E}_1 + \mathcal{E}_2$,

$$\mathcal{E}_2^\pm(W) := (1/2)\sum_{r=0}^{s-1} c^{-r-1}\langle(\chi_+K_+(\partial_x) + \chi_-K_-(\partial_x))\partial_x^r W, \partial_x^r W\rangle, \qquad (4.26)$$

where operator $K_\pm(\partial_x)$ is defined by

$$\widehat{K_\pm(\partial_x f)}(\xi) := iK_\pm(\xi)\hat{f}(\xi) \qquad (4.27)$$

with $K_\pm(\cdot)$ as in (3.53), $\hat{g}$ denoting Fourier transform of g, and calculating the $\mathcal{E}_2$ contribution as in the argument of Proposition 3.20, we obtain

$$(d/dt)\mathcal{E}(W(t)) \leq \sum_{r=0}^{s} c^{-r}\left(-\frac{C_*}{2}\langle\partial_x^r w^I, \alpha\Theta\partial_x^r w^I\rangle - \frac{\theta}{2}\langle\partial_x^{r+1} w^{II}, \alpha\partial_x^{r+1} w^{II}\rangle\right)$$

$$- \theta\sum_{r=1}^{s} c^{-r}\langle(\chi_+ + \chi_-)\partial_x^r w^I, \alpha\Theta\partial_x^r w^I\rangle$$

$$+ \sum_{r=1}^{s} c^{-r}\left(CC_*\zeta\langle\partial_x^r w^I, \alpha\partial_x^r w^I\rangle\right) + C(C_*)|W|_{L^2}^2$$

$$\leq \sum_{r=0}^{s} c^{-r}\left(-\frac{\theta}{2}\langle\partial_x^r w^I, \alpha\partial_x^r w^I\rangle - \frac{\theta}{2}\langle\partial_x^{r+1} w^{II}, \alpha\partial_x^{r+1} w^{II}\rangle\right)$$

$$+ C(C_*)|W|_{L^2}^2$$

(4.28)

for C_* sufficiently large that $C_*\Theta \gg \theta$ for $|x| \leq M$, and ζ sufficiently small that $C_*\zeta \ll \theta$, where $\mathcal{E}^{1/2}(W)$ is equivalent to $|W|_{H^s}$.

Using the fact that α is bounded everywhere above and below, by definition 4.18 together with integrability of Θ, along with equivalence of $\mathcal{E}^{1/2}(W)$ and $|W|_{H^s}$, we obtain from (4.28) that

$$(d/dt)\mathcal{E}(W(t)) \leq -\theta_1|W(t)|_{H^s}^2 + C_2|W(t)|_{L^2}^2$$
$$\leq -\theta_2\mathcal{E}(W(t)) + C_2|W(t)|_{L^2}^2.$$

(4.29)

Applying Gronwall's inequality, similarly as in the proof of Proposition 3.22, we thus obtain $e^{\theta_2 t}\mathcal{E}|_0^t \leq C\int_0^t e^{\theta_2 s}|W(s)|_{L^2}^2\, ds$ and thereby the result.   ●

Remark 4.4 Note that (4.2) gives pointwise-in-time control on the $H^s(x)$ norm of W in terms of a time-weighted $L^2(t)$ average of the $L^2(x)$ norm. We will take advantage of this fact in the linearized analysis of Section 4.3, where we establish L^2-time-averaged high-frequency estimates by a simple Parseval argument, thus avoiding the complicated pointwise bounds of [MaZ.4, Z.4].

4.2. Linearized estimate. We next establish the following estimate on the linearized inhomogeneous problem

$$U_t - LU = f_1 + \partial_x f_2, \qquad U(0) = U_0, \tag{4.30}$$

where L is defined as in (3.8), assuming the low-frequency bounds of [Z.3].

Proposition 4.5 *Given (A1)–(A2), (H0)–(H5), assumption 2.8, and strong spectral, structural, and strong refined dynamical stability, the solution $U(t) := \int_0^t e^{L(t-s)}f(s)ds$ of (4.30) satisfies*

$$\int_0^T e^{-8\theta_1(T-s)}|U(s)|_{L^2}^2\,ds \le$$

$$C(1+T)^{-(d-1)/2+2\epsilon}|U_0|_{L^1}^2 + Ce^{-2\theta_1 T}\left(|U_0|_{H^1}^2 + |LU_0|_{H^1}^2\right)$$

$$+ C\int_0^T e^{-2\theta_1(T-s)}\left(|f_1(s)|_{H^1}^2 + |\partial_x f_2|_{H^1}^2\right)ds \tag{4.31}$$

$$+ C\left(\int_0^T (1+T-s)^{-(d-1)/4+\epsilon}|f_1(s)|_{L^1}\,ds\right)^2$$

$$+ C\left(\int_0^T (1+T-s)^{-(d-1)/4+\epsilon-1/2}|f_2(s)|_{L^1}\,ds\right)^2$$

for any fixed $\epsilon > 0$, for some $C = C(\epsilon) > 0$ sufficiently large, for any U_0, f_1, f_2 such that the righthand side is well-defined. In the case of strong (inviscid, not refined) dynamical stability, we may take $\epsilon = 0$ and $C > 0$ fixed.

Remark 4.6 Somewhat stronger linearized bounds were obtained in [Z.4] by a more detailed analysis (pointwise in time, with no loss in regularity, and for the initial-data rather than the zero-data inhomogeneous problem). However, these will suffice for our argument and are much easier to obtain.

4.2.1. High- and mid-frequency resolvent bounds. Our first step is to estimate solutions of resolvent equation (3.20).

Proposition 4.7 *(High-frequency bound) Given (A1)–(A2), (H0)–(H1),*

$$|(\lambda - L_{\tilde{\xi}})^{-1}|_{\hat{H}^1(x_1)} \le C \quad \text{for } |(\tilde{\xi},\lambda)| \ge R \text{ and } \Re\lambda \ge -\theta, \tag{4.32}$$

for some R, $C > 0$ sufficiently large and $\theta > 0$ sufficiently small, where $|\cdot|_{\hat{H}^1}$ denotes $\hat{H}^1$ operator norm, $|f|_{\hat{H}^1} := |(1 + |\partial_{x_1}| + |\tilde{\xi}|)f|_{L^2}$.

Proof. A Laplace–Fourier transformed version (now with respect to $(t, \tilde{x})$) of nonlinear estimate (4.24), $s = 1$, carried out on the linearized equations (3.8) written in W-coordinates, $W := (\partial W/\partial U)(\bar{U})U$, yields

$$\Re\lambda\left((1 + |\tilde{\xi}|^2)|W|^2 + |\partial_{x_1}W|^2\right) \le -\theta_1\left(|\tilde{\xi}|^2|W|^2 + |\partial_{x_1}W|^2\right)$$
$$+ C_1\left(|W|^2 + (1 + |\tilde{\xi}|^2)|W||f| + |\partial_{x_1}W||\partial_{x_1}f|\right) \tag{4.33}$$

for some $C_1 > 0$ sufficiently big and $\theta_1 > 0$ sufficiently small, where $|\cdot|$ denotes $|\cdot|_{L^2(x_1)}$ and

$$(\lambda - L_{\tilde{\xi}})U = f. \tag{4.34}$$

We omit the proof, which is just the translation into frequency domain of the proof of Proposition 4.1, carried out in the much simpler, linearized setting. For related calculations, see the proof of Proposition 3.6 or the proof of (K3) in Lemma 3.18.

Rearranging, we have

$$(\mathbb{R}\lambda + \theta_1)|(1 + |\tilde{\xi}| + |\partial_{x_1}|)W|^2 \leq C_1|(1 + |\tilde{\xi}| + |\partial_{x_1}|)f|^2 + C_1|W|^2. \qquad (4.35)$$

On the other hand, taking the imaginary part of the L^2 inner product of u against $\lambda u = f + Lu$ we have also the standard estimate

$$\begin{aligned}
\Im\lambda|u|_{L^2}^2 &= \Im\langle u, f + Lu\rangle \\
&\leq |u|_{L^2}|f|_{L^2} + |\langle u, Lu\rangle| \qquad (4.36) \\
&\leq C|f|_{L^2}^2 + C^{-1}|u|_{L^2}^2 + |u|_{H^1}^2,
\end{aligned}$$

or, taking the Fourier transform in $\tilde{x}$, with U the Fourier transform of u,

$$\Im\lambda|W|^2 \leq C_2|f|^2 + C_2^{-1}|W|^2 + |(|\tilde{\xi}| + |\partial_x|)W|^2. \qquad (4.37)$$

Summing (4.35) and (4.37) and dividing by factor $|(1 + |\tilde{\xi}| + |\partial_{x_1}|)|W|$, we obtain $|W|_{\hat{H}^1(x_1)} \leq C_3|f|_{\hat{H}^1(x_1)}$, and thus the a priori bound

$$|U|_{\hat{H}^1(x_1)} \leq C|(\lambda - L_{\tilde{\xi}})U|_{\hat{H}^1(x_1)},$$

with $\theta := \theta_1/2$ (i.e, $\mathbb{R}\lambda + \theta_1 \geq \theta_1/2 > 0$) and $C := 2C_1/\theta$, for $|(\tilde{\xi}, \lambda)| \geq 2(C_1 + C_2^{-1})$ (i.e., sufficiently large that $(C_1 + C_2^{-1})|W|$ may be absorbed in the lefthand side). Observing that the adjoint operator satisfies a corresponding a priori bound, yielding by duality invertibility of $(\lambda - L_{\tilde{\xi}})$, we obtain (4.32). •

Exercise 4.8 *1. For $E := \langle u, Su\rangle \sim |u|_{H^s}^2$, S without loss of generality symmetric and real-valued, show more generally that damping estimate*

$$E_t \leq -2\theta E + C|u|_{L^2}^2 \qquad (4.38)$$

$\theta > 0$, for $u_t = Lu$, which implies $E(t) \leq e^{-2\theta}E(0) + C\int_0^t e^{-2\theta(t-s)}|u(s)|_{L^2}^2 ds$, is equivalent to

$$\mathbb{R}\langle Su, Lu\rangle \leq -\theta\langle Su, u\rangle + C|u|_{L^2}^2. \qquad (4.39)$$

Provided $LU \leq C|u|_{H^s}^2$ (so that $|\lambda||u|_{L^2} \leq C|u|_{H^s} + |f|_{L^2}$ for $\lambda u = Lu + f$), show that (4.39) implies also the high-frequency resolvent estimate

$$|(\lambda - L)^{-1}f|_{H^s} \leq C_2(\mathbb{R}\lambda + \theta)^{-1}|f|_{H^s}$$

for $\mathbb{R}\lambda \geq -\theta$ and $|\lambda|$ sufficiently large. (Hint: establish an a priori estimate for $(\lambda - L)u = f$, concluding invertibility of $(\lambda - L)$ as usual by a corresponding estimate on the adjoint.) This may be viewed as a weak analog of the Lumer–Phillips Theorem described in Remark 6.19 (global resolvent bounds and C, C_1, $\theta = 0$). For a more complete version, see Exercise 6.28.

2. *Show using the identity operator $Lu = u$ that the converse does not hold. (As a bounded multiplication operator, L satisfies uniform high-frequency resolvent bounds $|(\lambda - L)^{-1}f|_{H^1} \leq C|\lambda|^{-1}|f|_{H^1}$ for $|\lambda|$ sufficiently large , but $\langle u, SLu \rangle = \langle u, Su \rangle \not\leq C|u|^2_{L^2}$ for any $C > 0$ for $\langle u, Su \rangle \sim |u|^2_{H^2}$, so damping does not hold.) Thus, the principle described in part 1 is a one-way version of the Lumer–Phillips Theorem.*

Proposition 4.9 *(Mid-frequency bound) Given (A1)–(A2), (H0)–(H1), and strong spectral stability (2.14),*

$$|(\lambda - L_{\tilde{\xi}})^{-1}|_{\hat{H}^1(x_1)} \leq C \quad \text{for } R^{-1} \leq |(\tilde{\xi}, \lambda)| \leq R \text{ and } \Re\lambda \geq -\theta, \qquad (4.40)$$

for any $R > 0$, and $C = C(R) > 0$ sufficiently large and $\theta = \theta(R) > 0$ sufficiently small, where $|\cdot|_{\hat{H}^1}$ is as defined in Proposition 4.7.

Proof. This follows by compactness of the set of frequencies under consideration together with the fact that the resolvent $(\lambda - L_{\tilde{\xi}})^{-1}$ is analytic with respect to $\hat{H}^1$ in $(\tilde{\xi}, \lambda)$ for λ in the resolvent set $\rho_{\hat{H}^1}(L)$ of $L_{\tilde{\xi}}$ with respect to $\hat{H}^1$ once we establish that λ lies in the $\rho_{\hat{H}^1}(L_{\tilde{\xi}})$ whenever $R^{-1} \leq |(\tilde{\xi}, \lambda)| \leq R$ and $\Re\lambda \geq -\theta$. Here, the $\hat{H}^1$-resolvent set $\rho_{\hat{H}^1}(L_{\tilde{\xi}})$ is defined as the set of λ such that $(\lambda - L_{\tilde{\xi}})$ has a bounded inverse with respect to the $\hat{H}^1$-operator norm [Kat, Yo, Pa], and similarly for $\rho_{L^2}(L_{\tilde{\xi}})$, to which we refer elsewhere simply as $\rho(L_{\tilde{\xi}})$; see Definition 6.3, Appendix A. Analyticity in λ on the resolvent set of $(\lambda - L)^{-1}$ holds for general operators L (Lemma 6.5, Appendix A), while analyticity in $\tilde{\xi}$ on the resolvent set of $(\lambda - L_{\tilde{\xi}})^{-1}$ follows using standard properties of asymptotically constant-coefficient ordinary differential operators [He, Co, ZH] from analyticity in $\tilde{\xi}$ of the limiting, constant-coefficient symbol together with convergence at integrable rate of the coefficients of $L_{\tilde{\xi}}$ to their limits as $x_1 \to \pm\infty$. Alternatively, we may establish this directly for θ sufficiently small, using the smoothing properties induced by (A1)–(A2), (H0)–(H2); see Exercises 6.6.1–2, Appendix A.

Under assumptions (A1)–(A2), (H0)–(H2), $\rho_{\hat{H}^1}(L_{\tilde{\xi}})$ and $\rho_{L^2}(L_{\tilde{\xi}})$ agree on the set $R^{-1} \leq |(\tilde{\xi}, \lambda)| \leq R$, $\Re\lambda \geq -\theta$, provided θ is chosen sufficiently small relative to R^{-1}. For, standard considerations yield that, on the "domain of consistent splitting" comprised of the intersection of the rightmost components of the resolvent sets of the limiting, constant-coefficient operators $L_{\tilde{\xi}}^{\pm}$ as $x_1 \to \pm\infty$, the spectra of the asymptotically constant-coefficient ordinary differential operators $L_{\tilde{\xi}}$ with respect to $\hat{H}^s$, for any $0 \leq s \leq q(d)$, $q(d)$ as defined in (H0) (indeed, with respect to any reasonable norm), consist entirely of isolated L^2 eigenvalues, i.e., $\lambda - L_{\tilde{\xi}}$ is Fredholm in each $\hat{H}^s$; for further discussion, see [He, GZ, ZH, ZS, Z.3, Z.4] or Section 5, below. By (K3), the domain of consistent splitting includes the domain under consideration for θ sufficiently small, giving $\rho_{\hat{H}^1}(L_{\tilde{\xi}}) = \rho_{L^2}(L_{\tilde{\xi}})$ as claimed. Alternatively, we may establish this by the direct but more special argument that (4.35) together with $|(\lambda - L_{\tilde{\xi}})^{-1}|_{L^2(x_1)} \leq C_1$ implies $|(\lambda - L_{\tilde{\xi}})^{-1}|_{\hat{H}^1(x_1)} \leq C_2$.

With this observation, the result follows by the strong spectral stability assumption 2.12, since analyticity of the resolvent in $(\tilde{\xi}, \lambda)$ implies that $\{(\tilde{\xi}, \lambda) : \lambda \in \rho_{L^2}(L_{\tilde{\xi}})\}$ is open, and therefore contains an open neighborhood of $\{(\tilde{\xi}, \lambda) : \Re\lambda \geq 0\} \setminus \{(0,0)\}$. (Recall, $\sigma_{L^2}(L) := \rho_{L^2}(L)^c$.) •

Remarks 4.10 1. As the argument of Proposition 4.9 suggests, uniform mid-frequency bounds are essentially automatic in the context of asymptotically constant - coefficient ordinary differential operators, following from compactness/ continuity of the resolvent alone.

2. Resolvent bounds (4.32) for λ with negative real part correspond roughly to time-exponential decay in high-frequency modes, similarly as in (4.2). Indeed, a uniform bound $|(\lambda - L_{\tilde{\xi}})^{-1}|_{\hat{H}^1} \leq C$ for all $\tilde{\xi}$, $\Re\lambda \geq -\theta < 0$, or equivalently $|(\lambda - L)^{-1}|_{H^1} \leq C$ for all $\Re\lambda \leq -\theta < 0$ would imply time-exponential linearized stability $|e^{Lt}|_{H^1} \leq C(\tilde{\theta})e^{-\tilde{\theta}t}$ for any $\tilde{\theta} < \theta$, by a general Hilbert-space theorem of Prüss [Pr]; see Remark 6.23, Appendix A. Contrasting with estimate (3.20) corresponding to local well-posedness, what we have done here is to use the additional structure (A2), (H1) to push γ_0 into the negative half-plane. Note that we made no stability assumptions in the statement of Proposition 4.7.

3. High-frequency linearized resolvent bounds may alternatively be obtained by Kreiss symmetrizer techniques applying to a much more general class of systems; see [GMWZ.4]. In particular, we may dispense with the symmetry assumptions in (A1)–(A2) connected with energy estimates using integration by parts, replacing them everywhere with sharp, spectral conditions of hyperbolicity/parabolicity. By corresponding pseudodifferential estimates, it might be possible to recover also the nonlinear time-evolutionary estimate (4.7) for this more general class of equations. This would be a very interesting direction for future investigation.

4.2.2. Splitting the solution operator. Letting L as usual denote the linearized operator defined in (3.8), define a "low-frequency cutoff"

$$S_1(t)f(x) :=$$

$$\int_{|\tilde{\xi}|^2 \leq \theta_1 + \theta_2} \oint_{\Re\lambda = \theta_2 - |\tilde{\xi}|^2 - |\Im\lambda|^2 \geq -\theta_1} e^{i\tilde{\xi}\cdot\tilde{x} + \lambda t}(\lambda - L_{\tilde{\xi}})^{-1}\hat{f}(x_1, \tilde{\xi})\,d\tilde{\xi}d\lambda \quad (4.41)$$

of the solution operator e^{Lt} described in (3.12), similarly as in the constant-coefficient argument carried out in Appendix B, where $\theta_1, \theta_2 > 0$ are to be determined later. From Propositions 4.7–4.9, we obtain the following decompositions justifying this analogy.

Corollary 4.11 *Given (A1)–(A2), (H0)–(H1), and strong spectral stability, let $|e^{Lt}| \leq Ce^{\gamma_0 t}$ denote the C^0 semigroup on L^2 generated by L, with domain*

$\mathcal{D}(L) := \{U : U, LU \in L^2\}$. Then, for arbitrary $\theta_2 > 0$, $\theta_1 > 0$ sufficiently small relative to θ_2, and all $t \geq 0$,

$$e^{Lt} f(x) = S_1(t)f(x)$$
$$+ \text{P.V.} \int_{-\theta_1 - i\infty}^{-\theta_1 + i\infty} \int_{\mathbb{R}^{d-1}} \mathbb{I}_{|\tilde{\xi}|^2 + |\Im\lambda|^2 \geq \theta_1 + \theta_2} e^{i\tilde{\xi}\cdot\tilde{x} + \lambda t} (\lambda - L_{\tilde{\xi}})^{-1} \hat{f}(x_1, \tilde{\xi}) \, d\tilde{\xi} d\lambda$$

(4.42)

for all $f \in \mathcal{D}(L) \cap H^1$, where $\mathbb{I}_P$ denotes the indicator function for logical proposition P (one for P true and zero for P false) and $\hat{f}$ denotes Fourier transform of f. Likewise, for $0 \leq t \leq T$,

$$\int_0^t e^{L(t-s)} f(s) \, ds = \int_0^T S_1(t-s)f(s) \, ds$$
$$+ \text{P.V.} \int_{-\theta_1 - i\infty}^{-\theta_1 + i\infty} \int_{\mathbb{R}^{d-1}} \mathbb{I}_{|\tilde{\xi}|^2 + |\Im\lambda|^2 \geq \theta_1 + \theta_2} e^{i\tilde{\xi}\cdot\tilde{x} + \lambda t} (\lambda - L_{\tilde{\xi}})^{-1} \widehat{\widehat{f^T}}(x_1, \tilde{\xi}, \lambda) \, d\tilde{\xi} d\lambda$$

(4.43)

for all $f \in L^1([0,T]; \mathcal{D}(L) \cap H^1(x))$, where $\hat{\hat{f}}$ denotes Laplace–Fourier transform of f and truncation f^T is as in (3.14).

Proof. Using analyticity on the resolvent set $\rho(L)$ of the resolvent $(\lambda - L)^{-1}$ (Lemma 6.5, Appendix A), the bound

$$|(\lambda - L)^{-1} f| \leq C(|(\lambda - L)^{-1}||Lf| + |f|)|\lambda|^{-1}$$

(4.44)

coming from resolvent identity (6.4) (Lemma 6.7), and the results of Propositions 4.7–4.9, we may deform the contour in (3.22) using Cauchy's Theorem to obtain

$$e^{Lt} f(x) = \oint_{\Re\lambda = \theta_2 - |\Im\lambda|^2; |\Im\lambda|^2 \leq \theta_1 + \theta_2} e^{\lambda t} (\lambda - L)^{-1} f(x) \, d\lambda$$
$$+ \text{P.V.} \int_{-\theta_1 - i\infty}^{-\theta_1 + i\infty} \mathbb{I}_{|\Im\lambda|^2 \geq \theta_1 + \theta_2} e^{\lambda t} (\lambda - L)^{-1} f(x) \, d\lambda$$
$$= \oint_{\Re\lambda = \theta_2 - |\Im\lambda|^2; |\Im\lambda|^2 \leq \theta_1 + \theta_2} \int_{\tilde{\xi} \in \mathbb{R}^{d-1}} e^{i\tilde{\xi}\cdot\tilde{x} + \lambda t} (\lambda - L_{\tilde{\xi}})^{-1} \hat{f}(x_1, \tilde{\xi}) \, d\tilde{\xi} d\lambda$$
$$+ \text{P.V.} \int_{-\theta_1 - i\infty}^{-\theta_1 + i\infty} \mathbb{I}_{|\Im\lambda|^2 \geq \theta_1 + \theta_2} \int_{\tilde{\xi} \in \mathbb{R}^{d-1}} e^{i\tilde{\xi}\cdot\tilde{x} + \lambda t} (\lambda - L_{\tilde{\xi}})^{-1} \hat{f}(x_1, \tilde{\xi}) \, d\tilde{\xi} d\lambda.$$

(4.45)

Using exercise 4.12 below, we may exchange the order of integration to write the first term of the last line as

$$\int_{\tilde{\xi} \in \mathbb{R}^{d-1}} \oint_{\Re\lambda = \theta_2 - |\Im\lambda|^2; |\Im\lambda|^2 \leq \theta_1 + \theta_2} e^{i\tilde{\xi}\cdot\tilde{x} + \lambda t} (\lambda - L_{\tilde{\xi}})^{-1} \hat{f}(x_1, \tilde{\xi}) \, d\tilde{\xi} d\lambda. \quad (4.46)$$

Deforming individual contours in the second integral, using analyticity of $(\lambda - L_{\tilde{\xi}})^{-1}$ on the resolvent set $\rho(L_{\tilde{\xi}})$ together with the results of Propositions

(4.7)–(4.9), and exchanging orders of integration using the result of Exercise 4.12, we may rewrite this further as

$$\int_{|\tilde{\xi}|^2 \leq \theta_1 + \theta_2} \oint_{\Re\lambda = \theta_2 - |\Im\lambda|^2 - |\tilde{\xi}|^2} e^{i\tilde{\xi}\cdot\tilde{x} + \lambda t}(\lambda - L_{\tilde{\xi}})^{-1}\hat{f}(x_1, \tilde{\xi})\,d\tilde{\xi}d\lambda$$

$$+ \int_{\tilde{\xi}\in\mathbb{R}^{d-1}} \int_{-\theta_1 - i\sqrt{\theta_1 + \theta_2}}^{-\theta_1 + i\sqrt{\theta_1 + \theta_2}} \mathbb{I}_{|\Im\lambda|^2 + |\tilde{\xi}|^2 \geq \theta_1 + \theta_2} e^{i\tilde{\xi}\cdot\tilde{x} + \lambda t}(\lambda - L_{\tilde{\xi}})^{-1}\hat{f}(x_1, \tilde{\xi})\,d\tilde{\xi}d\lambda$$

$$= \int_{|\tilde{\xi}|^2 \leq \theta_1 + \theta_2} \oint_{\Re\lambda = \theta_2 - |\Im\lambda|^2 - |\tilde{\xi}|^2} e^{i\tilde{\xi}\cdot\tilde{x} + \lambda t}(\lambda - L_{\tilde{\xi}})^{-1}\hat{f}(x_1, \tilde{\xi})\,d\tilde{\xi}d\lambda$$

$$+ \int_{-\theta_1 - i\sqrt{\theta_1 + \theta_2}}^{-\theta_1 + i\sqrt{\theta_1 + \theta_2}} \int_{\tilde{\xi}\in\mathbb{R}^{d-1}} \mathbb{I}_{|\Im\lambda|^2 + |\tilde{\xi}|^2 \geq \theta_1 + \theta_2} e^{i\tilde{\xi}\cdot\tilde{x} + \lambda t}(\lambda - L_{\tilde{\xi}})^{-1}\hat{f}(x_1, \tilde{\xi})\,d\tilde{\xi}d\lambda.$$

$$(4.47)$$

Substituting back into (4.45) and recombining terms, we obtain (4.42). A similar computation, together with the observation that

$$\widehat{\widehat{f^T}}(x_1, \tilde{\xi}, \lambda) := \int_0^T e^{-\lambda s}\hat{f}(x_1, \tilde{\xi}, s)\,ds \tag{4.48}$$

is analytic in $(\tilde{\xi}, \lambda)$ as the absolutely convergent integral of an analytic integrand, yields (4.43).   ●

Exercise 4.12 *For* $g \in H^1(x)$, *show using Parseval's identity and the Fourier transform definition* $|g|_{H^s(x)} := |(1 + |\xi|^2)^{s/2}\hat{g}(\xi)|_{L^2(\xi)}$ *that*

$$\int_{|\tilde{\xi}| \geq L} |\hat{g}|_{L^2(x_1)}^2\,d\tilde{\xi} \leq C(1 + L^2)^{-1})|g|_{H^1(x)}^2 \tag{4.49}$$

converges uniformly to zero as $L \to +\infty$, *where* $\hat{g}(x_1, \tilde{\xi})$ *denotes the Fourier transform of* g *with respect to* $\tilde{x}$.

Remark 4.13 Defining the "high-frequency cutoff" $S_2(t) := e^{Lt} - S_1(t)$, we have $\int_0^t e^{L(t-s)}\,ds = \int_0^t S_1(t-s)\,ds + \int_0^t S_2(t-s)\,ds$, where, by (4.43),

$$\int_0^t S_2(t-s)f(s)\,ds = \int_t^T S_1(t-s)f(s)\,ds$$

$$+ \text{P.V.} \int_{-\theta_1 - i\infty}^{-\theta_1 + i\infty} \int_{\mathbb{R}^{d-1}} \mathbb{I}_{|\tilde{\xi}|^2 + |\Im\lambda|^2 \geq \theta_1 + \theta_2} e^{i\tilde{\xi}\cdot\tilde{x} + \lambda t}(\lambda - L_{\tilde{\xi}})^{-1}\widehat{\widehat{f^T}}(x_1, \tilde{\xi}, \lambda)\,d\tilde{\xi}d\lambda.$$

$$(4.50)$$

Note that $S_1(t) = -S_2(t) \neq 0$ for $t < 0$, in violation of causality. "Causality error" $\int_t^T S_1(t-s)f(s)\,ds$ is the price for fixing the cutoff T in the second term of (4.50).

4.2.3. Proof of the main estimate. Now, assume the following low-frequency estimate of [Z.3], to be discussed further in Section 5. Similar results hold for profiles of nonclassical, over- or undercompressive type, as described in [Z.3, Z.4].

Proposition 4.14 ([Z.3][11]) *Under the assumptions of Proposition 4.5,*

$$|S_1(t)(f_1 + \partial_x f_2)|_{L^2(x)} \leq C(1+t)^{-(d-1)/4+\epsilon}|f_1|_{L^1(x)}$$
$$+ C(1+t)^{-(d-1)/4+\epsilon-1/2}|f_2|_{L^1(x)} \tag{4.51}$$

for all $t \geq 0$, for any fixed $\epsilon > 0$, for some $C = C(\epsilon) > 0$ sufficiently large, for any $f_1 \in L^1(x)$ and $f_2 \in L^1(x)$. In the case of strong (inviscid, not refined) dynamical stability, we may take $\epsilon = 0$ and $C > 0$ fixed.

Corollary 4.15 *Under the assumptions of Proposition 4.5,*

$$V(x,t) := \int_0^t S_1(T-t)f(x,s)ds \tag{4.52}$$

satisfies

$$\int_0^T e^{-8\theta_1(T-t)}|V(t)|_{L^2(x)}^2 \, dt \leq C\Big(\int_0^T (1+T-s)^{-(d-1)/4+\epsilon-1/2}|f(s)|_{L^1} \, ds \Big)^2 \tag{4.53}$$

for all $T \geq 0$, for any $f(x,t) \in L^1([0,T]; L^1(x))$.

Proof. Using (4.51) and the inequality

$$e^{-2\theta_1(T-t)}(1+t-s)^{-(d-1)/4+\epsilon-1/2} \leq C(1+T-s)^{-(d-1)/4+\epsilon-1/2} \tag{4.54}$$

for $0 \leq s \leq t \leq T$, we obtain

$$\int_0^T e^{-8\theta_1(T-t)}|V(t)|_{L^2(x)}^2 \, dt = \int_0^T e^{-8\theta_1(T-t)} \Big(\int_0^t S_1(t-s)|f(s)|_{L^1(x)}ds \Big)^2 dt$$
$$\leq C \int_0^T e^{-8\theta_1(T-t)} \Big(\int_0^t (1+t-s)^{-(d-1)/4+\epsilon-1/2}|f(s)|_{L^1(x)}ds \Big)^2 dt$$
$$\leq C \int_0^T e^{-4\theta_1(T-t)} \Big(\int_0^T (1+T-s)^{-(d-1)/4+\epsilon-1/2}|f(s)|_{L^1(x)}ds \Big)^2 dt$$
$$\leq C_2 \Big(\int_0^T (1+T-s)^{-(d-1)/4+\epsilon-1/2}|f(s)|_{L^1(x)}ds \Big)^2. \tag{4.55}$$

$\bullet$

Lemma 4.16 *Under the assumptions of Proposition 4.9,*

$$|S_1(t)f|_{L^2(x)} \leq Ce^{-\theta_1 t}|f|_{H^1} \tag{4.56}$$

for all $t < 0$, for any $f \in H^1(x)$.

Proof. Immediate, from representation (4.41) and Proposition 4.9. $\bullet$

Corollary 4.17 *Under the assumptions of Proposition 4.9,*

$$W_1(x,t) := \int_t^T S_1(T-t)f(x,s)ds \tag{4.57}$$

satisfies

$$\int_0^T e^{-8\theta_1(T-t)}|W_1(t)|^2_{L^2(x)}\, dt \le C \int_0^T e^{-2\theta_1(T-t)}|f(t)|^2_{H^1(x)}\, dt \tag{4.58}$$

for all $T \ge 0$, for any $f(x,t) \in L^2([0,T], H^1(x))$.

Proof. Using the triangle inequality, and (4.56), we obtain

$$\int_0^T e^{-8\theta_1(T-t)}|W_1(t)|^2_{L^2(x)}\, dt =$$

$$\int_0^T e^{-8\theta_1(T-t)}\left|\int_t^T S_1(t-s)f(s)ds\right|^2_{L^2(x)} dt$$

$$\le C \int_0^T e^{-8\theta_1(T-t)} \int_t^T e^{-2\theta_1(t-s)}|f(s)|^2_{H^1(x)}\, dsdt \tag{4.59}$$

$$\le C \int_0^T e^{-4\theta_1(T-t)} \int_0^T e^{-2\theta_1(T-t)}|f(s)|^2_{H^1(x)}\, dsdt$$

$$\le C_2 \int_0^T e^{-2\theta_1(T-t)}|f(s)|^2_{H^1(x)}\, ds.$$

$\bullet$

Lemma 4.18 *Under the assumptions of Proposition 4.9,*

$$W_2(x,t) :=$$

$$\text{P.V.} \int_{-\theta_1-i\infty}^{-\theta_1+i\infty} \int_{\mathbb{R}^{d-1}} \mathbb{I}_{|\tilde\xi|^2+|\Im\lambda|^2 \ge \theta_1+\theta_2}\, e^{i\tilde\xi\cdot\tilde x+\lambda t}(\lambda - L_{\tilde\xi})^{-1}\widehat{\widehat{f^T}}(x_1,\tilde\xi,\lambda)\, d\tilde\xi d\lambda \tag{4.60}$$

satisfies

$$\int_0^T e^{-2\theta_1(T-t)}|W_2(t)|^2_{L^2(x)}\, dt \le C \int_0^T e^{-2\theta_1(T-t)}|f(t)|^2_{H^1(x)}\, dt \tag{4.61}$$

for all $T \ge 0$, for any $f(x,t) \in L^2([0,T], H^1(x))$.

Proof. Using the relation

$$\mathcal{L}h(\lambda) = \mathcal{F}(e^{-\Re\lambda t}h)(-\Im\lambda) \tag{4.62}$$

between Fourier transform $\mathcal{F}h$ and Laplace transform $\mathcal{L}h$ of a function $h(t)$, we have

$$\mathcal{F}e^{\theta_1 t}W_2(t)(-k) = \mathcal{L}W(-\theta_1 + ik) = (\lambda - L)^{-1}\mathcal{L}f(-\theta_1 + ik). \qquad (4.63)$$

Together with Parseval's identity and Propositions 4.7–4.9, this yields

$$
\begin{aligned}
\int_0^T e^{2\theta_1 t}\big|W_2(t)\big|_{L^2(x)}^2\, dt &= \big|e^{\theta_1 t}W_2(t)\big|_{L^2(x,[0,T])}^2 = \\
&\int_{-\theta_1-i\infty}^{-\theta_1+i\infty} \Big|\mathbb{I}_{|\tilde\xi|^2+|\Im\lambda|^2\geq\theta_1+\theta_2}(\lambda - L_{\tilde\xi})^{-1}\widehat{\widetilde{f^T}}\Big|_{L^2(x_1,\tilde\xi)}^2\, d\lambda \\
&\leq C\int_{-\theta_1-i\infty}^{-\theta_1+i\infty}\big|\widehat{\widetilde{f^T}}\big|_{H^1(x_1,\tilde\xi)}^2\, d\lambda \\
&= C\int_0^T e^{2\theta_1 t}\big|f(t)\big|_{H^1(x)}^2\, dt
\end{aligned}
\qquad (4.64)
$$

yielding the claimed estimate upon multiplication by $e^{-2\theta_1 T}$. $\qquad\bullet$

Lemma 4.19 *Under the assumptions of Proposition 4.9,*

$$W_0(x,t) :=$$

$$\text{P.V.}\int_{-\theta_1-i\infty}^{-\theta_1+i\infty}\int_{\mathbb{R}^{d-1}}\mathbb{I}_{|\tilde\xi|^2+|\Im\lambda|^2\geq\theta_1+\theta_2}e^{i\tilde\xi\cdot\tilde x+\lambda t}(\lambda - L_{\tilde\xi})^{-1}\widehat{U_0}(x_1,\tilde\xi)\, d\tilde\xi d\lambda \qquad (4.65)$$

satisfies

$$\int_0^T e^{-2\theta_1(T-t)}|W_0(t)|_{L^2(x)}^2\, dt \leq Ce^{-2\theta_1 T}\big(|U_0|_{H^1(x)}^2 + |LU_0|_{H^1}^2\big) \qquad (4.66)$$

for all $T\geq 0$, for U_0, $LU_0 \in H^1(x,[0,T])$.

Proof. Without loss of generality taking $\Re\lambda \neq 0$, we have by (4.62), Parseval's identity, Propositions 4.7–4.9, and the resolvent identity

$$(\lambda - L_{\tilde\xi})^{-1}\widehat{U_0} = \lambda^{-1}\big((\lambda - L_{\tilde\xi})^{-1}L_{\tilde\xi}\widehat{U_0} + U_0\big) \qquad (4.67)$$

((6.4), Appendix A), that

$$
\begin{aligned}
\int_0^\infty e^{2\theta_1 t}\big|W_0(t)\big|_{L^2(x)}^2\, dt &= \big|e^{\theta_1 t}W_0(t)\big|_{L^2(x,t)}^2 = \\
&\int_{-\theta_1-i\infty}^{-\theta_1+i\infty}\Big|\mathbb{I}_{|\tilde\xi|^2+|\Im\lambda|^2\geq\theta_1+\theta_2}(\lambda - L_{\tilde\xi})^{-1}\widehat{U_0}\Big|_{L^2(x_1,\tilde\xi)}^2\, d\lambda \\
&\leq C\int_{-\theta_1-i\infty}^{-\theta_1+i\infty}|\lambda|^{-2}\, d\lambda\big(|L_{\tilde\xi}\widehat{U_0}|_{\hat H^1(x_1,\tilde\xi)} + |U_0|_{L^2(x_1,\tilde\xi)}\big) \\
&\leq C\big(|LU_0|_{H^1(x)} + |U_0|_{L^2(x)}\big),
\end{aligned}
\qquad (4.68)
$$

from which the claimed estimate follows upon multiplication by $e^{-2\theta_1 T}$. $\qquad\bullet$

Proof of Proposition 4.5. Immediate, combining the estimates of Proposition 4.14, Corollaries 4.15 and 4.17, and Lemmas 4.19 and 4.18, and using representation

$$U(T) = e^{LT}U_0 + \int_0^T e^{L(T-t)}(f_1 + \partial_x f_2)\, dt =: S_1(T)U_0 + V + W_0 + W_1 + W_2$$

(4.69)

coming from (4.42)–(4.43), (4.50).   ●

4.3. Nonlinear stability argument. We are now ready to establish the result of Theorem 2.2, assuming for the moment the bounds asserted in Proposition 4.14 on the low-frequency solution operator $S_1(t)$. Define the nonlinear perturbation

$$U := \tilde{U} - \bar{U},$$

(4.70)

where $\tilde{U}$ denotes a solution of (1.2) with initial data $\tilde{U}_0$ close to $\bar{U}$. Then, Taylor expanding about $\bar{U}$, we may rewrite (1.2) as

$$U_t - LU = \partial_x Q(U, \partial_x U) =: \partial_x f_2, \qquad U(0) = U_0,$$

(4.71)

where

$$|Q(U, \partial_x U)| \leq C(|U||\partial_x U| + |U|^2),$$
$$|\partial_x Q(U, \partial_x U)| \leq C(|U||\partial_x^2 U| + |\partial_x U|^2 + |U||\partial_x U|),$$
$$|\partial_x^2 Q(U, \partial_x U)| \leq C(|U||\partial_x^3 U| + |\partial_x U||\partial_x^2 U| + |\partial_x U|^2),$$

(4.72)

so long as U remains uniformly bounded.

Lemma 4.20 *So long as U remains uniformly bounded,*

$$|f_2(t)|_{L^1(x)} \leq C|U(t)|^2_{H^1(x)},$$
$$|\partial_x f_2(t)|_{H^1(x)} \leq C|U(t)|^2_{H^{[d/2]+2}(x)}.$$

(4.73)

Proof. Immediate, by (4.72) and $|\partial_x U|_{L^\infty(x)} \leq |U|_{H^{[d/2]+2}(x)}$.   ●

Proof of Theorem 1.18. Set $8\theta_1 := \theta_2$, $\theta_2 > 0$ as in the statement of Proposition 4.1, and define

$$\zeta(t) := \sup_{0 \leq \tau \leq t} |U(\tau)|_{L^2}(1+\tau)^{\frac{(d-1)}{4}-\epsilon},$$

(4.74)

where $\epsilon > 0$ is as in the statement of Theorem 2.2. By local well-posedness in H^s, Proposition 1.16, and the standard principle of continuation, there exists a solution $U \in H^s(x)$, $U_t \in H^{s-2}(x) \subset L^2(x)$ on the open time-interval for which $|U|_{H^s}$ remains bounded. On this interval, ζ is well-defined and continuous.

Let $[0, T)$ be the maximal interval on which $|U|_{H^s(x)}$ remains bounded by some fixed, sufficiently small constant $\delta > 0$. By Proposition 4.1,

$$|U(t)|^2_{H^s} \leq C|U(0)|^2_{H^s} e^{-\theta t} + C \int_0^t e^{-\theta_2(t-\tau)}|U(\tau)|^2_{L^2}\, d\tau$$

$$\leq C_2\left(|U(0)|^2_{H^s} + \zeta(t)^2\right)(1+\tau)^{-(d-1)/2+2\epsilon}. \qquad (4.75)$$

Combining (4.31) with the bounds of Lemma 4.20, we thus obtain

$$\int_0^t e^{-\theta_2(t-\tau)}|U(\tau)|^2_{L^2}\, d\tau \leq C(1+t)^{-(d-1)/4+\epsilon}|U_0|^2_{L^1}$$

$$+ Ce^{-2\theta_1 t}\left(|U_0|_{L^2} + |LU_0|_{H^1}\right)$$

$$+ C \int_0^t e^{-2\theta_1(t-\tau)}|\partial_x f_2(\tau)|^2_{H^1}\, d\tau$$

$$+ C\left(\int_0^t (1+t-\tau)^{-(d-1)/4+\epsilon-1/2}|f_2(\tau)|_{L^1}\, d\tau\right)^2$$

$$\leq C(1+t)^{-(d-1)/2+2\epsilon}|\tilde{U}_0 - \bar{U}|^2_{L^1 \cap H^3}$$

$$+ C\zeta(t)^4\left(\int_0^t (1+t-\tau)^{-(d-1)/4+\epsilon-1/2}(1+\tau)^{-(d-1)/2+2\epsilon}\, d\tau\right)^2$$

$$\leq C_2\left(|\tilde{U}_0 - \bar{U}|^2_{L^1 \cap H^3} + C\zeta(t)^4\right)(1+\tau)^{-(d-1)/2+2\epsilon} \qquad (4.76)$$

for $d \geq 2$ and $\epsilon = 0$, or $d \geq 3$ and ϵ sufficiently small (exercise).

Applying Proposition 4.1 a second time, we obtain

$$|U(t)|^2_{H^s} \leq Ce^{-\theta_2 t}|\tilde{U}_0 - \bar{U}|^2_{H^s} + C \int_0^t e^{-\theta_2(t-\tau)}|U(\tau)|^2_{L^2}\, d\tau$$

$$\leq C_2\left(|\tilde{U}_0 - \bar{U}|^2_{L^1 \cap H^s} + C\zeta(t)^4\right)(1+\tau)^{-(d-1)/2+2\epsilon} \qquad (4.77)$$

and therefore

$$\zeta(t) \leq C\left(|\tilde{U}_0 - \bar{U}|_{L^1 \cap H^s} + C\zeta(t)^2\right) \qquad (4.78)$$

so long as $|U|_{H^s}$ and thus $|U|$ remains uniformly bounded.

From (4.78), it follows by continuous induction (exercise) that

$$\zeta(t) \leq 2C|U(0)|_{L^1 \cap H^s} \qquad (4.79)$$

for $|U(0)|_{L^1 \cap H^s}$ sufficiently small, and thus also

$$|U(t)|_{H^s} \leq 2C(1+t)^{-(d-1)/4+\epsilon}|U(0)|_{L^1 \cap H^s} \qquad (4.80)$$

as claimed, on the maximal interval $[0, T)$ for which $|U|_{H^s} < \delta$. In particular, $|U(t)|_{H^s} < \delta/2$ for $|U(0)|_{L^1 \cap H^s}$ sufficiently small, so that $T = +\infty$, and we obtain both global existence and H^s decay at the claimed rate. Applying (1.13), we obtain the same bound for $|U(t)|_{L^\infty}$, and thus, by L^p interpolation (formula (6.43), Appendix B), for $|U(t)|_{L^p}$, all $2 \leq p \leq \infty$.   ●

Together with the prior results of [Z.3], the results of this section complete the analysis. We'll review those prior results in the low-frequency analysis of the next section.

Remarks 4.21 1. Sharp rates of decay in L^p, $2 \leq p \leq \infty$ may be obtained by a further, bootstrap argument, as in the related analysis of [Z.4].

2. The approach followed here, yielding high-frequency estimates entirely from resolvent bounds, also greatly simplifies the analysis of the one-dimensional case, replacing the detailed pointwise high-frequency estimates of [MaZ.3]. Pointwise low-frequency bounds for the moment are still required; however, a new method of shock-tracking introduced in [GMWZ.3] might yield an alternative approach based on resolvent bounds. This would be an interesting direction for further investigation. At the same time, the new approach yields slightly more general results: specifically, we may drop the assumption that the hyperbolic convection matrix $A_*(\bar{U}(x))$ have constant multiplicity with respect to x, and we may reduce the regularity requirement on the coefficients from C^5 to C^3 and the regularity of the perturbation from $L^1 \cap H^3$ to $L^1 \cap H^2$.[12]

3. Kawashima-type energy estimates are available also in the "dual", relaxation case, at least for small-amplitude shocks. Thus, we immediately obtain by the methods of this article, together with prior results of [Z.3], a corresponding result of nonlinear multi-dimensional stability of small-amplitude relaxation fronts. In the one-dimensional case, an energy estimate has recently been established for large-amplitude profiles in [MaZ.5]; in combination with the methods of this article, this recovers the large-amplitude nonlinear stability result of [MaZ.5] without the restrictive hypothesis of constant multiplicity of relaxation characteristics. Besides being mathematically appealing, this improvement is important for applications to moment-closure models.

4. The analysis so far, with the exception of the large-amplitude version of the "magic" energy estimate (4.2), is all "soft". The novelty lies, rather, in the argument structure, which for the first time successfully integrates Parseval- and semigroup-type (i.e., direct, pointwise-in-time) bounds in the analysis of long-time viscous shock stability. See [KK] for an interesting earlier approach based entirely on Parseval's identity and Hausdorff–Young's inequality, which may be used in cases of sufficiently fast decay to establish $\int_0^\infty |U(s)|_{L^2}^2 \, ds < +\infty$; in the shock-wave setting, this occurs for dimensions $d = 3$ and higher [GMWZ.2]. For further discussion/comparison, see Remark 5.3 below. More generally, our mid- and high-frequency analysis could be viewed as addressing the larger problem of obtaining time-algebraic decay estimates from spectral

---

[12] More precisely, to C^4 and $L^1 \cap H^3$ by the arguments of this article alone. To obtain the weaker regularity stated, we must substitute for the uniform $H^1 \to H^1$ high- and mid-frequency resolvent bounds of Propositions 4.7 and 4.9 the corresponding $L^2 \to L^2$ bounds obtained in [MaZ.3] by pointwise methods or in [GMWZ.4] by symmetrizer estimates.

information in the absence of either a spectral gap or sectorial-type resolvent bounds; see Remark 6.23.

Note in particular that our conclusions are independent of the particular setting considered here, depending only on the availability of nonlinear smoothing-type energy estimates favorable in the highest derivative. Thus, our methods may be of use in the study of delicate stability phenomena arising in other equations. Moreover, they are truly multi-dimensional, in the sense that they do not intrinsically depend on planar structure/decoupling of Fourier modes, so could be applied also to the case of a background wave $\bar{u}$ varying also in the $\tilde{x}$ directions: for example to traveling waves in a channel $(x_1, \tilde{x}) \in \mathbb{R} \times X$, $X \subset \mathbb{R}^{d-1}$ bounded.

Indeed, there is no requirement that the resolvent equation even be posable as an (possibly infinite-dimensional) ODE. Thus our methods may be of use in interesting situations for which this standard assumption does not hold: in particular, *stability of irrational-speed semi-discrete traveling-waves for non-upwind schemes* (see discussion, treatment of the upwind case in [B, BHR]), or *stability of shock profiles for Boltzmann equations*. Of course, in each of the mentioned cases (fully multi-dimensional, semidiscrete, and kinetic equations), there remains the problem of carrying out a suitable low-frequency linearized analysis; nonetheless, it is a substantial reduction of the problem.

5 Low Frequency Analysis/Completion of Proofs

We complete our analysis by carrying out the remaining, low-frequency analysis, establishing Propositions 4.14 and 2.14, and Theorem 2.1. Our main tools in this endeavor are the conjugation lemma, Corollary 3.13, by which we reduce the resolvent equation to a constant-coefficient equation with implicitly determined transmission condition, and the Evans function, through which we in effect obtain asymptotics for the resulting, unknown transmission condition in the low-frequency limit (Proposition 2.14).

5.1. Bounds on $S_1(t)$. Continuing in our backwards fashion, we begin by establishing the deferred low-frequency bounds cited in Proposition 4.14 of the previous section under appropriate resolvent bounds, thus reducing the remaining analysis to a detailed study of the resolvent and eigenvalue equations, to be carried out in the remainder of this section. The following low-frequency estimates were obtained in [Z.3] by pointwise estimates on the resolvent kernel; we sketch a different proof in Section 5.4, based on degenerate symmetrizer estimates as in [GMWZ.1, W].

Proposition 5.1 ([Z.3, Z.4]) *Assuming (A1)–(A3), (H0)–(H5), for $\rho := |(\tilde{\xi}, \lambda)|$ sufficiently small, and*

$$\Re\lambda = -\theta(|\tilde{\xi}|^2 + |\Im\lambda|^2)$$

for θ sufficiently small, there holds

$$|(L_{\tilde{\xi}} - \lambda)^{-1}\partial_{x_1}^\beta f|_{L^p(x_1)} \le C\gamma_1\gamma_2\rho^{(1-\alpha)|\beta|-1}|f|_{L^1(x_1)} \tag{5.1}$$

for all $2 \le p \le \infty$, $0 \le |\beta| \le 1$, where

$$\gamma_1(\tilde{\xi}, \lambda) := \begin{cases} 1 \text{ in case of strong (uniform) dynamical stability,} \\ 1 + \sum_j [\rho^{-1}|\Im\lambda - i\tau_j(\tilde{\xi})| + \rho]^{-1} \text{ otherwise,} \end{cases} \tag{5.2}$$

$$\gamma_2(\tilde{\xi}, \lambda) := 1 + \sum_{j,\pm} [\rho^{-1}|\Im\lambda - \eta_j^\pm(\tilde{\xi})| + \rho]^{-t}, \qquad 0 < t < 1, \tag{5.3}$$

$\eta_j(\cdot)$ is as in (H5) and the zero-level set of $\Delta(\tilde{\xi}, i\tau)$ for $\tilde{\xi}$, τ real is given by $\cup_{j,\tilde{\xi}}(\tilde{\xi}, i\tau_j(\tilde{\xi}))$, and

$$\alpha := \begin{cases} 0 \text{ for Lax or overcompressive case,} \\ 1 \text{ for undercompressive case.} \end{cases} \tag{5.4}$$

(For a definition of overcompressive and undercompressive types, see, e.g., Section 1.2, [Z.4].) More precisely, $t := 1-1/K_{max}$, where $K_{max} := \max K_j^\pm = \max s_j^\pm$ is the maximum among the orders of all branch singularities $\eta_j^\pm(\cdot)$, $s_j^\pm$ and $\eta_j^\pm$ defined as in (H5); in particular, $t = 1/2$ in the (generic) case that only square-root singularities occur.

Remark 5.2 Rewriting the $L^1 \to L^2$ resolvent bound (5.1) as

$$|(L_{\tilde{\xi}} - \lambda)^{-1}f|_{L^2(x_1)} \le \frac{C\gamma_2|f|_{L^1(x_1)}}{\sqrt{\Re\lambda + \rho^2}}, \tag{5.5}$$

we may recognize it as essentially a second-order correction of the corresponding $L^2 \to L^2$ resolvent bound

$$|(L_{\tilde{\xi}} - \lambda)^{-1}f|_{L^2(x_1)} \le \frac{C|f|_{L^2(x_1)}}{\Re\lambda} \tag{5.6}$$

of the inviscid theory, which in turn is approximately the condition for a bounded C^0 semigroup (Appendix A). The singular factor γ_2 appearing in the numerator of (5.5), and the square root in the denominator, are new effects connected with the fact that the bound is taken between L^1 and L^2 rather than in L^2 alone; for further discussion, see [Z3].

Proof of Proposition 4.14. For simplicity, we restrict to the Lax-type, uniformly dynamically stable case; other cases are similar. By analyticity of the resolvent on the resolvent set, we may deform the contour in (4.14) to obtain

$$S_1(t)f(x) =$$

$$\int_{|\tilde{\xi}|^2 \le \theta_1+\theta_2} \oint_{\Re\lambda = \frac{-\theta_1}{\theta_1+\theta_2}(|\tilde{\xi}|^2 - |\Im\lambda|^2) \ge -\theta_1} e^{i\tilde{\xi}\cdot\tilde{x}+\lambda t}(\lambda - L_{\tilde{\xi}})^{-1}\hat{f}(x_1, \tilde{\xi})\,d\tilde{\xi}d\lambda,$$

$$\tag{5.7}$$

where $\theta_j > 0$ are as in (4.41), and Corollary 4.11, with θ_1 sufficiently small in relation to θ_2. Bounding

$$|\tilde{f}|_{L^\infty(\xi', L^1(x_1))} \leq |f|_{L^1(x_1, \tilde{x})} = |f|_1 \tag{5.8}$$

using Hausdorff–Young's inequality, and appealing to the $L^1 \to L^p$ resolvent estimates of Proposition 5.1 with $\theta := \frac{-\theta_1}{\theta_1 + \theta_2} > 0$ taken suffiently small, we may thus bound

$$|\hat{u}(x_1, \xi', \lambda)|_{L^2(x_1)} \leq |f|_1 b(\tilde{\xi}, \lambda), \tag{5.9}$$

for $\hat{u} := (L_{\tilde{\xi}} - \lambda)f$ and $\rho := |\tilde{\xi}| + |\lambda|$ sufficiently small, where

$$b(\tilde{\xi}, \lambda) := \rho^{-1} \gamma_2(\tilde{\xi}, \lambda). \tag{5.10}$$

Denoting by $\Gamma(\tilde{\xi})$ the arc

$$\Re \lambda = \frac{-\theta_1}{\theta_1 + \theta_2}(|\tilde{\xi}|^2 + |\Im \lambda|^2) \geq -\theta_1 \tag{5.11}$$

and using in turn Parseval's identity, Fubini's Theorem, the triangle inequality, and our $L^1 \to L^2$ resolvent bounds, we may estimate

$$|S_1(t)f|_{L^2(x_1, \tilde{x})}(t) = \left(\int_{x_1} \int_{\tilde{\xi} \in \mathbb{R}^{d-1}} \left| \oint_{\lambda \in \tilde{\Gamma}(\tilde{\xi})} e^{\lambda t} \hat{u}(x_1, \tilde{\xi}, \lambda) d\lambda \right|^2 d\tilde{\xi} \, dx_1 \right)^{1/2}$$

$$= \left(\int_{\tilde{\xi} \in \mathbb{R}^{d-1}} \left| \oint_{\lambda \in \tilde{\Gamma}(\tilde{\xi})} e^{\lambda t} \hat{u}(x_1, \tilde{\xi}, \lambda) d\lambda \right|_{L^2(x_1)}^2 d\tilde{\xi} \right)^{1/2}$$

$$\leq \left(\int_{\tilde{\xi} \in \mathbb{R}^{d-1}} \left| \oint_{\lambda \in \tilde{\Gamma}(\tilde{\xi})} |e^{\lambda t}| |\hat{u}(x_1, \tilde{\xi}, \lambda)|_{L^2(x_1)} d\lambda \right|^2 d\tilde{\xi} \right)^{1/2}$$

$$\leq |f|_1 \left(\int_{\tilde{\xi} \in \mathbb{R}^{d-1}} \left| \oint_{\lambda \in \tilde{\Gamma}(\tilde{\xi})} e^{\Re \lambda t} b(\tilde{\xi}, \lambda) d\lambda \right|^2 d\tilde{\xi} \right)^{1/2}, \tag{5.12}$$

from which we readily obtain the claimed bound on $|S_1(t)f|_2$ using (5.10) and the definition of γ_2. Specifically , parametrizing $\Gamma(\tilde{\xi})$ by

$$\lambda(\tilde{\xi}, k) = ik - \theta_1(k^2 + |\tilde{\xi}|^2), \qquad k \in \mathbb{R},$$

and observing that in nonpolar coordinates

$$\rho^{-1} \gamma_2 \leq \left[(|k| + |\tilde{\xi}|)^{-1}(1 + \sum_{j \geq 1} \left(\frac{|k - \tau_j(\tilde{\xi})|}{\rho} \right)^{\frac{1}{s_j} - 1} \right]$$

$$\leq \left[|k| + |\tilde{\xi}|)^{-1}(1 + \sum_{j \geq 1} \left(\frac{|k - \tau_j(\tilde{\xi})|}{\rho} \right)^{\varepsilon - 1} \right], \tag{5.13}$$

where $\varepsilon := \frac{1}{\max_j s_j}$ ($0 < \varepsilon < 1$ chosen arbitrarily if there are no singularities), we obtain a contribution bounded by

$$C|f|_1 \left(\int_{\tilde{\xi} \in \mathbb{R}^{d-1}} \left| \int_{-1}^{+1} e^{-\theta(k^2 + |\tilde{\xi}|^2)t} (\rho)^{-1} \gamma_2 dk \right|^2 d\tilde{\xi} \right)^{1/2}$$

$$\leq C|f|_1 \int_{\tilde{\xi} \in \mathbb{R}^{d-1}} \left(e^{-2\theta|\tilde{\xi}|^2 t} |\tilde{\xi}|^{-2\varepsilon} \left| \int_{-\infty}^{+\infty} e^{-\theta|k|^2 t} |k|^{\varepsilon - 1} dk \right|^2 d\tilde{\xi} \right)^{1/2}$$

$$+ C \sum_{j \geq 1} |f|_1 \int_{\tilde{\xi} \in \mathbb{R}^{d-1}} \left(e^{-2\theta|\tilde{\xi}|^2 t} |\tilde{\xi}|^{-2\varepsilon} \left| \int_{-\infty}^{+\infty} e^{-\theta|k|^2 t} |k - \tau_j(\tilde{\xi})|^{\varepsilon - 1} dk \right|^2 d\tilde{\xi} \right)^{1/2}$$

$$\leq C|f|_1 \int_{\tilde{\xi} \in \mathbb{R}^{d-1}} \left(e^{-2\theta|\tilde{\xi}|^2 t} |\tilde{\xi}|^{-2\varepsilon} \left| \int_{-\infty}^{+\infty} e^{-\theta|k|^2 t} |k|^{\varepsilon - 1} dk \right|^2 d\tilde{\xi} \right)^{1/2}$$

$$\leq C|f|_1 t^{-(d-1)/4},$$

$$(5.14)$$

yielding the asserted bound for $t \geq 1$; for $t \leq 1$ on the other hand, we obtain the asserted uniform bound by local integrability together with boundedness of the region of integration. Derivative bounds, $\beta = 1$, follow similarly.   ●

Remark 5.3 In the proof of Proposition 4.14, we have used in a fundamental way the semigroup representation and the autonomy of the underlying equations, specifically in the parabolic deformation of contours revealing the stabilizing effect of diffusion. The resulting bounds are not available through Parseval's identity or Hausdorff–Young's inequality, corresponding to the choice of straight-line contours parallel to the imaginary axis. Such "parabolic," semigroup-type estimates do not immediately translate to the small-viscosity context, for which the linearized equations are variable-coefficient in time. Moreover, it is not clear in this context how one could obtain resolvent bounds between different norms, since standard pseudo- or paradifferential techniques are based on Fourier decompositions respecting L^2 but not necessarily other norms. The efficient accounting of diffusive effects in this setting is an interesting and fundamental problem that appears to be of general mathematical interest [MeZ.1].

Remark 5.4 As noted in the acknowledgements, all estimates on the linearized solution operator, both on $S_1(t)$ and $S_2(t)$, have been obtained through resolvent bounds alone. These could be obtained in principle by a variety of methods, an observation that may be useful in the more general situations (e.g., semidiscrete or Boltzmann shock profiles) described in Remark 4.21.4.

5.2. Link to the inviscid case. It remains to study the resolvent (resp. eigenvalue) equation

$$(L_{\tilde{\xi}} - \lambda)U = \begin{cases} f, \\ 0, \end{cases} \tag{5.15}$$

where

$$(L_{\tilde{\xi}} - \lambda)U = \overbrace{(B^{11}U')' - (A^1 U)' - i\sum_{j\neq 1} A^j \xi_j U}^{L_0 U}$$
$$+ i\sum_{j\neq 1} B^{j1}\xi_j U' + i\sum_{k\neq 1}(B^{1k}\xi_k U)' - \sum_{j,k\neq 1} B^{jk}\xi_j\xi_k U - \lambda U. \tag{5.16}$$

Up to this point our analysis has been rather general; from here on, we make extensive use of the property, convenient for the application of both Evans function and Kreiss symmetrizer techniques, that the (5.15) may be written as an ODE

$$W' - \mathbb{A}(\tilde{\xi}, \lambda, x_1)W = \begin{cases} F, \\ 0, \end{cases} \tag{5.17}$$

in the phase variable $W := \begin{pmatrix} U \\ B_{II}^{11} U' \end{pmatrix} \in \mathbb{C}^{n+r}$, where $F \sim f$. This is a straight-forward consequence of block-diagonal structure, (A1), and invertibility of $\tilde{A}_{11}^1$, (H1); see [MaZ.3, Z.4] for further details. (There is ample reason to relax this condition, however; see Remark 4.21.4.) Our first task, carried out in this subsection, is to make contact with the inviscid case.

5.2.1. Normal modes. By the conjugation lemma, Corollary 3.13, we may reduce (5.16) by a change of coordinates $W = P_\pm Z_\pm$, $f = P_\pm \tilde{f}$, with $P_\pm \to I$ as $x_1 \to \pm\infty$, P analytic in $(\tilde{\xi}, \lambda)$, to a pair of constant-coefficient equations

$$Z'_\pm - \mathbb{A}_\pm(\tilde{\xi}, \lambda)Z_\pm = \begin{cases} \tilde{f} \\ 0, \end{cases} \tag{5.18}$$

on the half-lines $x_1 \gtrless 0$, coupled by the implicitly determined transmission conditions

$$P_+ Z_+(0) - P_- Z_-(0) = 0 \tag{5.19}$$

at the boundary $x_1 = 0$, where

$$\mathbb{A}_\pm(\tilde{\xi}, \lambda) := \lim_{x_1 \to \pm\infty} \mathbb{A}(\tilde{\xi}, \lambda, x_1) : \tag{5.20}$$

that is, a system that at least superficially resembles that arising in the inviscid theory. The following lemma generalizes a standard result of Hersch [H] in the inviscid theory; see Exercise 2.10.

Lemma 5.5 ([ZS, Z.3]) *Assuming (A1)–(A2), (H1) (and, implicitly, existence of a profile), the matrices $\mathbb{A}_\pm$ have no center subspace on*

$$\Lambda := \{\lambda : \Re\lambda > \frac{-\theta(|\tilde{\xi}|^2 + |\Im\lambda|^2)}{1 + |\tilde{\xi}|^2 + |\Im\lambda|^2}\}, \tag{5.21}$$

for $\theta > 0$ sufficiently small; moreover, the dimensions of their stable and unstable subspaces agree, summing to full rank $n + r$.

Proof. The fundamental modes of (5.18) are of form $e^{\mu x}V$, where μ, V satisfy the *characteristic equation*

$$\left[\mu^2 B_\pm^{11} + \mu(-A_\pm^1 + i\sum_{j\neq 1} B_\pm^{j1}\xi_j + i\sum_{k\neq 1} B^{1k}\xi_k) \right.$$
$$\left. -(i\sum_{j\neq 1} A^j\xi_j + \sum_{jk\neq 1} B^{jk}\xi_j\xi_k + \lambda I) \right] \mathbf{v} = 0. \tag{5.22}$$

The existence of a center manifold thus corresponds with existence of solutions $\mu = i\xi_1$, V of (5.22), ξ_1 real, i.e., solutions of the *dispersion relation*

$$(-\sum_{j,k} B_\pm^{jk}\xi_j\xi_k - i\sum_j A_\pm^j\xi_j - \lambda I)\mathbf{v} = 0. \tag{5.23}$$

But, $\lambda \in \sigma(-B^{\xi\xi} - iA_\pm^\xi)$ implies, by (3.52), that

$$\Re\lambda \leq -\theta_1|\xi|^2/(1+|\xi|^2).$$

Noting that for low frequencies $|\Im\lambda| = \mathcal{O}(|\xi|)$, we thus obtain

$$\Re\lambda \leq \frac{-\theta(|\tilde{\xi}|^2 + |\Im\lambda|^2)}{1+|\tilde{\xi}|^2 + |\Im\lambda|^2},$$

in contradiction of (5.21).

Nonexistence of a center manifold, together with connectivity of the set described in (5.21), implies that the dimensions of stable and unstable manifolds at $+\infty/-\infty$ are constant on Λ. Taking $\lambda \to +\infty$ along the real axis, with $\tilde{\xi} \equiv 0$, we find that these dimensions sum to the full dimension $n + r$ as claimed. For, Fourier expansion about $\xi_1 = \infty$ of the one-dimensional ($\tilde{\xi} = 0$) dispersion relation (see, e.g., Appendix A.4 [Z.4]) yields $n - r$ "hyperbolic" modes

$$\lambda_j = -i\xi_1 a_j^* + \dots, \quad j = 1, \dots, n-r,$$

where a_j^* denote the eigenvalues of

$$A_*^1 := A_{11}^1 - (b_2^{11})^{-1}b_1^{11}A_{12}^1$$

and r "parabolic" modes

$$\lambda_{n-r+j} = -b_j\xi_1^2 + \dots, \quad j = 1, \dots r,$$

where b_j denote the eigenvalues of b_2^{11}; here, we have suppressed the $\pm$ indices for readability. Inverting these relationships to solve for $\mu := i\xi_1$, we find, for $\lambda \to \infty$, that there are $n - r$ hyperbolic roots $\mu_j \sim -\lambda/a_j^*$, and r parabolic roots $\mu_{n-r+j}^\pm \sim \sqrt{\lambda/b_j}$. By assumption (H1), $\det A_*^1(x_1) \neq 0$ for all x_1, and so the former yield a fixed number $k/(n - r - k)$ of stable/unstable roots, independent of x_1, and thus of $\pm$. Likewise, $(\tilde{H}1(\text{i}))$ implies that the latter yields r stable, r unstable roots. (Note: we have here used the existence of a connecting profile: i.e., equality is not a consequence of algebraic structure alone.) Combining, we find the desired consistent splitting, with $(k+r)/(n-k)$ stable/unstable roots at both $\pm\infty$.   •

Corollary 5.6 *Assuming (A1)–(A2) and (H1), on the set Λ defined in (5.21), there exists an analytic choice of basis vectors $V_j^\pm$, $j = 1, \ldots k$ (resp. $V_j^\pm$, $j = k+1, \ldots, n+r$) spanning the stable (resp. unstable) subspace of $\mathbb{A}_\pm$.*

Proof. By spectral separation of stable and unstable subspaces, the associated (group) eigenprojections are analytic. The existence of analytic bases then follows by a standard result of Kato; see [Kat], pp. 99–102.   •

Definition 5.7 For $x_1 \gtrless 0$, define *normal modes* for the variable-coefficient ODE (5.17) as

$$W_j^\pm(x_1) := P_\pm e^{\mathbb{A}_\pm x_1} V_j^\pm. \qquad (5.24)$$

In particular, $W_1^+, \ldots, W_k^+$ span the manifold of solutions of the eigenvalue equation decaying as $x_1 \to +\infty$, and $W_{k+1}^+, \ldots, W_{n+r}^+$ span the manifold of solutions of the eigenvalue equation decaying as $x_1 \to -\infty$.

5.2.2. Low-frequency asymptotics. For comparison with the inviscid case, it is convenient to introduce polar coordinates

$$(\tilde{\xi}, \lambda) =: (\rho\tilde{\xi}_0, \rho\lambda_0), \qquad (5.25)$$

$\rho \in \mathbb{R}^1$, $(\tilde{\xi}_0, \lambda_0) \in \mathbb{R}^{d-1} \times \{\Re\lambda \geq 0\} \setminus \{(0,0)\}$, and consider $W_j^\pm$, $V_j^\pm$ as functions of $(\rho, \tilde{\xi}_0, \lambda_0)$. With this notation, the resolvent (resp. eigenvalue) equation (5.16) may be viewed as a singular perturbation as $\rho \to 0$ of the corresponding inviscid equation. The following results quantify this observation, separating normal modes into slow-decaying modes asymptotic to those of the inviscid theory (the "slow manifold" in the singular perturbation limit) and fast-decaying transient modes associated with the viscous regularization.

Lemma 5.8 ([K, Me.4]) *Let there hold (A1), (H2), and (H4), or, more generally, $\sigma(df^\xi(u_\pm))$ real, semisimple, and of constant multiplicity for $\xi \in \mathbb{R}^d \setminus \{0\}$ and $\det df^1 \neq 0$. Then, the vectors $\{r_1^-, \cdots, r_{n-i_-}^-\}, \{r_{i_++1}^+, \cdots, r_n^+\}$ defined as in (2.6)–(2.8), Section 2.2 (and thus also $\Delta(\tilde{\xi}, \lambda)$) may be chosen to be homogeneous degree zero (resp. one), analytic on $\tilde{\xi} \in \mathbb{R}^{d-1}$, $\Re\lambda > 0$ and continuous at the boundary $\tilde{\xi} \in \mathbb{R}^{d-1} \setminus \{0\}$, $\Re\lambda = 0$.*

Proof. See Exercises 4.23–4.24 and Remark 4.25, Section 4.5.2 of [Z.3] for a proof in the general case (three alternative proofs, based respectively on [K], [CP], and [ZS]). In the case of main interest, when (H5) holds as well, this will be established through the explicit computations of Section 5.4.   •

Lemma 5.9 ([ZS, Z.3, MeZ.2]) *Under assumptions (A1) – (A2), (H0) – (H4), the functions $V_j^\pm$ may be chosen within groups of r "fast", or "viscous" modes bounded away from the center subspace of coefficient $\mathbb{A}_\pm$, analytic in $(\rho, \tilde\xi_0, \lambda_0)$ for $\rho \geq 0$, $\tilde\xi_0 \in \mathbb{R}^{d-1}$, $\Re\lambda_0 \geq 0$, and n "slow", or "inviscid" modes approaching the center subspace as $\rho \to 0$, analytic in $(\rho, \tilde\xi_0, \lambda_0)$ for $\rho > 0$, $\tilde\xi_0 \in \mathbb{R}^{d-1}$, $\Re\lambda_0 \geq 0$ and continuous at the boundary $\rho = 0$, with limits*

$$V_j^\pm(0, \tilde\xi_0, \lambda_0) = \begin{pmatrix} (A_\pm^1)^{-1} r_j^\pm(\tilde\xi_0, \lambda_0) \\ 0 \end{pmatrix}, \tag{5.26}$$

$r_j^\pm$ *defined as in (2.6)–(2.8).*

Proof. We carry out here the simpler case $\Re\lambda_0 > 0$. In the case of interest, that (H5) also holds, the case $\Re\lambda = 0$ follows by the detailed computations in Section 5.4; see Remark 5.17. For the general case (without (H5)), see [Me.2].

Substituting $U = e^{\mu x_1} \mathbf{v}$ into the limiting eigenvalue equations written in polar coordinates, we obtain the polar characteristic equation,

$$\left[\mu^2 B_\pm^{11} + \mu\left(-A_\pm^1 + i\rho \sum_{j\neq 1} B_\pm^{j1}\xi_j + i\rho \sum_{k\neq 1} B^{1k}\xi_k\right) \right.$$
$$\left. - \left(i\rho \sum_{j\neq 1} A^j\xi_j + \rho^2 \sum_{jk\neq 1} B^{jk}\xi_j\xi_k + \rho\lambda I\right) \right] \mathbf{v} = 0, \tag{5.27}$$

where for notational convenience we have dropped subscripts from the fixed parameters $\tilde\xi_0$, λ_0. At $\rho = 0$, this simplifies to

$$\left(\mu^2 B_\pm^{11} - \mu A_\pm^1\right) \mathbf{v} = 0,$$

which, by the analysis in Appendix A.2 of the linearized traveling-wave ordinary differential equation $\left(\mu B_\pm^{11} - A_\pm^1\right)\mathbf{v} = 0$, has n roots $\mu = 0$, and r roots $\Re\mu \neq 0$. The latter, "fast" roots correspond to stable and unstable subspaces, which extend analytically as claimed by their spectral separation from other modes; thus, we need only focus on the bifurcation as ρ varies near zero of the n-dimensional center manifold associated with "slow" roots $\mu = 0$.

Positing a first-order Taylor expansion

$$\begin{cases} \mu = 0 + \mu^1\rho + o(\rho), \\ \mathbf{v} = \mathbf{v}^0 + \mathbf{v}^1\rho + o(\rho), \end{cases} \tag{5.28}$$

and matching terms of order ρ in 5.27, we obtain

$$\left(-\mu^1 A_\pm^1 - i\sum_{j\neq 1} A^j\xi_j - \lambda I\right)\mathbf{v}^0 = 0, \tag{5.29}$$

or equivalently $-\mu_1$ is an eigenvalue of $(A^1)^{-1}(\lambda + iA^{\tilde\xi})$ with associated eigenvector $\mathbf{v}^0$.

For $\Re\lambda > 0$, $(A^1)^{-1}(\lambda + iA^{\tilde\xi})$ has no center subspace. For, substituting $\mu^1 = i\xi_1$ in 5.29, we obtain $\lambda \in \sigma(iA^\xi)$, pure imaginary, a contradiction. Thus, the stable/unstable spectrum *splits* to first order, and we obtain the desired analytic extension by standard matrix perturbation theory, though not in fact the analyticity of individual eigenvalues μ.   •

Remark 5.10 The first-order approximation 5.29 is exactly the matrix perturbation problem arising in the inviscid theory [K, Me.1].

Corollary 5.11 *Denoting $W = (U, b_1^{11}u^I + b_2^{11}u^{II})$ as above, we may arrange at $\rho = 0$ that all $W_j^\pm$ satisfy the linearized traveling-wave ODE*

$$(B^{11}U')' - (A^1U)' = 0,$$

with constant of integration

$$B^{11}U_j^{\pm'} - A^1U_j^\pm \equiv \begin{cases} 0, & \text{for fast modes,} \\ r_j^\pm, & \text{for slow modes,} \end{cases} \tag{5.30}$$

$r_j^\pm$ as above, with fast modes analytic at the $\rho = 0$ boundary and independent of $(\tilde\xi_0, \lambda_0)$ for $\rho = 0$, and slow modes continuous at $\rho = 0$.

Proof. Immediate   •

5.2.4. The Evans function and its low-frequency limit. We now complete the analogy with the inviscid case, introducing the Evans function D and establishing its relation to the Lopatinski determinant Δ in the limit as frequency goes to zero, Proposition 2.14.

Definition 5.12 Following the standard construction of, e.g., [E.1, E.2, E.3, E.4, E.5, AGJ, PW, GZ, ZS], we define on the set Λ an *Evans function*

$$\begin{aligned} D(\tilde\xi, \lambda) &:= \det\left(W_1^+, \ldots, W_k^+, W_{k+1}^-, \ldots, W_N^-\right)_{|x=0,\lambda} \\ &= \det\left(P_+V_1^+, \ldots, P_+V_k^+, P_-V_{k+1}^-, \ldots, P_-V_N^-\right)_{|x=0,\lambda}, \end{aligned} \tag{5.31}$$

measuring the (solid) angle of intersection between the manifolds of solutions of the eigenvalue equation decaying as $x_1 \to +\infty$ and $x_1 \to -\infty$, respectively, where $P_\pm$, $V_j^\pm$, $W_j^\pm$ are as in Definition 5.7.

Evidently, D vanishes if and only if there exists an exponentially decaying solution of the eigenvalue equation, i.e., λ is an eigenvalue of $L_{\tilde\xi}$, or, equivalently, there exists a solution of the boundary-value problem (5.18) satisfying boundary condition (5.19). That is, D is the analog for the viscous problem of the Lopatinski determinant Δ for the inviscid one.

Proof of Proposition 2.14. We carry out the proof in the Lax case only. The proofs in the under- and overcompressive cases are quite similar; see [ZS]. We are free to make any analytic choice of bases, and any nonsingular choice of coordinates, since these affect the Evans function only up to a nonvanishing analytic multiplier which does not affect the result. Choose bases $W_j^\pm$ as in Lemma 5.11, $W = (U, z_2')$, $z_2 = b_1^{11} u^I + b_2^{11} u^{II}$. Noting that $L_0 \bar{U}' = 0$, by translation invariance, we have that $\bar{U}'$ lies in both Span $\{U_1^+, \cdots, U_K^+\}$ and Span $(U_{K+1}^-, \cdots, U_{n+r}^-)$ for $\rho = 0$, hence without loss of generality

$$U_1^+ = U_{n+r}^- = \bar{U}', \tag{5.32}$$

independent of $\tilde{\xi}$, λ. (Here, as usual, "$'$" denotes $\partial/\partial x_1$).

More generally, we order the bases so that

$$W_1^+, \ldots, W_k^+ \text{ and } W_{n+r-k-1}^-, \cdots, W_{n+r}^-$$

are fast modes (decaying for $\rho = 0$) and

$$W_{k+1}^+, \ldots, W_K^+ \text{ and } W_{K+1}^-, \ldots, W_{n+r-k-2}^-$$

are slow modes (asymptotically constant for $\rho = 0$), fast modes analytic and slow modes continuous at $\rho = 0$ (Corollary 5.11).

Using the fact that U_1^+ and U_{n+r}^- are analytic, we may express

$$\begin{aligned} U_1^+(\rho) &= U_1^+(0) + U_{1,\rho}^+ \rho + o(\rho), \\ U_{n+r}^-(\rho) &= U_{n+r}^-(0) + U_{n+r,\rho}^- \rho + o(\rho), \end{aligned} \tag{5.33}$$

Writing out the eigenvalue equation

$$\begin{aligned} (B^{11} w') &= (A^1 w') - i\rho \sum_{j \neq 1} B^{j1} \xi_j w' \\ &\quad - i\rho \Big(\sum_{k \neq 1} B^{1k} \xi_k w \Big)' + i\rho \sum_{j \neq 1} A^j \xi_j w + \rho \lambda w \\ &\quad - \rho^2 \sum_{j,k \neq 1} B^{jk} \xi_j \xi_k w, \end{aligned} \tag{5.34}$$

in polar coordinates, we find that $Y^+ := U_{1,\rho}^+(0)$ and $Y^- := U_{n+r,\rho}^-(0)$ satisfy the variational equations

$$\begin{aligned} (B^{11} Y')' &= (A^1 Y') - \Big[i \sum_{j \neq 1} B^{j1} \xi_j \bar{U}' \\ &\quad + i \Big(\sum_{k \neq 1} B^{1k} \xi_k \bar{U}' \Big) + i \sum_{j \neq 1} A^j \xi_j \bar{U}' + \lambda \bar{U}' \Big], \end{aligned} \tag{5.35}$$

with boundary conditions $Y^+(+\infty) = Y^-(-\infty) = 0$.

Integrating from $+\infty$, $-\infty$ respectively, we obtain therefore

$$
\begin{aligned}
B^{11}Y\pm' - A^1Y\pm' = {}& if^{\tilde{\xi}}(\bar{U}) - iB^{1\tilde{\xi}}(\bar{U})\bar{U}' \\
& - iB^{\tilde{\xi}1}(\bar{U})\bar{U}' + \lambda\bar{U} - \left[if^{\tilde{\xi}}(u_\pm) + \lambda u_\pm\right],
\end{aligned}
\tag{5.36}
$$

hence $\tilde{Y} := (Y^- - Y^+)$ satisfies

$$
B^{11}\tilde{Y}' - A^1\tilde{Y} = i[f^{\tilde{\xi}}] + \lambda[u].
\tag{5.37}
$$

By (A1) together with (H1), $\begin{pmatrix} A^1_{11} & A^1_{12} \\ b^{11}_1 & b^{11}_2 \end{pmatrix}$ is invertible, hence

$$
\begin{aligned}
(U, z_2') \to {}& (z_2, -z_1, z_2' + (A^1_{21} - b^{11'}_1, A^1_{22} - b^{11'}_2)U) \\
& = (z_2, B^{11}U' - AU)
\end{aligned}
\tag{5.38}
$$

is a nonsingular coordinate change, where $\begin{pmatrix} z_1 \\ z_2 \end{pmatrix} := \begin{pmatrix} A^1_{11} & A^1_{12} \\ b^{11}_1 & b^{11}_2 \end{pmatrix} U$.

Fixing $\tilde{\xi}_0$, λ_0, and using $W^+_1(0) = W^-_{n+r}(0)$, we have

$$
\begin{aligned}
D(\rho) = {}& \det\left(W^+_1(0) + \rho W^+_{1_\rho}(0) + o(\rho), \cdots, W^+_K(0) + o(1),\right. \\
& \left. W^-_{K+1}(0) + o(1), \cdots, W^-_{n+r}(0)\rho W^-_{n+r_\rho}(0) + o(\rho)\right) \\
= {}& \det\left(W^+_1(0) + \rho W^+_{1_\rho}(0) + o(\rho), \cdots, W^+_K(0) + o(1),\right. \\
& \left. W^-_{K+1}(0) + o(1), \cdots, \rho\tilde{Y}(0) + o(\rho)\right) \\
= {}& \det\left(W^+_1(0) \cdots, W^+_K(0)\, W^-_{K+1}(0) \cdots, \rho\tilde{Y}(0)\right) + o(\rho)
\end{aligned}
$$

Applying now (5.38) and using (5.30) and (5.37), we obtain

$$
D(\rho) = C \det \left(
\begin{array}{l}
\overbrace{z^+_{2,1} \cdots z^+_{2,k},}^{\text{fast}} \quad \overbrace{*, \cdots, *, *, \cdots, *,}^{\text{slow}} \\
0, \cdots, 0, \quad r^+_{i_+ + 1}, \cdots, r^+_n, r^-_1, \cdots, r^-_{n - i_-} \\[1em]
\overbrace{z^-_{2, n+r-k-1}, \cdots, z^-_{2, n+r-1},}^{\text{fast}} \qquad * \\
0, \cdots, 0, \qquad\qquad i[f^{\tilde{\xi}}(u)] + \lambda[u]
\end{array}
\right)_{\big|_{x_1 = 0}} + o(\rho)
$$

$$
= \gamma\Delta(\tilde{\xi}, \lambda) + o(\rho)
$$

as claimed, where

$$\gamma := C \det \left(z_{2,1}^+, \cdots, z_{2,k}^+, z_{2,n+r-k-1}^-, \cdots, z_{2,n+r-1}^- \right)_{x_1=0}.$$

Noting that $\{z_{2,1}^+, \cdots, z_{2,k}^+\}$ and $\{z_{2,n+r-k-1}^-, \cdots, z_{2,n+r}^-\}$ span the tangent manifolds at $\bar{U}(\cdot)$ of the stable/unstable manifolds of traveling wave ODE (3.2) at U_+/U_-, respectively, with $z_{2,1}^+ = z_{2,n+r}^- = (b_1^{11}, b_2^{11})\bar{U}'$ in common, we see that γ indeed measures transversality of their intersection; moreover, γ is constant, by Corollary 5.11. $\qquad\bullet$

Remark 5.13 The proof of Proposition 2.14 may be recognized as a generalization of the basic Evans function calculation pioneered by Evans [E.4], relating behavior near the origin to geometry of the phase space of the traveling wave ODE and thus giving an explicit link between PDE and ODE dynamics. The corresponding one-dimensional result was established in [GZ]; for related calculations, see, e.g., [J, AGJ, PW].

5.3. Spectral bounds and necessary conditions for stability. Using Proposition 2.14, we readily obtain the stated necessary conditions for stability, Theorem 2.1. Define the *reduced Evans function* as

$$\bar{\Delta}(\tilde{\xi}, \lambda) := \lim_{\rho \to 0} \rho^{-\ell} D(\rho\tilde{\xi}, \rho\lambda). \tag{5.39}$$

By the results of the previous section, the limit $\bar{\Delta}$ exists and is analytic, with

$$\bar{\Delta} = \gamma\Delta, \tag{5.40}$$

for shocks of pure type (indeed, such a limit exists for all types). Evidently, $\bar{\Delta}(\cdot, \cdot)$ is homogeneous, degree ℓ.[13]

Lemma 5.14 ([ZS]) *Let $\bar{\Delta}(0, 1) \neq 0$. Then, near any root $(\tilde{\xi}_0, \lambda_0)$ of $\bar{\Delta}(\cdot, \cdot)$, there exists a continuous branch $\lambda(\tilde{\xi})$, homogeneous degree one, of solutions of*

$$\bar{\Delta}(\tilde{\xi}, \lambda(\tilde{\xi})) \equiv 0 \tag{5.41}$$

defined in a neighborhood V of $\tilde{\xi}_0$, with $\lambda(\tilde{\xi}_0) = \lambda_0$. Likewise, there exists a continuous branch $\lambda_*(\tilde{\xi})$ of roots of

$$D(\tilde{\xi}, \lambda_*(\tilde{\xi})) \equiv 0, \tag{5.42}$$

defined on a conical neighborhood $V_{\rho_0} := \{\tilde{\xi} = \rho\tilde{\eta} : \tilde{\eta} \in V, 0 < \rho < \rho_0\}$, $\rho_0 > 0$ sufficiently small, "tangent" to $\lambda(\cdot)$ in the sense that

$$|\lambda_*(\tilde{\xi}) - \lambda(\tilde{\xi})| = o(|\tilde{\xi}|) \tag{5.43}$$

as $|\tilde{\xi}| \to 0$, for $\tilde{\xi} \in V_{\rho_0}$.

---

[13] Here, and elsewhere, homogeneity is with respect to the positive reals, as in most cases should be clear from the context. Recall that Δ (and thus $\bar{\Delta}$) is only defined for real $\tilde{\xi}$, and $\Re\lambda \geq 0$.

Proof. Provided that $\bar{\Delta}(\tilde{\xi}_0, \cdot) \not\equiv 0$, the statement 5.41 follows by Rouche's Theorem, since $\bar{\Delta}(\tilde{\xi}_0, \cdot)$ are a continuous family of analytic functions. But, otherwise, restricting λ to the positive real axis, we have by homogeneity that

$$0 = \lim_{\lambda \to +\infty} \bar{\Delta}(\tilde{\xi}_0, \lambda) = \lim_{\lambda \to +\infty} \bar{\Delta}(\tilde{\xi}_0/\lambda, 1) = \bar{\Delta}(0, 1),$$

in contradiction with the hypothesis. Clearly we can further choose $\lambda(\cdot)$ homogeneous degree one, by homogeneity of $\bar{\Delta}$. Similar considerations yield existence of a branch of roots $\bar{\lambda}(\tilde{\xi}, \rho)$ of the family of analytic functions

$$g^{\tilde{\xi}, \rho}(\lambda) := \rho^{-\ell} D(\rho\tilde{\xi}, \rho\lambda), \tag{5.44}$$

for ρ sufficiently small, since $g^{\tilde{\xi},0} = \bar{\Delta}(\tilde{\xi}, \cdot)$. Setting $\lambda_*(\tilde{\xi}) := |\tilde{\xi}|\bar{\lambda}(\tilde{\xi}/|\tilde{\xi}|, |\tilde{\xi}|)$, we have

$$D(\tilde{\xi}, \lambda_*(\tilde{\xi})) = |\tilde{\xi}|^\ell g^{\tilde{\xi}/|\tilde{\xi}|, |\tilde{\xi}|}(\lambda_*) \equiv 0,$$

as claimed. "Tangency," in the sense of 5.43, follows by continuity of $\bar{\lambda}$ at $\rho = 0$, the definition of λ_*, and the fact that $\lambda(\tilde{\xi}) = |\tilde{\xi}|\lambda(\tilde{\xi}/|\tilde{\xi}|)$, by homogeneity of $\bar{\Delta}$. $\bullet$

Proof of Theorem 2.1. Weak spectral stability is clearly necessary for viscous stability. For, unstable (L^p) spectrum of $L_{\tilde{\xi}}$ (necessarily point spectrum, by Lemma 5.5 combined with the conjugation lemma, Corollary 3.13) corresponds to unstable (L^p) essential spectrum of the operator L for $p < \infty$, by a standard limiting argument (see e.g. [He, Z.1]), and unstable point spectrum for $p = \infty$. This precludes $L^p \to L^p$ stability, by the generalized Hille–Yosida theorem, (Proposition 6.11, Appendix A). Moreover, standard spectral continuity results [Ka, He, Z.2] yield that instability, if it occurs, occurs for a band of $\tilde{\xi}$ values, from which we may deduce by inverse Fourier transform the exponential instability of (3.8) for test function initial data $U_0 \in C_0^\infty$, with respect to any L^p, $1 \le p \le \infty$.

Thus, it is sufficient to establish that failure of weak refined dynamical stability implies failure of weak spectral stability, i.e., existence of a zero $D(\tilde{\xi}, \lambda) = 0$ for $\xi \in \mathbb{R}^{d-1}$, $\Re\lambda > 0$. Failure of weak inviscid stability, or $\Delta(\tilde{\xi}, \lambda) = 0$ for $\tilde{\xi} \in \mathbb{R}^{d-1}$, $\Re\lambda > 0$, implies immediately the existence of such a root, by tangency of the zero-sets of D and Δ at the origin, Lemma 5.14. Therefore, it remains to consider the case that weak inviscid stability holds, but there exists a root $D(\xi, i\tau)$ for ξ, τ real, at which Δ is analytic, $\Delta_\lambda \ne 0$, and $\beta(\xi, i\tau) < 0$, where β is defined as in (2.17).

Recalling that $D(\rho\tilde{\xi}, \rho\lambda)$ vanishes to order ℓ in ρ at $\rho = 0$, we find by L'Hopital's rule that

$$(\partial/\partial\rho)^{\ell+1} D(\rho\tilde{\xi}, \rho\lambda)|_{\rho=0, \lambda=i\tau} = (1/\ell!)(\partial/\partial\rho)g^{\tilde{\xi}, i\tau}(0)$$

and

$$(\partial/\partial\lambda) D(\rho\tilde{\xi}, \rho i\tau)|_{\rho=0, \lambda=i\tau} = (1/\ell!)(\partial/\partial\lambda)g^{\tilde{\xi}, i\tau}(0),$$

where $g^{\tilde{\xi},\rho}(\lambda) := \rho^{-\ell}D(\rho\tilde{\xi}, \rho\lambda)$ as in 5.44, whence

$$\beta = \frac{(\partial/\partial\rho)g^{\tilde{\xi},\lambda}(0)}{(\partial/\partial\lambda)g^{\tilde{\xi},\lambda}(0)}, \tag{5.45}$$

for β defined as in Definition 2.16, with $(\partial/\partial\lambda)g^{\tilde{\xi},\lambda}(0) \neq 0$.

By the (analytic) Implicit Function Theorem, therefore, $\lambda(\tilde{\xi}, \rho)$ is analytic in $\tilde{\xi}, \rho$ at $\rho = 0$, with

$$(\partial/\partial\rho)\,\lambda(\tilde{\xi}, 0) = -\beta, \tag{5.46}$$

where $\lambda(\tilde{\xi}, \rho)$ as in the proof of Lemma 5.14 is defined implicitly by $g^{\tilde{\xi},\lambda}(\rho) = 0$, $\lambda(\tilde{\xi}, 0) := i\tau$. We thus have, to first order,

$$\lambda(\tilde{\xi}, \rho) = i\tau - \beta\rho + \mathcal{O}(\rho^2). \tag{5.47}$$

Recalling the definition $\lambda_*(\tilde{\xi}) := |\tilde{\xi}|\bar{\lambda}(\tilde{\xi}/|\tilde{\xi}|, |\tilde{\xi}|)$, we have then, to second order, the series expansion

$$\lambda_*(\rho\tilde{\xi}) = i\rho\tau - \beta\rho^2 + \mathcal{O}(\rho^3), \tag{5.48}$$

where $\lambda_*(\tilde{\xi})$ is the root of $D(\tilde{\xi}, \lambda) = 0$ defined in Lemma 5.14. It follows that there exist unstable roots of D for small $\rho > 0$ unless $\Re\beta \geq 0$.   •

5.4. Low-frequency resolvent estimates. It remains to establish the low-frequency resolvent bounds of Proposition 5.1. Accordingly, we restrict attention to arcs

$$\Gamma^{\tilde{\xi}} : \Re\lambda = \theta(|\tilde{\xi}|^2 + |\Im\lambda|^2), \quad 0 < |(\tilde{\xi}, \Im\lambda)| \leq \delta, \tag{5.49}$$

with $\theta > 0$ and δ taken sufficiently small.

5.4.1. Second-order perturbation problem. We begin by deriving a second-order, viscous correction of the central matrix perturbation problem 5.29 underlying the inviscid stability analysis [K, M.1, M.2, M.3, Me.1]. Introduce the curves

$$(\tilde{\xi}, \lambda)(\rho, \tilde{\xi}_0, \tau_0) := (\rho\tilde{\xi}_0, \rho i\tau_0 - \theta_1\rho^2), \tag{5.50}$$

where $\tilde{\xi}_0 \in \mathbb{R}^{d-1}$ and $\tau_0 \in \mathbb{R}$ are restricted to the unit sphere $S^d : |\tilde{\xi}_0|^2 + |\tau_0|^2 = 1$. Evidently, as $(\tilde{\xi}_0, \tau_0, \rho)$ range in the compact set $S^d \times [0, \delta]$, $(\tilde{\xi}, \lambda)$ traces out the surface $\cup_{\tilde{\xi}}\Gamma^{\tilde{\xi}}$ of interest.

Making as usual the Ansatz $U =: e^{\mu x_1}\mathbf{v}$, and substituting $\lambda = i\rho\tau_0 - \theta_1\rho^2$ into 5.27, we obtain the characteristic equation

$$\left[\mu^2 B_{\pm}^{11} + \mu(-A_{\pm}^1 + i\rho\sum_{j\neq 1}B_{\pm}^{j1}\xi_{0_j} + i\rho\sum_{k\neq 1}B^{1k}\xi_{0_k})\right.$$
$$\left. -(i\rho\sum_{j\neq 1}A^j\xi_{0_j} + \rho^2\sum_{jk\neq 1}B^{jk}\xi_{0_j}\xi_{0_k} + (\rho i\tau_0 - \theta_1\rho^2)I)\right]\mathbf{v} = 0. \tag{5.51}$$

Note that this agrees with 5.27 up to second order in ρ, hence the (first-order) matrix bifurcation analysis of Lemma 5.9 applies for any fixed θ. We focus on "slow", or "inviscid" modes $\mu \sim \rho$.

Positing the Taylor expansion

$$\begin{cases} \mu = 0 + \mu^1 \rho + \cdots, \\ \mathbf{v} = \mathbf{v}^0 + \cdots \end{cases} \tag{5.52}$$

(or Puisieux expansion, in the case of a branch singularity) as before, and matching terms of order ρ in (5.51), we obtain

$$(-\mu^1 A_\pm^1 - i \sum_{j \neq 1} A^j \xi_{0_j} - i\tau_0 I)\mathbf{v} = 0, \tag{5.53}$$

just as in 5.29, or equivalently

$$[(A^1)^{-1}(i\tau_0 + iA^{\tilde{\xi}_0}) - \alpha_0 I]\mathbf{v} = 0, \tag{5.54}$$

with $\mu_1 =: -\alpha_0$, which can be recognized as the equation occurring in the inviscid theory on the imaginary boundary $\lambda = i\tau_0$.

In the inviscid stability theory, solutions of (5.53) are subcategorized into "elliptic" modes, for which α_0 has nonzero real part, "hyperbolic" modes, for which α_0 is pure imaginary and locally analytic in $(\tilde{\xi}, \tau)$, and "glancing" modes lying on the elliptic–hyperbolic boundary, for which α_0 is pure imaginary with a branch singularity at $(\tilde{\xi}_0, \tau_0)$. Elliptic modes admit a straightforward treatment, both in the inviscid and the viscous theory; however, hyperbolic and glancing modes require a more detailed matrix perturbation analysis.

Accordingly, we now restrict to the case of a pure imaginary eigenvalue $\alpha_0 =: i\xi_{0_1}$; here, we must consider quadratic order terms in ρ, and the viscous and inviscid theory part ways. Using $\mu = -i\rho\xi_{0_1} + o(\rho)$, we obtain at second order the modified equation:

$$[(A^1)^{-1}(i\tau_0 + \rho(B^{\xi_0\xi_0} - \theta_1) + iA^{\tilde{\xi}_0}) - \tilde{\alpha}I]\tilde{\mathbf{v}} = 0, \tag{5.55}$$

where $\tilde{\alpha}$ is the next order correction to $\alpha \sim -\mu/\rho$, and $\tilde{\mathbf{v}}$ the next order correction to $\mathbf{v}$. (Note that this derivation remains valid near branch singularities, since we have only assumed continuity of μ/ρ and not analyticity at $\rho = 0$). Here, $B^{\xi_0\xi_0}$ as usual denotes $\sum B^{jk}\xi_{0_j}\xi_{0_k}$, where $\xi_0 := (\xi_{0_1}, \tilde{\xi}_0)$. Equation (5.55) generalizes the perturbation equation

$$[(A^1)^{-1}(i\tau_0 + \gamma + iA^{\tilde{\xi}_0}) - \tilde{\alpha}I]\tilde{\mathbf{v}} = 0, \tag{5.56}$$

$\gamma := \Re\lambda \to 0^+$, arising in the inviscid theory near the imaginary boundary $\lambda = i\tau_0$ [K, Me.4].

Note that τ_0 is an eigenvalue of A^{ξ_0}, as can be seen by substituting $\alpha_0 = i\xi_{0_1}$ in (5.54), hence $|\tau_0| \leq C|\xi_0|$ and therefore (since clearly also $|\xi_0| \geq |\tilde{\xi}_0|$)

$$|\xi_0| \geq (1/C)(|(\tilde{\xi}_0, \tau_0)| = 1/C. \tag{5.57}$$

Thus, for B positive definite, and θ_1 sufficiently small, perturbation

$$\rho(B^{\xi_0 \xi_0} - \theta_1),$$

roughly speaking, enters (5.55) with the same sign as does γI in (5.56), and the same holds true for semidefinite B under the genuine coupling condition (1.20); see (K1), Lemma 3.18. Indeed, for identity viscosity $B^{jk} := \delta_k^j I$, (5.55) reduces for fixed $(\tilde{\xi}_0, \tau_0)$ exactly to (5.56) for θ_1 sufficiently small, by the rescaling $\rho \to \rho/(|\tilde{\xi}_0|^2 - \theta_1)$. This motivates the improved viscous resolvent bounds of Proposition 5.1; see Remark 5.5.

5.4.2. Matrix bifurcation analysis. The matrix bifurcation analysis for (5.55) goes much as in the inviscid case, but with additional technical complications due to the presence of the additional parameter ρ. The structural hypothesis (H5), however, allows us to reduce the calculations somewhat, at the same time imposing additional structure to be used in the final estimates of Section 5.4.3.

For hyperbolic modes, $(\tilde{\xi}_0, \tau_0)$ bounded away from the set of branch singularities $\cup(\tilde{\xi}, \eta_j(\tilde{\xi}))$, we may treat (5.55) as a continuous family of single-variable matrix perturbation problem in ρ, indexed by $(\tilde{\xi}_0, \tau_0)$; the resulting continuous family of analytic perturbation series will then yield uniform bounds by compactness. For glancing modes, $(\tilde{\xi}_0, \tau_0)$ near a branch singularity, on the other hand, we must vary both ρ and $(\tilde{\xi}_0, \tau_0)$, in general a complicated multi-variable perturbation problem. Using homogeneity, however, and the uniform structure assumed in (H5), this can be reduced to a two-variable perturbation problem that again yields uniform bounds. For, noting that $\eta_j(\tilde{\xi}) \equiv 0$, we find that $\tilde{\xi}_0$ must be bounded away from the origin at branch singularities; thus, we may treat the direction $\tilde{\xi}_0/|\tilde{\xi}_0|$ as a fixed parameter and vary only ρ and the ratio $|\tau_0|/|\tilde{\xi}_0|$. Alternatively, relaxing the restriction of $(\tilde{\xi}_0, \tau_0)$ to the unit sphere, we may fix $\tilde{\xi}_0$ and vary ρ and τ_0, obtaining after some rearrangement the rescaled equation

$$[(A^1)^{-1}(i\tau_0 + [i\sigma + \rho(B^{\xi_0 \xi_0} - \theta_1(|\tilde{\xi}_0|^2 + |\tau_0 + \sigma|^2)] + iA^{\tilde{\xi}_0}) - \tilde{\alpha}I]\tilde{v} = 0, \tag{5.58}$$

where σ denotes variation in τ_0.

Proposition 5.15 ([Z.3, Z.4]) *Under the hypotheses of Theorem 2.2, let $\alpha_0 = i\xi_{0_1}$ be a pure imaginary root of the inviscid equation (5.54) for some given $\tilde{\xi}_0$, τ_0, i.e. $\det(A^{\xi_0} + \tau_0) = 0$. Then, associated with the corresponding root $\tilde{\alpha}$ in (5.55), we have the following behavior, for some fixed ϵ, $\theta > 0$ independent of $(\tilde{\xi}_0, \tau_0)$:*

(i) For $(\tilde{\xi}_0, \tau_0)$ bounded distance ϵ away from any branch singularity $(\tilde{\xi}, \eta_j(\tilde{\xi}))$ involving α, η_j as defined in (H5), the root $\tilde{\alpha}(\rho)$ in (5.55) such that $\tilde{\alpha}(0) := \alpha_0$ bifurcates smoothly into m roots $\tilde{\alpha}_1, \ldots, \tilde{\alpha}_m$, where m is the dimension of $\ker(A^{\xi_0} + \tau_0)$, satisfying

$$\Re \tilde{\alpha}_j \geq \theta \rho \ or \ \Re \tilde{\alpha}_j \leq -\theta \rho; \tag{5.59}$$

moreover, there is an analytic choice of eigenvectors spanning the associated group eigenspace $\mathcal{V}$, *in which coordinates the restriction* $\mathbb{A}_{\mathcal{V}}^{\pm}$ *of the limiting coefficient matrices for (5.17) satisfy*

$$\theta \rho^2 \leq \Re \mathbb{A}_{\mathcal{V}}^{\pm} \leq C\rho^2 \ or \ -C\rho^2 \leq \Re \mathbb{A}_{\mathcal{V}}^{\pm} \leq -\theta \rho^2, \tag{5.60}$$

$C > 0$, *in accordance with (5.59), for* $0 < \rho \leq \epsilon$.

(ii) *For* $(\tilde{\xi}_0, \tau_0)$ *lying at a branch singularity* $(\tilde{\xi}, \eta_j(\tilde{\xi}))$ *involving* α, *the root* $\tilde{\alpha}(\rho, \sigma)$ *in (5.58) such that* $\tilde{\alpha}(0,0) = \alpha_0$ *bifurcates (nonsmoothly) into* m *groups of* s *roots each:*

$$\{\tilde{\alpha}_1^1, \dots, \tilde{\alpha}_s^1\}, \dots, \{\tilde{\alpha}_1^m, \dots, \tilde{\alpha}_s^m\}, \tag{5.61}$$

where m *is the dimension of* $\ker(A^{\xi_0} + \tau_0)$ *and* s *is some positive integer, such that, for* $0 \leq \rho \leq \epsilon$ *and* $|\sigma| \leq \epsilon$,

$$\tilde{\alpha}_k^j = \alpha + \pi_k^j + o(|\sigma| + |\rho|)^{1/s}, \tag{5.62}$$

and, in appropriately chosen (analytically varying) coordinate system, the associated group eigenspaces $\mathcal{V}^j$ *are spanned by*

$$\mathbf{v}_k^j = \begin{pmatrix} 0 \\ \vdots \\ 0 \\ e_k \\ \Pi^j e_k \\ (\Pi^j)^2 e_k \\ \vdots \\ (\Pi^j)^{s-1} e_k \\ 0 \\ \vdots \\ 0 \end{pmatrix} + o(|\sigma| + |\rho|)^{1/s}, \tag{5.63}$$

where

$$\pi_k^j := \varepsilon^j i (p\sigma - iq_k\rho)^{1/s}, \tag{5.64}$$

$\varepsilon := 1^{1/s}$, *for* $p(\tilde{\xi}_0)$ *real-valued and uniformly bounded both above and away from zero and* q_k *denoting the eigenvalues of* $Q \in \mathbb{C}^{m \times m}$ *such that* $\operatorname{sgn} pQ \geq \theta > 0$,

$$\Pi^j := \varepsilon^j i (\sigma p I_m - iQ\rho)^{1/s} \tag{5.65}$$

(well-defined, by definiteness of p, Q), e_k *denote the standard basis elements in* $\mathbb{C}^m$, *and, moreover, the restrictions* $\mathbb{A}_{\mathcal{V}^j}^{\pm}$ *of the limiting coefficient matrices for (5.17) to the invariant subspaces* $\mathcal{V}_{\pm}^j$ *are of form* $\Pi^j + o(|\sigma| + |\rho|)^{1/s}$, *satisfying*

$$\theta\rho\Re\tilde\pi^j \le \Re A_{\mathcal V}^{\pm} \le C\rho\Re\tilde\pi^j, \ \ or \ -C\rho\Re\tilde\pi^j \le \Re A_{\mathcal V}^{\pm} \le -\theta\rho\Re\tilde\pi^j, \tag{5.66}$$

$\tilde\pi^j := i(\sigma - i\rho)^{j/s}$, $C > 0$, for $0 < \rho \le \epsilon$.

Proof. We here carry out the proof in the much simpler strictly hyperbolic case, which permits a direct and relatively straightforward treatment. A proof of the general case is given in Appendix C. In this case, the dimension of $\ker(A^{\xi_0} + \tau_0)$ is one, hence m is simply one. Let $l(\tilde\xi_0, \tau_0)$ and $r(\tilde\xi_0, \tau_0)$ denote left and right zero eigenvectors of $(A^{\xi_0} + \tau_0) = 1$, spanning co-kernel and kernel, respectively; these are necessarily real, since $(A^{\xi_0} + \tau_0)$ is real. Clearly r is also a right (null) eigenvector of $(A^1)^{-1}(i\tau_0 + iA^{\xi_0})$, and lA^1 a left eigenvector.
    Branch singularities are signalled by the relation

$$lA^1r = 0, \tag{5.67}$$

which indicates the presence of a single Jordan chain of generalized eigenvectors of $(A^1)^{-1}(i\tau_0 + iA^{\xi_0})$ extending up from the genuine eigenvector r; we denote the length of this chain by s.

Observation 5.16 ([Z.3]) Bound (3.52) implies that

$$lB^{\xi_0\xi_0}r \ge \theta > 0, \tag{5.68}$$

uniformly in $\tilde\xi$.

Proof of Observation. In our present notation, (3.52) can be written as

$$\Re\sigma(-iA^{\xi_0} - \rho B^{\xi_0\xi_0}) \le -\theta_1\rho, \tag{5.69}$$

for all $\rho > 0$, some $\theta_1 > 0$. (Recall: $|\xi_0| \ge \theta_2 > 0$, by previous discussion). By standard matrix perturbation theory [Kat], the simple eigenvalue $\gamma = i\tau_0$ of $-iA^{\xi_0}$ perturbs analytically as ρ is varied around $\rho = 0$, with perturbation series

$$\gamma(\rho) = i\tau_0 - \rho lB^{\xi_0\xi_0}r + o(\rho). \tag{5.70}$$

Thus,

$$\Re\gamma(\rho) = -\rho lB^{\xi_0\xi_0}r + o(\rho) \le -\theta_1\rho, \tag{5.71}$$

yielding the result.   •
    In case (i), $\tilde\alpha(0) = \alpha$ is a simple eigenvalue of $(A^1)^{-1}(i\tau_0 + iA\tilde\xi_0)$, and so perturbs analytically in (5.55) as ρ is varied around zero, with perturbation series

$$\tilde\alpha(\rho) = \alpha + \rho\mu^1 + o(\rho), \tag{5.72}$$

where $\mu^1 = \tilde l(A^1)^{-1}\tilde r$, $\tilde l, \tilde r$ denoting left and right eigenvectors of $(A^1)^{-1}(i\tau_0 + A\tilde\xi_0)$. Observing by direct calculation that $\tilde r = r$, $\tilde l = lA^1/lA^1r$, we find that

$$\mu^1 = lB^{\xi_0\xi_0}r/lA^1r \tag{5.73}$$

is real and bounded uniformly away from zero, by Observation 5.16, yielding the result (5.59) for any fixed $(\tilde{\xi}_0, \tau_0)$, on some interval $0 \leq \rho \leq \epsilon$, where ϵ depends only on a lower bound for μ^1 and the maximum of $\gamma''(\rho)$ on the interval $0 \leq \rho \leq \epsilon$. By compactness, we can therefore make a uniform choice of ϵ for which (5.59) is valid on the entire set of $(\tilde{\xi}_0, \tau_0)$ under consideration. As $A_{\mathcal{V}}^{\pm}$ are scalar for $m = 1$, (5.60) is in this case identical to (5.59).

In case (ii), $\tilde{\alpha}(0, 0) = \alpha$ is an s-fold eigenvalue of $(A^1)^{-1}(i\tau_0 + iA\tilde{\xi}_0)$, corresponding to a single $s \times s$ Jordan block. By standard matrix perturbation theory, the corresponding s-dimensional invariant subspace (or "total eigenspace") varies analytically with ρ and σ, and admits an analytic choice of basis with arbitrary initialization at $\rho, \sigma = 0$ [Kat]. Thus, by restricting attention to this subspace we can reduce to an s-dimensional perturbation problem; moreover, up to linear order in ρ, σ, the perturbation may be calculated with respect to the fixed, initial coordinization at $\rho, \sigma = 0$.

Choosing the initial basis as a real, Jordan chain reducing the restriction (to the subspace of interest) of $(A^1)^{-1}(i\tau_0 + iA\tilde{\xi}_0)$ to i times a standard Jordan block, we thus reduce (5.58) to the canonical problem

$$\left(iJ + i\sigma M + \rho N - (\tilde{\alpha} - \alpha) \right) \mathbf{v}_I = 0, \tag{5.74}$$

where

$$J := \begin{pmatrix} 0 & 1 & 0 & \cdots & & 0 \\ 0 & 0 & 1 & 0 & \cdots & \\ 0 & 0 & 0 & 1 & \cdots & \\ \vdots & \vdots & \vdots & \vdots & & \vdots \\ 0 & 0 & 0 & \cdots & & 0 \end{pmatrix}, \tag{5.75}$$

$\mathbf{v}_I$ is the coordinate representation of $\mathbf{v}$ in the s-dimensional total eigenspace, and M and N are given by

$$M := \tilde{L}(A^1)^{-1}\tilde{R} \tag{5.76}$$

and

$$N := \tilde{L}(A^1)^{-1}(B^{\xi_0\xi_0} - \theta_1)\tilde{R}, \tag{5.77}$$

respectively, where $\tilde{R}$ and $\tilde{L}$ are the initializing (right) basis, and its corresponding (left) dual.

Now, we have only to recall that, as may be readily seen by the defining relation

$$\tilde{L}(A^1)^{-1}(i\tau_0 + iA\tilde{\xi}_0)\tilde{R} = J, \tag{5.78}$$

or equivalently $(A^1)^{-1}(i\tau_0 + iA\tilde{\xi}_0)\tilde{R} = \tilde{R}J$ and $\tilde{L}(A^1)^{-1}(i\tau_0 + iA\tilde{\xi}_0) = J\tilde{L}$, the first column of $\tilde{R}$ and the last row of $\tilde{L}$ are genuine left and right eigenvectors $\tilde{r}$ and $\tilde{l}$ of $(A^1)^{-1}(i\tau_0 + iA\tilde{\xi}_0)$, hence without loss of generality

$$\tilde{r} = r, \quad \tilde{l} = plA^1 \tag{5.79}$$

as in the previous (simple eigenvalue) case, where p is an appropriate nonzero real constant. Applying again Observation 5.16, we thus find that the crucial $s, 1$ entries of the perturbations M, N, namely p and $pl(B^{\xi_0\xi_0} - \theta_1)r =: q$, respectively, are real, nonzero and of the same sign. Recalling, by standard matrix perturbation theory, that this entry when nonzero is the only significant one, we have reduced finally (modulo $o(|\sigma| + |\rho|)^{1/s}$) errors) to the computation of the eigenvalues/eigenvectors of

$$
i
\begin{pmatrix}
0 & 1\,0\cdots & 0 \\
0 & 0\,1\,0 & \cdots \\
0 & 0\,0\,1 & \cdots \\
\vdots & \vdots\,\vdots\,\vdots & \vdots \\
p\sigma - iq\rho\,0\,0 & \cdots & 0
\end{pmatrix},
\tag{5.80}
$$

from which results (5.62)–(5.64) follow by an elementary calculation, for any fixed $(\tilde{\xi}_0, \tau_0)$, and some choice of $\epsilon > 0$; as in the previous case, the corresponding global results then follow by compactness. Finally, bound (5.66) follows from (5.62) and (5.64) by direct calculation. (Note that the addition of further $\mathcal{O}(|\sigma| + |\rho|)$ perturbation terms in entries other than the lower lefthand corner of (5.80) does not affect the result. Note also that $\mathbb{A}^{\pm}_{\mathcal{V}j}$ are scalar in the strictly hyperbolic case $m = 1$, and $\mathcal{V}^j_{\pm}$ are simply eigenvectors of $\mathbb{A}_{\pm}$.) This completes the proof in the strictly hyperbolic case.   ●

Remark 5.17 The detailed description of hyperbolic and glancing modes given in Proposition 5.15 readily yield the result of Lemma 5.9 in the deferred case $\Re\lambda_0 = 0$, under the additional hypothesis (H5) (exercise).

5.4.3. Main estimates.
Combining the Evans-function estimates of Proposition 2.14 (and, in the case that refined but not uniform dynamical stability holds, also those established in the course of the proof of Proposition 2.1) with the matrix perturbation analysis of Proposition 5.15, we have all of the ingredients needed to carry out the basic $L^1 \to L^2$ resolvent estimates of Proposition 5.1. In particular, in the uniformly dynamically stable case, they may be obtained quite efficiently by Kreiss symmetrizer estimates generalizing those of the inviscid theory. We refer to [GMWZ.1] or the article of M. Williams [W] in this volume for a presentation of the argument in the strictly hyperbolic, Laplacian viscosity case. With the results of Proposition 5.15, a block version of the same argument applies in the general case, substituting invariant subspaces $\mathcal{V}^j$ for individual eigenvectors $\mathcal{V}$. We omit the details, which are beyond the scope of this article.

In the refined, but not uniformly dynamically stable case, the estimates may be obtained instead as in [Z.3, Z.4] by detailed pointwise estimates on the resolvent kernel, using the second-order Evans function estimates carried out in the course of the proof of Proposition 2.1 and the explicit representation formula for the resolvent kernel of an ordinary differential operator [MaZ.3, Z.4]. We refer to [Z.3, Z.4] for an account of these more complicated arguments.

5.4.4. Derivative estimates. Improved derivative estimates, $|\beta| = 1$, may now easily be obtained by a method introduced by Kreiss and Kreiss [KK] in the one-dimensional case; see [GMWZ.1] or the notes of M. Williams [W] in this volume. Specifically, given a differentiated source $f = \partial_{x_1} g$, in resolvent equation $(L_{\tilde{\xi}} - \lambda)U = f$, consider first the *auxiliary equation*

$$L_0 W = (B^{11} W_{x_1})_{x_1} - (A^1 W)_{x_1} = \partial_{x_1} g. \tag{5.81}$$

Using the conservative (i.e., divergence-form) structure of L_0, we may integrate (5.81) from $-\infty$ to x_1 to obtain a reduced ODE

$$z_2' - \alpha(x_1)z_2 = \begin{cases} 0, \\ G, \end{cases} \qquad W = \Phi(z_2) \tag{5.82}$$

analogous to that obtained in [KK, GMWZ.1] for the strictly parabolic case, where $z_2 := B_{II}^{11} W$ and $G \sim g$, $W \sim z_2$. More precisely, the inhomogeneous version $z_2' - \alpha(x_1)z = 0$ is the linearization about $\bar{U}$ of the (integrated) traveling-wave ODE (3.5), from which we obtain by transversality $\gamma \neq 0$ that the solution of (5.82) is unique modulo $\bar{U}'$. That a solution exists follows easily from the fact (3.7) relating the signatures of $\alpha_{\pm}$ to those of $A_{\pm}^1$ (a consequence of transversality, together with our assumptions on the profile; see Remark 3.5), together with standard arguments for asymptotically constant ODE as in, e.g., [He, Co, CL]; see, in particular, the argument of section 10.2, [GMWZ.1] in the strictly parabolic case, for which α reduces to A^1. See also [Go.2, KK], or the article [W] by M. Williams in this volume.

Indeed, imposing an additional condition

$$\langle \ell, W \rangle = 0, \tag{5.83}$$

where ℓ is any constant vector satisfying $\langle \ell, \bar{U}' \rangle = \ell \cdot [U] \neq 0$, we have [He, Co, CL, GMWZ.1, Go.2, KK, W] the bound $|z_2|_{W^{1,p}} \leq C|G|_{L^p}$ for any p, yielding in particular

$$|W|_{L^1} + |B^{11} W_{x_1}|_{L^1} \leq C|g|_{L^1}. \tag{5.84}$$

The reduction to form (5.82) goes similarly as the reduction of the nonlinear traveling-wave ODE in Section 3.1; we leave this as an exercise.

Setting now $U = W + Y$, and substituting the auxiliary equation into the eigenvalue equation, we obtain equation

$$\begin{aligned}(L_{\tilde{\xi}} - \lambda)Y &= \mathcal{O}(\rho)(|W|_{L^1} + |B^{11} W_{x_1}|_{L^1}) \\ &= \mathcal{O}(\rho|g|_{L^1}),\end{aligned} \tag{5.85}$$

for the residual Y, from which we obtain the desired bound from the basic $L^1 \to L^2$ estimate of Section 5.4.3 above.

Alternatively, one may obtain the same bounds as in [Z.3, Z.4] by direct computation on the original resolvent equation. Both methods are based ultimately on the fact that, for Lax-type shocks, the only L^1 time-invariants of solutions of the linearized equations are those determined by conservation of mass. This property is shared by over- but not undercompressive shocks, hence the degraded bounds in the latter case; for further discussion, see [LZ.2, Z.2]. This completes the proof of Proposition 5.1, and the analysis.

6 Appendices

6.1 Appendix A: Semigroup Facts

Definition 6.1 Given a Banach space X, and a linear operator $L : \mathcal{D}(L) \subset X \to X$, we say that L is *densely defined* in X if $\mathcal{D}(L)$ is dense in X. We say that L is *closed* if $u_n \to u$ and $x_n := Lu_n \to x$ (with respect to $|\cdot|_X$) for $u_n \in \mathcal{D}(L)$ and $x_n \in X$ implies that $u \in \mathcal{D}(L)$ and $Lu = x$. $\mathcal{D}(L)$ is called the *domain* of L. We define associated domains $\mathcal{D}(L^n)$ by induction as the set of $x \in \mathcal{D}(L^{n-1})$ such that $L^{n-1}x \in \mathcal{D}(L)$.

Exercises 6.2 *1. If L is closed, show that $\mathcal{D}(L)$ is a Banach space under the canonical norm $|u|_{\mathcal{D}(L)} := |u|_X + |Lu|_X$, i.e., each Cauchy sequences $u_n \in \mathcal{D}(L)$ has a limit $u \in \mathcal{D}(L)$. (First note that u_n and $x_n := Lu_n$ are Cauchy with respect to $|\cdot|_X$, hence have limits u, x in X.) With this choice of norm, $L : \mathcal{D}(L) \to X$ is a bounded operator, hence (trivially) closed in the usual, Banach space sense.*

2. Show that $L - \lambda$ is closed if and only if is L.

3. Show that $|u| \le C|Lu|$ for L closed implies that range(L) *is closed.*

Definition 6.3 Given a Banach space X, and a closed, densely defined linear operator $L : \mathcal{D}(L) \subset X \to X$, the *resolvent set* $\rho(L)$, written $\rho_X(L)$ when we wish to identify the space, is defined as the set of $\lambda \in \mathbb{C}$ for which $(\lambda - L)$ has a bounded inverse $(\lambda - L)^{-1} : X \to \mathcal{D}(L)$. The operator $(\lambda - L)^{-1}$ is called the *resolvent* of L. The *spectrum* $\sigma(L)$ of L, written $\sigma_X(L)$ when we wish to identify the space, is defined as the complement of the resolvent set, $\sigma(L) := \rho(L)^c$.

Exercise 6.4 *If L is densely defined and $\lambda \in \rho(L)$ for some λ, show that $\mathcal{D}(L^n) = $ range$(\lambda - L)^{-n}$ is dense in X. (Show by induction that each $\mathcal{D}(L^{n+1})$ is dense in $\mathcal{D}(L^n)$.)*

Lemma 6.5 *For a closed, densely defined operator L on Banach space X, the resolvent operator $(\lambda - L)^{-1}$ is analytic in λ with respect to $|\cdot|_X$ for λ in the resolvent set $\rho(L)$.*

Proof. For λ sufficiently near $\lambda_0 \in \rho(L)$, we may expand

$$(\lambda - L) = (\lambda_0 - L)^{-1}\left(I - (\lambda_0 - \lambda)(\lambda_0 - L)^{-1}\right)^{-1}$$

$$= (\lambda_0 - L)^{-1} \sum_{j=0}^{\infty} \left((\lambda_0 - \lambda)(\lambda_0 - L)^{-1}\right)^j, \tag{6.1}$$

using $|(\lambda_0 - \lambda)(\lambda - L)^{-1}| \leq C|\lambda_0 - \lambda|$ by the definition of resolvent set and the Neumann expansion $(I - T)^{-1} = \sum_{j=0}^{\infty} T^j$ for $|T|$ sufficiently small.   •

Exercise 6.6 *1. Let $L(\alpha)$ be a family of closed, densely defined operators on a single Banach space X, such that L is "relatively analytic" in α with respect to $(L(\alpha_0) - \lambda_0)$, $\lambda_0 \in \rho(L(\alpha_0))$, in the sense that $L(\alpha)(\lambda_0 - L(\alpha_0))^{-1}$ is analytic in α with respect to $|\cdot|_X$. Show that the family of resolvent operators $(\lambda - L(\alpha))^{-1}$ is analytic in (α, λ) in a neighborhood of (α_0, λ_0).*

2. Including in (4.35) formerly discarded beneficial w^{II} terms in the energy estimates from which it derives, we obtain the sharpened version

$$(\Re\lambda + \theta_1)\Big(|W|_{\dot{H}^1} + |\partial_{x_1}^2 w^{II}|\Big) \leq C_1\Big(|f|_{\dot{H}^1} + C_1|W|\Big), \tag{6.2}$$

revealing smoothing in w^{II} for $\Re\lambda > -\theta_1$. Show that this implies analyticity in $\tilde{\xi}$ with respect to $|\cdot|_{\dot{H}^1}$ of $L_{\tilde{\xi}}(\lambda_0 - L_{\tilde{\xi}_0})^{-1}$ (note: $\partial_{\tilde{\xi}}L_{\tilde{\xi}}$ is a continuous-coefficient first-order differential operator, for which the derivative falls only on w^{II} components), and conclude that $(\lambda - L_{\tilde{\xi}})^{-1}$ is analytic in $(\tilde{\xi}, \lambda)$ for $\lambda \in \rho_{\dot{H}^1}(L_{\tilde{\xi}})$ and $\Re\lambda \geq -\theta_1$ with $\theta_1 > 0$ sufficiently small.

Lemma 6.7 (Resolvent identities) *For λ, μ in the resolvent set of a closed, densely defined operator $L : \mathcal{D}(L) \to X$ on a Banach space X,*

$$(\lambda - L)^{-1}(\mu - L)^{-1} = \frac{(\mu - L)^{-1} - (\lambda - L)^{-1}}{\lambda - \mu} = (\mu - L)^{-1}(\lambda - L)^{-1} \tag{6.3}$$

on X and $L(\lambda - L)^{-1} = \lambda(\lambda - L)^{-1} - I = (\lambda - L)^{-1}L$ on $\mathcal{D}(L)$: in particular,

$$(\lambda - L)^{-1}u = \lambda^{-1}\big(u + (\lambda - L)^{-1}Lu\big) \quad \text{for } u \in \mathcal{D}(L). \tag{6.4}$$

Proof. Rearranging $(\mu - L)(\mu - L)^{-1} = I$, we obtain $(\lambda - L)(\mu - L)^{-1} = (\lambda - \mu)(\mu - L)^{-1}I$, from which the first equality of (6.3) follows upon application of $(\lambda - L)^{-1}$ from the left, and the second by symmetry. Rearranging defining relation $(\lambda - L)(\lambda - L)^{-1} = I = (\lambda - L)^{-1}(\lambda - L)$, we obtain the second assertion, whereupon (6.4) follows by further rearrangement after multiplication by λ^{-1}.   •

Exercise 6.8 *Assuming that the resolvent set is open, recover the result of Lemma 6.5 directly from the definition of derivative, using resolvent identity (6.3) to establish differentiability, $(d/d\lambda)(\lambda - L)^{-1} = -(\lambda - L)^{-2}$.*

Definition 6.9 ([Pa]) A C^0 semigroup on Banach space X is a family of bounded operators $T(t)$ satisfying the properties (i) $T(0) = I$, (ii) $T(t + s) = T(t)T(s)$ for every t, $s \geq 0$, and (iii) $\lim_{t\to 0+} T(t)x = x$ for all $x \in X$. The generator L of the semigroup is defined as $Lx = \lim_{t\to 0+}(T(t)x - x)/t$ on the domain $\mathcal{D}(L) \subset X$ for which the limit exists. We write $T(t) = e^{Lt}$. Every C^0 semigroup satisfies $|e^{Lt}| \leq Ce^{\gamma_0 t}$ for some γ_0, C; see [Pa], Theorem 2.2.

Remark 6.10 For a C^0 semigroup, $(d/dt)e^{Lt}f = Le^{Lt}f = e^{Lt}Lf$ for all $f \in \mathcal{D}(L)$ and $t \geq 0$; see [Pa], Theorem 2.4(c). Thus, e^{Lt} is the solution operator for initial-value problem $u_t = Lu$, $u(0) = f$, justifying the exponential notation.

Proposition 6.11 (Generalized Hille–Yosida theorem) *An operator* L: $\mathcal{D}(L) \to X$ *is the generator of a* C^0 *semigroup* $|e^{Lt}| \leq Ce^{\gamma_0 t}$ *on* X *with domain* $\mathcal{D}(L)$ *if and only if: (i) it is closed and densely defined, and (ii)* $\lambda \in \rho(L)$ *and* $|(\lambda - L)^{-k}| \leq C|\lambda - \gamma_0|^{-k}$ *for sufficiently large real* λ, *in which case also* $|(\lambda - L)^{-k}| \leq C|\Re\lambda - \gamma_0|^{-k}$ *for all* $\Re\lambda > \gamma_0$.

Proof. ($\Rightarrow$) By the assumed exponential decay, the Laplace transform $\hat{T} := \int_0^\infty e^{-\lambda s}T(s)\,ds$ is well-defined for $\Re\lambda > \gamma_0$, with $|\hat{T}(\lambda)| \leq C|\Re\lambda - \gamma_0|^{-1}$. By the properties of the solution operator, we have also $\hat{T}L = L\hat{T} = \widehat{LT}$ on $\mathcal{D}(L)$, as well as $\widehat{LT} = \widehat{\partial_t T} = \lambda\hat{T} - I$. Thus, $(\lambda - L)\hat{T} = \hat{T}(\lambda - L) = I$ on $\mathcal{D}(L)$. But, also,

$$h^{-1}(T(h) - I)\hat{T} = h^{-1}(e^{\lambda h} - I)\hat{T} - h^{-1}\int_0^h e^{-\lambda(s-h)}T(s)\,ds \qquad (6.5)$$

approaches $\lambda\hat{T} - I$ as $h \to 0^+$, yielding $\hat{T}X \subset \mathcal{D}(L)$ and $L\hat{T} = \lambda\hat{T} - I$ on X, by definition. Thus, $\lambda \in \rho(L)$ for $\Re\lambda > \gamma_0$, and $\hat{T} = (\lambda - L)^{-1}$, whence $|(\lambda - L)^{-k} \leq C|\Re\lambda - \gamma_0|^{-k}$ for $k = 1$. The bound for general k follows from the computation $\hat{T}^k = \mathcal{L}(\underbrace{T * T \cdots * T}_{k \text{ times}}) = \mathcal{L}(t^k T/k!)$, where $\mathcal{L}$ denotes Laplace transform and "$*$" convolution, $g * h(t) := \int_0^t g(t-s)h(s)\,ds$, together with $\int_0^\infty e^{-z}z^k/k!\,dz \equiv 1$ for all $k \geq 0$ (exercise).

($\Leftarrow$) Without loss of generality, take γ_0 0. For real numbers z, $e^z = (e^{-z})^{-1} = \lim_{n\to\infty}(1 - z/n)^{-n}$ gives a stable approximation for $z < 0$. This motivates the introduction of approximants

$$T_n := (I - Lt/n)^{-n}, \qquad (6.6)$$

which, by resolvent bound (ii), satisfies $|T_n| \leq C$. Restricting to the dyadic approximants T_{2^j}, and using the elementary difference formula $a^n - b^n = (a^{n-1} + a^{n-2}b + \cdots + b^{n-1})(a - b)$ for commuting operators a, b together with bound (ii), we find for $x \in \mathcal{D}(L^2)$ that

$$|(T_{2^{j+1}} - T_{2^j})x| \leq 2^j C^2 \left| \left((I - Lt/2^{j+1})^{-2} - (I - Lt/2^j)^{-1} \right)x \right|$$

$$= 2^j C^2 \left| (I - Lt/2^{j+1})^{-2}(I - Lt/2^j)^{-1} \right| t^2 |L^2 x| 2^{-2j-2} \quad (6.7)$$

$$\leq C^4 t^2 |L^2 x| 2^{-j-2}$$

and thus the sequence converges geometrically to a limit Tx for all $x \in \mathcal{D}(L^2)$. Since $\mathcal{D}(L^2)$ is dense in X (Exercise 6.4), and the T_n are uniformly bounded, this implies convergence for all x, defining a bounded operator $T(t)$.

Clearly, $T(0) = T_n(0) \equiv I$. Also, $T_n x \to x$ as $t \to 0$ for $x \in \mathcal{D}(L^n)$, whence we obtain $Tx \to x$ for all x by uniform convergence of the dyadic approximants and density of convergence of the approximants and density of each $\mathcal{D}(L^n)$ in X. Finally, for $x \in \mathcal{D}(L)$, we have $(d/dt)(I - Lt/n)x = -Lx/n$, from which we obtain $(d/dt)T_n x = -n(I - Lt/n)^{-n-1}(-L/n)x = L_n(t)T_n x$, where $L_n(t) := L(I - Lt/n)^{-1}$ is bounded for bounded t (exercise). Observing that $|LT_n(t)x| = |T_n Lx| \leq C|Lx|$ for all t, we find for $x \in \mathcal{D}(L^2)$ that

$$e_n(r,t) := u(t) - v(t) = T_n(r+t)x - T_n(t)T_n(r)x, \qquad (6.8)$$

$u' = L_n(s)u$, $v' = L_n(r+s)v$, $u(0) = v(0) := T_n(r)x$, satisfies

$$e_n' = L_n(s+r)e_n + f_n(r,s), \quad e(0) = 0, \qquad (6.9)$$

$f_n(r,s) := (L_n(s) - L_n(s+r))T_n(r+s)x$, where

$$
\begin{aligned}
|f_n(r,s)| &= \left| \left((I - Ls/n)^{-1} - (I - L(s+r)/n)^{-1} \right) T_n Lx \right| \\
&= |(I - Ls/n)^{-1}(I - L(s+r)/n)^{-1} T_n(r/n)L^2 x| \qquad (6.10) \\
&\leq |L^2 x| r/n
\end{aligned}
$$

goes to zero uniformly in n for bounded r. Expressing

$$e_n(r,t) = \int_0^t T_n(t+r)T_n(t+r-s)^{-1} f_n(r,s)\, ds \qquad (6.11)$$

using Duhamel's principle/uniqueness of solutions of bounded-coefficient equations, and observing (Exercise 6.13 below) that $|T_n(t+r)T_n(t+r-s)^{-1}| \leq C$, we thus have $|e_n(r,t)| \leq Ct|L^2 x| r/n \to 0$ uniformly in n for bounded r, t, verifying semigroup property (ii) Definition 6.9 in the limit for $x \in \mathcal{D}(L^2)$, and thus, by continuity, for all $x \in X$.   ●

Remark 6.12 ([Pa]) The approximants $T_n(t)$ correspond to the finite - difference approximation obtained by first-order implicit Euler's method with mesh t/n. In the language of numerical analysis, boundedness of $|T_n|$ corresponds to "$\mathcal{A}$-stability" of the scheme, i.e., suitability for "stiff" ODE.

Exercise 6.13 *Show by direct computation that*

$$T_n(r+s)T_n(s)^{-1} = \left(\left(\frac{r}{r+s} \right)(I - L(r+s)/n)^{-1} + \left(\frac{s}{r+s} \right) \right)^n. \qquad (6.12)$$

Assuming resolvent bound (ii), and thus $|(I - L(r+s)/n)^{-k}| \leq C$, show that $|T_n(r+s)T_n(s)^{-1}| \leq C$, using binomial expansion and the fact that resolvents commute.

Exercise 6.14 *Show by careful expansion of*

$$T_{n+1} - T_n = \left((I - Lt/(n+1))^{-(n+1)} - (I - Lt/(n+1))^{-n}\right)$$
$$+ \left((I - Lt/(n+1))^{-n} - (I - Lt/n))^{-n}\right)$$

(6.13)

that $|T_{n+1}x - T_nx| \leq C(|L^2x| + |Lx| + |x|)/n^2$, so that the entire sequence $\{T_n\}$ is convergent for $x \in \mathcal{D}(L^2)$.

Exercise 6.15 *Show directly, using the same Neumann expansion argument used to prove analyticity of the resolvent that $|(\lambda - L)^{-1}| \leq (\lambda - \gamma_0)^{-1}$ for sufficiently large real λ implies $|(\lambda - L)^{-1}| \leq (\Re\lambda - \gamma_0)^{-1}$ for all $\Re\lambda > \gamma_0$.*

Definition 6.16 We say that a linear operator L is dissipative if it satisfies an a priori estimate

$$|\lambda - \gamma_0|^k |U|_X \leq C|(\lambda - L)^k U|_X \tag{6.14}$$

for all real $\lambda > \lambda_0$. Condition 6.14 together with range$(\lambda - L) = X$ is equivalent to condition (ii) of Proposition 6.11.

Corollary 6.17 *An operator $L : \mathcal{D}(L) \to X$ is the generator of a C^0 semigroup $|e^{Lt}| \leq Ce^{\gamma_0 t}$ on X with domain $\mathcal{D}(L)$ if and only if: (i) it is closed and densely defined, and (ii') both L and L^* are dissipative, where $L^* : X^* \to \mathcal{D}(L)^*$ defined by $\langle L^*v, u\rangle := \langle v, Lu\rangle$ denotes the adjoint of L.*

Proof. See Corollary 4.4 of [Pa], or exercise 6.20, below.   ●

Corollary 6.18 (Generalized Lumer–Phillips theorem)
A densely defined operator $L : \mathcal{D}(L) \to X$ generates a C^0 semigroup $|e^{Lt}| \leq Ce^{\gamma_0 t}$ on X with domain $\mathcal{D}(L)$ if and only if L is dissipative and range$(\lambda_0 - L) = X$ for some real $\lambda_0 > \gamma_0$. In particular, for a densely defined dissipative operator L with constants C, γ_0, the real ray $(\gamma_0, +\infty)$ consists either entirely of spectra, or entirely of resolvent points.

Proof. See Theorem 4.3 of [Pa], or Exercise 6.21 below.   ●

Remark 6.19 In the contractive case, $C = 1$, dissipativity is equivalent to $\Re\langle v, u\rangle \leq 0$ for some $v \in X^*$ such that $\langle v, u\rangle = |u||v|$; see Theorem 4.2, [Pa]. If X is a Hilbert space, contractive dissipativity $(C = 1)$ of both L and its adjoint L^* reduce to the single condition $\Re\langle u, Lu\rangle \leq 0$, which is therefore necessary and sufficient that L generate a C^0 contraction semigroup; see Exercise 6.22. This corresponds to $(d/dt)(1/2)|u|^2 \leq 0$, motivating our terminology.

Exercise 6.20 *For a closed operator L, show that (ii) of Proposition 6.11 is equivalent to (ii') of Proposition 6.17. The forward direction follows by the general facts that $|A| = |A^*|$ and $(A^{-1})^* = (A^*)^{-1}$. The reverse follows by closure of range$(\lambda - L)$ (Exercise 6.2.3), the Hahn–Banach Theorem, and the observation that $0 = \langle f, (\lambda - L)u\rangle = \langle(\lambda - L)^*f, u\rangle$ for all $u \in \mathcal{D}(L)$ implies $f = 0$, which together yield range$(\lambda - L) = X$.*

Exercise 6.21 *Let L be a dissipative operator with constants C, γ_0, such that* range($\lambda_0 - L$) $= X$ *for some real $\lambda_0 > \gamma_0$.*

   1. Using Exercise 6.2.3, show that L is closed.

   2. If $(\lambda_1 - L)u_1 = (\lambda_2 - L)u_2$ for $\lambda_j > \gamma_0$, show that $|u_1 - u_2| \leq C|\lambda_1 - \gamma_0|^{-1}|\lambda_1 - \lambda_2||u_2|$.

   3. Using the result of 2, show that range($\lambda_n - L$) $= X$ *for $\lambda_n \to \lambda > \gamma_0$ implies* range($\lambda - L$) $= X$. *(Show that $\{u_j\}$ is Cauchy for $(\lambda_j - L)u_j := x$, then use closure of L.) Conclude that* range($\lambda - L$) $= X$ *for all real $\lambda > \gamma_0$, by the fact that $\rho(L)$ is open.*

Exercise 6.22 *For a densely defined linear operator L on a Hilbert space X, and $u \in \mathcal{D}(L)$, show by direct inner-product expansion that $|(\lambda - L)u|^2 \geq |(\lambda - \gamma_0)u|^2$ for real $\lambda > \gamma_0$ if and only if $\langle u, Lu \rangle \leq \gamma_0|u|^2$, if and only if $\langle u, L^*u \rangle \leq \gamma_0|u|^2$.*

Remark 6.23 Proposition 6.11 includes the rather deep stability estimate $|e^{Lt}| \leq Ce^{\gamma_0 t}$ converting global spectral information to a sharp rate of linearized time-exponential decay. For a Hilbert space, a useful alternative criterion for exponential decay $|e^{LT}| \leq Ce^{\gamma_0 t}$ has been given by Prüss [Pr]: namely, that, for some $\gamma < \gamma_0$, $\{\lambda : \Re\lambda \geq \gamma\} \subset \rho(L)$, and $|(\lambda - L)^{-1} \leq M$ on $\Re\lambda = \gamma$. A useful observation of Kapitula and Sandstede [KS, ProK] is that, in the case that there exist isolated, finite-multiplicity eigenvalues of L in $\{\lambda : \Re\lambda \geq \gamma\} \subset \rho(L)$, the same result may be used to show exponential convergence to the union of their associated eigenspaces.

However, these tools are not available in the case, as here, that essential spectrum of L approaches the contour $\Re\lambda = \gamma_0$ ($\gamma_0 = 0$ in our case). In this situation, one may take the alternative approach of direct estimation using the inverse Laplace transform formula, given just below. However, notice that we do not get from this "local" formula the Hille-Yosida bound, which is not implied by behavior on any single contour $\Re\lambda = \gamma$ (recall that $\lambda \to \infty$ in the proof of Proposition 6.11). Indeed, the spectral resolution formula holds under much weaker bounds than required for existence of a semigroup; see Exercise 6.26. A theme of this article is that we can nonetheless get somewhat weaker bounds by similarly simple criteria, involving exponential decay with loss of derivatives, and that, provided we have available a nonlinear damping estimate analogous to (3.66), we can use these to close a nonlinear argument in which the deficiencies of our linearized bounds disappear. The result is effectively a "Prüss-type" bound on the high-frequency part of the solution operator, analogous to the bounds obtained by Kapitula and Sandstede by spectral decomposition in the case that slow- and fast-decaying modes are spectrally separated.

Moreover, a damping bound (3.66) obtained by standard energy estimates implies uniform high-frequency resolvent bounds, Exercise 4.8. Thus, *an energy estimate (3.66) alone is sufficient to yield exponential control of the high-frequency part of the solution operator,* a (one-directional) high-frequency version of the Lumer–Phillips Theorem of Remark 6.19. For a simple, linear version of this principle, see (6.29), Exercise 6.28 below.

Proposition 6.24 *Let $L : \mathcal{D}(L) \to X$ be a closed, densely defined operator on Banach space X, generating a C^0 semigroup $|e^{Lt}| \le Ce^{\gamma_0 t}$: equivalently, satisfying resolvent bound (3.16). Then, for $f \in \mathcal{D}(L)$,*

$$e^{Lt}f = \text{P.V.} \int_{\gamma-i\infty}^{\gamma+i\infty} e^{\lambda t}(\lambda - L)^{-1} f \, d\lambda \qquad (6.15)$$

for any $\gamma > \gamma_0$, with convergence in $L^2([0,\infty); X)$. For $f, Lf \in \mathcal{D}(L)$, (6.15) converges pointwise for $t > 0$, with uniform convergence on compact subintervals $t \in [\epsilon, 1/\epsilon]$ at a rate depending only on the bound for $|f|_X + |LF|_X + |L(Lf)|_X$.

Proof. By the bound $|e^{Lt}| \le Ce^{\gamma_0 t}$, the Laplace transform $\hat{u}$ of $u(t) := e^{Lt}f$ is well-defined for $\Re\lambda > \gamma_0$, with

$$u(t) = \text{P.V.} \int_{\gamma-i\infty}^{\gamma+i\infty} e^{\lambda t} \hat{u}(\lambda) \, d\lambda \qquad (6.16)$$

for $\gamma > \gamma_0$, by convergence of Fourier integrals on L^2, and the relation between Fourier and Laplace transform (see (4.62)). Moreover, $u(t) \in C^1([0,\infty); X)$ by Remark 6.10, with $u_t = Lu$ and $|u_t|, |Lu| \le Ce^{\gamma_0 t}$, whence $\widehat{u_t} = \lambda\hat{u} - u(0) = \lambda\hat{u} - f$ and also $\widehat{Lu} = L\hat{u}$. Thus, $(\lambda - L)\hat{u} = f$, giving (6.15) by (6.16) and invertibility of $\lambda - L$ on $\Re\lambda > \gamma_0$.

Next, suppose that $f, Lf \in \mathcal{D}(L)$, without loss of generality taking $\gamma \ne 0$.[14] Expanding

$$(\lambda - L)^{-1}f = \lambda^{-2}\big((\lambda - L)^{-1}L \cdot Lf + Lf\big) + \lambda^{-1}f \qquad (6.17)$$

using (6.4), we may split the righthand side of (6.15) into the sum of an integral

$$\text{P.V.} \int_{\gamma-i\infty}^{\gamma+i\infty} e^{\lambda t}\lambda^{-2}\big((\lambda - L)^{-1}L \cdot Lf + Lf\big) \, d\lambda \qquad (6.18)$$

that is absolutely convergent on bounded time-intervals $t \in [0, \delta^{-1}]$ and the integral P.V. $\int_{\gamma-i\infty}^{\gamma+i\infty} e^{\lambda t}\lambda^{-1} \, d\lambda f = f$, uniformly convergent on compact intervals $t \in [\delta, \delta^{-1}]$; see Exercise 6.27 below.

Note that (6.18) converges to

$$\text{P.V.} \int_{\gamma-i\infty}^{\gamma+i\infty} \lambda^{-2}\big((\lambda - L)^{-1}L \cdot Lf + Lf\big) \, d\lambda = 0 \qquad (6.19)$$

as $t \to 0$, directly verifying semigroup property (iii). (Taking $\gamma \to +\infty$ shows that the lefthand side is zero.) For an alternative proof, see Corollary 7.5 of [Pa]. •

---

[14] The restriction $\gamma \ne 0$ may be removed by considering the shifted semigroup $\tilde{T}(t) := e^{-\gamma_0 t}e^{Lt}$. (Exercise: verify that $\tilde{T}$ is generated by $L - \gamma_0$.)

Proposition 6.25 *Let $L : \mathcal{D}(L) \to X$ be a closed, densely defined operator on Banach space X, generating a C^0 semigroup $|e^{Lt}| \leq Ce^{\gamma_0 t}$: equivalently, satisfying resolvent bound (3.16). Then, for $f \in L^1([0,T]; \mathcal{D}(L))$ or $f \in L^2([0,T]; X)$,*

$$\int_0^t e^{L(t-s)} f(s)\, ds = \text{P.V.} \int_{\gamma-i\infty}^{\gamma+i\infty} e^{\lambda t} (\lambda - L)^{-1} \widehat{f^T}(\lambda)\, d\lambda \qquad (6.20)$$

for any $\gamma > \gamma_0$, $0 \leq t \leq T$, with convergence in $L^2([0,T]; X)$, where $\hat{g}$ denotes Laplace transform of g and $f^T(x,s) := f(x,s)$ for $0 \leq s \leq T$ and zero otherwise. For f, $Lf \in L^1([0,T]; \mathcal{D}(L))$ or $f \in L^q([0,T]; \mathcal{D}(L))$, $q > 1$, the convergence is pointwise, and uniform for $t \in [0,T]$.

Proof. Equivalence and convergence in $L^2([0,T]; X)$ for $f \in L^1([0,T]; \mathcal{D}(L))$ follows similarly as in the proof of Proposition 6.15, observing that $u := \int_0^t e^{L(t-s)} f(s)\, ds$ satisfies $u_t - Lu = f^T$ for $0 \leq t \leq T$ with $u(0) = 0$, and $\hat{u}$, $\hat{f}$ are well-defined, with $(\lambda - L)\hat{u} = \hat{f}^T$. Likewise, we may obtain pointwise convergence for f, $Lf \in L^1([0,T]; \mathcal{D}(L))$ using expansion (6.17) and the Hausdorff–Young inequality

$$|\hat{g}^T(\lambda)| = \left| \int_0^T e^{-\lambda s} g^T(s) ds \right| \leq (1 + e^{-\Re \lambda T}) |g|_{L^1(t)} \qquad (6.21)$$

to obtain a uniformly absolutely convergent integral plus the uniformly convergent integral

$$\text{P.V.} \int_{\gamma-i\infty}^{\gamma+i\infty} \lambda^{-1} e^{\lambda t} \widehat{f^T}(\lambda)\, d\lambda = \int_0^T f(s) ds\, \text{P.V.} \int_{\gamma-i\infty}^{\gamma+i\infty} \lambda^{-1} e^{\lambda(t-s)} \widehat{f^T}(\lambda)\, d\lambda$$

$$= \int_0^t f(s) ds; \qquad (6.22)$$

see Exercise 6.27 below.

Convergence in L^2 for $f \in L^2([0,T]; X)$ follows using Parseval's identity,

$$\text{P.V.} \int_{\gamma-i\infty}^{\gamma+i\infty} |\widehat{g^T}(\lambda)|^2\, d\lambda = \int_0^T |e^{-\Re \lambda s} g(s)|^2\, ds \leq (1 + |e^{-\Re \lambda T}|)^2 |g|_{L^2[0,T]}^2, \qquad (6.23)$$

together with the fact that $|e^{\lambda t} (\lambda - L)^{-1}| \leq C(\gamma)$ for $\Re \lambda = \gamma$. We may then conclude equivalence by a limiting argument, using continuity with respect to f of the left-hand side and equivalence in the previous case $f \in L^1([0,T]; \mathcal{D}(L))$. A similar estimate together with expansion (6.17) yields pointwise convergence for $f \in L^q([0,T]; \mathcal{D}(L))$, $q > 1$, using Hölder's inequality $|\lambda^{-1} \widehat{g^T}|_{L^2(\lambda)} \leq C |\lambda_{L^q}^{-1}| \widehat{g^T}|_{L^p}$ and the Hausdorff–Young inequality $|\widehat{g^T}|_{L^p} \leq C |g|_{L^2[0,T]}$, where $1/p + 1/q = 1$. $\bullet$

Exercise 6.26 *Let $L : \mathcal{D}(L) \to X$ be a closed, densely defined operator on Banach space X, satisfying resolvent bound (3.16) (equivalently, generating a C^0 semigroup $|e^{Lt}| \le Ce^{\gamma_0 t}$).*

1. For $f \in \mathcal{D}(L)$, show by judicious deformation of the contour that P.V. $\int_{\gamma-i\infty}^{\gamma+i\infty} e^{\lambda t} (\lambda - L)^{-1} f \, d\lambda = 0$ for any $\gamma > \gamma_0$ and $t < 0$, with uniform pointwise convergence on compact intervals $[-\epsilon, -\epsilon^{-1}]$, $\epsilon > 0$, at a rate depending only on the bound for $|f|_X + |LF|_X$. Indeed, this holds under the resolvent growth bound $|(\lambda - L)^{-1}| \le C(1 + |\lambda|)^r$ for $\Re\lambda \ge \gamma$, for any $r < 1$.

2. For $f \in \mathcal{D}(L)$, show without reference to semigroup theory that the righthand side of (6.15) converges in $L^2(x, t)$ to a weak solution in the sense of [Sm] of initial value problem $u_t = Lu$, $u(0) = f$, using Parseval's identity and resolvent identity (6.4) together with the fact that distributional derivatives and limits commute. (By moving γ to $\omega > 0$, show that the $L^2(t; X)$ norm of $e^{-\omega t}(e^{Lt} f - f) = \int e^{(\lambda-\omega)t} \lambda^{-1} (\lambda - L)^{-1} L f \, d\lambda$ is order $(\omega - \gamma_0)^{-1}$, yielding a trace at $t = 0$.) Indeed, this holds under the resolvent growth bound $|(\lambda - L)^{-1}| \le C(1 + |\lambda|)^r$ for $\Re\lambda = \gamma$, for any $r < 1/2$.

Exercise 6.27 *Show that P.V. $\int_{\gamma-i\infty}^{\gamma+i\infty} z^{-1} e^z \, dz$ is uniformly bounded, independent of γ, and converges uniformly to a Heaviside function*

$$H(\gamma) = \begin{cases} 1 & \gamma > 0 \\ 0 & \gamma < 0 \end{cases}$$

for γ bounded away from zero.

Exercise 6.28 *Let L be such that $|Lf|_{L^2} \le C|f|_{H^s}$ satisfy energy estimate $E_t \le -2\theta E + C|u|_{L^2}^2$, $\theta > 0$, for $u_t = Lu$, $E := \langle u, Su \rangle \sim |u|_{H^s}^2$, S without loss of generality symmetric and real-valued, or equivalently (Exercise 4.8)*

$$\mathbb{R}\langle Su, Lu \rangle \le -\theta \langle Su, u \rangle + C|u|_{L^2}^2. \tag{6.24}$$

Then,

$$E(t) \le e^{-2\theta} E(0) + C \int_0^t e^{-2\theta(t-s)} |u(s)|_{L^2}^2 ds \tag{6.25}$$

and, by Exercise 4.8,

$$|(\lambda - L)^{-1} f|_{H^s} \le C_2 (\mathbb{R}\lambda + \theta)^{-1} |f|_{H^s} \tag{6.26}$$

for $\mathbb{R}\lambda \ge -\theta$ and $|\lambda| \ge R > 0$ sufficiently large.

Define the high-frequency part of solution operator $e^L t$ by

$$S_2(t)f := \left(\int_{\gamma-i\infty}^{\gamma-iR} + \int_{\gamma+iR}^{\gamma+i\infty} \right) e^{\lambda t} (\lambda - L)^{-1} f \, d\lambda. \tag{6.27}$$

(i) Using (6.17), $|Lf|_{L^2} \le |f|_{H^s}$ and (6.26), show that

$$e^{\lambda t}(\lambda - L)^{-1}f = \overbrace{e^{\lambda t}\lambda^{-1}f}^{integral\ zero} + \overbrace{e^{-\theta t}O(|\lambda|^{-2}|f|_{H^{3s}})}^{abs.\ integrable},$$

where $O(\cdot)$ refers to $|\cdot|_{H^s}$ norm, hence

$$|S_2(t)f|_{H^s} \le Ce^{-\theta t}|f|_{H^{3s}}. \tag{6.28}$$

(ii) For $\tilde{S} := \partial_x^{2s}S\partial_x^{2s} + C\partial_x^s S\partial_x^s + C^2S$, $C > 0$ sufficiently large, show that $\mathbb{R}\langle \tilde{S}u, Lu \rangle \le -\theta\langle \tilde{S}u, u \rangle + C|u|_{L^2}^2$, hence

$$|e^{Lt}f|_{H^{3s}} \le e^{-2\theta}|f|_{H^{3s}}^2 + C\int_0^t e^{-2\theta(t-s)}|e^{Ls}f|_{L^2}^2 ds.$$

Combining with (6.28), conclude exponential control of S_2 by $S_1 := e^{Lt} - S_2$:

$$|e^{Lt}f|_{H^{3s}}^2 \le e^{-2\theta}|f|_{H^{3s}}^2 + C\int_0^t e^{-2\theta(t-s)}|S_1(s)f|_{L^2}^2 ds. \tag{6.29}$$

The implication $(6.24) \Rightarrow (6.26)$–$(6.29)$ is a high-frequency version of (one direction of) the Lumer–Phillips Theorem described in Remark 6.19.

Definition 6.29 A C^0 evolutionary system on Banach space X is a family of bounded operators $U(s,t)$, $s \le t$, satisfying the properties (i) $U(s,s) = I$, (ii) $U(s, t + \tau) = U(s,t)U(t, t + \tau)$ for every $s \le t$, $\tau \ge 0$, and (iii) $\lim_{t \to s+} U(s,t)x = x$ for all $x \in X$. The instantaneous generators $L(s)$ of the system are defined as $L(s)x = \lim_{t \to s+}(U(s,t)x - x)/(t - s)$ on the domain $\mathcal{D}(L(s)) \subset X$ for which the limit exists.

Remark 6.30 For a C^0 evolutionary system,

$$(d/dt)U(s,s)f = L(s)U(s,s)f$$

for all $f \in \mathcal{D}(L(s))$; see [Pa], p. 129. Thus, if there is a common dense subspace $Y \subset \mathcal{D}(L(t))$ for all t that it is invariant under the flow $U(s,t)$, then, for $f \in Y$, $U(s,t)$ is a solution operator for initial-value problem

$$u_t = L(t)u, \quad u(s) = f,$$

generalizing the notion of semigroup to the nonautonomous case.

Proposition 6.31 Let $L(s) : \mathcal{D}(L(s)) \to X$ be a family of closed, densely defined operators on Banach space X, each generating a contraction semigroup $|e^{L(s)t}| \le e^{\gamma_0(s)t}$. Suppose also that there exists a Banach space $Y \subset \cap_s \mathcal{D}(L(s))$ contained in the common intersection of their domains, dense in X, invariant under the flow of each semigroup $e^{L(s)t}$, on which $e^{L(s)t}$ generates a semigroup also in the Y-norm, and for which $L(s) : Y \to X$ is continuous in s with respect to the $X \to Y$ operator norm. Then, $L(s)$ generates a C^0 evolutionary system on X with domains $\mathcal{D}(L(s))$.

Proof (Sketch of proof). Substituting for $L(t)$ the piecewise constant approximations $L^n(t) := L(k/n)$ for $t \in [k/n, (k+1)/n)$, we obtain a sequence of approximate evolutionary systems $U^n(s,t)$ satisfying the uniform bound $|U(s,t)| \le e^{\gamma_0(t-s)}$ for $s \le t$. For $x \in Y \subset \mathcal{D}(L(s))$ for all s, the sequence $U^n(s,t)x$ may be shown to be Cauchy by a Duhamel argument similar to the one used in the proof of Proposition 6.11. See Theorem 3.1 of [Pa] for details.

●

Remark 6.32 There exist more general versions applying to systems for which the generators are not contractive, but stability in this case is difficult to verify. See [Pa], Section 3, for further discussion.

6.2 Appendix B: Proof of Proposition 1.21

Exercise 6.33 ([Kaw]) *(Optional) 1. Using (3.52), show that*

$$|e^{P_-(i\xi)t}| \le Ce^{-\theta|\xi|^2 t/(1+|\xi|^2)}, \tag{6.30}$$

where $P_-(i\xi) := -i\left(\sum_j i\xi_j A^j - \sum_{j,k} \xi_i \xi_j B^{jk}\right)_-$.

2. Denoting by $S(t)$ the solution operator of the constant-coefficient problem

$$U_t + \sum_j A^j_- U_{x_j} = \sum_{j,k}(B^{jk}_- U_{x_k})_{x_j}, \tag{6.31}$$

we have $\widehat{S(t)U_0} = e^{P_-(i\xi)t}\hat{U}_0$, where ^ denotes Fourier transform in x. Decomposing

$$S = S_1 + S_2, \tag{6.32}$$

where $\widehat{S_j(t)U_0} = \chi_j(\xi)e^{P_-(i\xi)t}\hat{U}_0$, where $\chi_1(\xi)$ is one for $|\xi| \le 0$ and zero otherwise, and $\chi_2 = 1 - \chi_1$, show, using Parseval's identity $|f|_{L^2(x)} = |\hat{f}|_{L^2(\xi}$ and the Hausdorff–Young inequality $|\hat{f}|_{L^\infty(\xi)} \le |f|_{L^1(x)}$, that

$$|S_1(t)\partial_x^\alpha f|_{L^2} \le C(1+t)^{-d/4-|\alpha|/2}|f|_{L^1}, \quad |\alpha| \ge 0, \tag{6.33}$$

and

$$|S_2(t)f|_{L^2} \le Ce^{-\theta t}|f|_{L^2}. \tag{6.34}$$

Proof of Proposition 1.21. Taylor expanding about $U = U_-$, we may rewrite (1.2) as

$$U_t + \sum_j A^j_- U_{x_j} = \sum_{j,k}(B^{jk}_- U_{x_k})_{x_j} = \partial_x Q(U, \partial_x U), \tag{6.35}$$

where $|Q(U, \partial_x U| \le C|U||\partial_x U|$ and $|\partial_x Q(U, \partial_x U| \le C(|U||\partial_x^2 U| + |\partial_x U|^2)$ so long as $|U|$ remains uniformly bounded. Using Duhamel's principle (variation of constants), we may thus express the solution of (1.2) as

$$U(t) = S(t)U_0 + \int_0^t S(t-s)\partial_x Q(U, \partial U_x)(s)\, ds, \tag{6.36}$$

where $S(\cdot) = S_1(\cdot) + S_2(\cdot)$ is the solution operator discussed in Exercise 6.33.

Defining

$$\zeta(t) := \sup_{0 \le s \le t} |U(s)|_{H^s}(1+s)^{d/4}, \tag{6.37}$$

and using (6.33) and (6.34), we may thus bound

$$|U(t)|_{L^2} \le |S_1(t)U_0|_{L^2} + |S_2(t)U_0|_{L^2} + \int_0^t |S_1(t-s)\partial_x Q(U, \partial U_x)(s)|_{L^2}\, ds$$

$$+ \int_0^t |S_2(t-s)\partial_x Q(U, \partial U_x)(s)|_{L^2}\, ds$$

$$\le C(1+t)^{-d/4}|U_0|_{L^1} + Ce^{-\theta t}|U_0|_{L^2}$$

$$+ C\int_0^t (1+t-s)^{-d/4-1/2}|Q(U, \partial U_x)(s)|_{L^1}\, ds$$

$$+ C\int_0^t e^{-\theta(t-s)}|\partial_x Q(U, \partial U_x)(s)|_{L^2}\, ds$$

$$\le C((1+t)^{-d/4}|U(0)|_{L^1 \cap L^2}$$

$$+ C\zeta(t)^2 \int_0^t (1+t-s)^{-d/4-1/2}(1+s)^{-d/2}\, ds$$

$$\le C(1+t)^{-d/4}(|U(0)|_{L^1 \cap L^2} + \zeta(t)^2) \tag{6.38}$$

so long as $|U|_{H^s}$ remains uniformly bounded. Applying now (3.77), we obtain (exercise)

$$|U(t)|_{H^s}^2 \le Ce^{-\theta t}|U(0)|_{H^s}^2 + C\int_0^t e^{-\theta(t-s)}|U(s)|_{L^2}^2\, ds$$

$$\le Ce^{-\theta t}|U(0)|_{H^s}^2 + C(1+t)^{-d/2}(|U(0)|_{L^1 \cap L^2} + \zeta(t)^2)^2 \tag{6.39}$$

$$\le C(1+t)^{-d/2}(|U(0)|_{L^1 \cap H^s} + \zeta(t)^2)^2,$$

and thus

$$\zeta(t) \le C(|U(0)|_{L^1 \cap H^s} + \zeta(t)^2). \tag{6.40}$$

From (6.40), it follows by continuous induction (exercise) that

$$\zeta(t) \le 2C|U(0)|_{L^1 \cap H^s} \tag{6.41}$$

for $|U(0)|_{L^1 \cap H^s}$ sufficiently small, and thus

$$|U(t)|_{H^s} \le 2C(1+t)^{-d/4}|U(0)|_{L^1 \cap H^s} \tag{6.42}$$

as claimed. Applying (1.13), we obtain the same bound for $|U(t)|_{L^\infty}$, and thus for $|U(t)|_{L^p}$, all $2 \le p \le \infty$, by the L^p interpolation formula

$$|f|_{L^{p*}} \leq |f|_{L^{p_1}}^{\beta} |f|_{L^{p_2}}^{1-\beta} \tag{6.43}$$

for all $1 \leq p_1 < p_* < p_2$, where $\beta := p_1(p_2 - p_*)/p_*(p_2 - p_1)$ is determined by $1/p_* = \beta/p_1 + (1-\beta)/p_2$ (here applied between L^2 and L^∞, i.e., with $p_1 = 2$, $p_* = p$, $p_2 = \infty$). ●

6.3 Appendix C: Proof of Proposition 5.15

In this appendix, we complete the proof of Proposition 5.15 in the general *symmetrizable, constant-multiplicity* case; here, we make essential use of recent results of Métivier [Me.4] concerning the spectral structure of matrix $(A^1)^{-1}(i\tau + iA^{\tilde{\xi}})$. Without loss of generality, take A^ξ *symmetric*; this may be achieved by the change of coordinates $A^\xi \to \tilde{A}_0^{1/2} A^\xi \tilde{A}_0^{-1/2}$.

With these assumptions, the kernel and co-kernel of $(A^{\xi_0} + \tau_0)$ are of fixed dimension m, not necessarily equal to one, and are spanned by a common set of zero-eigenvectors $r_1, \dots, r_m$. Vectors $r_1, \dots, r_m$ are necessarily right zero-eigenvectors of $(A^1)^{-1}(i\tau_0 + iA^{\xi_0})$ as well. Branch singularities correspond to the existence of one or more Jordan chains of generalized zero-eigenvectors extending up from genuine eigenvectors in their span, which by the argument of Lemma 3.4 is equivalent to

$$\det(r_j^t A^1 r_k) = 0. \tag{6.44}$$

In fact, as pointed out by Métivier [Me.4], the assumption of constant multiplicity implies considerable additional structure.

Observation 6.34 ([Me.2]) Let $(\tilde{\xi}_0, \tau_0)$ lie at a branch singularity involving root $\alpha_0 = i\xi_{0_1}$ in (5.54), with τ_0 an m-fold eigenvalue of A^{ξ_0}. Then, for $(\tilde{\xi}, \tau)$ in the vicinity of $(\tilde{\xi}_0, \tau_0)$, the roots α bifurcating from α_0 in (5.54) consist of m copies of s roots $\alpha_1, \dots, \alpha_s$, where s is some fixed positive integer.

Proof of Observation. Let $a(\tilde{\xi}, \alpha)$ denote the unique eigenvalue of A^ξ lying near $-\tau_0$, where, as usual, $-i\xi_1 := \alpha$; by the constant multiplicity assumption, $a(\cdot, \cdot)$ is an analytic function of its arguments. Observing that

$$\det[(A^1)^{-1}(i\tau + iA^{\tilde{\xi}}) - \alpha] = \det i(A^1)^{-1} \det(\tau + A^\xi)$$
$$= e(\tilde{\xi}, \tau, \alpha)(\tau + a(\tilde{\xi}, \alpha))^m, \tag{6.45}$$

where $e(\cdot, \cdot, \cdot)$ does not vanish for $(\tilde{\xi}, \tau, \alpha)$ sufficiently close to $(\tilde{\xi}_0, \tau_0, \alpha_0)$, we see that the roots in question occur as m-fold copies of the roots of

$$\tau + a(\tilde{\xi}, \alpha) = 0. \tag{6.46}$$

But, the lefthand side of (6.46) is a family of analytic functions in α, continuously varying in the parameters $(\tilde{\xi}, \tau)$, whence the number of zeroes is constant near $(\tilde{\xi}_0, \tau_0)$. ●

Observation 6.35 ([Z.3])   The matrix $(r_j^t A^1 r_k)$, j, $k = 1, \ldots, m$ is a real multiple of the identity,

$$(r_j^t A^1 r_k) = (\partial a / \partial \xi_1) I_m, \tag{6.47}$$

where $a(\xi)$ denotes the (unique, analytic) m-fold eigenvalue of A^ξ perturbing from $-\tau_0$.

More generally, if

$$(\partial a / \partial \xi_1) = \cdots = (\partial^{s-1} a / \partial \xi_1^{s-1}) = 0, \quad (\partial^s a / \partial \xi_1^s) \neq 0 \tag{6.48}$$

at ξ_0, then, letting $r_1(\tilde{\xi}), \ldots, r_m(\tilde{\xi})$ denote an analytic choice of basis for the eigenspace corresponding to $a(\tilde{\xi})$, orthonormal at $(\tilde{\xi}_0, \tau_0)$, we have the relations

$$(A^1)^{-1}(\tau_0 + A^{\xi_0})r_{j,p} = r_{j,p-1}, \tag{6.49}$$

for $1 \leq p \leq s - 1$, and

$$(r_{j,0}^t A^1 r_{k,p-1}) = p! \, (\partial^p a / \partial \xi_1^p) I_m, \tag{6.50}$$

for $1 \leq p \leq s$, where

$$r_{j,p} := (-1)^p p(\partial^p r_j / \partial \xi_1^p). \tag{6.51}$$

In particular,

$$\{r_{j,0}, \ldots, r_{j,s-1}\}, \quad j = 1, \ldots, m \tag{6.52}$$

is a right Jordan basis for the total zero eigenspace of $(A^1)^{-1}(\tau_0 + A^{\xi_0})$, for which the genuine zero-eigenvectors $\tilde{l}_j$ of the dual, left basis are given by

$$\tilde{l}_j = (1/s! (\partial^s a / \partial \xi_1^s)) A^1 r_j. \tag{6.53}$$

Proof of Observation. Considering A^ξ as a matrix perturbation in ξ_1, we find by standard spectral perturbation theory that the bifurcation of the m-fold eigenvalue τ_0 as ξ_1 is varied is governed to first order by the spectrum of $(r_j^t A^1 r_k)$. Since these eigenvalues in fact do not split, it follows that $(r_j^t A^1 r_k)$ has a single eigenvalue. But, also, $(r_j^t A^1 r_k)$ is symmetric, hence diagonalizable, whence we obtain result (6.47).

Result (6.50) may be obtained by a more systematic version of the same argument. Let $R(\xi_1)$ denote the matrix of right eigenvectors

$$R(\xi_1) := (r_1, \ldots, r_m)(\xi_1). \tag{6.54}$$

Denoting by

$$a(\xi_1 + h) =: a^0 + a^1 h + \cdots + a^p h^p + \ldots \tag{6.55}$$

and

$$R(\xi_1 + h) =: R^0 + R^1 h + \cdots + R^p h^p + \ldots \tag{6.56}$$

the Taylor expansions of functions $a(\cdot)$ and $R(\cdot)$ around ξ_0 as ξ_1 is varied, and recalling that

$$A^\xi = A^{\xi_0} + hA^1, \tag{6.57}$$

we obtain in the usual way, matching terms of common order in the expansion of the defining relation $(A - a)R = 0$, the heirarchy of relations:

$$(A^{\xi_0} - a^0)R^0 = 0,$$

$$(A^{\xi_0} - a^0)R^1 = -(A^1 - a^1)R^0,$$

$$(A^{\xi_0} - a^0)R^2 = -(A^1 - a^1)R^1 + a^2 R^0, \tag{6.58}$$

$$\vdots$$

$$(A^{\xi_0} - a^0)R^p = -(A^1 - a^1)R^{p-1} + a^2 R^{p-2} + \cdots + a^p R^0.$$

Using $a^0 = \cdots = a^{s-1} = 0$, we obtain (6.49) immediately, from equations $p = 1, \ldots, s-1$, and $R^p = (1/p!)(r_{1,p}, \ldots, r_{m,p})$. Likewise, (6.50), follows from equations $p = 1, \ldots, s$, upon left multiplication by $L^0 := (R^0)^{-1} = (R^0)^t$, using relations $L^0(A^{\xi_0} - a^0) = 0$ and $a^p = (\partial^p a/\partial \xi_1^p)/p!$.

From (6.49), we have the claimed right Jordan basis. But, defining $\tilde{l}_j$ as in (6.53), we can rewrite (6.50) as

$$(\tilde{l}_j^t r_{k,p-1}) = \begin{cases} 0 & 1 \leq p \leq s-1, \\ I_m, & p = s; \end{cases} \tag{6.59}$$

these ms criteria uniquely define $\tilde{l}_j$ (within the ms-dimensional total left eigenspace) as the genuine left eigenvectors dual to the right basis formed by vectors $r_{j,p}$ (see also exercise just below).   •

Observation 6.35 implies in particular that Jordan chains extend from *all* or *none* of the genuine eigenvectors $r_1, \ldots, r_m$, with common height s. As suggested by Observation 6.34 (but not directly shown here), this uniform structure in fact persists under variations in $\tilde{\xi}, \tau$, see [Me.4]. Observation 6.35 is a slightly more concrete version of Lemma 2.5 in [Me.4]; note the close similarity between the argument used here, based on successive variations in basis r_j, and the argument of [Me.4], based on variations in the associated total projection.

With these preparations, the result goes through essentially as in the strictly hyperbolic case. Set

$$p := 1/(s!(\partial^s a/\partial \xi_1^s)) \tag{6.60}$$

and define

$$pR^t B \tilde{\xi}_0, \tilde{\xi}_0 R =: Q \tag{6.61}$$

Note, as claimed, that $p \neq 0$ by assumption $(\partial^s a/\partial \xi_1^s) \neq 0$ in Observation 6.35, and $sgn(p)Q > 0$ by (K1), Proposition 3.18.

Thus, working in the Jordan basis defined in Observation 6.35, we find similarly as in the strictly hyperbolic case that the matrix perturbation problem (5.58) reduces to an $ms \times ms$ block-version

$$\big(iJ + i\sigma M + \rho N - (\tilde{\alpha} - \alpha)\big)\mathbf{v}_{I_j} = 0 \tag{6.62}$$

of (6.2) in the strictly hyperbolic case, where

$$J := \begin{pmatrix} 0 & I_m & 0 & \cdots & 0 \\ 0 & 0 & I_m & 0 & \cdots \\ 0 & 0 & 0 & I_m & \cdots \\ \vdots & \vdots & \vdots & \vdots & \vdots \\ 0 & 0 & 0 & \cdots & 0 \end{pmatrix}, \tag{6.63}$$

denotes the standard block-Jordan block, and the lower-lefthand block of $i\sigma M + \rho N$ is $\sigma p I_m - i\rho Q \sim |\sigma| + |\rho|$. To lowest order $O(|\sigma| + |\rho|)^{1/s}$, therefore, the problem reduces to the computation of eigenvectors and eigenvalues of the perturbed block-Jordan block

$$diag\{i \begin{pmatrix} 0 & I_m & 0 & \cdots & 0 \\ 0 & 0 & I_m & 0 & \cdots \\ 0 & 0 & 0 & I_m & \cdots \\ \vdots & & \vdots & \vdots & \vdots & \vdots \\ \sigma p I_m - i\rho Q & 0 & 0 & \cdots & 0 \end{pmatrix}\}, \tag{6.64}$$

from which results (5.62)–(5.66) follow as before by standard matrix perturbation theory; see, e.g., Section 2.2.4, *Splitting of a block-Jordan block* of [Z.4]. (Note that the simple eigenvalue case $s = 1$ follows as a special case of the block-Jordan block computation, with $\sigma \equiv 0$.) This completes the proof of Proposition 5.15 in the general case.

References

[BE]   A.A. Barmin and S.A. Egorushkin, *Stability of shock waves.* Adv. Mech. 15 (1992) No. 1–2, 3–37.

[B]   S. Benzoni–Gavage, *Stability of semi-discrete shock profiles by means of an Evans function in infinite dimensions.* J. Dynam. Differential Equations 14 (2002), no. 3, 613–674.

[BHR] S. Benzoni–Gavage, P. Huot, and F. Rousset, *Nonlinear stability of semidiscrete shock waves.* preprint (2003).

[BRSZ] S. Benzoni–Gavage, D. Serre, and K. Zumbrun, *Generic types and transitions in hyperbolic initial-boundary-value problems.* Proc. Roy. Soc. Edinburgh Sect. A 132 (2002) 1073–1104.

[BSZ] S. Benzoni–Gavage, D. Serre, and K. Zumbrun, *Alternate Evans functions and viscous shock waves.* SIAM J. Math. Anal. 32 (2001), 929–962.

[BiB.1] S. Bianchini and A. Bressan, *BV solutions for a class of viscous hyperbolic systems.* Indiana Univ. Math. J. 49 (2000), no. 4, 1673–1713.

[BiB.2] S. Bianchini and A. Bressan, *The vanishing viscosity limit for a class of viscous hyperbolic systems.* preprint (2002).

[Bl]   A.M. Blokhin, *Strong discontinuities in magnetohydrodynamics.* Translated by
       A. V. Zakharov. Nova Science Publishers, Inc., Commack, NY, 1994. x+150
       pp. ISBN: 1-56072-144-8.

[BT.1] A. Blokhin and Y. Trakhinin, *Stability of strong discontinuities in fluids and
       MHD.* in Handbook of mathematical fluid dynamics, Vol. I, 545–652, North-
       Holland, Amsterdam, 2002.

[BT.2] A.M. Blokhin and Y. Trakhinin, *Stability of fast parallel MHD shock waves
       in polytropic gas.* Eur. J. Mech. B Fluids 18 (1999) 197–211.

[BT.3] A.M. Blokhin and Y. Trakhinin, *Stability of fast parallel and transversal
       MHD shock waves in plasma with pressure anisotropy.* Acta Mech. 135 (1999).

[BT.4] A.M. Blokhin and Y. Trakhinin, *Hyperbolic initial-boundary value problems
       on the stability of strong discontinuities in continuum mechanics.* Hyperbolic
       problems: theory, numerics, applications, Vol. I (Zürich, 1998), 77–86, Inter-
       nat. Ser. Numer. Math., 129, Birkhäuser, Basel, 1999.

[BTM.1] A.M. Blokhin, Y. Trakhinin, and I.Z. Merazhov, *On the stability of shock
       waves in a continuum with bulk charge.* (Russian) Prikl. Mekh. Tekhn. Fiz. 39
       (1998) 29–39; translation in J. Appl. Mech. Tech. Phys. 39 (1998) 184–193.

[BTM.2] A.M. Blokhin, Y. Trakhinin, and I.Z. Merazhov, *Investigation on stability
       of electrohydrodynamic shock waves.* Matematiche (Catania) 52 (1997) 87–114
       (1998).

[Bo]   G. Boillat, *On symmetrization of partial differential systems.* Appl. Anal. 57
       (1995) 17–21.

[B]   A. Bressan, *Center manifold notes.* lecture notes (2002).

[BDG]   T.J. Bridges, G. Derks, and G. Gottwald, *Stability and instability of solitary
       waves of the fifth-order KdV equation: a numerical framework.* Phys. D 172
       (2002), no. 1-4, 190–216.

[Br.1] L. Brin, *Numerical testing of the stability of viscous shock waves,* Doctoral
       thesis, Indiana University (1998).

[Br.2] L. Q. Brin, *Numerical testing of the stability of viscous shock waves.* Math.
       Comp. 70 (2001) 235, 1071–1088.

[BrZ]   L. Brin and K. Zumbrun, *Analytically varying eigenvectors and the stability
       of viscous shock waves.* Mat. Contemp. (2003).

[CF]   R. Courant and K.O. Friedrichs, *Supersonic flow and shock waves,* Springer–
       Verlag, New York (1976) xvi+464 pp.

[CP]   J. Chazarain-A. Piriou, *Introduction to the theory of linear partial differen-
       tial equations,* Translated from the French. Studies in Mathematics and its
       Applications, 14. North-Holland Publishing Co., Amsterdam-New York, 1982.
       xiv+559 pp. ISBN: 0-444-86452-0.

[E1]   J.W. Evans, *Nerve axon equations: I. Linear approximations.* Ind. Univ. Math.
       J. 21 (1972) 877–885.

[E2]   J.W. Evans, *Nerve axon equations: II. Stability at rest.* Ind. Univ. Math. J.
       22 (1972) 75–90.

[E3]   J.W. Evans, *Nerve axon equations: III. Stability of the nerve impulse.* Ind.
       Univ. Math. J. 22 (1972) 577–593.

[E4]   J.W. Evans, *Nerve axon equations: IV. The stable and the unstable impulse.*
       Ind. Univ. Math. J. 24 (1975) 1169–1190.

[FMé]   J. Francheteau and G. Métivier, *Existence de chocs faibles pour des systémes
       quasi-linéaires hyperboliques multidimensionnels.* C.R.AC.Sc. Paris, 327 Série
       I (1998) 725–728.

[Fre.1] H. Freistühler, *Some results on the stability of Non-classical shock waves.* J. Partial Diff. Eqs. 11 (1998), 23–38.

[Fre.2] H. Freistühler, *Dynamical stability and vanishing viscosity: A case study of a nonstrictly hyperbolic system of conservation laws.* Comm. Pure Appl. Math. 45 (1992) 561–582.

[Fre.3] H. Freistühler, *A short note on the persistence of ideal shock waves.* Arch. Math. (Basel) 64 (1995), no. 4, 344–352.

[FreL] H. Freistühler and T.-P. Liu, *Nonlinear stability of overcompressive shock waves in a rotationally invariant system of viscous conservation laws.* Commun. Math. Phys. 153 (1993) 147–158.

[FreS] H. Freistühler and P. Szmolyan, *Spectral stability of small shock waves.* Arch. Ration. Mech. Anal. 164 (2002) 287–309.

[FreZ] H. Freistühler and K. Zumbrun, *Examples of unstable viscous shock waves.* unpublished note, Institut für Mathematik, RWTH Aachen, February 1998.

[F.1] K.O. Friedrichs, *Symmetric hyperbolic linear differential equations.* Comm. Pure and Appl. Math. 7 (1954) 345–392.

[F.2] K.O. Friedrichs, *On the laws of relativistic electro-magneto-fluid dynamics.* Comm. Pure and Appl. Math. 27 (1974) 749–808.

[FL] K.O. Friedrichs and P. Lax, *Systems of conservation equations with a convex extension.* Proc. nat. Acad. Sci. USA 68 (1971) 1686–1688.

[GZ] R. Gardner and K. Zumbrun, *The Gap Lemma and geometric criteria for instability of viscous shock profiles.* Comm. Pure Appl. Math. 51 (1998), no. 7, 797–855.

[Ge] I.M. Gelfand, *Some problems in theory of quasilinear equations.* Am. Mat. Soc. Trans., Ser. 2, 29 (1963) 295–381.

[Gi] D. Gilbarg, *The existence and limit behavior of the one-dimensional shock layer.* Amer. J. Math. 73, (1951). 256–274.

[God] P. Godillon, *Stabilité linéaire des profils pour les systémes avec relaxation semi-linéaire.* Phys. D 148 (2001) 289–316.

[G] S.K. Godunov, *An interesting class of quasilinear systems.* Sov. Math. 2 (1961) 947–948.

[Go.1] J. Goodman, *Nonlinear asymptotic stability of viscous shock profiles for conservation laws.* Arch. Rational Mech. Anal. 95 (1986), no. 4, 325–344.

[Go.2] J. Goodman, *Remarks on the stability of viscous shock waves,* in: Viscous profiles and numerical methods for shock waves. (Raleigh, NC, 1990), 66–72, SIAM, Philadelphia, PA, (1991).

[Go.3] J. Goodman, *Stability of viscous scalar shock fronts in several dimensions.* Trans. Amer. Math. Soc. 311 (1989), no. 2, 683–695.

[GM] J. Goodman and J.R. Miller, *Long-time behavior of scalar viscous shock fronts in two dimensions.* J. Dynam. Differential Equations 11 (1999), no. 2, 255–277.

[GX] J. Goodman and Z. Xin, *Viscous limits for piecewise smooth solutions to systems of conservation laws.* Arch. Rat. Mech. Anal. 121 (1992), no. 3, 235–265.

[GrR] E. Grenier-F. Rousset, *Stability of one-dimensional boundary layers by using Green's functions.* Comm. Pure Appl. Math. 54 (2001), 1343–1385.

[GMWZ.1] O. Gues, G. Métivier, M. Williams, and K. Zumbrun, *Multidimensional viscous shocks I: degenerate symmetrizers and long time stability.* preprint (2002).

[GWMZ.2] O. Gues, G. Métivier, M. Williams, and K. Zumbrun, *Multidimensional viscous shocks II: the small viscosity limit.* to appear, Comm. Pure and Appl. Math. (2004).

[GWMZ.3] O. Gues, G. Métivier, M. Williams, and K. Zumbrun, *A new approach to stability of multidimensional viscous shocks.* preprint (2003).

[GMWZ.4] O. Gues, G. Métivier, M. Williams, and K. Zumbrun, *Navier–Stokes regularization of multidimensional Euler shocks.* in preparation.

[H] R. Hersh, *Mixed problems in several variables.* J. Math. Mech. 12 (1963) 317-334.

[He] D. Henry, *Geometric theory of semilinear parabolic equations,* Lecture Notes in Mathematics, Springer–Verlag, Berlin (1981), iv + 348 pp.

[HoZ.1] D. Hoff and K. Zumbrun, *Pointwise Green's function bounds for multidimensional scalar viscous shock fronts.* J. Differential Equations 183 (2002) 368-408.

[HoZ.2] D. Hoff and K. Zumbrun, *Asymptotic behavior of multi-dimensional scalar viscous shock fronts.* Indiana Univ. Math. J. 49 (2000) 427-474.

[HuZ] J. Humpherys and K. Zumbrun, *Spectral stability of small amplitude shock profiles for dissipative symmetric hyperbolic–parabolic systems.* Z. Angew. Math. Phys. 53 (2002) 20-34.

[JLy.1] K. Jenssen and G. Lyng, *Evaluation of the Lopatinski determinant for multidimensional Euler equations.* Appendix to [Z.3] (see below).

[JLy.2] K. Jenssen and G. Lyng, *Low frequency multi-dimensional stability of viscous detonation waves,* in preparation.

[Jo] F. John, *Formation of singularities in one-dimensional nonlinear wave propagation.* Comm. Pure Appl. Math. 27 (1974), 377-405.

[J] C.K.R.T. Jones, *Stability of the travelling wave solution of the FitzHugh–Nagumo system.* Trans. Amer. Math. Soc. 286 (1984), 431-469.

[KS] T. Kapitula and B. Sandstede, *Stability of bright solitary-wave solutions to perturbed nonlinear Schrdinger equations.* Phys. D 124 (1998), no. 1-3, 58-103.

[Kat] T. Kato, *Perturbation theory for linear operators.* Springer–Verlag, Berlin Heidelberg (1985).

[Kaw] S. Kawashima, *Systems of a hyperbolic–parabolic composite type, with applications to the equations of magnetohydrodynamics.* thesis, Kyoto University (1983).

[KSh] S. Kawashima and Y. Shizuta, *On the normal form of the symmetric hyperbolic-parabolic systems associated with the conservation laws.* Tohoku Math. J. 40 (1988) 449-464.

[KlM] S. Klainerman and A. Majda, *Formation of singularities for wave equations including the nonlinear vibrating string.* Comm. Pure Appl. Math. 33 (1980), no. 3, 241-263.

[K] H.O. Kreiss, *Initial boundary value problems for hyperbolic systems.* Comm. Pure Appl. Math. 23 (1970) 277-298.

[KK] G. Kreiss and H.O. Kreiss, *Stability of systems of viscous conservation laws.* Comm. Pure Appl. Math. 51 (1998), no. 11–12, 1397-1424.

[KL] G. Kreiss and M. Liefvendahl, *Numerical investigation of examples of unstable viscous shock waves.* Hyperbolic problems: theory, numerics, applications, Vol. I, II (Magdeburg, 2000), 613-621, Internat. Ser. Numer. Math., 140, 141, Birkhäuser, Basel, 2001.

[La]   P.D. Lax, *Hyperbolic systems of conservation laws and the mathematical the-ory of shock waves.* Conference Board of the Mathematical Sciences Regional Conference Series in Applied Mathematics, No. 11. Society for Industrial and Applied Mathematics, Philadelphia, Pa., 1973. v+48 pp.

[La.2]   P.D. Lax, *Hyperbolic systems of conservation laws. II.* Comm. Pure Appl. Math. 10 1957 537–566.

[L.1]   T.-P. Liu, *Pointwise convergence to shock waves for viscous conservation laws.* Comm. Pure Appl. Math. 50 (1997), no. 11, 1113–1182.

[L.2]   T.-P. Liu, *The entropy condition and the admissibility of shocks.* J. Math. Anal. Appl. 53 (1976) 78–88.

[L.3]   T.-P. Liu, *Nonlinear stability and instability of overcompressive shock waves,* in: *Shock induced transitions and phase structures in general media.* 159–167, IMA Vol. Math. Appl., 52, Springer, New York, 1993.

[L.4]   T.-P. Liu, *Interactions of Hyperbolic waves,* in: *Viscous profiles and numerical methods for shock waves.* ed: M. Shearer, Philadelphia, SIAM (1991) 66–72.

[LZ.1]   T.P. Liu and K. Zumbrun, *Nonlinear stability of an undercompressive shock for complex Burgers equation.* Comm. Math. Phys. 168 (1995), no. 1, 163–186.

[LZ.2]   T.P. Liu and K. Zumbrun, *On nonlinear stability of general undercompressive viscous shock waves.* Comm. Math. Phys. 174 (1995), no. 2, 319–345.

[LZe]   T.-P. Liu and Y. Zeng, *Large time behavior of solutions for general quasilinear hyperbolic–parabolic systems of conservation laws.* AMS memoirs 599 (1997).

[Ly]   G. Lyng, *One-dimensional stability of combustion waves.* Thesis, Indiana University (2003).

[LyZ.1] G. Lyng and K. Zumbrun, *A stability index for detonation waves in Majda's model for reacting flow.* preprint (2003).

[LyZ.2] G. Lyng and K. Zumbrun, *On the one-dimensional stability of viscous strong detonation waves.* preprint (2003).

[M.1]   A. Majda, *The stability of multi-dimensional shock fronts – a new problem for linear hyperbolic equations.* Mem. Amer. Math. Soc. 275 (1983).

[M.2]   A. Majda, *The existence of multi-dimensional shock fronts.* Mem. Amer. Math. Soc. 281 (1983).

[M.3]   A. Majda, *Compressible fluid flow and systems of conservation laws in several space variables.* Springer-Verlag, New York (1984), viii+ 159 pp.

[MP]   A. Majda and R. Pego, *Stable viscosity matrices for systems of conservation laws.* J. Diff. Eqs. 56 (1985) 229–262.

[MaZ.1] C. Mascia and K. Zumbrun, *Pointwise Green's function bounds and stability of relaxation shocks.* Indiana Univ. Math. J. 51 (2002), no. 4, 773–904.

[MaZ.2] C. Mascia and K. Zumbrun, *Stability of shock profiles of dissipative symmetric hyperbolic–parabolic systems.* preprint (2001).

[MaZ.3] C. Mascia and K. Zumbrun, *Pointwise Green's function bounds for shock profiles with degenerate viscosity.* Arch. Rational Mech. Anal., to appear.

[MaZ.4] C. Mascia and K. Zumbrun, *Stability of large-amplitude shock profiles of hyperbolic–parabolic systems.* Arch. Rational Mech. Anal., to appear.

[MaZ.5] C. Mascia and K. Zumbrun, *Stability of large-amplitude shock profiles for general relaxation systems.* preprint (2003).

[MeP]   R. Menikoff and B. Plohr, *The Riemann problem for fluid flow of real materials.* Rev. Modern Phys. 61 (1989), no. 1, 75–130.

[Mé.1]   G. Métivier, *Interaction de deux chocs pour un système de deux lois de conservation, en dimension deux d'espace.* Trans. Amer. Math. Soc. 296 (1986) 431–479.

[Mé.2] G. Métivier, *Stability of multidimensional shocks*. Advances in the theory of shock waves, 25–103, Progr. Nonlinear Differential Equations Appl., 47, Birkhäuser Boston, Boston, MA, 2001.

[Mé.3] G.Métivier. *The Block Structure Condition for Symmetric Hyperbolic Problems*. Bull. London Math.Soc., 32 (2000), 689–702

[Mé.4] G.Métivier. *Stability of multidimensional weak shocks*, Comm. Partial Diff. Equ. 15 (1990) 983–1028.

[MéZ.1] G.Métivier-K.Zumbrun, *Viscous Boundary Layers for Noncharacteristic Nonlinear Hyperbolic Problems*. preprint (2002).

[MéZ.2] G.Métivier-K.Zumbrun, *Symmetrizers and continuity of stable subspaces for parabolic–hyperbolic boundary value problems*. to appear, J. Discrete. Cont. Dyn. Systems (2004).

[MéZ.3] G.Métivier-K.Zumbrun, *Hyperbolic boundary value problems for symmetric systems with variable multiplicities*, in preparation.

[Pa] A. Pazy, *Semigroups of linear operators and applications to partial differential equations*. Applied Mathematical Sciences, 44, Springer-Verlag, New York-Berlin, (1983) viii+279 pp. ISBN: 0-387-90845-5.

[P] R.L. Pego, *Stable viscosities and shock profiles for systems of conservation laws*. Trans. Amer. Math. Soc. 282 (1984) 749–763.

[PW] R. L. Pego-M.I. Weinstein, *Eigenvalues, and instabilities of solitary waves*. Philos. Trans. Roy. Soc. London Ser. A 340 (1992), 47–94.

[PZ] R. Plaza and K. Zumbrun, *An Evans function approach to spectral stability of small-amplitude viscous shock profiles*. to appear, J. Discrete and Continuous Dynamical Systems.

[ProK] K. Promislow and N. Kutz, *Bifurcation and asymptotic stability in the large detuning limit of the optical parametric oscillator*. Nonlinearity 13 (2000) 675–698.

[Pr] J. Prüss, *On the spectrum of C_0-semigroups*. Trans. Amer. Math. Soc. 284 (1984), no. 2, 847–857.

[Ra] Lord Rayleigh (J.W. Strutt), *On the stability, or instability, of certain fluid motions, II,* Scientific Papers, 3 (1887) 2–23, Cambridge University Press.

[Sat] D. Sattinger, *On the stability of waves of nonlinear parabolic systems*. Adv. Math. 22 (1976) 312–355.

[Si] T. Sideris, *Formation of singularities in solutions to nonlinear hyperbolic equations*. Arch. Rational Mech. Anal. 86 (1984), no. 4, 369–381.

[Sm] J. Smoller, *Shock waves and reaction–diffusion equations*. Second edition, Grundlehren der Mathematischen Wissenschaften, Fundamental Principles of Mathematical Sciences, 258. Springer-Verlag, New York, 1994. xxiv+632 pp. ISBN: 0-387-94259-9.

[SX] A. Szepessy and Z. Xin, *Nonlinear stability of viscous shock waves*. Arch. Rat. Mech. Anal. 122 (1993) 53–103.

[SZ] A. Szepessy and K. Zumbrun, *Stability of rarefaction waves in viscous media*. Arch. Rat. Mech. Anal. 133 (1996) 249–298.

[T] M.Taylor. *Partial Differential Equations*III. Applied Mathematical Sciences 117, Springer, 1996.

[vN] J. von Neumann, *Collected works*. Vol. VI: Theory of games, astrophysics, hydrodynamics and meteorology. General editor: A. H. Taub. A Pergamon Press Book The Macmillan Co., New York 1963 x+538 pp. (1 plate).

[W] M. Williams, *Stability of multidimensional viscous shock waves*. C.I.M.E. summer school notes (2003).

[Yo] W.-A. Yong, *Singular perturbations of first-order hyperbolic systems*. PhD Thesis, Universität Heidelberg (1992).

[Y]   K. Yosida, *Functional analysis*. Reprint of the sixth (1980) edition, Classics in Mathematics, Springer-Verlag, Berlin, 1995, xii+501 pp. ISBN: 3-540-58654-7.

[Yu] S.-H. Yu. *Zero-dissipation limit of solutions with shocks for systems of hyperbolic conservation laws*. Arch. Ration. Mech. Anal. 146 (1999), no. 4, 275–370.

[Z.1] K. Zumbrun, *Stability of viscous shock waves*. Lecture Notes, Indiana University (1998).

[Z.2] K. Zumbrun, *Refined Wave–tracking and Nonlinear Stability of Viscous Lax Shocks*. Methods Appl. Anal. 7 (2000) 747–768.

[Z.3] K. Zumbrun, *Multidimensional stability of planar viscous shock waves*. Advances in the theory of shock waves, 307–516, Progr. Nonlinear Differential Equations Appl., 47, Birkhuser Boston, Boston, MA, 2001.

[Z.4] K. Zumbrun, *Stability of large-amplitude shock waves of compressible Navier-Stokes equations*. for Handbook of Fluid Dynamics, preprint (2003).

[ZH] K. Zumbrun and P. Howard, *Pointwise semigroup methods and stability of viscous shock waves*. Indiana Mathematics Journal V47 (1998), 741–871.

[ZH.e] K. Zumbrun-P. Howard, *Errata to: "Pointwise semigroup methods, and stability of viscous shock waves" [Indiana Univ. Math. J. 47 (1998), no. 3, 741–871.* Indiana Univ. Math. J. 51 (2002), no. 4, 1017–1021.

[ZS] K. Zumbrun and D. Serre, *Viscous and inviscid stability of multidimensional planar shock fronts*. Indiana Univ. Math. J. 48 (1999) 937–992.

A

Tutorial on the Center Manifold Theorem

Alberto Bressan
Department of Mathematics, Penn State University
University Park, Pa. 16802 U.S.A.
bressan@math.psu.edu

A.1 Review of Linear O.D.E's

Let A be an $n \times n$ matrix and consider the Cauchy problem for a linear system of O.D.E's with constant coefficients

$$\dot{x} = Ax, \qquad x(0) = \bar{x}. \tag{1.1}$$

The explicit solution (see [P]) can be written as

$$x(t) = e^{tA}\bar{x}, \qquad e^{tA} \doteq \sum_{k=0}^{\infty} \frac{t^k A^k}{k!}. \tag{1.2}$$

If $B = R^{-1}AR$ for some invertible matrix R, then

$$e^{tA} = e^{tRBR^{-1}} = Re^{tB}R^{-1}.$$

The actual computation of the exponential matrix e^{tA} can thus be carried out by reducing A to a more convenient canonical form B, and then computing e^{tB}. We give here an illustration of this procedure.

Example 1. Assume that A is a 6×6 matrix, with

$$\det(\zeta I - A) = (\zeta - \lambda)(\zeta - \mu)^3 (\zeta - (\alpha + i\beta))(\zeta - (\alpha - i\beta)),$$

so that λ is a simple real eigenvalue, μ is a multiple eigenvalue and $\alpha \pm i\beta$ are a pair of complex conjugate eigenvalues. Assume that the geometric multiplicity of μ is 1. Then there exists an invertible matrix R that reduces A to the canonical form

$$B = R^{-1}AR = \begin{pmatrix} \lambda & 0 & 0 & 0 & 0 & 0 \\ 0 & \mu & 1 & 0 & 0 & 0 \\ 0 & 0 & \mu & 1 & 0 & 0 \\ 0 & 0 & 0 & \mu & 0 & 0 \\ 0 & 0 & 0 & 0 & \alpha & -\beta \\ 0 & 0 & 0 & 0 & \beta & \alpha \end{pmatrix}$$

In this case one has

$$e^{tB} = \begin{pmatrix} e^{\lambda t} & 0 & 0 & 0 & 0 & 0 \\ 0 & e^{\mu t} & te^{\mu t} & (t^2/2)e^{\mu t} & 0 & 0 \\ 0 & 0 & e^{\mu t} & te^{\mu t} & 0 & 0 \\ 0 & 0 & 0 & e^{\mu t} & 0 & 0 \\ 0 & 0 & 0 & 0 & e^{\alpha t}\cos\beta t & -e^{\alpha t}\sin\beta t \\ 0 & 0 & 0 & 0 & e^{\alpha t}\sin\beta t & e^{\alpha t}\cos\beta t \end{pmatrix}.$$

We say that a subspace $V \subset \mathbb{R}^n$ is **invariant** for the flow of (1.1) if $x \in V$ implies $e^{At}x \in V$ for all $t \in \mathbb{R}$. A natural way to decompose the space $\mathbb{R}^n$ as the sum of three invariant subspaces is now described.

Consider the eigenvalues of A, i.e. the zeroes of the polynomial $p(\zeta) \doteq \det(\zeta I - A)$. These are finitely many points in the complex plane (fig. 1).

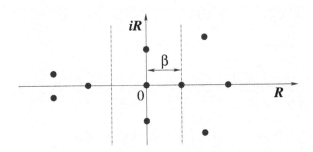

Fig. A.1.

The space $\mathbb{R}^n$ can then be decomposed as the sum of a stable, an unstable and a center subspace, respectively spanned by the (generalized) eigenvectors corresponding to eigenvalues with negative, positive and zero real part. We thus have

$$\mathbb{R}^n = V^s \oplus V^u \oplus V^c$$

with continuous projections

$$\pi_s : \mathbb{R}^n \mapsto V^s, \qquad \pi_u : \mathbb{R}^n \mapsto V^u, \qquad \pi_c : \mathbb{R}^n \mapsto V^c,$$

$$x = \pi_s x + \pi_c x + \pi_u x \,.$$

These projections commute with A and hence with the exponential e^{At} as well:

$$\pi_s e^{At} = e^{At} \pi_s \,, \qquad \pi_u e^{At} = e^{At} \pi_u \,, \qquad \pi_c e^{At} = e^{At} \pi_c \,. \tag{1.3}$$

In particular, these subspaces are invariant for the flow of (1.1). Defining the **spectral gap** of A as

$$\beta \doteq \min \left\{ |\mathrm{Re}\,\lambda| \,; \quad \lambda \text{ is an eigenvalue with non-zero real part} \right\} \tag{1.4}$$

(see fig. 1), the following key estimates hold. For every $\varepsilon \in \,]0, \beta[$ there exists a constant C_ε such that

$$\begin{aligned}
\left\| e^{At} \pi_s \right\| &\leq C_\varepsilon e^{-(\beta - \varepsilon)t} & t &\geq 0, \\
\left\| e^{At} \pi_u \right\| &\leq C_\varepsilon e^{(\beta - \varepsilon)t} & t &\leq 0, \\
\left\| e^{At} \pi_c \right\| &\leq C_\varepsilon e^{\varepsilon |t|} & t &\in \mathbb{R}.
\end{aligned} \tag{1.5}$$

A.2 Statement of the Center Manifold Theorem

Consider a nonlinear O.D.E. having an equilibrium point at the origin, say

$$\dot{x} = f(x) \,, \tag{2.1}$$

where $f : \mathbb{R}^n \mapsto \mathbb{R}^n$ is a smooth function with $f(0) = 0$. The trajectory of (2.1) taking the initial value $x(0) = y$ will be denoted as

$$t \mapsto x(t) \doteq \tilde{x}(t, y) \,. \tag{2.2}$$

Calling $A = Df(0)$ the Jacobian matrix of f at the origin, we can write (2.1) in the form

$$\dot{x} = Ax + g(x) \,, \tag{2.3}$$

where $g(0) = 0$, $Dg(0) = 0$. It is reasonable to expect that, in a small neighborhood of the origin, the flow of (2.1) should look like the flow of the corresponding linearized system (1.1). The main result in this direction (fig. 2) is the famous

Theorem (Hartman-Grobman). *Let f be smooth. If all the eigenvalues of the matrix $A \doteq Df(0)$ have non-zero real part, then the flows of (1.1) and (2.1) are equivalent. More precisely, there exists a homeomorphism φ of a neighborhood $\mathbb{N}$ of the origin onto another neighborhood of the origin such that*

$$e^{At} \varphi(y) = \varphi\big(\tilde{x}(t, y)\big)$$

for all y, t such that $y, \tilde{x}(t, y) \in \mathbb{N}$.

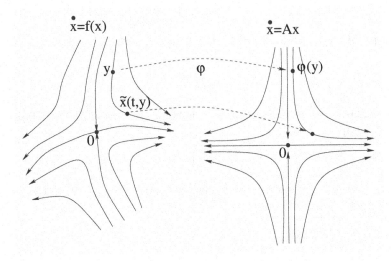

Fig. A.2.

For a proof, see [P]. This theorem settles the case where the center subspace vanishes: $V^c = \{0\}$. The Center Manifold Theorem, on the other hand, applies to the case where V^c is nontrivial. In essence, it says that near the origin all the interesting dynamics takes place on an invariant manifold $\mathcal{M}$, tangent to the center subspace V^c. Its main usefulness lies in this dimensional reduction: instead of studying a flow on the entire space $\mathbb{R}^n$, one can then restrict the analysis to a "center manifold" having the same dimension as V^c.

Center Manifold Theorem. *Let $f : \mathbb{R}^n \mapsto \mathbb{R}^n$ be a vector field in $\mathbb{C}^{k+1}$ (here $k \geq 1$), with $f(0) = 0$. Consider the matrix $A = Df(0)$, and let V^s, V^u, V^c be the corresponding stable, unstable, center subspaces. Then there exists $\delta > 0$ and a local center manifold $\mathcal{M}$ with the following properties.*

(i) There exists a $\mathbb{C}^k$ function $\phi : V^c \mapsto \mathbb{R}^n$ with $\pi_c \, \phi(x_c) = x_c$ such that

$$\mathcal{M} = \left\{ \phi(x_c) \; ; \quad x_c \in V^c, \quad |x_c| < \delta \right\}. \tag{2.4}$$

(ii) The manifold $\mathcal{M}$ is locally invariant for the flow of (2.1), i.e. $x \in \mathcal{M}$ implies $\tilde{x}(t, x) \in \mathcal{M}$ for $|t|$ small.

(iii) $\mathcal{M}$ is tangent to V^c at the origin.

(iv) Every globally bounded orbit remaining in a suitably small neighborhood of the origin is entirely contained inside $\mathcal{M}$.

(v) Given any trajectory such that $x(t) \to 0$ as $t \to +\infty$, there exists $\eta > 0$ and a trajectory $t \mapsto y(t) \in \mathcal{M}$ on the center manifold such that

$$e^{\eta t} \, |x(t) - y(t)| \to 0 \qquad \text{as} \quad t \to +\infty. \tag{2.5}$$

Remarks. By (i), the manifold $\mathcal{M}$ is parametrized by points on the center subspace V^c. In particular, it has the same dimension as V^c. The invariance property (ii) means that the vector field f is tangent to $\mathcal{M}$ at every point $x \in \mathcal{M}$. By (v), every solution which approaches the origin as $t \to +\infty$ can be described as an exponentially small perturbation of some trajectory on the center manifold. An entirely similar statement holds for solutions which approach the origin as $t \to -\infty$. The proof will show that in (2.5) one can choose any constant $\eta \in \,]0, \beta[$ smaller than the spectral gap of A.

A.3 Proof of the Center Manifold Theorem

The proof, mainly following [V], will be given in several steps. Throughout the following, the Landau notation $\mathcal{O}(1)$ will be used to indicate a quantity depending only on the vector field f, whose absolute value remains uniformly bounded.

A.3.1 Reduction to the Case of a Compact Perturbation.

Set $g(x) \doteq f(x) - Ax$. As a first step we show that, by using a cutoff function, one can assume that g has compact support and that its $\mathbb{C}^1$ norm is arbitrarily small. Indeed, let $\rho : \mathbb{R} \mapsto [0,1]$ be a smooth, even function with compact support, such that

$$\rho(\zeta) = \begin{cases} 1 & if \quad |\zeta| \leq 1, \\ 0 & if \quad |\zeta| \geq 2. \end{cases}$$

For $\varepsilon > 0$ small, define the truncated function

$$g_\varepsilon(x) \doteq \rho\big(|x|/\varepsilon\big) \, g(x) \,.$$

Observing that

$$\big|g(x)\big| = \mathcal{O}(1) \cdot |x|^2 \,, \qquad \big|Dg(x)\big| = \mathcal{O}(1) \cdot |x| \,,$$

we obtain

$$\begin{aligned}
\big\|g_\varepsilon\big\|_{\mathbb{C}^1} &\leq \sup_{|x|<2\varepsilon} \Big\{\big|g_\varepsilon(x)\big| + \big|Dg_\varepsilon(x)\big|\Big\} \\
&\leq \sup_{|x|<2\varepsilon} \Big\{\big|g(x)\big| + \varepsilon^{-1}\big|\rho'(|x|/\varepsilon)\big|\,\big|g(x)\big| + \big|Dg(x)\big|\Big\} \\
&= \mathcal{O}(1) \cdot \varepsilon \,.
\end{aligned}$$

By possibly replacing g with g_ε, we can thus assume that $g \in \mathbb{C}_c^{k+1}$ and that $\|g\|_{\mathbb{C}^1}$ is as small as we like. With these assumptions we shall prove the existence of a global center manifold:

$$\mathcal{M} = \big\{\phi(x_c) \,; \quad x_c \in V^c\big\}, \tag{3.1}$$

parametrized by the whole subspace V^c, without any restriction on the size of x_c.

In the general case, the corresponding local properties (i)–(v) can then be easily obtained, observing that g_ε coincides with g for $|x| \le \varepsilon$.

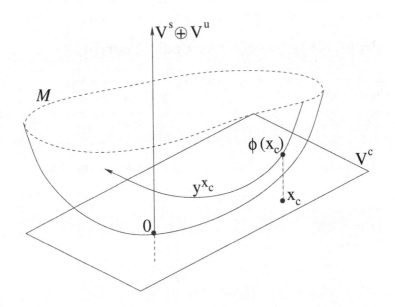

Fig. A.3.

A.3.2 Characterization of the Global Center Manifold.

We now come to the actual construction of points on the center manifold (fig. 3). Let $\beta > 0$ be the spectral gap of the matrix A, as in (1.4). Fix any number $\eta \in \,]0, \beta[$ and define the space of functions with "slow growth"

$$Y_\eta \doteq \left\{ y : \mathbb{R} \mapsto \mathbb{R}^n \; ; \; \|y(\cdot)\|_\eta \doteq \sup_t e^{-\eta|t|}|y(t)| < \infty \right\}.$$

Of course, this implies

$$|y(t)| \le e^{\eta|t|}\|y\|_\eta . \tag{3.2}$$

For every $x_c \in V^c$ we seek a trajectory $y(\cdot) \in Y_\eta$ such that $\pi_c y(0) = x_c$. The following arguments, based on the Contraction Mapping Principle, will show the existence and uniqueness of such a trajectory.

Any solution $t \mapsto y(t)$ of (2.3) can be represented by the "variation of constant formula"

$$y(t) = e^{A(t-t_0)} y(t_0) + \int_{t_0}^{t} e^{A(t-\tau)} g(y(\tau)) \, d\tau. \tag{3.3}$$

We can decompose (3.3) as a sum of its center, stable and unstable components. Notice that here we can choose different starting times in connection with different components:

$$
\begin{aligned}
y(t) = {} &\pi_c \left(e^{A(t-t_c)} y(t_c) + \int_{t_c}^{t} e^{A(t-\tau)} g(y(\tau)) \, d\tau \right) \\
&+ \pi_s \left(e^{A(t-t_s)} y(t_s) + \int_{t_s}^{t} e^{A(t-\tau)} g(y(\tau)) \, d\tau \right) \\
&+ \pi_u \left(e^{A(t-t_u)} y(t_u) + \int_{t_u}^{t} e^{A(t-\tau)} g(y(\tau)) \, d\tau \right).
\end{aligned}
\tag{3.4}
$$

We now choose $t_c = 0$ and let $t_s \to -\infty$ while $t_u \to +\infty$. Recalling that e^{At} commutes with all three projections π_c, π_s, π_u, from (3.4) and the assumptions $y(0) = x_c$, $y(\cdot) \in Y_\eta$ we obtain

$$
\begin{aligned}
y(t) = {} &e^{At} x_c + \int_{0}^{t} e^{A(t-\tau)} \pi_c g(y(\tau)) \, d\tau \\
&+ \int_{-\infty}^{t} e^{A(t-\tau)} \pi_s \, g(y(\tau)) \, d\tau - \int_{t}^{\infty} e^{A(t-\tau)} \pi_u \, g(y(\tau)) \, d\tau.
\end{aligned}
\tag{3.5}
$$

Indeed, for fixed t, as $t_u \to \infty$ by (1.5) and (3.2) we have

$$\lim_{t_u \to \infty} \left| e^{A(t-t_u)} \pi_u \, y(t_u) \right| \leq \lim_{t_u \to \infty} C_\varepsilon e^{(\beta-\varepsilon)(t-t_u)} e^{\eta |t_u|} \|y\|_\eta = 0.$$

Similarly, as $t_s \to -\infty$ we have

$$\lim_{t_s \to -\infty} \left| e^{A(t-t_s)} \pi_u \, y(t_s) \right| \leq \lim_{t_s \to -\infty} C_\varepsilon e^{-(\beta-\varepsilon)(t-t_s)} e^{\eta |t_s|} \|y\|_\eta = 0.$$

Remark. The representation (3.3) is useful in connection with a Cauchy problem, with data assigned at time $t = t_0$. On the other hand, one can regard (3.5) as representing the solution to a three-point boundary value problem. We are here assigning the center component $\pi_c y(0) = x_c$ at time $t = 0$, while (the asymptotic behavior of) the stable component $\pi_s \, y$ is prescribed at time $t = -\infty$ and (the asymptotic behavior of) the unstable component $\pi_u \, y$ is prescribed at time $t = +\infty$.

A.3.3 Construction of the Center Manifold.

For each $x_c \in V^c$, a unique solution $y(\cdot) \in Y_\eta$ of (3.4) will be obtained by the contraction mapping principle. Define the map $\Gamma : V^c \times Y_\eta \mapsto Y_\eta$ by setting

$$\Gamma(x_c, y)(t) = e^{At}x_c + \int_0^t e^{A(t-\tau)}\pi_c g(y(\tau))\, d\tau$$
$$+ \int_{-\infty}^t e^{A(t-\tau)}\pi_s g(y(\tau))\, d\tau - \int_t^\infty e^{A(t-\tau)}\pi_u g(y(\tau))\, d\tau.$$
(3.6)

To show that $\Gamma(x_c, y) \in Y_\eta$ we use the inequalities (1.5), choosing $\varepsilon \doteq \eta$ to estimate the center component and $\varepsilon \doteq \beta - \eta$ to estimate the stable and unstable components.

$$\left|\Gamma(x_c, y)(t)\right| \le C_\eta e^{\eta|t|}|x_c| + \int_0^t C_\eta e^{\eta|t-\tau|}\|g\|_{C^0}\, d\tau$$
$$+ \int_{-\infty}^t C_{\beta-\eta} e^{-\eta(t-\tau)}\|g\|_{C^0}\, d\tau + \int_t^\infty C_{\beta-\eta} e^{\eta(t-\tau)}\|g\|_{C^0}\, d\tau$$
$$= C\|g\|_{C^0} e^{\eta|t|}$$

for a suitable constant C. We claim that, for every fixed x_c, the map $y \mapsto \Gamma(x_c, y)$ is a strict contraction. Indeed, call $\delta_0 \doteq \|y_1 - y_2\|_\eta$. By (3.2) this implies

$$\left|y_1(t) - y_2(t)\right| \le \delta_0 e^{\eta|t|}, \qquad \left|g(y_1(t)) - g(y_2(t))\right| \le \delta_0 e^{\eta|t|}\|g\|_{C^1}$$

for all $t \in \mathbb{R}$. Therefore

$$\left|\Gamma(x_c, y_1)(t) - \Gamma(x_c, y_2)(t)\right|$$
$$\le \int_0^t C_\varepsilon e^{\varepsilon|t-\tau|}\delta_0 e^{\eta|\tau|}\|g\|_{C^1}\, d\tau + \int_{-\infty}^t C_\varepsilon e^{-(\beta-\varepsilon)(t-\tau)}\delta_0 e^{\eta|\tau|}\|g\|_{C^1}\, d\tau$$
$$+ \int_t^\infty C_\varepsilon e^{(\beta-\varepsilon)(t-\tau)}\delta_0 e^{\eta|\tau|}\|g\|_{C^1}\, d\tau \le C' \cdot \delta_0 \|g\|_{C^1} e^{\eta|t|}$$

for some constant C' independent of y_1, y_2. Assuming that $\|g\|_{C^1} \le 1/2C'$ we thus have

$$\left\|\Gamma(x_c, y_1) - \Gamma(x_c, y_2)\right\|_\eta \le \frac{1}{2}\|y_1 - y_2\|_\eta,$$
(3.7)

proving our claim. By the Contraction Mapping Theorem (see the Appendix), for each $x_c \in V^c$ the map $y \mapsto \Gamma(x_c, y)$ has a unique fixed point $y^{x_c} \in Y_\eta$, which provides a solution to (3.5). Since Γ is Lipschitz continuous (namely: linear) w.r.t. the variable x_c, it follows that the map $x_c \mapsto y^{x_c}$ is Lipschitz continuous.

For every $x_c \in V^c$ we now set

$$\phi(x_c) \doteq y^{x_c}(0)$$
(3.8)

and define the manifold $\mathcal{M}$ in terms of (3.1). By the previous analysis, the map $\phi : V^c \mapsto \mathbb{R}^n$ is Lipschitz continuous. By (3.5) it is clear that $\pi_c\phi(x_c) = \pi_c y^{x_c}(0) = x_c$.

A.3.4 Proof of the Invariance Property (ii).

To show that $\mathcal{M}$ is invariant for the flow of (2.3), fix any point $x_0 \in \mathcal{M}$. By construction, the trajectory starting at x, which we denote as $t \mapsto \tilde{x}(t, x_0)$, lies in Y_η. Fix any time t_1. To prove that the point $x_1 \doteq \tilde{x}(t_1, x_0)$ also lies on $\mathcal{M}$ we need to show that the trajectory $t \mapsto \tilde{x}(t, x_1)$ lies in Y_η. But this is clear because

$$\left|\tilde{x}(t, x_1)\right| = \left|\tilde{x}(t + t_1, x_0)\right| \leq C\, e^{\eta|t+t_1|} \leq \left(C\, e^{\eta|t_1|}\right) e^{\eta|t|}.$$

A.3.5 Proof of (iv).

By construction, every trajectory having slow growth at $\pm\infty$, i.e. with $y \in Y_\eta$, is entirely contained in the center manifold $\mathcal{M}$. This is certainly the case for all globally bounded trajectories.

A.3.6 Proof of the Tangency Property (iii).

Since $g(0) = 0$, the function $y(t) \equiv 0$ is (trivially) a globally bounded solution. Hence, by (iv), the manifold $\mathcal{M}$ contains the origin.

To prove that $\mathcal{M}$ is tangent to V^c at the origin, for any $x_c \in V^c$ consider the function $y(t) \doteq e^{At}x_c$. Since $g \neq 0$, we don't expect this to be a solution of (2.3). However (see the Appendix), by the contraction property (3.7), the distance between y and the unique fixed point y^{x_c} can be estimated as twice the distance between y and its first iterate:

$$\left\|y - y^{x_c}\right\|_\eta \leq 2\left\|y - \Gamma(x_c, y)\right\|_\eta. \tag{3.9}$$

We now observe that

$$\left|g(y(\tau))\right| \leq \left|y(\tau)\right|^2 \|g\|_{\mathbb{C}^2} \leq \left(C_\varepsilon e^{\varepsilon|\tau|}|x_c|\right)^2 \|g\|_{\mathbb{C}^2}.$$

By the definition of Γ at (3.6) it now follows

$$
\begin{aligned}
\left|y(t) - \Gamma(x_c, y)(t)\right| &\leq \int_0^t C_\varepsilon e^{\varepsilon|t-\tau|}\left(C_\varepsilon e^{\varepsilon|\tau|}|x_c|\right)^2 \|g\|_{\mathbb{C}^2}\, d\tau \\
&+ \int_{-\infty}^t C_\varepsilon e^{-(\beta-\varepsilon)(t-\tau)}\left(C_\varepsilon e^{\varepsilon|\tau|}|x_c|\right)^2 \|g\|_{\mathbb{C}^2}\, d\tau \\
&+ \int_t^\infty C_\varepsilon e^{(\beta-\varepsilon)(t-\tau)}\left(C_\varepsilon e^{\varepsilon|\tau|}|x_c|\right)^2 \|g\|_{\mathbb{C}^2}\, d\tau \\
&\leq C\,|x_c|^2 e^{\eta|t|}
\end{aligned}
$$

for some constant C independent of $|x_c|$. Therefore

$$\left|y(0) - y^{x_c}(0)\right| \leq \|y - y^{x_c}\|_\eta \leq 2\left\|y - \Gamma(x_c, y)\right\|_\eta \leq 2C|x_c|^2. \tag{3.10}$$

Recalling that $y(0) = x_c$, $y^{x_c}(0) = \phi(x_c)$, an easy consequence of (3.10) is

$$\lim_{x_c \to 0} \frac{|\phi(x_c) - x_c|}{|x_c|} = 0.$$

Hence the manifold $\mathcal{M}$ is tangent to V^c at the origin.

A.3.7 Proof of the Asymptotic Approximation Property (v).

Let $x : [0, +\infty[\mapsto \mathbb{R}^n$ be a solution of (2.1) which approaches the origin as $t \to +\infty$. We extend $x(\cdot)$ to a bounded function $x^*(\cdot)$ defined on the whole real line by setting

$$x^*(t) = \begin{cases} x(t) & if \quad t \geq 0, \\ x(0) & if \quad t < 0. \end{cases}$$

Notice that x^* provides a globally bounded solution to

$$\dot{x}^*(t) = Ax^* + g(x^*) + \varphi(t) \qquad \varphi(t) = \begin{cases} 0 & if \quad t > 0, \\ -Ax(0) - g(x(0)) & if \quad t < 0. \end{cases}$$

Therefore, x^* can be represented by the "variation of constant formula"

$$x^*(t) = e^{A(t-t_0)} \pi_s x^*(t_0) + \int_{t_0}^{t} e^{A(t-\tau)} \pi_s g(x^*(\tau)) \, d\tau + \int_{t_0}^{t} e^{A(t-\tau)} \pi_s \varphi(\tau) \, d\tau$$

$$+ e^{A(t-t_1)} \pi_{cu} x^*(t_1) + \int_{t_1}^{t} e^{A(t-\tau)} \pi_{cu} g(x^*(\tau)) \, d\tau + \int_{t_1}^{t} e^{A(t-\tau)} \pi_{cu} \varphi(\tau) \, d\tau .$$

$$(3.11)$$

Here and in the sequel, $\pi_{cu} = \pi_c + \pi_u$ denotes the projection on the center-unstable space $V^c \oplus V^u$.

Consider now the space of functions

$$Z_\eta \doteq \left\{ z : \mathbb{R} \mapsto \mathbb{R}^n ; \quad \|z(\cdot)\|_\eta \doteq \sup_t e^{\eta t} |z(t)| < \infty \right\}.$$

We claim that there exists a function $z \in Z_\eta$ such that $y = x^* + z \in Y_\eta$ is a global solution of (2.3), contained in the center manifold $\mathcal{M}$. Recalling (3.11), for any choice of t_0, t_1 such a function $z(\cdot)$ should provide a solution to the integral equation

$$z(t) = -\pi_s x^*(t) + e^{A(t-t_0)} \pi_s \big(x^*(t_0) + z(t_0)\big)$$

$$+ \int_{t_0}^t e^{A(t-\tau)} \pi_s g\big(x^*(\tau) + z(\tau)\big) \, d\tau$$

$$- \pi_{cu} x^*(t) + e^{A(t-t_1)} \pi_{cu} \big(x^*(t_1) + z(t_1)\big)$$

$$+ \int_{t_1}^t e^{A(t-\tau)} \pi_{cu} g\big(x^*(\tau) + z(\tau)\big) \, d\tau$$

$$= e^{A(t-t_0)} \pi_s z(t_0) + \int_{t_0}^t e^{A(t-\tau)} \pi_s \Big[g\big(x^*(\tau) + z(\tau)\big) - g\big(x^*(\tau)\big) \Big] \, d\tau$$

$$- \int_{t_0}^t e^{A(t-\tau)} \pi_s \varphi(\tau) \, d\tau + e^{A(t-t_1)} \pi_{cu} z(t_1)$$

$$+ \int_{t_1}^t e^{A(t-\tau)} \pi_{cu} \Big[g\big(x^*(\tau) + z(\tau)\big) - g\big(x^*(\tau)\big) \Big] \, d\tau$$

$$- \int_{t_1}^t e^{A(t-\tau)} \pi_{cu} \varphi(\tau) \, d\tau \, .$$

Letting $t_0 \to -\infty$ and $t_1 \to +\infty$ we obtain

$$z(t) = \int_{-\infty}^t e^{A(t-\tau)} \pi_s \Big[g\big(x^*(\tau) + z(\tau)\big) - g\big(x^*(\tau)\big) \Big] \, d\tau$$

$$- \int_{-\infty}^t e^{A(t-\tau)} \pi_s \varphi(\tau) \, d\tau$$

$$- \int_t^{\infty} e^{A(t-\tau)} \pi_{cu} \Big[g\big(x^*(\tau) + z(\tau)\big) - g\big(x^*(\tau)\big) \Big] \, d\tau \qquad (3.12)$$

$$+ \int_t^{\infty} e^{A(t-\tau)} \pi_{cu} \varphi(\tau) \, d\tau$$

$$\doteq \Lambda(z)(t) \, .$$

Recalling that $\varphi(\tau) = 0$ for $\tau > 0$ and using the basic inequalities (1.5), we see that the map $\Lambda : Z_\eta \mapsto Z_\eta$ is a strict contraction, provided that the norm $\|g\|_{C^1}$ is suitably small. Therefore Λ admits a unique fixed point $z \in Z_\eta$, which satisfies (3.12). Since x^* is globally bounded and $z \in Z_\eta \subset Y_\eta$, it is clear that $y \doteq x^* + z \in Y_\eta$, hence it represents a trajectory contained in the center manifold. For all $t > 0$ we now have

$$|x(t) - y(t)| = |z(t)| \le e^{-\eta t} \|z\|_\eta \, .$$

This implies (2.5) for any smaller choice of the exponent η.

A.3.8 Smoothness of the Center Manifold.

To complete the proof, it remains to show that the map $x_c \mapsto \phi(x_c)$ is k times continuously differentiable. This fact would easily follow from the implicit function theorem, if we could prove that $\Gamma : V^c \times Y_\eta \mapsto Y_\eta$ is a C^k map.

Unfortunately this is not true. Indeed, for any non-trivial function $g \in C_c^\infty$, the substitution operator $y \mapsto G(y)$ defined by

$$G(y)(t) \doteq g\big(y(t)\big) \tag{3.13}$$

is not differentiable as a map from Y_η into itself.

Example 2. Let $g : \mathbb{R} \mapsto \mathbb{R}$ be a smooth function with compact support, such that

$$g(x) = x^2 \qquad \text{for } |x| \le \varepsilon.$$

If the map $G : Y_\eta \mapsto Y_\eta$ in (3.13) were differentiable at the origin $0 \in Y_\eta$, its differential could only be the identically zero map. However, consider the sequence of functions

$$y_n(t) \doteq \begin{cases} \varepsilon & \text{if } t \in [n,\, n+1], \\ 0 & \text{otherwise.} \end{cases}$$

This is mapped into the sequence

$$G(y_n)(t) \doteq \begin{cases} \varepsilon^2 & \text{if } t \in [n,\, n+1], \\ 0 & \text{otherwise.} \end{cases}$$

By the definition of the norm on the space Y_η, as $n \to \infty$ one has

$$\|y_n\|_\eta = \sup_t \, e^{-\eta|t|} y_n(t) = \varepsilon\, e^{-\eta n} \to 0.$$

We now have

$$\lim_{n \to \infty} \frac{\|G(y_n)\|_\eta}{\|y_n\|_\eta} = \lim_{n \to \infty} \frac{\varepsilon^2\, e^{-\eta n}}{\varepsilon\, e^{-\eta n}} = \varepsilon \ne 0,$$

showing that the zero linear map cannot be the differential of G at the origin.

To overcome the difficulty pointed out by the previous example, one can observe that G becomes k times differentiable if viewed as a map from a smaller space $Y_{\eta'}$ (with a stronger norm) into a Y_η. The proof of the regularity of the center manifold $\mathcal{M}$ strongly relies on this fact.

Lemma. Let $g \in C^{k+1}$ and assume $0 < \eta' < (k+1)\eta' \le \eta$. Then the substitution operator G at (3.13) is k times differentiable as a map from $Y_{\eta'}$ into Y_η.

Proof. We begin by recalling Taylor's formula (see [D], p.190)

$$g(y + z) = T_k g(y, z) + R_k(y, z),$$

where

$$T_k g(y, z) \doteq \sum_{j=0}^{k} \frac{D^j g(y)}{j!} z^{[j]},$$

and

$$R_k(y, z) = \left(\int_0^1 \frac{(1-\xi)^k}{k!} D^{k+1} g(y + \xi z) \, d\xi \right) z^{[k+1]}.$$

The j-th derivative of a function is here written as a multilinear symmetric operator, while $z^{[j]} = z \otimes \cdots \otimes z$ denotes the tensor product of j factors all equal to z. In order to prove that the map $y \mapsto G(y)$ is in $\mathbb{C}^k(Y_{\eta'}, Y_\eta)$, we need to check that

$$\left\| G(y+z) - T_k G(y, z) \right\|_\eta = \sup_t e^{-\eta|t|} \left| g(y(t)+z(t)) - T_k g(y(t), z(t)) \right| = \mathcal{O}(1) \cdot \|z\|_{\eta'}^{k+1}.$$

This is clear because

$$e^{-\eta|t|} \left| R_k(y(t), z(t)) \right| \leq e^{-\eta|t|} \cdot \frac{1}{k!} \|g\|_{\mathbb{C}^{k+1}} \|z(t)\|^{k+1}$$

$$\leq e^{-\eta|t|} \cdot \frac{1}{k!} \|g\|_{\mathbb{C}^{k+1}} e^{(k+1)\eta'|t|} \|z\|_{\eta'}^{k+1} \leq \frac{1}{k!} \|g\|_{\mathbb{C}^{k+1}} \|z\|_{\eta'}^{k+1},$$

provided that $(k+1)\eta' \leq \eta$.   •

Corollary. *For every* $j = 1, \ldots, k$, *the operator* Γ *defined at (3.6) is a* $\mathbb{C}^\ell$ *map from* $V^c \times Y_{\eta'}$ *into* Y_η, *provided that* $2\ell\eta' \leq \eta$.

Indeed, Γ can be written in the form

$$\Gamma(x, y) \doteq Sx + K \circ G(y), \tag{3.14}$$

where

$$(Sx)(t) \doteq e^{At} x, \tag{3.15}$$

$$(Kv)(t) \doteq \int_0^t e^{A(t-\tau)} \pi_c v(\tau) \, d\tau + \int_{-\infty}^t e^{A(t-\tau)} \pi_s v(\tau) \, d\tau - \int_t^\infty e^{A(t-\tau)} \pi_u v(\tau) \, d\tau. \tag{3.16}$$

Since both $S : V^c \mapsto Y_\eta$ and $K : Y_\eta \mapsto Y_\eta$ are continuous linear mappings, by the previous lemma it follows that Γ is a $\mathbb{C}^\ell$ map from $V^c \times Y_{\eta'}$ into Y_η, provided that $(\ell + 1)\eta' \leq \eta$.   •

We now resume the proof of the main theorem. By induction, define the sequence of mappings $y_\nu : V^c \mapsto Y_\eta$, such that

$$y_0(x) \equiv 0 \qquad y_\nu(x) \doteq \Gamma(x, y_{\nu-1}) \qquad \nu \geq 1.$$

In particular, one has

$$y_0(x)(t) = 0 \qquad y_1(x)(t) \doteq e^{At} x.$$

By the argument at (3.7), we already know that the sequence y_ν converges pointwise to the function $x_c \mapsto y^{x_c}$ uniformly for $x_c \in V^c$ on bounded sets. We now show that the same is true also for all derivatives, up to order k. Recalling that $\phi(x_c) \doteq y^{x_c}(0)$, this will show that $\phi \in C^k$, completing the proof.

Fix $\eta \in]0, \beta[$ and consider the numbers $0 < \eta_0 < \eta_1 < \ldots < \eta_k = \eta < \beta$, defined by

$$\eta_j \doteq e^{2j-2k}\eta \qquad j = 0, 1, \ldots, k.$$

By the previous Corollary, Γ is then a mapping of class C^ℓ from $V^c \times Y_{\eta_i}$ into Y_{η_j}, provided that $i + \ell \leq j$. For convenience, the higher derivatives of a map $y : V^c \mapsto Y_\eta$ will be denoted as $D^j y \doteq d^j y/dx^{[j]}$. We recall that these are elements of the space of j-linear mappings $L^j(V^c, Y_\eta)$.

As $\nu \to \infty$, the convergence of the sequence of derivatives $D^j y_\nu$ will be proved by induction on j. Assume that, for all $i = 0, 1, \ldots, j-1$, the sequence of derivatives

$$x \mapsto D^i y_\nu(x)$$

converges uniformly on bounded sets, in the space of maps

$$V^c \mapsto L^i(V^c; Y_{\eta_i}).$$

We claim that the sequence of j-derivatives also converges. To see this, we first compute the derivatives of the composite map $x \mapsto \Gamma(x, y(x))$. Differentiating (3.14) several times, one finds

$$\frac{d}{dx}\Gamma(x, y(x)) = S + K \circ DG\, Dy,$$

$$\frac{d^2}{dx^{[2]}}\Gamma(x, y(x)) = K \circ \left[DG\, D^2 y + D^2 G\,(Dy \otimes Dy)\right]$$

$$\frac{d^3}{dx^{[3]}}\Gamma(x, y(x)) = K \circ \left[DG\, D^3 y + 3D^2 G\,(Dy \otimes D^2 y) + D^3 G\,(Dy \otimes Dy \otimes Dy)\right]$$

By induction (see [B]), it is clear that the j-th derivative of (3.14) has an expression of the form

$$\frac{d^j}{dx^{[j]}}\Gamma(x, y(x)) = K \circ \left[DG\, D^j y + \Phi_j(y, Dy, \ldots, D^{j-1} y)\right]$$

for a suitable function Φ_j involving only derivatives of lower order. The inductive assumption now guarantees that the sequence of mappings

$$x \mapsto \Phi_j\big(y_\nu(x), Dy_\nu(x), \ldots, D^{j-1} y_\nu(x)\big)$$

converges for all $x \in V^c$, uniformly on bounded sets. On the other hand, if $\|g\|_{C^1}$ is small enough, for every $y \in Y_\eta$ the operator

$$\psi \mapsto \left[K \circ DG(y)\right]\psi$$

is a strict contraction in the space $L^j(V^c, Y_{\eta_j})$. An application of the Contraction Mapping Theorem (see the Appendix) now yields the convergence of the sequence $D^j y_\nu$, uniformly on bounded sets. This completes the proof of the theorem. •

Assuming that the norm $\|g\|_{C^1}$ is sufficiently small, we proved the existence and uniqueness of a GLOBAL center manifold consisting of all trajectories $t \mapsto y(t)$ having "slow growth" at $\pm\infty$, so that $y \in Y_\eta$. In the general case, we could still prove the existence of a LOCAL center manifold, defined in a neighborhood of the origin. However, one should be aware that this local center manifold may not be unique, because its construction depends on the choice of the cut-off function.

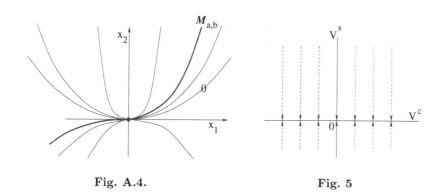

Fig. A.4.   **Fig. 5**

Example 3. Consider the system (fig. 4)

$$\dot{x}_1 = -x_1^3, \qquad\qquad \dot{x}_2 = -x_2. \qquad\qquad (3.17)$$

Linearizing at the origin, one obtains the system (fig. 5)

$$\begin{pmatrix} \dot{x}_1 \\ \dot{x}_2 \end{pmatrix} = \begin{pmatrix} 0 & 0 \\ 0 & -1 \end{pmatrix} \begin{pmatrix} x_1 \\ x_2 \end{pmatrix}.$$

The corresponding stable, center and unstable subspaces are

$$V^s = \{(0, x_2); \ x_2 \in \mathbb{R}\}, \qquad V^c = \{(x_1, 0); \ x_1 \in \mathbb{R}\}, \qquad V^u = \{(0,0)\}.$$

By explicitly solving the equation

$$\frac{dx_1}{dx_2} = \frac{x_1^3}{x_2}$$

we find that, for any choice of $a, b \in \mathbb{R}$, the following manifold $\mathcal{M}_{a,b}$ is invariant w.r.t. the flow of (3.17):

$$\mathcal{M}_{a,b} \doteq \left\{ \big(x, \psi(x)\big) \right\} \qquad \psi(x) = \begin{cases} a\,e^{-1/2x^2} & if \quad x < 0, \\ 0 & if \quad x = 0, \\ b\,e^{-1/2x^2} & if \quad x > 0. \end{cases}$$

Notice that every $\mathcal{M}_{a,b}$ is smooth, and tangent to V^c at the origin, In this case, we thus have a continuum family of center manifolds.

A.4 The Contraction Mapping Theorem

For reader's convenience, we prove here the fixed point theorem which was used throughout these notes.

Contraction Mapping Theorem. *Let* X, Y *be Banach spaces and let* $\Gamma :$ $X \times Y \mapsto Y$ *be a continuous mapping which is a strict contraction in the* y *variable, i.e.*

$$\big\|\Gamma(x,y) - \Gamma(x,y')\big\| \leq \kappa\|y - y'\| \qquad for\ all \quad x \in X, \quad y, y' \in Y, \quad (A.1)$$

for some constant $\kappa < 1$. *Then the following holds.*

(i) For every $x \in X$, *there exists a unique* $y(x) \in Y$ *such that*

$$y(x) = \Gamma\big(x,\, y(x)\big). \qquad (A.2)$$

(ii) For every $x \in X$, $y \in Y$ *one has*

$$\big\|y - y(x)\big\| \leq \frac{1}{1 - \kappa}\big\|y - \Gamma(x,y)\big\|. \qquad (A.3)$$

(iii) If Γ *is Lipschitz continuous w.r.t.* x, *say*

$$\big\|\Gamma(x,y) - \Gamma(x',y)\big\| \leq L\,\|x - x'\| \qquad for\ all \quad x, x' \in X, \quad y \in Y, \quad (A.4)$$

then the same is true of the map $x \mapsto y(x)$. *Namely*

$$\big\|y(x) - y(x')\big\| \leq \frac{L}{1 - \kappa}\,\|x - x'\|. \qquad (A.5)$$

(iv) Consider any convergent sequence $x_\nu \to \bar{x}$ *in* X. *Then for every* $y_0 \in Y$ *the sequence of iterates*

$$y_{\nu+1} \doteq \Gamma(x_\nu, y_\nu)$$

converges to the point $\bar{y} \doteq y(\bar{x})$.

Proof. Fix any point $y \in Y$. For each x, consider the sequence

$$y_0 \doteq y, \quad y_1 \doteq \Gamma(x, y_0), \qquad \cdots \qquad y_{\nu+1} \doteq \Gamma(x, y_\nu), \quad \cdots$$

By induction, for every $\nu \geq 0$ one checks that

$$\|y_{\nu+1} - y_\nu\| \leq \kappa \|y_\nu - y_{\nu-1}\| \leq \kappa^\nu \|y_1 - y_0\| = \kappa^\nu \|\Gamma(x,y) - y\|. \qquad (A.6)$$

Since $\kappa < 1$, the sequence y_ν is Cauchy and converges to some limit point, which we call $y(x)$. The continuity of Γ now implies

$$y(x) = \lim_{\nu \to \infty} y_{\nu+1} = \lim_{\nu \to \infty} \Gamma(x, y_\nu) = \Gamma\left(x, \lim_{\nu \to \infty} y_\nu\right) = \Gamma(x, y(x)).$$

Hence (A.2) holds. The uniqueness of the fixed point $y(x)$ is proved observing that, if $y_1 = \Gamma(x, y_1)$ and $y_2 = \Gamma(x, y_2)$, then (A.1) implies

$$\|y_1 - y_2\| = \|\Gamma(x, y_1) - \Gamma(x, y_2)\| \leq \kappa \|y_1 - y_2\|.$$

Hence $y_1 = y_2$.

To prove (A.3), it suffices to observe that (A.6) implies

$$\|y - y(x)\| \leq \sum_{\nu=0}^\infty \|y_{\nu+1} - y_\nu\| \leq \sum_{\nu=0}^\infty \kappa^\nu \|\Gamma(x,y) - y\| = \frac{\|\Gamma(x,y) - y\|}{1 - \kappa}.$$

Toward a proof of (A.5), we use (A.3) with $y \doteq y(x')$ and obtain

$$\|y(x') - y(x)\| \leq \frac{1}{1-\kappa} \|y(x') - \Gamma(x, y(x'))\|$$
$$= \frac{1}{1-\kappa} \|\Gamma(x', y(x')) - \Gamma(x, y(x'))\| \leq \frac{L}{1-\kappa} \|x' - x\|.$$

To prove the remaining statement (iv), observe that the quantities $\varepsilon_\nu \doteq \|\Gamma(x_\nu, \bar{y}) - \Gamma(\bar{x}, \bar{y})\|$ converge to zero as $\nu \to \infty$. Moreover, the contraction property implies

$$\begin{aligned}
\|y_{\nu+1} - \bar{y}\| &= \|\Gamma(x_\nu, y_\nu) - \Gamma(\bar{x}, \bar{y})\| \\
&\leq \|\Gamma(x_\nu, y_\nu) - \Gamma(x_\nu, \bar{y})\| + \|\Gamma(x_\nu, \bar{y}) - \Gamma(\bar{x}, \bar{y})\| \qquad (A.7) \\
&\leq \kappa \|y_\nu - \bar{y}\| + \varepsilon_\nu.
\end{aligned}$$

¿From (A.7) we deduce

$$\|y_\nu - \bar{y}\| \leq \kappa^\nu \|y_0 - \bar{y}\| + \sum_{j=1}^\nu \kappa^{\nu-j} \varepsilon_j$$

and therefore

$$\limsup_{\nu \to \infty} \|y_\nu - \bar{y}\| \leq \limsup_{\nu \to \infty} \frac{\varepsilon_\nu}{1 - \kappa} = 0.$$

References

[B] A. Bressan, High order approximation of implicitly defined maps, *Annali Mat. Pura Appl.*, (IV) **137** (1984), 163–173.

[D] J. Diéudonne, *Foundations of Modern Analysis*, Academic Press, New York, 1969.

[P] L. Perko, *Differential Equations and Dynamical Systems*, Third Edition, Springer-Verlag, 2001.

[V] A. Vanderbauwhede, Centre manifolds, normal forms and elementary bifurcations, *Dynamics Reported, Vol. 2* (1989), 89–169.

List of Participants

1. Amadori Debora
   University of L' Aquila
   amadori@univaq.it

2. Ancona Fabio
   University of Bologna
   ancona@ciram.unibo.it

3. Bellettini Giovanni
   University of Roma II
   belletti@mat.uniroma2.it

4. Berselli Luigi
   University of Pisa
   berselli@dma.unipi.it

5. Bianchini Stefano
   IAC CNR
   s.bianchini@iac.rm.cnr.it

6. Bobyleva Olga
   University of Mosca
   olga_1028@front.ru

7. Bressan Alberto (lecturer)
   Penn State University, USA
   bressan@math.psu.edu

8. Calvo Daniela
   University of Torino
   calvo@dm.unipi.it

9. Candela Simona
   University of L' Aquila
   candela@univaq.it

10. Cappiello Marco
    University of Torino
    marco@alpha01.dm.unito.it

11. Casella Emanuela
    University of Brescia
    casella@ing.unibs.it

12. Cavazzoni Rita
    University of Firenze
    cavazzoni@math.unifi.it

13. Chau Chun Dong
    University of Eindoven
    c.p.chau@student.tue.nl

14. Chesnokov Alexander
    University of Novosibirsk
    chesnokov@hydro.nsc.ru

15. Christoforou Cleopatra
    Brown University, USA
    cleo@math.brown.edu

16. Chugainova Anna
    University of Mosca
    annach@bk.ru

17. Coclite Giuseppe
    SISSA
    coclite@sissa.it

18. Colombo Rinaldo
    University of Brescia
    rinaldo@ing.unibs.it

19. Corrias Lucilla
    University of Evry
    corrias@maths.univ-evry.fr

20. Da Lio Francesca
    University of Torino
    dalio@dm.unito.it

21. De Falco Carlo
    University of Milano
    defalco@mat.unimi.it

22. Devarapu Anilkumar
    University of Madras
    anil@math.net

23. Di Francesco Marco
    University of L' Aquila
    difrance@univaq.it

24. Donatelli Donatella
    University of L' Aquila
    donatell@univaq.it

25. Dragoni Federica
    SNS Pisa
    f.dragoni@sns.it
26. Fonte Massimo
    SISSA *fonte@sissa.it*
27. Garavello Mauro
    SISSA *mgarav@sissa.it*
28. Gianazza Ugo
    University of Pavia
    gianazza@imati.cnr.it
29. Int'panhuis Peter
    University of Eindoven
    p.h.m.w.i.t.panhuis@student.tue.nl
30. Karch Grzegorz
    University of Wroklaw
    karch@math.uni.wroc.pl
31. Khattri Sanjay
    University of Bergen, Norvegia
    skhattri@neo.tamu.edu
32. Khe Alexander
    University of Novosibirsk
    alekhe@hydro.nsc.ru
33. Lucente Sandra
    University of Bari
    lucente@dm.uniba.it
34. Makarenkov Oleg
    University of Voronezh
    omakarenkov@mail.ru
35. Makarenkova Irina
    University of Voronezh
    makarenkova_irin@mail.ru
36. Marcati Pierangelo (editor)
    University of L'Aquila
    marcati@univaq.it
37. Marson Andrea
    University of Padova
    marson@math.unipd.it
38. Mascia Corrado
    University of Roma 1
    mascia@mat.uniroma1.it
39. Morando Alessandro
    University of Brescia
    morando@ing.unibs.it
40. Natalini Roberto
    IAC CNR
    r.natalini@iac.cnr.it
41. Novaga Matteo
    University of Pisa
    novaga@dm.unipi.it
42. Olech Michal
    University of Wroklaw
    s102507@math.uni.wroc.pl
43. Priuli Fabio Simone
    SISSA
    priuli@sissa.it

44. Rocca Elisabetta
    University of Pavia
    rocca@dimat.unipv.it
45. Rosini Massimiliano
    University of Napoli
    igfu2002@yahoo.it
46. Rossi Riccarda
    University of Pavia
    riccarda@dimat.unipv.it
47. Rozanova Olga
    University of Mosca
    rozanova@mech.math.msu.su
48. Rubino Bruno
    University of L' Aquila
    rubino@univaq.it
49. Sampalmieri Rosella
    University of L' Aquila
    sampalm@univaq.it
50. Savarè Giuseppe
    University of Pavia
    savare@imati.cnr.it
51. Serre Denis (lecturer)
    ENS Lions, France
    serre@umpa.ens-lyon.fr
52. Sfakianakis Nikolaos
    University of Creta
    sfaknikj@math.uoc.gr
53. Stefanelli Ulisse
    IMATI CNR, Pavia
    ulisse@imati.cnr.it
54. Terracina Andrea
    University of Roma 1
    terracin@mat.uniroma1.it
55. Timofeeva Galina
    Ural St University ofiv
    gtimofeeva@mail.ru
56. Vinh Pham Chi
    Nat University ofiv Hanoi
    pvc@maths.gla.ac.uk
57. Visciglia Nicola
    SNS, Pisa
    viscigli@mail.unipi.it
58. Wang Shu
    University of Vienna
    shu.wang@univie.ac.at
59. Williams Mark (lecturer)
    University of North Carolina
    Chapel Hill, USA
    williams@math.unc.edu
60. Zappacosta Stefano
    University of L' Aquila
    zappacos@univaq.it
61. Zumbrun Kevin (lecturer)
    Indiana University, Bloomington, USA
    kzumbrun@indiana.edu

LIST OF C.I.M.E. SEMINARS

Published by C.I.M.E

Published by Ed. Cremonese, Firenze

1966   39. Calculus of variations
       40. Economia matematica
       41. Classi caratteristiche e questioni connesse
       42. Some aspects of diffusion theory

1967   43. Modern questions of celestial mechanics
       44. Numerical analysis of partial differential equations
       45. Geometry of homogeneous bounded domains

1968   46. Controllability and observability
       47. Pseudo-differential operators
       48. Aspects of mathematical logic

1969   49. Potential theory
       50. Non-linear continuum theories in mechanics and physics and their applications
       51. Questions of algebraic varieties

1970   52. Relativistic fluid dynamics
       53. Theory of group representations and Fourier analysis
       54. Functional equations and inequalities
       55. Problems in non-linear analysis

1971   56. Stereodynamics
       57. Constructive aspects of functional analysis (2 vol.)
       58. Categories and commutative algebra

1972   59. Non-linear mechanics
       60. Finite geometric structures and their applications
       61. Geometric measure theory and minimal surfaces

1973   62. Complex analysis
       63. New variational techniques in mathematical physics
       64. Spectral analysis

1974   65. Stability problems
       66. Singularities of analytic spaces
       67. Eigenvalues of non linear problems

1975   68. Theoretical computer sciences
       69. Model theory and applications
       70. Differential operators and manifolds

Published by Ed. Liguori, Napoli

1976   71. Statistical Mechanics
       72. Hyperbolicity
       73. Differential topology

1977   74. Materials with memory
       75. Pseudodifferential operators with applications
       76. Algebraic surfaces

Published by Ed. Liguori, Napoli & Birkhäuser

1978   77. Stochastic differential equations
       78. Dynamical systems

1979   79. Recursion theory and computational complexity
       80. Mathematics of biology

Published by Springer-Verlag

Lecture Notes in Mathematics

For information about earlier volumes
please contact your bookseller or Springer
LNM Online archive: springerlink.com

Vol. 1764: A. Cannas da Silva, Lectures on Symplectic Geometry (2001)

Vol. 1765: T. Kerler, V. V. Lyubashenko, Non-Semisimple Topological Quantum Field Theories for 3-Manifolds with Corners (2001)

Vol. 1766: H. Hennion, L. Hervé, Limit Theorems for Markov Chains and Stochastic Properties of Dynamical Systems by Quasi-Compactness (2001)

Vol. 1767: J. Xiao, Holomorphic Q Classes (2001)

Vol. 1768: M. J. Pflaum, Analytic and Geometric Study of Stratified Spaces (2001)

Vol. 1769: M. Alberich-Carramiñana, Geometry of the Plane Cremona Maps (2002)

Vol. 1770: H. Gluesing-Luerssen, Linear Delay-Differential Systems with Commensurate Delays: An Algebraic Approach (2002)

Vol. 1771: M. Émery, M. Yor (Eds.), Séminaire de Probabilités 1967-1980. A Selection in Martingale Theory (2002)

Vol. 1772: F. Burstall, D. Ferus, K. Leschke, F. Pedit, U. Pinkall, Conformal Geometry of Surfaces in S^4 (2002)

Vol. 1773: Z. Arad, M. Muzychuk, Standard Integral Table Algebras Generated by a Non-real Element of Small Degree (2002)

Vol. 1774: V. Runde, Lectures on Amenability (2002)

Vol. 1775: W. H. Meeks, A. Ros, H. Rosenberg, The Global Theory of Minimal Surfaces in Flat Spaces. Martina Franca 1999. Editor: G. P. Pirola (2002)

Vol. 1776: K. Behrend, C. Gomez, V. Tarasov, G. Tian, Quantum Comohology. Cetraro 1997. Editors: P. de Bartolomeis, B. Dubrovin, C. Reina (2002)

Vol. 1777: E. García-Río, D. N. Kupeli, R. Vázquez-Lorenzo, Osserman Manifolds in Semi-Riemannian Geometry (2002)

Vol. 1778: H. Kiechle, Theory of K-Loops (2002)

Vol. 1779: I. Chueshov, Monotone Random Systems (2002)

Vol. 1780: J. H. Bruinier, Borcherds Products on O(2,1) and Chern Classes of Heegner Divisors (2002)

Vol. 1781: E. Bolthausen, E. Perkins, A. van der Vaart, Lectures on Probability Theory and Statistics. Ecole d' Eté de Probabilités de Saint-Flour XXIX-1999. Editor: P. Bernard (2002)

Vol. 1782: C.-H. Chu, A. T.-M. Lau, Harmonic Functions on Groups and Fourier Algebras (2002)

Vol. 1783: L. Grüne, Asymptotic Behavior of Dynamical and Control Systems under Perturbation and Discretization (2002)

Vol. 1784: L. H. Eliasson, S. B. Kuksin, S. Marmi, J.-C. Yoccoz, Dynamical Systems and Small Divisors. Cetraro, Italy 1998. Editors: S. Marmi, J.-C. Yoccoz (2002)

Vol. 1785: J. Arias de Reyna, Pointwise Convergence of Fourier Series (2002)

Vol. 1786: S. D. Cutkosky, Monomialization of Morphisms from 3-Folds to Surfaces (2002)

Vol. 1787: S. Caenepeel, G. Militaru, S. Zhu, Frobenius and Separable Functors for Generalized Module Categories and Nonlinear Equations (2002)

Vol. 1788: A. Vasil'ev, Moduli of Families of Curves for Conformal and Quasiconformal Mappings (2002)

Vol. 1789: Y. Sommerhäuser, Yetter-Drinfel'd Hopf algebras over groups of prime order (2002)

Vol. 1790: X. Zhan, Matrix Inequalities (2002)

Vol. 1791: M. Knebusch, D. Zhang, Manis Valuations and Prüfer Extensions I: A new Chapter in Commutative Algebra (2002)

Vol. 1792: D. D. Ang, R. Gorenflo, V. K. Le, D. D. Trong, Moment Theory and Some Inverse Problems in Potential Theory and Heat Conduction (2002)

Vol. 1793: J. Cortés Monforte, Geometric, Control and Numerical Aspects of Nonholonomic Systems (2002)

Vol. 1794: N. Pytheas Fogg, Substitution in Dynamics, Arithmetics and Combinatorics. Editors: V. Berthé, S. Ferenczi, C. Mauduit, A. Siegel (2002)

Vol. 1795: H. Li, Filtered-Graded Transfer in Using Noncommutative Gröbner Bases (2002)

Vol. 1796: J.M. Melenk, hp-Finite Element Methods for Singular Perturbations (2002)

Vol. 1797: B. Schmidt, Characters and Cyclotomic Fields in Finite Geometry (2002)

Vol. 1798: W.M. Oliva, Geometric Mechanics (2002)

Vol. 1799: H. Pajot, Analytic Capacity, Rectifiability, Menger Curvature and the Cauchy Integral (2002)

Vol. 1800: O. Gabber, L. Ramero, Almost Ring Theory (2003)

Vol. 1801: J. Azéma, M. Émery, M. Ledoux, M. Yor (Eds.), Séminaire de Probabilités XXXVI (2003)

Vol. 1802: V. Capasso, E. Merzbach, B. G. Ivanoff, M. Dozzi, R. Dalang, T. Mountford, Topics in Spatial Stochastic Processes. Martina Franca, Italy 2001. Editor: E. Merzbach (2003)

Vol. 1803: G. Dolzmann, Variational Methods for Crystalline Microstructure – Analysis and Computation (2003)

Vol. 1804: I. Cherednik, Ya. Markov, R. Howe, G. Lusztig, Iwahori-Hecke Algebras and their Representation Theory. Martina Franca, Italy 1999. Editors: V. Baldoni, D. Barbasch (2003)

Vol. 1805: F. Cao, Geometric Curve Evolution and Image Processing (2003)

Vol. 1806: H. Broer, I. Hoveijn. G. Lunther, G. Vegter, Bifurcations in Hamiltonian Systems. Computing Singularities by Gröbner Bases (2003)

Vol. 1807: V. D. Milman, G. Schechtman (Eds.), Geometric Aspects of Functional Analysis. Israel Seminar 2000-2002 (2003)

Vol. 1808: W. Schindler, Measures with Symmetry Properties (2003)

Vol. 1809: O. Steinbach, Stability Estimates for Hybrid Coupled Domain Decomposition Methods (2003)

Vol. 1810: J. Wengenroth, Derived Functors in Functional Analysis (2003)

Vol. 1811: J. Stevens, Deformations of Singularities (2003)

Vol. 1812: L. Ambrosio, K. Deckelnick, G. Dziuk, M. Mimura, V. A. Solonnikov, H. M. Soner, Mathematical Aspects of Evolving Interfaces. Madeira, Funchal, Portugal 2000. Editors: P. Colli, J. F. Rodrigues (2003)

Vol. 1813: L. Ambrosio, L. A. Caffarelli, Y. Brenier, G. Buttazzo, C. Villani, Optimal Transportation and its Applications. Martina Franca, Italy 2001. Editors: L. A. Caffarelli, S. Salsa (2003)

Vol. 1814: P. Bank, F. Baudoin, H. Föllmer, L.C.G. Rogers, M. Soner, N. Touzi, Paris-Princeton Lectures on Mathematical Finance 2002 (2003)

Vol. 1815: A. M. Vershik (Ed.), Asymptotic Combinatorics with Applications to Mathematical Physics. St. Petersburg, Russia 2001 (2003)

Vol. 1816: S. Albeverio, W. Schachermayer, M. Talagrand, Lectures on Probability Theory and Statistics. Ecole d'Eté de Probabilités de Saint-Flour XXX-2000. Editor: P. Bernard (2003)

Vol. 1817: E. Koelink, W. Van Assche (Eds.), Orthogonal Polynomials and Special Functions. Leuven 2002 (2003)

Vol. 1818: M. Bildhauer, Convex Variational Problems with Linear, nearly Linear and/or Anisotropic Growth Conditions (2003)

Vol. 1819: D. Masser, Yu. V. Nesterenko, H. P. Schlickewei, W. M. Schmidt, M. Waldschmidt, Diophantine Approximation. Cetraro, Italy 2000. Editors: F. Amoroso, U. Zannier (2003)

Vol. 1820: F. Hiai, H. Kosaki, Means of Hilbert Space Operators (2003)

Vol. 1821: S. Teufel, Adiabatic Perturbation Theory in Quantum Dynamics (2003)

Vol. 1822: S.-N. Chow, R. Conti, R. Johnson, J. Mallet-Paret, R. Nussbaum, Dynamical Systems. Cetraro, Italy 2000. Editors: J. W. Macki, P. Zecca (2003)

Vol. 1823: A. M. Anile, W. Allegretto, C. Ringhofer, Mathematical Problems in Semiconductor Physics. Cetraro, Italy 1998. Editor: A. M. Anile (2003)

Vol. 1824: J. A. Navarro González, J. B. Sancho de Salas, $\mathscr{C}^\infty$ – Differentiable Spaces (2003)

Vol. 1825: J. H. Bramble, A. Cohen, W. Dahmen, Multiscale Problems and Methods in Numerical Simulations, Martina Franca, Italy 2001. Editor: C. Canuto (2003)

Vol. 1826: K. Dohmen, Improved Bonferroni Inequalities via Abstract Tubes. Inequalities and Identities of Inclusion-Exclusion Type. VIII, 113 p, 2003.

Vol. 1827: K. M. Pilgrim, Combinations of Complex Dynamical Systems. IX, 118 p, 2003.

Vol. 1828: D. J. Green, Gröbner Bases and the Computation of Group Cohomology. XII, 138 p, 2003.

Vol. 1829: E. Altman, B. Gaujal, A. Hordijk, Discrete-Event Control of Stochastic Networks: Multimodularity and Regularity. XIV, 313 p, 2003.

Vol. 1830: M. I. Gil', Operator Functions and Localization of Spectra. XIV, 256 p, 2003.

Vol. 1831: A. Connes, J. Cuntz, E. Guentner, N. Higson, J. E. Kaminker, Noncommutative Geometry, Martina Franca, Italy 2002. Editors: S. Doplicher, L. Longo (2004)

Vol. 1832: J. Azéma, M. Émery, M. Ledoux, M. Yor (Eds.), Séminaire de Probabilités XXXVII (2003)

Vol. 1833: D.-Q. Jiang, M. Qian, M.-P. Qian, Mathematical Theory of Nonequilibrium Steady States. On the Frontier of Probability and Dynamical Systems. IX, 280 p, 2004.

Vol. 1834: Yo. Yomdin, G. Comte, Tame Geometry with Application in Smooth Analysis. VIII, 186 p, 2004.

Vol. 1835: O.T. Izhboldin, B. Kahn, N.A. Karpenko, A. Vishik, Geometric Methods in the Algebraic Theory of Quadratic Forms. Summer School, Lens, 2000. Editor: J.-P. Tignol (2004)

Vol. 1836: C. Năstăsescu, F. Van Oystaeyen, Methods of Graded Rings. XIII, 304 p, 2004.

Vol. 1837: S. Tavaré, O. Zeitouni, Lectures on Probability Theory and Statistics. Ecole d'Eté de Probabilités de Saint-Flour XXXI-2001. Editor: J. Picard (2004)

Vol. 1838: A.J. Ganesh, N.W. O'Connell, D.J. Wischik, Big Queues. XII, 254 p, 2004.

Vol. 1839: R. Gohm, Noncommutative Stationary Processes. VIII, 170 p, 2004.

Vol. 1840: B. Tsirelson, W. Werner, Lectures on Probability Theory and Statistics. Ecole d'Eté de Probabilités de Saint-Flour XXXII-2002. Editor: J. Picard (2004)

Vol. 1841: W. Reichel, Uniqueness Theorems for Variational Problems by the Method of Transformation Groups (2004)

Vol. 1842: T. Johnsen, A. L. Knutsen, K_3 Projective Models in Scrolls (2004)

Vol. 1843: B. Jefferies, Spectral Properties of Noncommuting Operators (2004)

Vol. 1844: K.F. Siburg, The Principle of Least Action in Geometry and Dynamics (2004)

Vol. 1845: Min Ho Lee, Mixed Automorphic Forms, Torus Bundles, and Jacobi Forms (2004)

Vol. 1846: H. Ammari, H. Kang, Reconstruction of Small Inhomogeneities from Boundary Measurements (2004)

Vol. 1847: T.R. Bielecki, T. Björk, M. Jeanblanc, M. Rutkowski, J.A. Scheinkman, W. Xiong, Paris-Princeton Lectures on Mathematical Finance 2003 (2004)

Vol. 1848: M. Abate, J. E. Fornaess, X. Huang, J. P. Rosay, A. Tumanov, Real Methods in Complex and CR Geometry, Martina Franca, Italy 2002. Editors: D. Zaitsev, G. Zampieri (2004)

Vol. 1849: Martin L. Brown, Heegner Modules and Elliptic Curves (2004)

Vol. 1850: V. D. Milman, G. Schechtman (Eds.), Geometric Aspects of Functional Analysis. Israel Seminar 2002-2003 (2004)

Vol. 1851: O. Catoni, Statistical Learning Theory and Stochastic Optimization (2004)

Vol. 1852: A.S. Kechris, B.D. Miller, Topics in Orbit Equivalence (2004)

Vol. 1853: Ch. Favre, M. Jonsson, The Valuative Tree (2004)

Vol. 1854: O. Saeki, Topology of Singular Fibers of Differential Maps (2004)

Vol. 1855: G. Da Prato, P.C. Kunstmann, I. Lasiecka, A. Lunardi, R. Schnaubelt, L. Weis, Functional Analytic Methods for Evolution Equations. Editors: M. Iannelli, R. Nagel, S. Piazzera (2004)

Vol. 1856: K. Back, T.R. Bielecki, C. Hipp, S. Peng, W. Schachermayer, Stochastic Methods in Finance, Bressanone/Brixen, Italy, 2003. Editors: M. Fritelli, W. Runggaldier (2004)

Vol. 1857: M. Émery, M. Ledoux, M. Yor (Eds.), Séminaire de Probabilités XXXVIII (2005)

Vol. 1858: A.S. Cherny, H.-J. Engelbert, Singular Stochastic Differential Equations (2005)

Vol. 1859: E. Letellier, Fourier Transforms of Invariant Functions on Finite Reductive Lie Algebras (2005)

Vol. 1860: A. Borisyuk, G.B. Ermentrout, A. Friedman, D. Terman, Tutorials in Mathematical Biosciences I. Mathematical Neurosciences (2005)

Vol. 1861: G. Benettin, J. Henrard, S. Kuksin, Hamiltonian Dynamics – Theory and Applications, Cetraro, Italy, 1999. Editor: A. Giorgilli (2005)

Vol. 1862: B. Helffer, F. Nier, Hypoelliptic Estimates and Spectral Theory for Fokker-Planck Operators and Witten Laplacians (2005)

Vol. 1863: H. Führ, Abstract Harmonic Analysis of Continuous Wavelet Transforms (2005)

Vol. 1864: K. Efstathiou, Metamorphoses of Hamiltonian Systems with Symmetries (2005)

Vol. 1865: D. Applebaum, B.V. R. Bhat, J. Kustermans, J. M. Lindsay, Quantum Independent Increment Processes I. From Classical Probability to Quantum Stochastic Calculus. Editors: M. Schürmann, U. Franz (2005)

Vol. 1866: O.E. Barndorff-Nielsen, U. Franz, R. Gohm, B. Kümmerer, S. Thorbjønsen, Quantum Independent Increment Processes II. Structure of Quantum Lévy Processes, Classical Probability, and Physics. Editors: M. Schürmann, U. Franz, (2005)

Vol. 1867: J. Sneyd (Ed.), Tutorials in Mathematical Biosciences II. Mathematical Modeling of Calcium Dynamics and Signal Transduction. (2005)

Recent Reprints and New Editions

4. Manuscripts should in general be submitted in English. Final manuscripts should contain at least 100 pages of mathematical text and should always include

   – a general table of contents;

   – an informative introduction, with adequate motivation and perhaps some historical remarks: it should be accessible to a reader not intimately familiar with the topic treated;

   – a global subject index: as a rule this is genuinely helpful for the reader.

   Lecture Notes volumes are, as a rule, printed digitally from the authors' files. We strongly recommend that all contributions in a volume be written in the same LaTeX version, preferably LaTeX2e. To ensure best results, authors are asked to use the LaTeX2e style files available from Springer's web-server at

   ftp://ftp.springer.de/pub/tex/latex/mathegl/mono.zip (for monographs) and
   ftp://ftp.springer.de/pub/tex/latex/mathegl/mult.zip (for summer schools/tutorials).

   Additional technical instructions, if necessary, are available on request from:

   lnm@springer-sbm.com.

5. Careful preparation of the manuscripts will help keep production time short besides ensuring satisfactory appearance of the finished book in print and online. After acceptance of the manuscript authors will be asked to prepare the final LaTeX source files (and also the corresponding dvi-, pdf- or zipped ps-file) together with the final printout made from these files. The LaTeX source files are essential for producing the full-text online version of the book. For the existing online volumes of LNM see:

   http://www.springerlink.com/openurl.asp?genre=journal&issn=0075-8434.

   The actual production of a Lecture Notes volume takes approximately 8 weeks.

6. Volume editors receive a total of 50 free copies of their volume to be shared with the authors, but no royalties. They and the authors are entitled to a discount of 33.3 % on the price of Springer books purchased for their personal use, if ordering directly from Springer.

7. Commitment to publish is made by letter of intent rather than by signing a formal contract. Springer-Verlag secures the copyright for each volume. Authors are free to reuse material contained in their LNM volumes in later publications: A brief written (or e-mail) request for formal permission is sufficient.

Addresses:

Professor J.-M. Morel, CMLA,
École Normale Supérieure de Cachan,
61 Avenue du Président Wilson, 94235 Cachan Cedex, France
E-mail: Jean-Michel.Morel@cmla.ens-cachan.fr

Professor F. Takens, Mathematisch Instituut,
Rijksuniversiteit Groningen, Postbus 800,
9700 AV Groningen, The Netherlands
E-mail: F.Takens@math.rug.nl

Professor B. Teissier, Institut Mathématique de Jussieu,
UMR 7586 du CNRS, Équipe "Géométrie et Dynamique",
175 rue du Chevaleret, 75013 Paris, France
E-mail: teissier@math.jussieu.fr

For the "Mathematical Biosciences Subseries" of LNM :
Professor P. K. Maini, Center for Mathematical Biology,
Mathematical Institute, 24-29 St Giles,
Oxford OX1 3LP, UK
E-mail : maini@maths.ox.ac.uk

Springer, Mathematics Editorial I, Tiergartenstr. 17,
69121 Heidelberg, Germany,
Tel.: +49 (6221) 487-8410
Fax: +49 (6221) 487-8355
E-mail: lnm@springer-sbm.com